AF255923

ENVIRONMENTAL DESIGN FOR PUBLIC PROJECTS

Edited by

David W. Hendricks, Evan C. Vlachos,
L. Scott Tucker, Joseph C. Kellogg

WATER RESOURCES PUBLICATIONS, LLC.
COLORADO, USA

For Information and Correspondence:

Water Resources Publications, LLC

P. O. Box 630026 • Highlands Ranch • Colorado 80163-0026 • USA

First printing 1975
Previous ISBN-10: 0-918334-08-X; ISBN-13: 978-0-918-334-08-4

Second Printing 2019

ENVIRONMENTAL DESIGN FOR PUBLIC PROJECTS

Edited by

**David W. Hendricks, Evan C. Vlachos,
L. cott Tucker, Joseph C. Kellogg**

ISBN-13 Number: 978-1-887201-95-7
Library of Congress Control Number: 2019940451

There is no need to elaborate on what is by now becoming obvious, namely, that public projects have caused many unintended changes on the environment.   Further, it is almost a truism at this point in time to repeat that interdisciplinary involvement and expertise in project planning is an imperative for the new era of professional practice now emerging.

The challenge of this work is not to further reiterate on these needs for reform, but to develop methods, procedures, and processes whereby implementation of reform is within the realm of professional practice.   More specifically, this means that relevant goals and values of society are identified and then effectively assimilated into the planning and design of large scale facilities.

Along these lines of aiding the trend toward interdisciplinary professional practice, the book has a threefold purpose. First, because of its structure and contents it is intended to sensitize the user as to the overall requirements for broad planning. Second, it attempts to serve as a brief text summarizing in a professional manner some of the fundamentals and working methods of other disciplines so that practitioners from the various disciplines will better know how to interface with one another.   And third, it hopefully can serve as a handbook for varied types of practitioners to glean some planning insight as to how--with methods, procedures and illustrative examples--a particular discipline area bears upon holistic planning and design.

Since we are at a stage--with the message for reform loud and clear--we must start translating this imperative into specific methods of practice.   This treatise is intended both as an expression of the holistic commitment to planning and a beginning of practical implementation.

William L. Slayton, Hon. AIA
Executive Vice-President
American Institute of Architects

Embarcadero Freeway in San Francisco, (Cushman, 1965) downloaded from website 15 June, 2019: Cushman-Aug-15-1965-fwy-offramps-and-ferry-bldg-brand-new-P14851.jpg. Referenced on page 23.

PREFACE

The calls for much needed and impending change in the practice of civil engineering have been sounding since the mid 1960's. Signs and guideposts for such change have become both visible and pronounced, through policies adopted by professional societies, the multiplicity of writings, and last but not least, the promulgation of such landmark and far-reaching legislative requirements as those incorporated into the National Environmental Policy Act of 1970.

Thus, an all encompassing holistic design ethic and the clarion for total environmental planning have been more than idle intellectual discussions since 1970. There is now an urgent call for a careful examination of the interrelationship of key elements affecting environmental equilibrium, namely, man, nature, culture, and technology, and for assimilation of these values into professional engineering practice. Furthermore, there is general agreement that the current environmental plight and concern are results of the essential indeterminancy, diffuse nature, and unpredictability of man's diverse actions. What appears in simplistic terms as "environmental crisis" or "environmental predicament" are facets of a much deeper and more extensive disequilibrium in the pursuit of a collective life that should aim towards both survival and fulfillment. The disharmony of parts and the disruption of the natural system are also manifestations of negative spillovers resulting from the lack of articulating shared goals, summarized under the general concept of *quality of life*.

It is with such concerns in mind, that the preparation of this work was undertaken. In its broad goals, it is both a focal point for bringing together diversified disciplines and approaches, and for acting as a catalyst for further thinking in the general area of environmental design. But beyond these broad implicit goals it is also intended as a working text and reference to facilitate the absorption of explicit methods of various disciplines into professional practice.

The present volume focuses upon environmental questions relating specifically to large scale engineering facilities-- dams, highways, power plants, transmission lines, etc.--which may be called individually *public projects*, and collectively the *physical infrastructure*. Two key questions are of central concern throughout the text: (1) what are the primary effects caused by the *physical presence* of a public project; and, (2) what are the secondary, or induced consequences resulting by the *functioning* of a public project. The last brings forward the corollary question as to how can unintended effects be anticipated and incorporated in design in order to mitigate or alleviate disequilibriums and negative consequences.

v

While it is not the purpose of the book to answer these
questions with an explicitly enumerated methodology, there is a
concerted effort throughout the volume in trying to provide a
sufficient understanding of associated disciplines and of how
they bear upon the environmental interactions with public projects. Thus, an expanded professional design of public projects
can include a wider variety of interlocking factors of both
natural and human environments. The central contention and hy-
pothesis of the volume is that, increasingly, civil engineering
projects will be planned, designed, and monitored by interdisciplinary teams.

Along these lines, a major objective of the book is to
provide a capability for interdisciplinary communication and
dialogue.   This can be partly accomplished by summarizing some
fundamental premises of related disciplines.   A second major
objective is to explicate in more consistent manner unintended
interactions between public projects and the natural and human
environments--as related to a variety of disciplines. Finally,
a third major objective is to better understand how a given
discipline can be absorbed into the design process and what
contributions may be made in the process of developing integrated, holistic planning.

Since engineering projects will continue to operate without waiting for the development of a formal science of environmental management, it was felt that if existing knowledge of
experienced practitioners could be put together into some
structured format, there could be an immediate professional
level capability to meet some of the environmental challenges
posed by public projects. The main tasks in accomplishing this
purpose then is first to have sufficient vision as to what
should comprise the best format or structure; and, second, to
mobilize an effort to achieve the necessary output.

As to how this effort was mobilized and motivated, special
attention is warranted - for its nature, content, and style is
very much a reflection of this development. Its nature is much
more than twenty assignments to twenty authors.   Indeed, for
many involved it has been a way of life for certain periods.

The book was conceived in 1970 by two of the Editors -
Hendricks and Tucker - as a manual for interdisciplinary professional practice in civil engineering.   The concept became a
project in the summer of 1971, with funding from the NSF-RANN
Program.   Because of the broad character and nature of the
proposed book, a "Project Management Board" (PMB) consisting of
the four editors, was formed in order to insure that the project effort would, so to speak, "fly in the right direction."
That all issues, questions and plans were broadly debated was
insured by the character traits of the individuals comprising
this "PMB". The PMB then undertook to form a "Project Advisory

Committee" (PAC) consisting originally of thirteen persons. This group was very carefully selected from among those disciplines which the PMB felt ought to be included in the contemplated book. Again, adequate debate and consideration of all facets of the project was insured by the character traits of the individuals comprising this group. The PMB met four times in formal group meetings with the PAC and communicated extensively with each member on an individual basis. Their advice and consideration of all issues and project strategy greatly amplified the visions of the PMB. The PAC was especially important in the selection of authors, and in amplifying the ideas of the PMB in determining the specific goals and content needed for each chapter. Thus the book was basically designed as a joint effort through considerable debate and deliberation. By January 1972 the authors were selected and joined the PMB and the PAC in a third meeting to discuss their initial chapter outlines and develop a rapport with the project.

An unusual feature of the project was a special "author's conference" held at Vail in June 1972. The authors wrote preliminary drafts of their respective chapters for scrutiny, debate, and revision at this conference. About forty practitioners from various disciplines, but all associated with actual engineering accomplishments, were invited to participate with the authors and the Advisory Committee in three hard days of vigorous debate, argument, and creative dialogue. Again this was another stage of amplification - and a "testing" stage. The tangible output from this conference was substantial and many useful liaisons were formed which proved helpful as the authors produced their final drafts. These completed chapters then became the final report submitted to the National Science Foundation in June 1973. This present book is a revision of that report. While there has been an effort to make the size more tractable for publication, the editors attempted to maintain a policy of preserving the basic integrity of each author's chapter. This was important, because although the Editors could have possibly boiled down the content into a more "efficient" presentation, the author's style as a professional, which contains some of the essential message, would have been lost.

The list of those who helped throughout the various phases of the project is long, and includes our authors, the Project Advisory Committee, and persons participating in the Vail Conference, but our gratitude to each is equally extensive. As a means of acknowledgement to the gracious and unfasting support, help, and insightful criticism we would like to express our particular appreciation to Frank Boerger, Henry Michel, Frank Montanari, John Peters, Jack Ward, Gardner Reynolds, P. H. McGauhey, Charles Notess, and Warren Bell.

Special mention should be given to the continuous support
and help from Donald C. Taylor, Research Manager for the American Society of Civil Engineers.   His many suggestions, helpful
liaisons and arrangements, and enthusiasm for the effort have
been invaluable in many ways.

Gail Richardson too deserves special mention for without
her continuous dedication to the effort - from beginning to
end - the project would have been both more difficult to manage
and diminished a measure by her pervasive cheerful spirit. Miss
Richardson handled the operational phases, taking initiative to
make sure all tasks were accomplished with perception and sensitivity.

The editors also appreciate the reviews of the NSF report
by various members of the ASCE Committee on Environmental Impacts, chaired by William Hedley, and the classroom use of that
report by Professor William Wisely (former Executive Director
of ASCE). All of these comments were considered in the revision
forming this book.

The financial support for the project upon which the book
is based was a grant (GI-29837) from the National Science Foundation, RANN Program, awarded to Colorado State University; the
senior editor was principal investigator. Dr. Ralph Long, Program Manager for the Advanced Technology Applications Program,
was project officer for NSF.   The development of this book was
the ultimate goal of the project supported by NSF.

David W. Hendricks
Evan C. Vlachos
L. Scott Tucker
Joseph C. Kellogg

Fort Collins, Colorado
February 1975

Editor's Note to the Second Printing ~ 2019

Environmental Design for Public Projects, supported by funding from the
NSF RANN Program, is as relevant today as when it was first published. The
principles presented in this book are both pertinent and applicable today in
helping civil engineers expand their perspective to assimilate contributions
from professionals from other disciplines in more holistic planning and design
of large-scale public projects. By the same token, professionals in other
disciplines are provided an introduction to selected topics in civil engineering
(such as water, transportation, and power). WRP is pleased to include this
book as part of our *Classic Resource Edition*.

viii

ADVISORY COMMITTEE*

ROBERT F. BAKER, Transportation Consultant, Bethesda, Maryland

DOUGLAS A. BENTON, Associate Professor of Management, Colorado State University, Fort Collins, Colorado

A. BRUCE BISHOP, Assistant Professor of Civil Engineering, Utah State University, Logan, Utah

FRANK C. BOERGER, Consulting Engineer, San Anselmo, California

CATHERINE BROWN, Freelance Editor and Writer, Ridgewood, New Jersey

WAYNE CAPRON, District Engineer, Colorado Highway Department, Greeley, Colorado

CARL CARLOZZI, Associate Professor of Resource Planning, University of Massachusetts, Amherst, Massachusetts

DESMOND M. CONNOR, Consulting Sociologist, D.M. Connor Development Services, Ottawa, Canada

C.L. GALLIMORE, Engineering and Construction Consultant, Omaha, Nebraska

R. BURNELL HELD, Professor of Economics and Recreation Resources, Colorado State University, Fort Collins, Colorado

WILLIAM T. HELM, Professor of Wildlife Resources, Utah State University, Logan, Utah

DAVID W. HENDRICKS, Associate Professor of Civil Engineering, Colorado State University, Fort Collins, Colorado

JOSEPH C. KELLOGG, President, Virodyne Corporation, Englewood, Colorado

WILLIAM M. KLEIN, Assistant Director, Missouri Botanical Gardens, St. Louis, Missouri

WALTER McCARTHY, Vice President, Management and Construction Services Corp., Minneapolis, Minnesota

P. H. McGAUHEY, Consulting Engineer, El Cerrito, California

---

*Authors are ex-officio members of the committee.

HENRY MICHEL, Partner, Parsons, Brinckerhoff, Quade and Douglas, New York, New York

DWAYNE C. NUZUM, Dean, College of Environmental Design, University of Colorado, Boulder, Colorado

JOHN C. PETERS, Environmental Specialist, U. S. Bureau of Reclamation, Denver, Colorado

LAWRENCE M. ROBERTSON, Consulting Engineer, Denver, Colorado

DAVID SEADER, Project Planner, Parsons, Brinckerhoff, Quade and Douglas, New York, New York

WILLIAM L. SLAYTON, Executive Vice President, American Institute of Architects, Washington, D. C.

DONALD C. TAYLOR, Research Manager, American Society of Civil Engineers, New York, New York

L. SCOTT TUCKER, Director, Urban Drainage and Flood Control District, Denver, Colorado

EVAN VLACHOS, Professor of Sociology, Colorado State University, Fort Collins, Colorado

MATT WALTON, Consulting Geologist, Englewood, Colorado

NORMAN A. WENGERT, Professor of Political Science, Colorado State University, Fort Collins, Colorado

J. LOUIS YORK, Environmental Consultant, Stearns-Roger Corporation, Denver, Colorado

OVERVIEW COMMITTEE*

DAVID AGGERHOLM, Chief, Long Range Planning, Institute for
Water Resources, Army Corps of Engineers, Alexandria, Virginia

HARVEY ATCHISON, Senior Landscape Architect-Administrator, Environmental Section, Colorado State Highway Dept., Denver,
Colorado

WARREN BELL, Civil Engineer, Edmonton, Alberta, Canada

BOB BREWSTER, Assistant to Regional Director, U. S. Bureau of
Reclamation, Salt Lake City, Utah

HENRY CAULFIELD, Professor of Political Science, Colorado State
University, Fort Collins, Colorado

JOHN CONGER, Senior Sociologist, Colorado Highway Department,
Denver, Colorado

GEORGE DARROW, Chairman, Montana Environmental Quality Council,
Billings, Montana

RICHARD deNEUFVILLE, Director, Civil Engineering Systems Laboratory, Massachusetts Institute of Technology, Cambridge,
Massachusetts

STEVEN DOLA, Engineering, Program Planning Group, Office of the
Secretary of the Army, Washington, D. C.

GLEN FULCHER, Chief, Division of Standards and Technology, U.S.
Bureau of Land Management, Denver, Colorado

J. WILLIAM GEISE, JR., Chief, Interagency Assistance and Evaluation Branch, Environmental Protection Agency, Denver, Colorado

SIGURD GRAVA, Professor of Architecture, Columbia University,
New York, New York

STUART HILL, Director, Environmental Affairs, Barton Aschmann
Associates, San Jose, California

DONALD A. JAMESON, Director, Regional Systems Program, Department of Forestry, Colorado State University, Fort Collins,
Colorado

---

*Convened with authors and Advisory Committee at 1972 Vail Con-
ference.

HENRY B. KEISER, President, Federal Publications Inc., Washington, D. C.

DuWAYNE A. KOCH, Chief, Systems Section, Economics Branch, Planning Division, Office of the Chief of Engineers, Washington, D. C.

R. ERNEST LEFFEL, Vice President, Camp, Dresser and McKee, Inc., Boston, Massachusetts

RICHARD MALES, Consulting Engineer, Environmental Engineers, Ltd., Cincinnati, Ohio

JOHN McWHORTER, Partner, King & King Attorneys, Washington, D.C.

F. W. MONTANARI, MANAGER, Denver Regional Office, Parsons, Brinckerhoff, Quade and Douglas, Denver, Colorado

VASILIS MORFOPOULOS, President, American Standards Testing Bureau, New York, New York

TABLE OF CONTENTS

PART I

THE CONTEXT OF PUBLIC PROJECTS

The overall goal of this part is to elucidate the context of public projects by juxtaposing demands of the past with increased requirements of today, and even more far-reaching demands of tomorrow. A central section discusses in detail the genesis and evolution of public projects and through examples in various area tries to shed light to the variegated requirements for interdisciplinary planning.

I

THE WORLD OF PUBLIC PROJECTS: TRADITIONS AND TRANSITIONS

Organized societies since antiquity have constructed large scale facilities for collective use. These facilities have included extensive road networks, numerous impressive dams and huge building complexes. The Roman edifices are probably best known and best organized as a system - though by no means are they the earliest. Such facilities provide the structural capacity for a society to metabolize - that is, to carry out the innumerable individual and collective tasks required of a viable functioning society.

The facilities which collectively comprise this structural capacity are called, herein, the *physical infrastructure*. Individual additions are called, *public projects*. In a modern industrial society these include at the "bulk level" : dams, canals, pipelines, levees, electric power generating plants, electric transmission lines, highways, airports sewage treatment plants, etc. Included in the continum at the microscale level of the internal city are: water distribution mains, sewage laterals, individual houses, city streets, etc.

The physical infrastructure of a modern industrial society is developed to such an extent that its various components often interfere with other systems in unintended ways. For example a new sewer line constructed with excess capacity to serve say a remote area provides a potential for a higher metabolism level - and hence further growth. An out of place building may disrupt the esthetic qualities and general character of an urban area. Or, the imposition of a dam on a stream may change completely the stream's ecology. The examples are of course endless. But from this brief insight it can be inferred that the unintended effects may be associated with either: (1) the new metabolic capacity provided by the facility, which may, because of waste or added populations, interfere with other systems, or (2) the spatial presence of the facility, which may interfere directly with other systems.

Traditionally, infrastructure facilities have been built by civil engineers who have been concerned largely with whether or not the facility is structurally sound, economically feasible, and fulfills its intended purpose. However, because of the increasingly serious disharmonies often induced it is becoming imperative that the unintended effects of infrastructure additions be anticipated, understood, and mitigated. If mitigating measures cannot be made acceptable, then informed choices can be made. Indeed this is the broad purpose of the National Environmental Policy Act - which gives a tangible reality to this goal.

To fulfill this larger goal in infrastructure development requires a broad understanding of the potential interactions with other systems, and a knowledge of how associated disciplines can be brought to bear in evaluating the interaction. This is required too in planning the societal role of the particular facility and in designing its configuration such that the interaction is within acceptable limits.

While these goals have been rather nebously recognized since the advent of the National Environmental Policy Act in 1970 there really is no science of environmental analysis, nor is there a formal professional approach to environmental assessment. An environmental assessment is largely structured about the format of an "impact statement" which tends to be a reactive approach, albeit highly valuable. One of the larger purposes of the Act is to use an interdisciplinary approach. Fulfillment of this, and other broad purposes of the Act, also constitutes a major purpose of this treatise, which is to provide a professional capability to plan and design physical infrastructure additions in the context of an emerging holistic planning ethos. This implies that engineers must understand the role of the physical infrastructure facility in society, and plan and design the facility in that context. This, of course, requires integration of a variety of disciplines in the planning and design processes as members of the interdisciplinary team.

All of this has been a strong trend since 1970. Most engineering firms, government engineering organizations, and state highway departments have reorganized themselves to respond to the new ethic. Not only has the trend been motivated by the National Environmental Policy Act but, in addition, numerous Executive Orders, federal regulations, state laws, and further reinforcing federal legislation have accelerated implementation of the new ethic.

What all the above imply is that formerly isolated disciplines will have to find ways to address some of these difficult problems by generating hard data followed by the application of systematic methods of analysis and finally by useful recommendations. This requires knowledge of the problem areas, of other disciplines, and of the laws and the political system that spawns the public projects, and of management skills necessary to effectively mobilize an interdisciplinary team.

Figure 1-1 depicts a holistic planning space which is indicative of the broad inclusive scope needed for planning and design of infrastructure additions (or "public projects"). Of the problem planes shown in the vertical axis, the utility satisfaction "plane" traditionally has been the major emphasis. Of the "discipline and profession" entries, "engineering" has been the operating plane. The result has been that only elements of one vector has been considered in traditional practice. This is illustrated graphically in Figure 1-2. Figure 1-1 on the other hand, shows the whole space in which problem solution should take place. Problems are many, and thus many discipline inputs are needed. Neither are problems isolated to just design or planning phases. Regional, national and global effects must be understood.

It is the purpose of this treatise to develop a beginning professional level capability for any discipline person to deal effectively with the whole problem space. The approach is to develop discipline fundamentals understandable to other disciplines and then to develop some application fundamentals delineating how each discipline can apply its principles to the appropriate problem planes. This should be done with sensitivity to planning, design, and operation. But the treatise is structured about selected *disciplines* and trusts to each author to adequately touch the relevant problem planes, with some discussion of scale (i.e., planning, design, operation).

The need for such a comprehensive and inclusive approach to design has not always been recognized in the development of large scale public facilities. It is, however, a crucial part of an emerging, but not as yet crystallized, *environmental ethic*, which has been shaping since about 1963; and it must be considered in that context in order to gain the proper perspective to deal with each specific problem as a part of a time continuum of orderly change vis a vis precipitous and uncoordinated action.

THE FRONTIER ETHIC

Traditions of development in the United States have their roots in the challenges of survival of the frontier. Customs of society, the focus of business, and the legal structure that

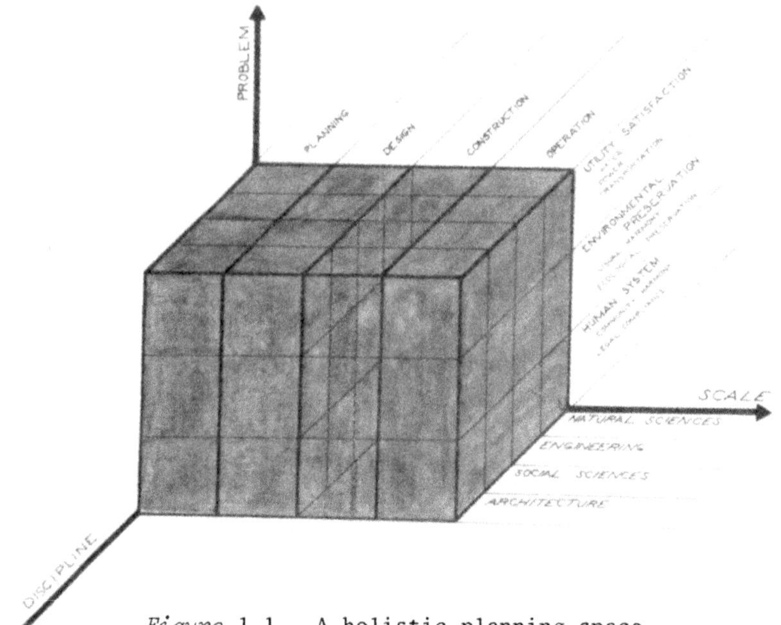

Figure 1-1. A holistic planning space.

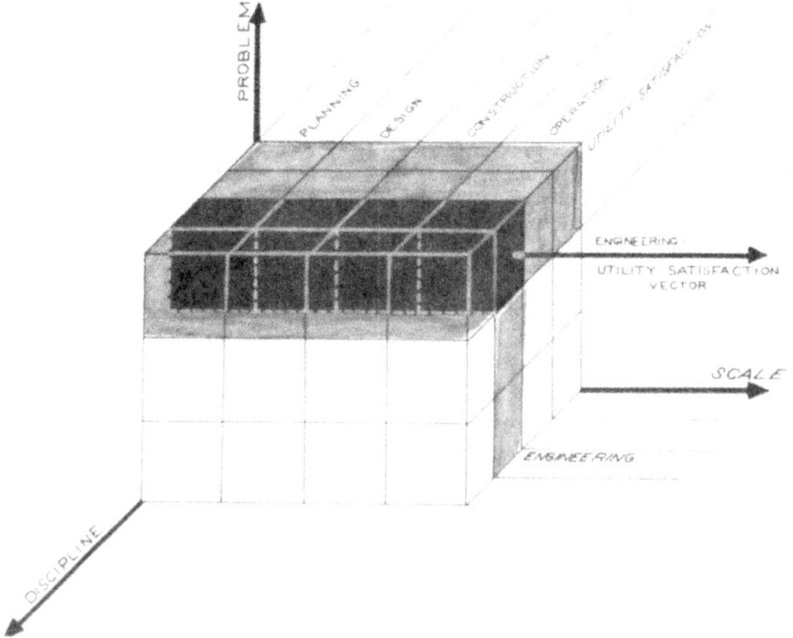

Figure 1-2.   Utility satisfaction vector of problem plane
and engineering of discipline and profession entries.

evolved all supported the ethic of unhindered development. The individuals who faced the sprawling continent generally were concerned with facilitating and promoting various types of development, the mastering of often harsh territory, and with an overall goal to develop a viable economy. All entrepreneurial forces coalesced and shaped an aggressive ethic whose primary objectives were exemplified in such terms as "development," "growth," "expansion," and economic prosperity. In this context, the function of government was to facilitate and support such forces, not to regulate, constrain, or otherwise interfere.

The pervasiveness of the frontier ethic is well documented not only in the historical records and the legends of fortunes made, but it has been circumscribed also in the appropriate legislation. Such legislation has been only one indicator of the spirit of the times--but it is probably the most important and the most tangible expression of shared social values. The various federal acts enumerated in Table 1-1 are representative of the type of social climate that has prevailed in the United States since early times; the lingering influence of such legislation survives strongly today.

Referring to Table 1-1, one can discern how the railroads, through land grants, became major economic forces controlling much of the development that did occur, such as the rise and fall of settlements, and the drawing of the economic map of the nation. The 1872 Mining Law, with the 1920 Mineral Leasing Act, has been the authorization for almost unlimited and virtually unregulated activity on public lands. Then, under the impetus of the 1902 Reclamation Act and with the strong presence of private irrigation companies, 43 million acres of land were under irrigation by 1969--a literal blooming of primarily arid land in seventeen Western States. This transformation has been the basis for a viable and agriculturally based economy for much of the West and the basis for the "urban oasis" that increasingly have come to characterize most states in the West. Figures 1-3 and 1-4 are the epitomy of the large scale projects and the robust testimony of man's control in an inhospitable area.

The major legal support bolstering the cause of reclamation and other private irrigation activities in the West has been the appropriation doctrine of beneficial use. Under such a doctrine, in-stream uses have no legal status; this has become the cause for serious conflicts emerging strongly around 1969 between those benefiting from the traditional uses of water requiring diversion and those wanting flowing streams for additional ecological purposes. This doctrine was ammended in some states in 1972 when in-stream uses were recognized as beneficial use.

Table 1-1. Federal Legislation or Administrative
Action Facilitating Development

Date Enacted	Name of Act	Key Provisions or Effects Related to Development
1824	Administrative Action	Corps of Engineers assigned responsibility to improve waterways
1852-1871	Railroad Land Grants	128 million acres to railroad use of which 91 million went to railroad corporations and 37 million acres went to states for benefit of the railroads
1862	Homestead Act	160 acre land grants to individuals 285,000,000 acres were patented under this act (1/3 of original public domain)
1872	1872 Mining Law	Permits unrestricted mineral exploration on all public lands and permitted complete control of all resources by a claimant
1877	Desert Land Act	Provided for sale of land at $1.25 per acre of 640 acres to settlers who would irrigate it within three years. Ten million acres were patented under the act
1902	Reclamation Act	Further advanced the amount of land irrigated
1908	Administrative Action	Inland Water Waterways Commission advances the doctrines of the river basin unit and multiple purpose water control projects
1920	Mineral Leasing Act	Provides for leasing by the Secretary of the Interior of coal, phosphate, sodium, oil and oil shale on public lands
1933	TVA Act	TVA was created

Figure 1-3. Roosevelt Dam, Salt River Project, Arizona, completed in 1911. (Courtesy Salt River Project).

The right of condemnation by state highway departments and public utility companies has allowed our basic physical infrastructure to develop without the continuing uncertainties over right-of-way acquisition. Thus, excluding this intrusion, the freedom to use one's property with only very few constraints has been a sacred tenet of the frontier ethic. Development on both public and private lands has proceeded with virtually no inhibitions.

There is no need to underline that the frontier ethic has served well in facilitating the transformation from a state of undeveloped frontier to a mature society. Our capital stock of land, water, minerals, and other basic resources has been utilized to create a viable economy and to provide a framework for the expansion necessary to permit industrial growth and development.

Figure 1-4. Morrow Point Dam, is a key feature of the Curecanti Unit of the Colorado River Storage Program. It was built on the Gunnison River near Montrose, Colorado, and provides long-term, carryover water storage and vitally needed hydroelectric power. The dam is 469 feet high, 740 feet long, 52 feet thick at the base, 12 feet thick at the crest, and contains 365,000 cubic yards of concrete. The powerplant located in the canyon wall underground consists of two generating units with a total capacity of 120,000 kilowatts of electricity. (Photograph courtesy of U.S. Bureau of Reclamation).

TRANSITION TOWARD A NEW ETHIC

Concern over unbridled resource exploitation first became a persistent and viable political force about the time of Theodore Roosevelt who with his forceful Forester, Gifford Pinchot (Chief of the U.S. Forest Service) led in the development of the conservation doctrine calling for "wise use" of resources. Although Roosevelt and Pinchot had much broader purposes in mind, the conservation movement was justified by many on the grounds that natural resources were important primarily as factors of production and as such needed protection, management and efficient utilization.

It would be inaccurate to state that the concern with the environment, or the "environmental movement" burst upon us suddenly in the 1960's as a result of sudden revolution or as a culmination of major social unrest. There has always been concern with unbridled exploitation, with the catastrophic possibilities of unchecked growth, and with the dangerous ramifications from a disregard of essential ecological constraints. But the larger and more nebulous problems of environmental quality did not come into critical public focus until early in 1960. In 1963, the latent, diffuse, and sporadic public concern was triggered into open debate throughout the United States and in Congress by Rachael Carson's *Silent Spring*. The environmental movement acquired its formal beginning as a persistent force and as a salient concern to stir public imagination.

Table 1-2 attempts to summarize key environmental legislation and the evolution towards more concerted efforts concerning the environment. At the same time, it should be noted that the so-called "environmental movement" has become much more widespread and all inclusive than the conservation movement--though the latter is the foundation of the former.

Table 1-2 is by no means inclusive. Left out are innumerable acts--mostly since 1967, such as the National Historic Preservation Act, Aircraft Noise Abatement Act, Fish and Wildlife Coordination Act (PL85-624), Water Resources Planning Act (PL89-80), ammended in 1971 by PL92-27, and section 4(f) of the Transportation Act of 1968, regulating the acquisition of highway right-of-way through parks and a host of similar acts. However, even with such a limited list of environmental acts, the trend exemplified in Table 1-2 is unmistakable. We are increasingly surrounded, starting with the broad concern on conservation about the turn of the century, by a strong legislative activity, and by toughening provisions reflecting both the mood of the times and the emergence of new social values concerning the relationship of man to his surrounding physical environment.

<u>*Table* 1-2. Federal Legislation Regulating Development</u>

Date Enacted	Name of Act	Key Provisions
1891	Forest Reserve Act	Established forest reserves--the precurser to the National Forests
1920	Federal Water Power Act	Created Federal Power Commission and authorized it to issue licences for navigation and power production from navigable waters
1932	Emergency Conservation Act	Provided employment on conservation and public works projects
1940	Ohio River Valley Sanitation Compact	For control and reduction of pollution in the streams of the Ohio River drainage basin
1948	Taft-Barkely Water Pollution Control Act PL80-845	Declared a pollution control policy* and provided aid to state agencies to control stream pollution
1952	Public Law 82-579	Financial provisions of PL80-845 extended to 1956
1956	Federal Water Pollution Control Act of 1956 PL84-660	Extended and strengthened 1948 law, encouraged interstate compacts and uniform state laws, financed research and technical assistance to states; grants for construction of treatment plants
1961	Federal Water Pollution Control Act Ammendments of 1961 PL87-88	Further strengthening of act by extending federal authority to control abatement of interstate waters; further financial authorizations
1965	Water Quality Act of 1965	Water quality standards required of each state; more construction grant monies
1965	Solid Waste Disposal Act of 1965, PL89-272	

Table 1-2. Federal Legislation Regulating Development (Con't)

1966	Clean Water Restoration	Comprehensive basin planning; acid mine water control; control of pollution in the Great Lakes; control of sewage from vessels; control of pollution by oil
1967	Air Quality Act of 1967 PL90-148	
1969	National Environmental Policy Act PL91-190	Council on Environmental Policy established; environmental impact statement required for all projects involving Federal government; declares national policy for environmental quality enhancement
1970	Clean Air Ammendments of 1970 PL-91-604	Vehicle emission control; control of stationary sources
1970	Resource Recovery Act of 1970 PL91-512	Encourages materials recycle
1972	Water Quality Act of 1972	18 billion for treatment plant construction; zero discharge by 1985

* Prior to this act the only three federal water pollution acts existed: (1) Rivers and Harbors Act of 1899; (2) Public Health Service Act of 1912; (3) Oil Pollution Act of 1924

A crucial point implicit in Table 1-2 is the initial focus and strong concern with water pollution. Water pollution is, indeed, tangible. Dead fish can be seen; sludge banks and floating debris are unsightly; hydrogen sulfide can be smelled, biochemical oxygen demand and coliforms, etc. can be measured. And so, because of its amenability for tangible discernment and explicit articulation and because of the undeniable existence of a problem, water pollution has been established early as a focal point of environmental quality concerns. Furthermore, water pollution is a known health hazard. This has a significance far beyond its face value in pollution control. Protection of public health has been the one sacred tenet of public policy that takes precedence over economic development. This has always been true, from early attempts to curb mortality in cities to all sources of morbidity plaguing our collective life. Thus, the disease potential of polluted water has been the basis for its control until 1965, when the Water Quality Act of 1965 considerably broadened the charter for clean water. Air pollution control was pursued initially on the basis of health impairment, then economic damage to buildings, crops, etc., and now the esthetics of clean air has a formal status in our value system. Briefly, the transformation has been from concern with survival to the quest towards fulfillment and happiness.

Since 1970, when the National Environmental Policy Act came into being, the scope of environmental quality concern has been considerably broadened beyond just pollution and health preoccupation. The clutter of signs in our urban areas, the sameness of the production line suburbs, the decay of the inner cities, the incessant noise of the mechanized and transistorized society, the community divisiveness caused by an urban freeway and its overwhelming presence, the lack of water in our streams, the criss-cross of our countryside with innumerable highways and utility corridors, and the carving up of all available pristine hinterland, with little escape or relief from the monotony of the suburbs, are continuous reminders and painful manifestations of a generally deteriorating environment, and of what popularly has come to be known as the environmental predicament.

Probably, among the biggest single handicaps in controlling use of our common property resources and of private and public lands on the scale needed for harmonious collective life lies exactly in the powerful traditions of the frontier ethic described above. Yet, despite our temporary despair and the surrounding signs of a deteriorating environment, there is little doubt that the transition to a new, more holistic ethic is under way. The National Environmental Policy Act is a culminating manifestation of new values and the first tangible evidence of things to come. This act itself offers a very broad charter and the imperative to plan and design along more holistic lines. It may be classified as reform legislation. Reform is insured by the procedural mechanisms of the Act assuring a confrontation of issues among interested parties. But more important, reform can also be brought about by the explicit ethos advocated in the Act, the requirement for multidisciplinary involvement, and the flexible language of many of its provisions. The last remains the key for the eventual success of the Act; its viability is not its dead letter, but the interpretation of the noble aspirations of a far-reaching legal document.

LIMITATIONS OF THE MARKET

It is important, however, at this point to emphasize strongly a crucial point that underlies all these larger concerns and the quest for a holistic, new thinking. All larger environmental concerns have some unifying characteristics. The most dominant relate to disutilities, or side effects. Further, it should not be forgotten, that development activities are not subject to control by private markets and that common property resources are vitally affected. While the private market deals with negotiables, both goods and services, that can be bartered in the market place, the effects of many private activities cannot be easily regulated by traditional market mechanisms. Sometimes, these externalities, such as by-products, pollution, litter, etc., can be put back into the market equation, or the price system, by certain strategic taxes or regulatory counter-incentives. Examples are taxes on waste contaminants in water or air, strategic subsidies for the encouragement of recycling and a series of laws and regulations establishing standards of various sorts.

It is clear, then, that some centralized planning and control of the use of common property resources is necessary if the public good is to be protected. This concept of *protection of common property* resources for environmental quality, although seemingly new, can be traced to laws and regulations all the way back to the enactment of seasons on wildlife hunting in the seventeenth century by some of the colonies.

It is of paramount importance to recognize, at this point, that while market forces provide the need for public projects (e.g. cars need highways) and that public projects may generate or reinforce market activity (e.g. highways encourage suburban housing developments) the use of common property resources for deleterious purposes--such as overloading air and water with wastes--is certainly beyond the realm of the market. In such a case the regulations that follows can be necessitated only through the evolution of an ethic and through political action leading to specific laws. The pollution control laws of the last decade, and other acts enumerated in Table 1-2 illustrate this point. Decisions creating these laws have been outside the market and are primarily an expression of political interests and of public pressure. Thus, markets for services and hardware in these areas are created only by laws, rather than as a response to traditional economic mechanisms. And the values being protected are essentially non-market. As such, they entail a ranking and a priority established through the political process. The law then becomes a nodal point for determining goods and services and for providing the impetus for new markets. An example: the opportunity to sell equipment for air and water pollution control and associated services comes about only through legal constraints imposed on the discharges.

MAN THE BUILDER

While the previous brief remarks have centered around old and emerging systems of values concerning technological inducements, and public projects, we need to turn our attention to the construction activities of man, who from time immemorial has attempted to manipulate his surrounding environment. At least since the Neolithic period (8000-2000 BC) man has built great edifices, to celebrate victories, to defy death, to provide shelter, or to guarantee the survival of his group. Environmental intervention in the form of some very large projects, has been with us for quite some time. What has changed is its tempo, rapidity, and extent.

The first large civil works projects to control water were in the Tigris-Euphrates Valley, where an extensive irrigated agriculture was developed. The largest project was the 400 foot wide Nahrwan Canal, 200 miles long; this facility included the usual ancillary features, (e.g., A diversion dam, sluice gates, elevated flumes, etc). By 1800 BC the entire Tigris Valley and a part of the Euphrates Valley was irrigated by an elaborate and effective physical infrastructure. The building, operation, and maintenance of these vast systems were made possible by a strong central government applying strict management control. The emergence of social organization was the key for the success of the project. Centralized wealth and management have always been necessary in order to accomplish large civil works projects.

Despite the extensive presence of such large structures, the pace of construction in antiquity was slow, and the way of life was people-centered rather than machine-centered which usually characterizes the needs of the industrial age. The edifices of antiquity are of immense significance in a utilitarian sense, but were more important as cultural expressions of the group; they are creations by man, rich in historic, cultural, and esthetic values. Each structure represents decades of human toil, untold suffering, and engineering ingenuity although power was limited by the human and animal energy that could be harnessed, the technology of building and the managerial skills needed to build large structures combined to provide impressive feats of construction.

By the time of Christ the technologies of building were advanced far enough to provide for a reasonable array of basic building techniques. The arch was used extensively and earthfill dams had existed for several millenia. A number of

building materials, including asphalt, existed. The techniques
of quarrying were well advanced, and large   stones   could   be
moved over long distances. Figure 1-5 attests to the Roman
technological capacity for road building. Roadbeds for modern
highways are not appreciably different in essential design
features.

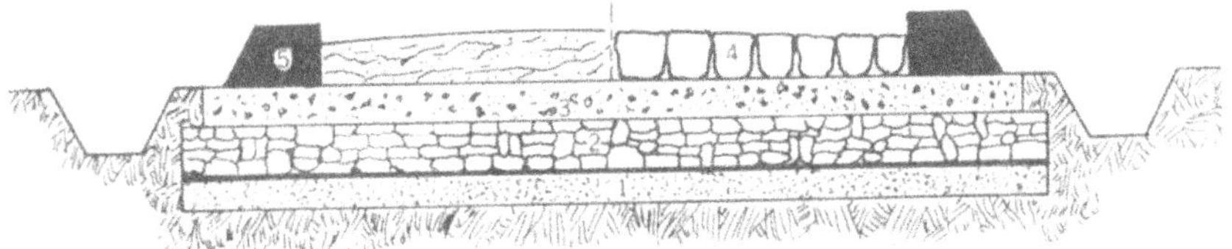

Figure 1-5. Roman via bed.

The Roman road system, viable and lasting because of such
technology, was extensive. Figure 1-6 gives some indication of
the thousands of kilometers included. The Romans also transported water over appreciable distances. Figure 1-7 is an
aqueduct; such structures were typical of their civilization.
Thus their supporting physical infrastructure was inpressive.

The historical periods that followed and a succession of
civilizations have each left their succinct and tangible im-
print. Both western and non-western societies have provided us
with a vast array of major buildings, structures, aquaducts,
dams, roads, fortifications, all attesting to the continuous
efforts to control the surrounding environment. The Industrial
Revolution, and the use of inanimate energy provided the major
breakthrough for a rapid acceleration of technology and an
increased domination of the surrounding landscape. The nature
of society has been altered from predominantly agricultural, to
a bustling urban-industrial millieu. We may see this transformation in the sequence of three figures. Figure 1-8 shows the
first attempt of man to control the rural environmental (in
this case irrigated agriculture in Colorado). The superimposi-
tion of the major Interstate on the rural hinterland, Figure1-9
shows another major transformation, as this public project
indicates not only control, but the opportunity for further
expansion. Finally, Figure 1-10 shows the total domination of
the massive urban agglomeration and a landscape shaped and
characterized by the arterial presence of a highway in its
midst.

Contrasted to other historical epochs, what is different
now is: (1) the more intense demand for amenities and development; (2) the ready availability of mechanical energy coupled
with the ease of procurement of building materials, as well as,
the improved technology of construction; (3) the diffuse

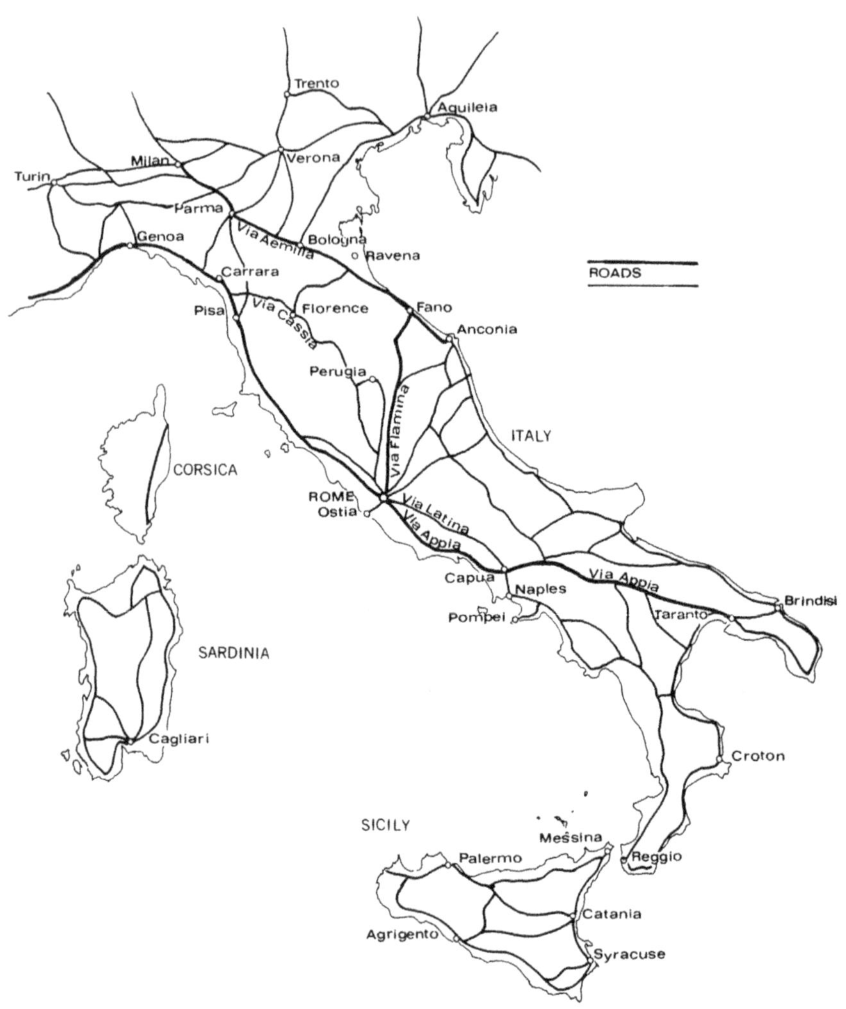

Figure 1-6. Road system of the Roman Empire

decision-making capability and means of financing among a multitude of public and private entities; and, (4) the rapid tempo of social change.

DEMANDS OF POPULATION AND AFFLUENCE

It is evident from the rate of population growth both in the United States and in the world (Figures 1-11 and 1-12) that there must be commensurate demand on resources, services, consumer goods, and large scale physical facilities. The

Figure 1-7. Aqueduct built by Romans crossing River Gand 12 miles from Nimes; a major feature in a 31 mile aqueduct system constructed in 63-13 BC in connection with the settling of some 8,000 colonials in the area (from Man the Builder, by Gosta E. Sandstrom, 1970, p. 85).

problem is compounded by the increases in per capita demand permitted by the increased affluence of the industrialized nations with resultant problems of waste production, and inefficient patterns of consumptive behavior. Population increase, therefore, and demands of increased affluence converge to create the essence of our problems, i.e., with the *additional* facilities needed, the *additional* resources demanded, and the attendant increased waste production. In an economic sense, the marginal social costs are now beginning in a number of cases to exceed the marginal social benefits of additional development.

THE DILEMMA OF THE INDUSTRIAL SOCIETY

Many people have built large structures over the centuries. Most of these structures, as a result of both planning and respect to the local culture, tend to blend harmoniously with the surrounding landscape. Many of these great edifices and buildings are now a part of the cultural heritage of the past. They are each unique and provide a rich community and landscape diversity from city to village, from city to city, and from village to countryside.

Figure 1-8. The rural landscape.

Figure 1-9. The rural hinterland with an interstate highway.

Figure 1-10. The urban landscape with a interstate highway.

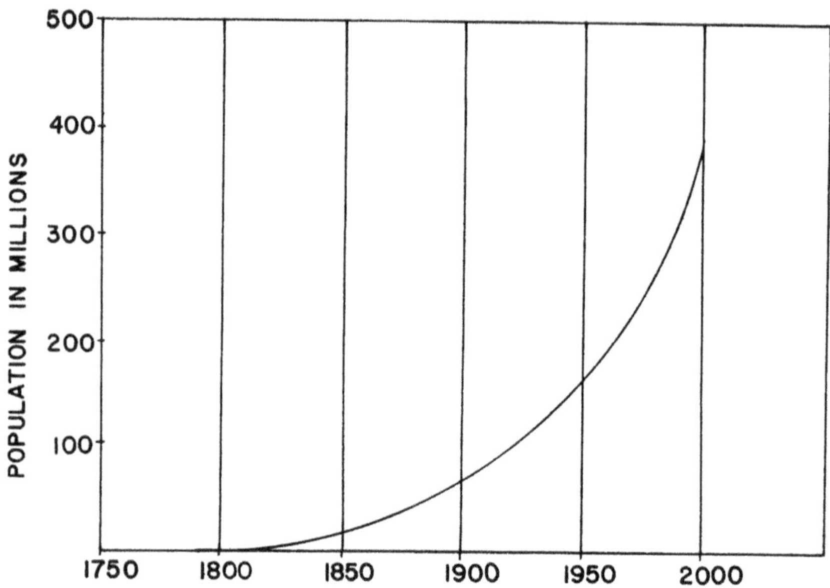

Figure 1-11.   Population growth of the United States.

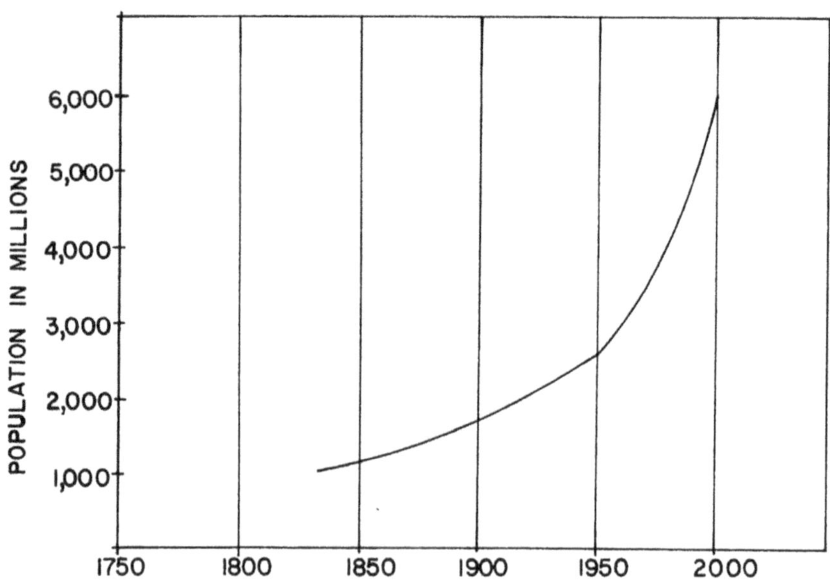

Figure 1-12.   Population growth of the World.

These structures perform not only the basic functions intended, but are integral parts of the life and culture of the region. The *rate* of development has been such that the traditions of architecture of the respective cultures have been mostly adhered to, and if nothing else, the minimum demands of growth have not stressed these systems to the point of instability and haphazard development. For example the charm and esthetic harmony of a French Village fulfills the esthic sensibilities of many foreign tourists. On the other hand some of the recent suburban expansions in the United States appeal only to demands for basic survival needs (e.g. shelter). To carry the theme further, two huge structures in San Francisco are interesting to contrast. The Golden Gate Bridge shown in Figure 1-13 is considered by most persons to compliment the natural beauty of the bay area and the unique urban character of San Francisco. On the other hand, the Embarcadero Freeway, shown in the preface section (page iv), was stopped because the disharmonies with the San Francisco character were too great. To harmonize function and utility with interacting values is the challenge. Or if harmony is not possible, reasonable alternatives must be proposed.

Figure 1-13. Golden Gate Bridge in San Francisco.

A number of themes can be developed to explain the general problem of environmental deterioration. One of the most succinct and appealing is found in Garrett Hardin's essay, *The Tragedy of the Commons*. In this article he supposes ten parties each grazing ten cows in a common pasture. One party decides that by adding an eleventh cow, he will increase his welfare by ten percent, while causing the pasture to deteriorate only one percent. As others decide on this strategy, namely to increase their own individual welfares, the pasture soon collapses. The concept can be applied to the use of any of our common property resources by individual interests.

Welfare economics - The field of study relating to common property goods is called "welfare economics". It is a relatively new branch of economics and it is frustrating to some because there is not much amenable to quantification.

In welfare economics a key ides is *value*. There are many types of value; health, esthetic, recreation, etc. If a good or service can be traded in the market it has *exchange value*. Exchange value has a property of being *cardinal*; therefore, the exchange value of various types of goods and services can be added. This permits a determination of profit or, in public financed projects, net benefits. This is the domain of conventional economics.

If a good or service has no exchange value, or if exchange value is not a fair indication of worth, then there must be a preference ranking; this characteristic is called ordinality. In making choices regarding public goods, the political process is the only means of ranking choices. This is why *public participation* has since 1972 become an important pahse of the decision making process.

The ideas of cardinality and ordinality are extremely important. For years there has been a delusion that a public project can be evaluated by benefit cost criterion along. In the 1972 Draft Guidelines of the Water Resources Council, benefit cost is only one criterion for decision making. The other criteria, which include environmental values, are difficult to evaluate because of their intangible characters. Despite the inherent difficulties with these new criterion it has been a great stride forward to identify and *formalize* the importance of intangible values.

The idea of *incremental benefits* and *incremental costs* is another useful idea (economists use the terms marginal benefits and marginal costs). The terms merely relate the changes

effected to some total. For example the social cost of daming
one stream for andromenous fish in the Northwest may be minor.
The social benefits from the power may be significant - especially if there is little total power development. If virtually
all streams have dams blocking andromenous fish runs (ignoring
for the moment that fish ladders will help circumvent the
problem) the social costs of daming the last remaining   free
flowing stream would be inordinate; the incremental social
benefits from the incremental power developed would be probably
small. This general concept, elaborated further in Chapter 9,
aids greatly in developing a rational approach to decision
making.

Diversity - A country to be rich in what it can offer its
citizens must have a large array of value attributes reflecting
the heterogeniety and mix of its population and the different
needs for fulfillment. A good example of a region having a
large diversity in value attributes is in the San Francisco
Bay Area in California. Culturally the city of San Francisco itself is rich beyond measure. But in addition the citizens have
unlimited *opportunities* for enjoyment of the natural environment both in the local area and within 200 miles in the High
Sierra's.

A major problem, however, is that many areas of the United
States are being homogenized. Using the San Francisco example
again, Lake Tahoe was formerly, until about 1955, one of the
most attractive pristine lakes in the world. Urbanization has
changed its character to such an extent, however, that   the
former exhilarating qualities have been destroyed.

EQUILIBRIUM

The idea of *equilibrium* provides a valuable conceptual
model for understanding the change process. It is borrowed from
the physical sciences.

First consider the idea of static equilibrium. This condition exists when all components of a system are at rest;
that is they do not change with time in any of their characteristics. The *state* of the system refers to its condition as
measured by certain indicators. Pressure, temperature,   and
volume specify the state of a gas. Any changes in any of these
conditions results in a new state. A *process* is the kinetic
quality of transition between states--such as a gas expanding.
Figure 1-14 illustrates in a general way. the *state* of a land
parcel may be altered by building. A lake may be changed in
state by pollution; indicators of the new state may be changed
in measured clarity; in plankton counts, or in dissolved oxygen.

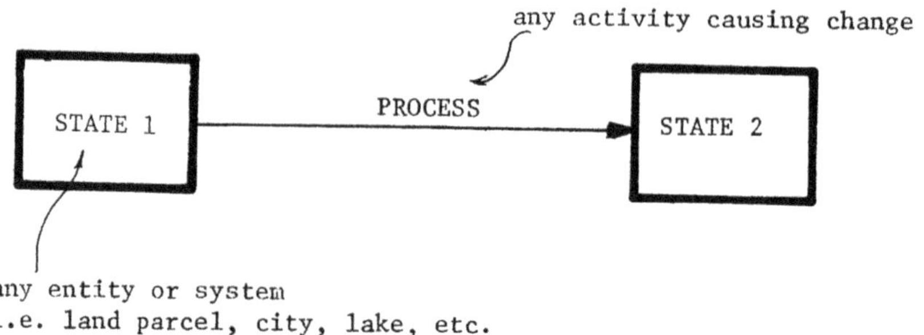

Figure 1-14. A process.

A system having input and output of materials is called an *open* system. The system being acted upon is in *dynamic* equilibrium as long as the rate of input and output do not change with time. Figure 1-15 defines a process. Figure 1-15 is a system in dynamic equilibrium. A city is in dynamic equilibrium. So is a household or a factory or any other entity. The *structure* of the system determines the relationship between input and output for a dynamic system. For an organized body - be it a factory or a city - can be made more effective, more efficient, or more effacious by appropriate structural changes which may be physical or non-physical.

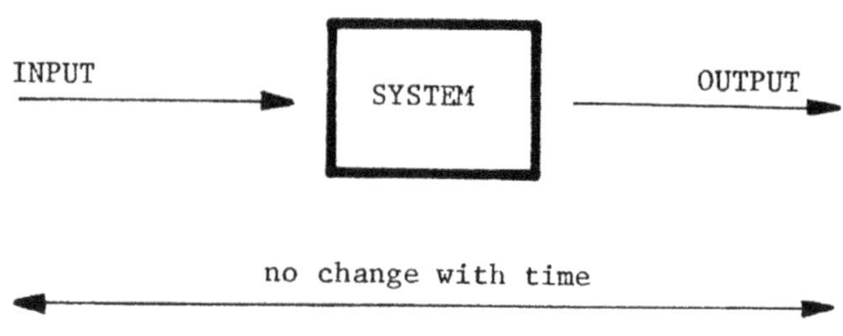

Figure 1-15. A system in dynamic equilibrium.

But regardless of the system, any perturbation on it by altering its structure or changing the input or output levels will alter that system in a myriad of diverse ways. Its *state* will change--and with it, *values*.

These are very simple concepts. They tell us that if a system is disturbed it will emerge to a new state. If a good

professional is present--say a fisheries biologist if a stream
is involved--or a sociologist if a human community is involved
--a few things may be deduced about the new state probable for
the stream or the community, the consequences of perturbation,
and the significance of change for *any* part of the system.

Figure 1-16 illustrates further showing the major components of a metabolizing system. The three broad categories shown
in Figure 1-16 are:

    (1)   resource extraction;
    (2)   metabolic activity within the societal system;
    (3)   waste production.

In a metabolizing system there are two key concerns, as
noted earlier; (1) how the *metabolism* of the system affects
other systems, and (2) how expansion of the physical infrastructure affects other system. These are *primary effects*. If
expansion of the physical infrastructure permits a greater
metabolic capacity, and this, in turn, affects another system,
then the effect on the other system caused by the facility may
be called a *secondary effect*.

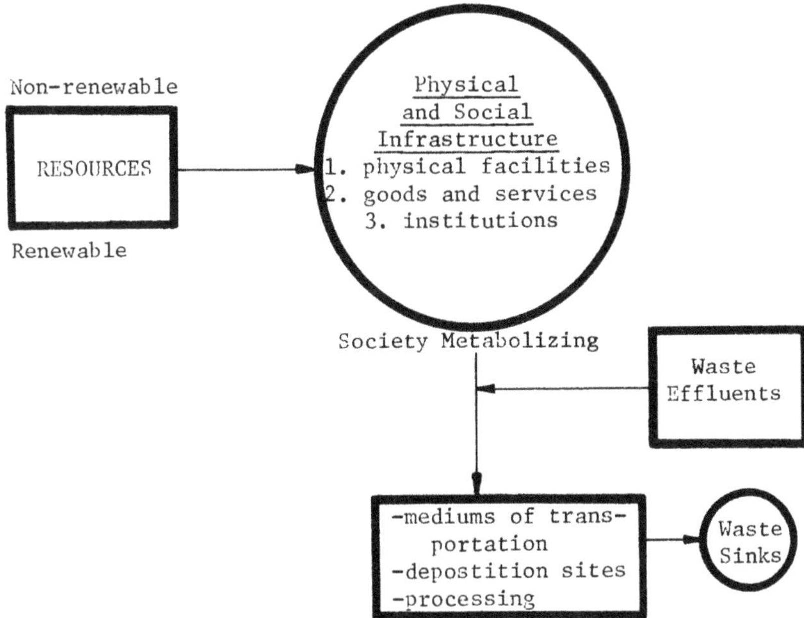

Figure 1-17. Elements of a metabolizing society.

It is the premise of this treatise that the needs for development activities requiring public support--public projects--will continue. And while the projects emanating from those perceived needs may or may not be advisable, we do need to develop planning and design approaches which will reach beyond the primary utilitarian functions of the facilities to be provided and incorporate, in a professional manner, the multitude of secondary effects caused. In other words, through a wider professional involvement, the undesirable *states* caused by the facility should be mitigated or avoided through the imaginative generation of alternatives in planning and through enlightened design.

This can be accomplished first by understanding the problems, the issues, the institutional arrangements, and the design requirements of the professionals and of the industries operating in the affected problem areas; transportation, water, and power are the specific problem areas circumscribed for this treatise. Then a knowledge of the affected environments---the earth's physical mantle, the biological communities, the human community, the esthetic--is needed in order to evaluate the disequilibriums that may occur and how to design in such a manner as to minimize the upsets and to harmonize parts wherever and whenever possible. For this purpose the disciplines representing these areas need to be delineated in both their substantive and operational characteristics. Their data requirements and methods of obtaining data, their analytical capabilities, and then the principles and discoveries underpinning the field of knowledge, and finally, the approaches in applying the respective knowledge to appropriate design of public projects, all must be delineated in a lucid professional to professional manner so that the other disciplines can effectively interface with those representing the affected environments. To effectively use this knowledge, a third area --the linkage mechanisms--must be delineated too. How to mobilize, coordinate, analyze, and to understand the means of facilitation and the limitations under the law must be shown in terms of management methods, systems analysis, economics and law, respectively. The central purposes of the chapters in this treatise is to provide first of all a sensitivity to the practitioner as to the various fields involved in any discussion of public projects' impacts and consequences, and then a springboard for the questions that need to be asked by various disciplines as to their contribution that they have to make in measuring spillovers of any technological intervention. Secondly, however, the purpose of such an exposition is also to alert the specialist in any given area as to what he can offer, or the types of challenges posed to him and the types of problems raised by practitioners in the field of public projects. What is envisioned, therefore, is a two-way interchange, namely a sensitivity of the practitioner as to the offerings of various fields, and, at the same time an awareness of the challenge, if not the urgent need and demand for providing an integrated framework that will bring together technical, social, economic, political, and cultural considerations in any attempt to solve the complex problems associated with natural and man-made activities within a given environment. In such a dialogue and interchange, there are common threads and understandings by both "hard" and "soft" scientists, and a recognition of the man-centered purpose of public projects. Similarly, emerging agreements must be emphasized concerning the secondary effects of public projects, and of the temporal perspective with potential long-range side effects.

Thus, we come again full circle to the original quest and purpose of this volume: to provide a professional level capability for holistic approaches on how to plan and design large scale facilities. To do this in the context of both fulfillment and survival is the basic challenge.

Acknowledgements

Figures 1-5 and 1-6 were reconstructed, by permission, from the book *Man the Builder*, by G. E. Sandstrom, published in 1970 by McGraw-Hill. Figure 1-7 was reproduced from that book also, and the book title inspired the section heading, Man the Builder in this chapter. The metabolism concept was inspired by an article by Abel Wolman, entitled, The Metabolism of Cities appearing in *Scientific American*, September 1965.

Professor P. H. McGauhey has thoughtfully criticized the chapter resulting in several important changes.

Glossary

Cardinal Value

> A value attribute is said to be cardinal if a number can be assigned as some measure of its value, which is additive with other substances or qualities. For example market value is a cardinal measure in dollars of a wide variety of goods and services. See also Chapter 8.

Ordinal Value

> A value attribute of some good or service or quality is ordinal in nature if it is *ranked* in comparison with other amenities. For example one may prefer playing tennis to bicycling. See also Chapter 8.

29

Indicator	Some attribute of an activity or substance which may be indicative of its condition and useful in making comparisons. Examples might be per capita income to reflect individual wealth of a country, gross national product to indicate the aggregate exchange value of all goods and services, coliform bacteria to indicate the possible presence of pathogenic bacteria in water.
State	Condition of a system at any given time described by the aggregate of all relevant indicators. Water quality, economic health of a region, recreation potential of a region, ecological condition, etc. all can be described by a set of indicators, usually extablished by the various professions.
Equilibrium	A system at rest is in *static* equilibrium; that none of its properties are changing with time. A system is in *dynamic* equilibrium if none of the inputs, outputs, or processes within the system change with time.
Process	The activity of a system changing from one state to another is a process.
Open System	A system having inputs and or outputs.
Social Benefits	Aggregate of all tangible and intangible benefits resulting from a project. Employment, income, recreation, etc. are examples.
Social Costs	Aggregate of all deleterious results of a project or activity. A freeway bisecting a community may upset patterns of communication. A dam blocking an andromenous fish run diminishes the fishing industry and sport fishing and could result in the demise of a species. These are all social costs.
Metabolism	The level or intensity of processing inputs and outputs to a system by that system is that system's metabolism. The term was adapted up from Abel Wolman's, The Metabolism of Cities, Scientific American, Sept. 1965.

Physical Infra-
structure

The aggregate of all facilities which allows a society to function as it does, including streets, highways railways, airports, canals, dams, power plants, transmission lines, telephone lines, etc. The *macro-scale physical infrastructure* allows the society to function in a gross or bulk sense; included are inter-city highways, transmission lines, power plants; large buildings, dams, canals, etc. The micro-scale physical infrastructure distributes from bulk parts of the system to individual units. Streets, power lines, sub-stations, water mains, sewers, dwelling houses, etc. comprise this portion of the system.

AUTHOR NOTES

This chapter represents the joint thinking of the four editors, D. W. Hendricks, E. Vlachos, J. C. Kellogg, and L. S. Tucker. Their concern was not only to define the scope of the total treatise, but to outline, in broad perspective, how professional practice is in a state of transition from the *survival* ethos of past years toward a more holistic *fulfillment* ethos, as a need for future.

D. W. Hendricks is Associate Professor of Civil Engineering at Colorado State University. His work has ranged broadly in teaching, research, and consulting in water resources and sanitary engineering to include traditional engineering, systems modeling, and environmental assessment. Since 1968 he has had a growing interest and involvement in interdisciplinary approaches to civil engineering. He has a doctorate in sanitary engineering and is a registered professional engineer.

Evan Vlachos is also the author of Chapter 6, where notes appear on his background.

L. Scott Tucker is the Excecutive Director of the Urban Drainage and Flood Control District of Metropolitan Denver, a regional service agency encompassing over thirty local governments. As a civil engineer, his area of speciality is in the area of urban water. He served three years as Deputy Program Director of the ASCE Urban Water Resource Research Program. He was also with Colorado State University for two years where he was an original principal in the project responsible for this treatise.

J. C. Kellog is President of Kellogg Corporation, an engineering management consulting firm located in Littleton, Colorado. Formerly he was Vice President of Al Johnson Construction Company of Minneapolis where he developed extensive construction and management experience on major projects through the United States. His most recent job, prior to starting his own firm in 1970, was Chief Engineer for the consortium of contractors building the Eisenhower Tunnel through Loveland Pass in Colorado.

2

THE EVOLVEMENT PROCESS

BY NORMAN I. WENGERT

The processes from which public projects evolve are tremendously complex, varying with the nature of the program, with the character of technology, with goals and with leadership, with the location of power and the distribution of influence, with the ability to "get things done," with public attitudes, understanding, expectations, perceptions and values, with the structure and organization of grovernment, and with societal institutions and institutional arrangements. And this is probably only a partial list! But in any case, to analyze all of the factors which shape the evolvement of public projects would require far more space than this chapter permits. Such an analysis would involve a review of the entire spectrum of societal decision making, and would necessitate case studies of many public and private sector programs and activities as related to the political process.

What is a Public Project?

The complexity is made more confusing to the practitioner since there is no fixed definition of either "public" or "project." The construction of a large dam by the Bureau of Reclamation, or the extension of a segment of Interstate Highway by a State Highway Department, are obviously public projects. But a generally applicable operational definition of the concept of public projects is more difficult to spell out. Although most electric power in the United States is generated by "investor owned" public utilities, their construction activities (e.g. the "Four Corners" power generation program) might logically be included in a definition of public projects. Perhaps an analogy may be drawn from the traditional legal definition of a "public utility," namely an activity "affected with the public interest," whether privately or publicly financed and managed.

But just as much difficulty is encountered with the term
"project." Congress has specifically authorized the Upper
Colorado River Project and the Columbia Basin Project. But a
stretch of highway a mile or two long is also called a project,
and the supervising engineer may bear the title of project
engineer. Obviously, how a project evolves is partly a matter
of defining the activity to which the term project is applied.
To illustrate, consider the previously cited cases; the Colorado
River Project has a history of many decades and a rich record
of political and social interaction; in contrast, a stretch of
highway through Nebraska may involve a very short and simple
decisional situation. But both may properly be designated
public projects from the viewpoint of this work.

What is One's Point of View?

Another variable of importance to a description of the way
in which public projects evolve concerns the point of view from
which the process is observed. Two perspectives deserve special
comment. One is the effect of one's hierarchical or authoritative position with respect to the project; the other is the
time perspective. An adviser to the Secretary of Transportation or to the head of the State highway agency will obviously
perceive the processes through which a project evolves differently from the newly graduated engineer working for the construction contractor, or the planner working for a city or
county. Basic policy decisions may be of interest to both but
do not directly involve their individual job responsibilities.
To both, moreover, most of the policy decisions have been made
before they become involved. Seeing the big picture at the
top of the hierarchy means, of course, that one is concerned
with longer term consequences and required, if not always competent, to assess meaningful alternatives. Men at the top may
be bogged down with details, but from time-to-time, at least,
they do deal with the formulation of new policies and programs
and thus they may participate in the initiation of a whole line
of public projects. In terms of the decisional time frame, it
makes a difference whether the practitioner is dealing with
preliminary, reconnaissance studies, comprehensive general
planning, detailed design, or actual construction. Clearly the
interests and perspectives of the policy planner are different
from those of the program implementer, and responsibilities
differ accordingly.

The Reclamation Act of 1902, which is the basic authorization for the program of the Bureau of Reclamation, owes a
part of its origin to the work of Frederick Haynes Newell who
was Chief of the Division of Hydrography of the U. S. Geological
Survey at the time the Act was passed. His input to irrigation
policy included preparing speech material for the President of

the United States, testifying before Congressional Committees, providing public information on the arid lands, etc. One popular encyclopedia lists him as the "founder of the U. S. Reclamation Service" of which he became the chief engineer. The act is often called "The Newlands Act" after the Nevada Senator who introduced it and pushed for its enactment. Yet if one were tracing the evolvement of this historic statute, one would have to begin at least twenty years before Newell or Newlands came on the scene. Newell graduated from Massachusetts Institute of Technology in 1885, and by that time John Wesley Powell had issued his famous report on the arid lands (1879) and had been jousting with Congress on public domain and irrigation policy for at least a half dozen years. And before Powell, the Mormon leaders had developed extensive experience with irrigation. The Reclamation Act of 1902 was, thus, the culmination of twenty or thirty years of experience and discussion. But strictly speaking, the Reclamation Act is not a public project, although it provided the authorization for many subsequent structures and river development plans, and its very existence has been crucial to evolvement of all the projects with which the Bureau of Reclamation has been concerned.

PUBLIC PROJECT DECISION MAKING

To consider how a public project (however one defines the term) evolves one needs to consider the context within which the set of decisions were made that resulted in authorization of the particular public project. This chapter, therefore, seeks to identify some of the more general factors which have been significant in the evolvement of many public projects, and also to suggest some distinctive differences among project evolvement in the substantive program fields with which this book deals (transportation, water development, and power generation). Its purpose is to describe the setting within which public projects are initiated and developed so that even the practitioner who may not be personally involved in policy or program decisions may yet understand the factors that influence project evolvement.

Contextual Elements

As the chapter proceeds, some of the important contextual elements will be identified and illustrated. But in a broad sense, contextual factors influencing public decisions may be catalogued to include the following:

> Human factors--the decision makers or actors, their personalities, their values, their roles, their power and influence as individuals, as officials in the decisional structure, as interest groups, etc.;

Scientific and technological factors--the state-of-the-art
and the competence of the practitioners; Physical
reality--the environment within which decisions are
made and action is to take place;
Social reality--the society, its institutions and organizations, its values and perceptions;
Political reality--government policies and structure,
formal and informal organizational arrangements,
constitutional authority and constraints;
Economic reality--costs and benefits, and also institutional arrangements for dealing with economic
dimensions;
Temporal reality--the time and circumstances under which
decisions are made or required;
Historical traditions and commitments--the evolution of
policies and programs over time.

These categories are not meant to be exhaustive, but
rather to suggest the complexities, not only of particular
project decisions but also of the background against which
such decisions are made and which condition action. Some of
these categories are discussed more fully in the following
pages.

Historical Trends

Public projects even in the fields of primary concern in
this treatise, have many purposes and develop in response to
many stimuli: economic development, resource conservation,
improved public services, protection of societal values, etc.
But increasingly, whatever the primary purpose, public projects
have environmental consequences and create environmental problems or result in environmental enhancement, and are evaluated
in these terms.

Changes in national attitudes with respect to the environment are results of four major developments, all of which
reached their fullest development in the Twentieth Century.
The first of these is the so-called conservation movement. The
second is the development of urban planning and land use control. The third is the resources programs of the New Deal with
their emphasis on economic revival and the attempt to merge
economic development and resource conservation. And the
fourth is the present concern with environmental preservation.

The conservation movement, articulated and brought to
public attention in the first decade of this century by
Theodore Roosevelt and Gifford Pinchot, was in fact the
fruition of years of effort on the part of a relatively small
group of scientists and civic leaders, intellectuals, and
academics who feared that the nation would soon run out of

basic resources, if it failed to develop policies of "wise use".
To this concern over resource depletion was added a negative
reaction to the mismanagement of resources by government and
private sector alike. Much of this concern focused on the
public domain and the national forests (first authorized in
1891).

There were many themes which converged in the conservation
movement, but the one common theme which united divergent
groups was their *willingness to use government to accomplish
social purposes* and to achieve what they considered to be "the
public interest." Although taken for granted today by conser-
vatives and liberals alike, this "pro-governmental bias" of the
first conservation movement represented a significant turning
point in American social philosophy.

The political effectiveness of the conservation movement
declined when it no longer benefited from the resourceful
leadership of Theodore Roosevelt and Gifford Pinchot, for there
just was not a sufficient bond of interest between the various
groups that made up the coalition once these forceful leaders
were out of positions of authority. Instead a variety of seg-
mented programs moved ahead.

As a part of the common agreement regarding the use of
government to accomplish societal ends, many of the participants in the conservation movement had sought control of resource waste and rationalization of resource development. This
became the "wise use" doctrine of the Forest Service, but the
point to note is that the emphasis was implicitly on *development* and not *preservation*. The Governors Conference in May,
1908[2] and the subsequent report of the National Conservation
Commission[3] compiled by the federal agencies, contained no suggestions of limiting growth. This is primarily a present-day
concern.

The conservation movement at the turn of the century,
premised on the use of government to preserve, protect, and
develop resources, and to punish the despoilers of these resources (particularly those involved in land fraud), had two
correlary consequences: *first* the expansion of the bureaucracy
through the creation of many new or expanded programs and
agencies, and *second*, an emphasis on professionalization, most
clearly evident in the development of forestry schools.

It is to be noted, too, that there was a clear technocratic and elitist bias in the early conservation movement in
that it sought to raise the professionally trained specialists
to positions of power and authority. At the same time, the
movement was influenced by Populist and Progressive values as
evidenced in its castigation of big business and "the trusts,"
and in its early sponsorship of public hydro-power development.

In positive terms, efficiency and wise use become important catch words, and were given major emphasis in 1912 when Theodore Roosevelt ran for the Presidency as the candidate of the "Bull Moose" party.

Urban planning and land use control had quite different origins. Essentially, they reflected a local governmental response to increasing urbanization and population growth, even as today these same forces demand even more effective planning and land use control. From 1900 to 1920 the population had grown by 50 million, from about 70 million to about 120 million. And for the first time urban population exceeded rural, a situation encouraged by industrialization and made possible by the automobile. This was also the period of tremendous immigration, but population growth, also reflected relatively high birth rates.

Urban zoning as a major technique of planning was experimented with in New York in 1916, and within ten years, encouraged by Herbert Hoover as Secretary of Commerce, the control of urban land use and of subdivision practices became widespread. The constitutionality of this new kind of government regulation was upheld in a case involving regulations of the City of Euclid, Ohio.[4] The most conservative Justice McReynolds on a very conservative court wrote the opinion declaring that urban zoning to regulate land rights was constitutional. The doctrines on which his opinion rested were primarily those of private nuisance control and police power action, with little attention being given to public interests in an attractive urban environment.

Partly because President Franklin D. Roosevelt was personally interested in conservation, partly because many of his associates (the so-called "Brain-Trust") were also conservation oriented, and partly because the mood of the times seemed to favor conservation, many New Deal programs were rationalized in terms of their conservation impact. But because of the Depression and the high rate of unemployment, public projects were regarded primarily in terms of the number of jobs they would produce and the stimulus they might give to the economy. The emphasis was that of "wise use, but the objective was economic growth and development.

Two programs were particularly symbolic of this era, the one involving substantial public works investment namely, TVA; and the other (which incidentally placed great emphasis on citizen participation) the program of the Soil Conservation Service and the State Soil Conservation Districts which it sponsored.[5] The appeal of the conservation rhetoric was evident in the fact that, after the initial statute to control agricultural production had been declared unconstitutional, Congress enacted

a very similar law justified primarily in terms of the public
interest in soil conservation on individual farms. The technique to be used to obtain this public goal was the expenditure
of public funds, paying farmers for following a variety of conservation practices.

Turning to the present, it is not at all clear what served
to stimulate the public concern for the environment and its
protection. Expressions of concern had been voiced contin-
ously, but they had not resulted in what seems to be a massive
shift in public attitudes and perceptions, nor did they lead to
dramatic political action in the interest of environmental protection. One can identify some of the stimuli, some of the
concepts, some of the expressions of concern and fear, but it
is difficult to indicate with any degree of assurance just
what triggered current attitudes and led to the increased
political activism to protect environmental values.

Very likely a combination of events and situations were
involved, and some of the dominant premises of the present
environmental movement may be identified. A new and dominant
theme is a lack of confidence in government agencies, in the
establishment, and in many (public program) goals which had for
many years been unchallenged. Perhaps the Viet Nam war contributed to this mood; perhaps our getting to the moon while
obvious evidence of environmental deterioration appeared all
around us, raised doubts as to National priorities. These and
many other stimuli probably contributed to the intensification
of environmental concerns.

For some politicians, as well as for some civic leaders
and for some academics, the current environmental movement has
provided ego rewards that cannot be minimized. But the most
significant and novel aspect of the new concern for environmental protection is the extent to which it reflects a lack of
confidence in government agencies and government programs, as
well as in private corporations and business leaders. Whereas
the earlier conservation movement had looked with hope and
confidence to governmental intervention in economic matters,
the present mood frequently expresses doubts both of the
motives and of the competence of government bureaucracy. Hence,
it was logical that those concerned for the environment should
emphasize *citizen participation* in planning and in other
aspects of the governmental process. While there are many
motives behind the quest for increased citizen participation,
much of it reflects a questioning of the ability of the government agencies, technicians and professional specialists to
respond to citizen interests. The objective of "power to the
people" is to make bureaucrat and technocrat accountable!

Citizen involvement and participation in the first conservation movement was minimal. When it occurred, its function was simply legitimation of decisions made by professional bureaucrats. The situation with respect to citizen participation in the New Deal period may be highlighted by the differences between the Soil Conservation Service program with its encouragement of Soil Conservation Districts, on one hand, and the farmer committee structure developed to control agricultural production, on the other. The committees were much more deeply rooted in the local political fabric, and they often possessed more interest and influence in the development of the programs with which they were involved. In many areas, in contrast, soil conservation districts tended to be dominated by the federal agency which had stimulated their creation. Moreover, the federal field empolyees of SCS took their orders from their superiors in regional offices and ultimately from Washington, rather than from locally elected farmer committees.

To summarize, among the significant changes in the late sixties, one was the grand disenchantment with many aspects of public programs, another was the demand for "maximum feasible participation." The latter was significant because it generated pressures to reexamine definitions of "public interest" from those which had been assumed by many professional public servants, and introduced new elements into the power structure. The demand for citizen participation often has a strong anti-intellectual bias, but it also frequently revealed citizen frustration with the consequences of overspecialization.

The lack of faith in the professionals, in the bureaucracy, in the government agencies, resulted in active interest groups more frequently challenging government decisions and in many cases in using the courts to try to stop certain kinds of action. As a consequence, many resource agencies were also forced to reexamine their relationships to "the public," and took a variety of steps to involve citizens more effectively in planning. In some cases, because of the nature of agency responsibility and because of deep rooted traditions, participation was conceived as simply a means for legitimation, in others it seems to promise better planning, and a more careful consideration of a wider range of inputs.

Another current theme, which supports the interest in environmental protection, is the renewed concern for "the future," and the fear that our rate of resource consumption will inevitably deprive future generations of the necessities of life. Often this concern is coupled with the fear that America is exploiting resources of the impoverished world for its own affluent way of life.

In short, the present environmental movement seeks to protect and conserve a wide range of environmental values and in the process it challenges growth of the economy as a proper public goal. At the same time, the current concern for environmental protection places great emphasis upon the consequences flowing from governmental action, and the result is that often the concern for environmental protection is negative -- against government programs -- and it is at this point that the use of law (both statutory and judicial) has become important.

Particular Stimuli to Action

Unlike human conception which occurs from a single act, the conception of public projects may have many different stimuli. In the simplest situation, a particular project may originate from a gleam in someone's eye, someone who has the persistence to promote the project until it reaches fruition. For twelve years, Senator George W. Norris fought for the development of the Tennessee River and from his efforts came the Tennessee Valley Authority in 1933. Others, of course, participated with him, and events prior to his taking an interest in the development of the Tennessee River are not without significance. But in a very real sense what became TVA began as a gleam in the eye of Senator Norris.

Another easily characterized situation is the crisis or cataclysm which stimulates public demands and ultimately may result in public projects. The flood control activities of the Army Corps of Engineers fall in this category.

Still another stimulus to public action results from the articulated demands of interest groups, pressure groups, civic or other leaders, or local communities. Not unrelated to this stimulus is that which comes from government agencies. Once the Reclamation Act was passed, it was to be expected that the Reclamation Service (later re-named the Bureau of Reclamation) would submit proposals for action to Congress. A good deal of stimulus to action comes from responses to "sunk costs," to precedents, to social (and bureaucratic) inertia. Thus, decision "a" requires decision "b" requires decision "c" and so on. Or, bordering on a kind of bureaucratic pathology, proposals are suggested and decisions made simply because of patterns established in prior years. The typical budget process perhaps illustrates this best, where items once in the budget are continued year after year. But even outside of budgeting, decisions based on precedents come naturally since they involve an obvious economy of effort.

Again, the purpose in listing some of these stimuli to action has not been to exhaust the possibilities, but to suggest the spectrum of alternatives, and to counter simplistic

views of how public projects are conceived--one of the most
frequently heard of which is that government simply responds to
the "felt needs" of the people. What people want and what they
need are not identical, although the rhetoric of political persuasion may involve convincing people that particular wants are
indeed needs. And this suggests the importance of leadership
in articulating demands, in formulating value goals, and in
seeking action.

Models of Decision Making

Popular discussions of decision making and some of the
literature on the subject, particularly with respect to the
evolvement of public projects, tends to suggest a level of
rationality both in procedure and in substance which in fact
does not exist. For a fuller understanding of how particular
projects may move from an idea in someone's head or a dimly
recognized need on the part of a leader or a segment of the
public to a full-blown public action commitment requires a more
sophisticated view of the kinds of forces and factors at work.

Typical of the simplistic view is the belief that a "felt
need" develops among citizens, is discussed and debated by them,
and then become crystalized as public wants (needs, or demands)
which lead to political responses which in turn lead to decisions for program action. Congress (the legislature) responding to the demands of people, passes a law, and the executive branch "executes" the law. Despite decades of criticism, it seems comfortable to many, following this conception
of decision-making, to draw a dichotomy between policy making
and project authorization, on the one hand, and administration
on the other. An example of this neat view of how public projects move to the action stage, has for years been presented by
the U.S. Army Corps of Engineers in its public relations pam-
phlets and is often referred to as "The Major Steps in the
Conception, Authorization and Construction of Civil Works Projects by the Corps of Engineers."

These so-called steps are designed to provide information
to interested citizens seeking help from the Corps. They are
significant both because they play an important part in the
administrative doctrines of a major project agency, and because
they obscure so much of reality.

It is useful to quote the steps here as a basis for comment, even though these procedures appear now to be in the process of change, partly as a result of the National Environmental Policy Act, and partly for other reasons.

    Step 1: Initiation of action by local interests.
    Step 2: Consultation by senator or representative with

the Public Works Committee [in order to get authorization to study the request of local interests. This is unique among Federal agencies, most having continuing authority for planning and program research.]

Step 3: Action by Public Works Committee [if it is convinced of the need for a review report].

Step 4: Assignment of investigation by Chief of Engineers [assuming that funds are appropriated for this purpose].

Step 5: Public hearings by division or district engineer [in order to ascertain the views and desires of local people with respect to the extent and character of the improvement desired].

Step 6: Investigation by division or district engineer [through field and office studies].

Step 7: Review by division engineer and issuance of public notice [so that those interested can file their views with the "board of engineers for rivers and harbors"].

Step 8: Review and hearing by board of engineers for rivers and harbors (or the Mississippi River Commission in the case of that stream).

Step 9: Preparation of proposed report by Chief of Engineers and review thereof by state and federal agencies.

Step 10: Transmittal of report to Bureau of the Budget [now Office of Management and Budget] for a determination of the relationship of the report to the program of the President.

Step 11: Transmittal of report to the Congress by the Chief of Engineers [complying with the resolution or act authorizing the investigation].

Step 12: Project authorization by the Congress [usually in the Omnibous River and Harbor and Flood Control Bills].

Step 13: Assurances of local cooperation [i.e. reaffirming the desire of local interest groups to support the project].

Step 14: Request for planning and construction funds [which may occur many years after authorization].

Step 15: Appropriation of planning and construction funds.

Step 16: Preparation of detailed plans for the project.

Step 17: Invitation to bid on the entire project or on parts thereof.

As subsequent discussion suggests, this view of the way in which public projects decisions are made ignores almost entirely the political interaction which is a necessary and inevitable part of the process. It overlooks leadership, and slights the interplay of power and influence, the conflict and competition, the bargaining and tradeoffs that are usually involved.

Harold Laswell many years ago defined politics as "who gets what, when and how" and to this should be added, "where". And project authorization necessarily involves politics in just this sense, if for no other reason than that there are never

sufficient funds available to undertake all of the projects
which may seem desirable or meritorious, and there is constant
competition for government benefits. So choices as to who gets
project money must be made! The demands made by important and
substantial interest and pressure groups are beyond the capacity of the system to satisfy. Thus, priorities must be set and
choices made. And *choosing is the essence of the political
process*. The question of project evolvement, then becomes *what
are the factors that influence or determine political choice?*

There are two dominant models of decision making. *One*
model places a heavy emphasis on step-by-step rationality, and
on the maximization of outcomes. This has been called "the
rational-comprehensive approach" because it emphasizes a logical progression, first, from a recognition and analysis of a
problem, through a review of all possible alternatives and a
balancing of the consequences of each alternative, and then,
finally to the selection of the *best possible solution*.

This essentially "technocratic" model, gives full play to
the specialist and specialized knowledge in-puts. It appeals
to engineers and other technically trained practitioners (including economists) just because it pushes aside the irration-
alities of human behavior, and the vagaries of individual desires and preferences. But as a result it is far from reality,
and it rests on an assumption, impossible to achieve, that it
is in fact possible to analyze all aspects of a problem and to
consider all possible alternatives.

The *second* model emphasizes incremental decision-making.
It is both descriptive and normative, finding considerable support among political scientists and sociologists, because it
can take human behavior and value differences into account.

The economist Anthony Downs[6] has pointed out that officials operate in a "realistic world, not a perfectly informed
world of traditional (economic) theory." And he has suggested
six limitations which to him make the rational-comprehensive
model of questionable validity:

1. Each decision maker can devote only a limited amount of
   time to any one decision.

2. Each decision maker can mentally weight and consider only a
   limited amount of information at one time ("span of attention").

3. The responsibilities of most officials require them to become involved in more activities than they can consider
   carefully and simultaneously; hence they must normally
   focus their attention on only part of their major concerns
   while the rest remain latent.

4. The amount of information initially available to every decision maker about each problem is only a small fraction of all the information potentially available on the subject.

5. Additional information bearing on any particular problem can usually be procured, but the cost of procurement and utilization may rise rapidly as the amount of data increases, and time pressures mount.

6. Important aspects of many problems involve information that cannot be procured at all especially concerning future events; hence many decisions must be made in the face of uncertainty.

Others, writing on administrative decision making, have emphasized such factors as the influence of the organizational context, the importance of outside pressures, and of current values and views, the constraints of traditions and customs, and a host of other socio-psychological factors including the personalities of decision makers. Not overlooked in this view is the bureaucracy and bureaucratic behavior, as well as the shear size of problems, and the complexity of the organization which may be involved in reaching a decision.

It is in the context of such analyses that "incrementalism" seems more nearly to apply to the real pattern of decision making. This is described by Charles E. Lindblom[7]

1. Rather than attempting a comprehensive survey and evaluation of all alternatives the decision maker focuses only on those policies which differ incrementally from existing policies. He prefers to move by small steps, rather than by drastic changes.

2. Only a relatively small number of policy alternatives are considered.

3. For each policy alternative, only a restricted number of important consequences are evaluated.

4. The problem confronting the decision maker is continually redefined, incrementally allowing for countless ends-means and means-ends adjustments which in effect make the problem more manageable.

5. Thus, there is no one decision or right solution, but a never-ending series of attacks on the issues at hand through serial analysis and evaluation which often involve a kind of trial and error approach.

6. As such, incremental decision making is described as remedial, geared more to the alleviation of present, concrete social imperfections than to the promotion of future long-range goals.

Often described as "muddling through," the incremental approach is modest about the correctness of any one solution, recognizing that many unforeseen independent variables may change results considerably. The incremental approach permits and even encourages compromises, adjustments, trade-offs, and revisions of views and values, and is not bothered by such accommodations, since it eschews absolutes. Herbert Simon has suggested that this pattern of decision making involves "satisficing," rather than optimizing, rejecting the latter for, among other reasons, its impossibly high information input requirements.

While the incremental approach may be criticized for being short range and skeptical of master plans and grand policies, it accommodates social-psychological factors in human behavior, and recognizes the variety of forces and factors that may shape public decisions. In most situations, incrementalism is tolerant of ambiguity and avoids the imperatives of resolving basic value-conflicts. It encourages continuous interchanges of ideas and a constant sifting and winnowing of proposals and programs, so necessary to the continuance of a diverse and pluralistic society.

But both the rational model and the incremental model should be recognized as models, i.e. simplifications of reality for analytical purposes. In either case, they do not describe the real world, but only part of it. Thus it is easy to argue that if one does not follow a "rational" approach to decision making, one must be irrational. Or as some economists have suggested, if costs exceed benefits (using an economic growth model) then the activity is clearly "uneconomic," and should not be undertaken.

Rationality in decision making is not a simple problem because what is rational is not automatically determined. In a thoughtful book on the subject, Paul Diesing differentiates among five types of rationality and the social conditions which apply to each type[8] His categories include technical, economic, social, legal, and political rationality, and he suggests that rationality is a *method of thought rather than a norm of thought;* it is instrumental, not ultimate!

It is in this sense that one who favors incremental decision making does not in fact reject rational inputs to decisions. Every intellectual, every professor, every specialist, is dedicated to increasing rational inputs into decisions through research, through the application of knowledge through systematic analysis and thought. Systems analysis and other logical techniques have advanced our ability to deal rationally with mahy problems that heretofore were unmanageable. And no

one would deny the appropriateness of using all the tools at
our command to reach wiser decisions. But the philosophical
incrementalist recognizes that at any one time that which we
know is so much less than that which we do not know. He in-
stinctively rejects Plato's Philosopher King or Orwell's Big
Brother. And there is persuasive evidence to suggest that the
unknown is like the universe--constantly expanding.

This subject of diversity in values, beliefs, preceptions,
and expectations cannot be dealt with without at the same time
recognizing that there are some who deplore the pluralism of
our society, who insist that no one has the right to be wrong,
and who would use the growing arsenal of psychological tools to
eliminate diversity, to socialize and condition the citizenry
to greater uniformity and hence harmony. This view is here rejected.

On the negative side, an incremental approach, which recognizes differences and tolerates ambiguity, also can contribute to avoiding decisions, to discounting the future effects
of present decisions, providing a convenient rationale for *not*
implementing action courses that seem imperative (planning).
This is the dilemma of our system; the challenge would seem to
be to develop a pragmatic merging of instrumental rationality
with an incremental outlook so as to have the best of both
worlds. It is easy to forget that the world has known few
benign dictators, and that dogmatic rationalism almost necessarily must override the often inarticulate values and views of
large segments of the populace. To deny their right to be
wrong, is to deny a part of their humanity.

Most American politicians are innately incremental, probably not for profound philosophical reasons, but because they
are often up against the complex values and preferences of a
pluralistic constituency. In this respect, and for somewhat
different reasons, American political leaders appear to be
somewhat different from their British counterparts. In a later
section of this chapter the fact that American political parties and politicians are not issue oriented or programmatic
will be discussed. The fact is that the American system most
often makes decisions incrementally. New government programs
rarely involve massive changes or drastically different actions.
Instead, changes are made in small steps to see how they work,
and to the extent possible, the options of changing or even
reversing directions are kept open.

This is one reason why the historic roots of particular
programs are important to understanding them. A recent example
is the Federal water pollution program begun in 1948 on a
modest basis, with no enforcement; revised in 1956 to include
modest enforcement provisions relying primarily on persuasion;

revised again in 1961, 1965, 1966, and 1970, each revision strengthening enforcement and broadening the scope of the Federal program. In the fall of 1972 Congress, over the President's veto, enacted a far-reaching new water quality program--the implications of which are not yet clear. In each enactment more funding was authorized and stronger administration accepted. From the beginning, vocal conservationists and environmentalists stressed that what Congress had done was not enough, and if the measures of effectiveness are reliable, perhaps they were right. But the politicians were right too, for they recognized the need to bring the public along in understanding the nature of the pollution problem and the need for action. Perhaps, they sensed that the technicians and the technology were not up to sweeping reforms and that gradualism permitted improvements in technology as well as increases in public acceptance.

The legislators were probably also aware that some of those urging actions were not particularly in that group which might be classed as "public regarding," but rather were promoting programs from which they would receive direct benefits, (e.g. consulting firms through the sale of their time and talents; equipment manufacturers through sale of equipment; contractors through construction contracts, etc.).

Program support may come from many sources and have many motives, just as many programs inevitably benefit others than those directly identified as beneficiaries in the discussion or debate at the time of programs authorizations. One need not adopt a conspiratorial theory of government to recognize that highway programs benefit the sellers of cement and construction equipment, as well as abutting landowners and employed labor. What is unfortunate is to overlook the fact that interests of this kind may also be influential in shaping public decisions.

Finally, it would seem appropriate to end this discussion of decision making by a reference to a wise comment of the late Justice Oliver Wendell Holmes who suggested (loosely paraphrased) that if one had sufficient power and was convinced of the rightness of his position, he had no choice but to override all opposition with force just because he was right. It is doubts about power and doubts about the rightness of a particular course, combined with a respect for the views and values of others, that reinforce incremental decision making in the American system.

INFLUENCING AND CONSTRAINING FACTORS

While a thorough review of all possible factors which might influence or constrain the evolution of public projects

is beyond the scope of this chapter, it is useful to comment on
some of those which seem currently to be of most significance
and interest, and which at the same time are often overlooked
or simply taken for granted.

The "People" and Citizen Participation

It is currently fashionable to emphasize the importance of
increasing citizen participation in planning and in administrative processes; some reference has already been made to
this subject. Because of its importance, a separate chapter is
included in this treatise.   The topic is relevant here because
citizen involvement and how it is structured vitally affects
the process of political decision making.

The literature of political philosophy is replete with
discussions of the relationship of the governors to the gov-
erned--and vice versa. Over the last several hundred years the
issues of democracy were issues of the relationships of citizens to government.   This was an important part of the American Revolution; it was an essential element of the French Revolution.   And many of the reforms in both Britain and America
in the 19th and 20th centuries represented responses to issues
of citizen/government relationships.

Yet most of this history sheds little light on why this
issue of citizen participation again became significant in the
early 1960's and continues so today.   For answers we must perhaps look not to history but to the more immediate social context of American politics in the decade just ended. No one can
fully explain the drives which have encouraged the demand for
more participation, although many have tried. Yet is is possible to identify what may be some of the contributing causes.
In part, the demand for involvement reflects widespread frustrations with the complexities of modern day life; in part, it
reflects the fact that from some points of view bureaucratic
action has indeed been high handed and has ignored important
values; and in part, it reflects the effort on the part of potential leaders to build new centers of power and to create viable support bases. While increasing citizen participation is
sound and desirable for many reasons, policy makers should realize that those who are most vociferous are not necessarily
those with the largest interests in particular decisions.   Democracy is not measured by decibels, nor equity and justice by
popularity polls.

Studies in recent decades have shown much about the character of public opinion. We have begun to understand the socialization process which shapes public opinions, attitudes,
perceptions and expectations. These investigations have pointed up the fact that on may issues there is a silent majority

and that those who articulate particular points of view do not
necessarily act as surrogates for the larger public or even for
any particular group within the public. They have also shown
that reason and rationality (even in terms of self interest)
may get short shrift?[9]

The decision maker confronting water project proposals
must realize, for example, that it is important not only to
know who supports and who opposes particular projects, but also
who benefits and who pays the costs. Weighing all factors is
the fundamental responsibility of the professional. He may
have to act, for example, in the face of strong bias, evident
particularly in water and highway planning, which often seems
to suggest that *only* the people living in a watershed or near
to the highway are the ones concerned with its development. In
many situations, too, the more vociferous speak for *now* and not
for the *future*.

Span of Attention and Time Horizons

Earlier it was suggested that decision makers are in reality unable to consider all possible alternatives. The same
basic human limitation affects the citizen public as well. It
is not simply lack of interest or perversity that prevents
meaningful public involvement--until the bulldozers have
started their work--but rather the fact that most people of
necessity can turn their attention only to those few topics
that at a particular time seem of dominant importance to them.
This is a basic psychological principle and it does no good to
excoriate the public for failing to use opportunities provided
to participate and to voice their interests. The hard fact is
that in many cases they do not or are not able to recognize
their interests until they become concrete and real, which
often means threatening. It must also be recognized that ordinary people probably do not think very far into the future.
Discounting the future is a common human trait. Together, the
limitations of span of attention and the inability to consider
long-range futures, make rational citizen decision making (i.e.
choices) difficult. Since decisions must be made, it is often
easier to make them in a highly personal frame of reference (e.
g. what's in it for me?). Moreover, even as ordinary people
suffer from these constraints, so also do politicians and
administrators.

Both individual and social psychology have accumulated
depressing evidence that man individually or collectively is
not the rational being he sometimes likes to think he is. This
is most evident in the political process where rhetoric may
substitute for thought and polemics for logical analysis.
Hence, it is not surprising that public relations and bureaucratic survival (legitimation) motives may all too often

characterize government behavior. There is little doubt that a
great deal of the concern for citizen participation is in fact
motivated by a desire for political support and legitimization
of agency activities and programs.

Power and influence. The loose jargon of everyday affairs
declares that *who* you know may be more important than *what* you
know. A quarter of a century ago Dale Carnegie authored a best
seller *How to Win Friends and Influence People*, which expressed
a practical approach to manipulating people, and views similar
to his have been common in the folklore of interpersonal relationships for many years. In a related way, research studies
on organizational theory have given emphasis to *informal* as
well as *formal* organization. It is through the informal structure that personal and personality factors most often come into
play. It is here where power and influence are often covertly
applied.

Theories of politics emphasizing the importance of groups,
and group interaction processes have been articulated at least
since Arthur Bently wrote in the first decade of this century.[9]
Earlier James Madison in *Federalist #10*[10] had formulated the
idea of group interaction in a democratic society and his ideas
have contributed to both folk wisdom and scholarly writings
about the American Political process ever since. For a time,
when reform was a dominant theme of American political action,
interest and pressure groups were regarded as evil and so
Kenneth Crawford could write a book entitled *The Pressure Boys*[11]
and Stuart Chase could deplore the "me first" philosophy of
Americans, criticizing the greed and selfishness which seemed
to motivate many interest and pressure groups.[12] Group theories
of politics, particularly after World War II, were thought to
offer a sufficient explanation of how the system worked, an explanation reinforced by economic theories of countervailing
power, such as those articulated by Kenneth Galbraith.[13]

That interest and pressure groups are important elements
of American political decision making hardly needs proof.   But
one major difficulty with group theories and concepts of countervailing power lies in the easy assumption that the sum of
expressed group positions represents, in fact, the total interest of society or of the community.   David Truman, one of
those responsible for the revival of group theory after World
War II, tried to avoid this pitfall by talking about "potential
groups," these being the groups which would arise when particular interests were violated.[14]

Group behavior remains an important factor in the development of public policy, but several decades of research and
analysis have begun to challenge group theory as a sufficient

explanation of the way in which the system operates. The critique of group theory by Mancur Olson[15] has been particularly effective in pointing out limitations of group explanations. Hence today such theories tend to be regarded as partial rather than an all-sufficient interpretation of how the system functions.

The Public Interest

The post-World War II era of Social Science, with its emphasis on individual behavior and its philosophical premises rooted deeply in logical positivism, saw a general decline in confidence in the concept of "the public interest." Although statutes continue to refer to the public interest, and administrative decisions are justified as implementing or seeking this objective, the trend has been to dismiss the concept of the public interest as unprovable. Instead, individuals are portrayed as self-seeking, dominated by personal interests, and trying as best they may to extract personal gain from the system. One result of weakening of the concept of the public interest has been to develop a crass cynicism with respect to public sector action. Only a few have continued to emphasize that, while it may not be possible to define the public interest in terms of eternal truths, the importance of the concept lies in the *search* for the public interest, in the requirement placed on government actors to rationalize and justify what they did or propose to do in terms of effects and consequences.

It is significant that this pragmatic emphasis has to some extent been restored to public policy evaluation in the new concern for the environment and for maintaining ecological balance. Perhaps the epitomy of this restatement is to be found in the National Environmental Policy Act of 1969, which establishes not a standard for a sound environment, but a process by which alternatives are explored and public decisions justified in terms of their impact on the human environment. The emphasis on the search for the public interest does not correct the fact that all humans necessarily view the world through their individual eyes, but what the new emphasis on the quality of the human environment provides is a procedure for assessing programs and actions and balancing position and negative consequences.

Although one may concede that all of us at some time define the public interest as that which coincides with our own views and interests, it is too restricted a view of human psychology to disregard the fact that individuals can be concerned for the welfare of others, that there are "public regarding" persons, and finally, that people may be educated, socialized and indoctrinated, with respect to civic virtues. Again the danger would seem to be that of oversimplification.

An example may serve to suggest at least a dim awareness of community values on the part of many citizens. No one likes to pay taxes, and many individuals and corporations use great skill to avoid them. Yet the fact is that taxes are enacted and often increased, even though it would clearly be in the power of the majority to abolish them. In states where the initiative and referendum is available to change constitutions and enact laws, taxes have not been eliminated or reduced; in fact, in some of those states the tax burdens are particularly heavy. No crass theory of simple self-interest can explain this fact. There would seem to be hope, therefore, that on some issues a sufficiently large number of citizens can be persuaded to seek out the public interest in certain lines of action, even when the immediate effects may be contrary to their own interests. One must note, in this connection, that there are also "public regarding" organizations--such as the League of Women Voters and the National Wildlife Federation, to mention but two. This is not saying that "public regarding" individuals or organizations are right, but simply that they are seeking public rather than individual or personal advantages.

Bureaucrats and Technocrats

Public projects, by definition, involve public agencies-- government organizations with responsibilities for carrying out project decisions and implementing public policies. For over fifty years (at least since Max Weber focused his attention on the subject), social scientists have been investigating bureaucracy and bureaucratic behavior, and formulating organizational theories. The practitioner working for or dealing with a public agency would do well to become familiar with the literature on these subjects. Some of what is popularly regarded as bureaucratic pathology may, in fact, be normal organizational behavior, reflecting size, and other aspects of agency structure and operating realities.

But in any case, whether particular ways of doing things are normal or pathological, agency practice and policy will shape the way in which public projects evolve. It is therefore necessary to understand administrative procedures as well as agency premises and goals: agency self-images as well as its traditions and biases.

Some examples: if one works for or deals with the U.S. Geological Survey, it is useful to understand the extent to which this agency is centralized, and how it regards its mission in highly scientific terms with traditions in this regard that trace back to the 19th century. Or if the agency is the Federal Aviation Administration, it is useful to know the extent to which private aviation interests and points of view

play an important role in its policy conceptions. If the a-
gency is the Corps of Engineers, it may be important to understand the intricate relationships between the commissioned
officers and the civilian employees, as well as the long-
standing and special relationships the Corps maintains with
Congress.

This is not the place to review the special characteristics of the many agencies involved in the program areas on
which this volume focuses. Nor can what is written here substitute for more detailed studies of bureaucratic behavior.
But it is important to recognize not only that decision makers,
being people, are motivated by the same views and interests
which motivate the human race generally, but also that organizational or bureaucratic social-psychology is important.
People functioning in organizations are influenced by those
organizations in particular as well as in general ways. An
understanding of these processes can be important to a grasp of
project evolvement.

Coordination

One of the most frequently encountered terms in decision
making is "coordination." Particularly in the Federal government, coordination is a dominant aspect of much bureaucratic
activity. This stress reflects, first, the tremendous size of
government and the many different programs and activities in
which it is engaged. For these reasons alone it is inevitable
that assignments to particular agencies relate to those of
other agencies, if they do not directly duplicate or overlap.
Second, the emphasis on coordination reflects an intellectual
concern for neatness and integrity, resting on the obvious fact
that in one sense everything relates to everything else.

At one time such a statement might have been regarded as
obvious, but completely non-operational. But the technique of
systems analysis, and information dissemination, storage, and
retrieval, together with a rapidly growing body of multidisciplinary research, availability of high speed computers, and
many other factors, have made it possible to consider interconnections and interrelatedness among numerous programs in
ways not possible before. As government becomes involved in
more and more aspects of the human endeavor, the opportunities,
if not the need, for better coordination become more apparent.

Four supermarkets on each of four corners at a highway
intersection are examples of free, competitive private enterprise; four post offices similarly situated would be vehemently
attacked as inefficient. The point is that the more government
undertakes, the more one will hear about the need for "coordination." And the more the tensions and other difficulties involved in coordination become apparent.

Symbolic of this emphasis upon coordination is the National Environmental Policy Act (NEPA), which requires a coordination of disciplinary approaches, a seeking out of interested and affected agencies, consultation with concerned members of the public, etc. While it does not directly deal with many of the fundamental dilemmas of coordination, it does recognize the responsibilities of the primary agency to choose and to decide. In this respect NEPA may be breaking new ground.

No one questions the desirability of effective information exchange, and often that is all that coordination represents. But the real problems of coordination stem not from lack of knowledge of "the facts" but from differing goals, differing problem perceptions, differing mandates, and differing clienteles. At the same time, there is commonly, a lack of authority to resolve conflicts and differences. For reasons that are deeply imbedded in the nature of bureaucracy, moreover, too often the term "coordination" is used to obscure the need for choice, for making decisions. And sometimes there seems to be a blind faith that simply exposing the facts and exchanging ideas will resolve all difficulties. In its richest meaning, coordination must include choices from among alternatives, decisions as to ends as well as means. And nowhere is this more important than with respect to public projects.

Coordination mechanisms have probably been most fully developed with respect to water projects. But thirty-five years of effort has not effectively resolved basic policy and procedural differences among the agencies responsible for water programs.

Geographic Politics

In a country as large as the United States, divided into independent states, with the states serving as a basis for legislative representation, it is not surprising that an important dimension of political choice involves distribution of costs and benefits among the several states and regions of the country. Both Congress and the executive branch are acutely aware of the need for spreading the benefits of public projects geographically. For example, Senator Vandenburg of Michigan frequently said in jest that he would not oppose TVA if Michigan could have a similar development project. A senator from a Southern coastal state is reputed to have said, in voting for the Reclamation Act of 1902, that since the coastal states had the benefits of the rivers and harbors·construction work of the Corps of Engineers, it was time to provide construction projects for the western states which did not have coast lines! Few put the issue so crassly, but there is little doubt that geographic politics has been and will continue to be important in the location of public projects.

The history of water projects provides many examples of the importance of geographical considerations. But other program areas have also been affected by geographic politics. The interstate highway program has established the same four lane standards across North Dakota, with few cities and sparse population, as for Pennsylvania or Missouri. But in addition to the simple areal factors, the existence of the Federal system operates to emphasize the importance of geographic constraints on decision making. Thus, for example, most Federal grant programs include grant formulas skewed to recognize the fact that the States exist. And for historical reasons, a strong emphasis on local factors (localism) continues to influence many aspects of political decision making -- from the obvious concern of politicians as to how public projects may contribute to their reelection, to the promotional enthusiasm of local chambers of commerce seeking markets and payrolls as boons flowing from project expenditures. As indicated elsewhere in this chapter, this situation is made more acute because so much of what the federal government undertakes is based on the constitutional spending power, and few things are more fluid or more readily divided on a geographic basis than dollars.

THE STRUCTURE AND PROCESS OF DECISION MAKING
FOR PUBLIC PROJECTS

Probably more than any other government in the world, the Government of the United States is characterized by a tremendous number of points of decision, virtually all of which in one way or another may shape and influence public projects. While most Americans take this situation for granted, the sheer numbers of governments in the structure literally boggle the mind of the foreign visitor. And although those involved with public projects in due course learn to work with the system, it is likely that few take the time to stand back and view this unbelievably complex structure.

The bare statistics begin to suggest how complex the structure in fact is. In addition to the Federal government and the fifty state governments, there are many, many local governments. Each of these, in turn, tend to separate legislative, executive, and judicial functions. All but one of the state legislatures have two houses, and both federal and state governments have dozens, if not hundreds, of departments and agencies in the executive structure. And the local level introduces chaos worse confounded!

There are a few more than 3000 counties in the U.S.; almost 250 metropolitan areas (SMSA's); over 600 urban places: 18,000 municipalities; and just over 17,000 townships. Citizens of the United States elect almost 522,000 officials, of which

508,000 are local, and the remainder state and federal. Not all of these officials, fortunately, are concerned with public projects, but a surprising number of them are.

Take highways, for example: the basic road system in the United States still remains largely a local government function. Of the 3.7 million miles of roads and highways in the United States, 2.3 million miles are under local control; 699 thousand are under state control; and 164 thousand are under federal control. The sources of funds is, of course, quite different, but this should not obscure the fact that a substantial amount of influence and control over road and highway decisions is lodged at the most local level--in road superintendents, city councils, public works departments, etc. In this maze of interlocking and shared responsibilities order and system is hard to find. It is no wonder that a review of these facts for some visiting Indonesian local officials resulted in an outburst from one of the officials: "And I thought Americans were efficient!"

Yet the practitioner involved with roads and highways has to learn to work in this maze. To be sure, the statistics tend to overstate the situation, since no one person, even those working for the Federal Highway Administration, would likely be involved with the whole structure. But certainly they must be aware of it, and more importantly, of how it may affect highway policies and programs. It is important to stress that in many ways this multiplex structure of government represents a dispersal of power and influence to local levels where in many cases it may be intimately related to successes and failures of particular officials when they run for office. It is clear that the basic strength of American political parties remains in the local communities, as will be pointed out below.

The simple and neat concepts of separation of powers and checks and balances in the American system have been pretty thoroughly demolished over the past fifty years. Clear distinctions among legislative, executive, and judicial functions just do not exist at either federal or state levels, although one can identify a kind of nodal clustering, with the executive and the executive agencies dominating the situation. But at the local level there is functional chaos. At all levels the organizational structure and patterns of authority and responsibility make the organizational specialist shudder.

BUREAUCRACY AND THE REALITIES OF EXECUTIVE CONTROL

Any conceptualization of the governmental system which identifies the "chief executive" and the executive function, tends also to imply that the executive in fact directs the administrative agencies of government. The analogies from business and industrial management come readily to mind. Yet one of the realities of American political life is that the Chief Executive (the President at the federal level, and governors and mayors at state and local levels) has only a limited control over the administrative establishment. We are accustomed to thinking of the interplay of politics in the legislative process, but tend to forget the tremendous political forces at work in the executive branch and among the many agencies of government.

A monolithic concept of the executive branch is very misleading. Among other things, it contributes to the myth that deep-seated and intractable issues of substance can be solved by reorganization. As Harold Seidman has said, "Where indeed is the commission or presidential task force with the self-restraint to forego proposing an organizational answer to the problems it cannot solve?[16]" Clinton Rossiter[17] quotes President Truman to the effect that the principle power possessed by a president is "to bring people in and try to persuade them to do what they ought to do without persuasion. That's what I spend most of my time doing. That's what the powers of the president amount to." Perhaps no American executive, business or governmental, can enforce his will simply by issuing an order, although simplistic critics continue to make this assumption. But the President, more than most American executives, is subjected to constant pulling and hauling from many power centers in and out of government. Some of this is a result of the pluralistic power structure of American society; some of it reflects an open and pluralistic society; some of it is a result simply of bureaucratic tendencies structure, and pathologies.

THE NETWORK OF INTERGOVERNMENTAL RELATIONSHIPS

The present American government was a response to a recognized need for a stronger central government. But the Constitution which went into effect in 1789 by no means abolished the states and the political power they represented. Altiough there is a great deal of handwringing about the gravitation of power to Washington, the states continue to exist and to exert tremendous influence on the way in which public programs are administered. Perhaps most important is the fact that much political power in the United States is local in character. As a result, federal programs must accommodate the fact of state and local interests. Undoubtedly the complex patterns of intergovernmental relationships which result are encouraged or reinforced by the fact that the national government remains a government of limited powers, and that its primary source of authority for acting has been the "spending power" through which local and state government conformance is often simply purchased.

At the same time, one would be blind to reality if one did not recognize the tremendous power and influence exercised by the federal government over state and local governments in how funds are spent. Congress has recently enacted a "revenue sharing" bill, which provides state and local governments with federal funds with a minimum of restrictions. But most federal funding of state and local projects continues to involve many substantive and procedural conditions. And while state and local officials may object to some of these restrictions, it is a rare situation in which they turn down federal funds because they do not want to meet the conditions involved.

In fiscal year 1972, federal grants to states and local governments amounted to almost forty billion dollars. Included in this total are two major public project programs, i.e., highways (over five billion) and water pollution control (about one billion) which together involved grants of over six billion dollars.

Substantive conditions include such things as design specifications in the case of both highways and sewage treatment plant construction. Procedural conditions include prohibitions against use of convict labor, against discrimination, and so forth. For certain programs coordination is required under the Intergovernmental Coordination Act of 1968 and its implementing regulations (Office of Management and Budget Circular A-95) which have instituted a new kind of intrastate regional planning.

Because of its sweeping language applying to all federal programs, the provisions of the National Environmental Policy Act of 1969 also apply to state activities financed in part from federal expenditures. As a result, for example, state highway departments are required to prepare impact statements with respect to construction of the interstate highway system. This provision reinforces provisions in the highway acts which relate to this subject.

REALITIES OF POLITICAL PARTY RESPONSIBILITY

Political scientists have for decades pointed out that American political parties are essentially "irresponsible." This harsh term emphasizes that American political parties are not programmatic or issue oriented, their primary function being to provide a system for the selection and election of office holders. National political parties provide a forum--an opportunity--for coalitions so necessary for success at election time. They are largely federations of state and local party organizations with few sanctions to discipline members.

Candidates, as well as citizens, choose their party affiliation on their own initiative without any requirement of loyalty or commitment. One result is that major parties contain within them a wide range of opinion, similarities between the two major parties often being greater than differences. Another consequence is that parties are not held accountable for programs.

Since parties have no programs, it is not possible to blame them for lack of accomplishment, nor to credit them with great success. It is not surprising, therefore, that into this policy and program vacuum have stepped interest or pressure groups and the bureaucracy. The active and open role of lobbies and pressure groups never fails to astound foreign observers, but even more surprising to many is the programmatic-political role of American bureaucrats. That drastic changes in what government does can originate with civil servants never ceases to amaze visitors from western democracies where the myth of civil service neutrality often comes close to reality.

ADMINISTRATIVE RESPONSIBILITY FOR PUBLIC PROJECTS: AN OVERVIEW

Major responsibility for public projects in the field of transportation, water, and energy lies with a variety of administrative agencies, both at the federal, state and local levels. Even though in certain program areas particular agencies have a dominant role, the situation is one generally characterized by a wide sharing of responsibility among many, many agencies--with competition and conflict not unusual.

A degree of centralization in the field of transportation at the federal level was achieved a few years ago through the creation of the Department of Transportation, and there is increasing interest in establishing similar agencies at the state level. But the existence of a single agency with the title "Department of Transportation" should not obscure the fact that responsibilities in the new department are widely diffused among a substantial number of powerful subordinate agencies; and some transportation activities remain outside the department.

It is not surprising that a major aspect of federal administration involves coordination, first, among federal agencies themselves, and second, between federal and state agencies. Without taking the variety of responsibilities associated with the different modes of transportation into account, it is nevertheless useful to indicate the organizational range of federal interests in the subject generally. Thus, the Treasury Department has an administrative responsibility for the Highway Trust Fund, the proceeds of which are accumulated from gasoline and other dedicated taxes. Presumably the Treasury also has an interest in the fiscal aspects of transportation, including such matters as gas taxes, oil depletion allowances, and taxes on airline tickets, as well as in local government decisions to build streets from the proceeds of tax-exempt bonds. The Department of Defense, through the U.S. Army Corps of Engineers, is a major factor in river and harbor improvements and thus plays an important role in water transportation. The Justice Department is concerned with antitrust problems, such as those raised in connection with recent railroad and airline consolidations.

Historically, as custodian of the public domain, the Interior Department was deeply involved in land grant policies important in the initial development of transcontinental railroads only 100 years ago. The Department of Agriculture is concerned with transportation as it relates to the movement of farm produce and the distribution of goods required by farmers. The Department of commerce is concerned with transportation as it affects the commerce of the nation. Its programs for regional development must deal with the transportation needs in particular regions. Strikes in transportation industries (e.g. those of longshoremen and railroad workers) involve the Labor department and several agencies dealing with labor disputes. The Department of Housing and Urban Development has an interest in urban transportation, both in terms of moving people and goods within the urban context, and in terms of mass transit as a key to reducing urban air pollution and achieving sound land use planning and development. The independent Environmental Protection Agency is involved with transportation as both a cause and a cure for air pollution. Since it regulates both rail and truck lines, the Interstate Commerce Commission is an important agency in the transportation picture. In terms of international trade and relationships, the Maritime Commission, and the Department of Commerce as well as the State Department, are important.

A similarly wide spectrum of agencies is concerned with water policies and projects. At the federal level those most directly concerned are the U. S. Army Corps of Engineers, the Bureau of Reclamation, and the Tennessee Valley Authority. Not unimportant is the Environmental Protection Agency with its pollution control programs involving authority for grants to state and local governments of several billion dollars per year. Agriculture, and Housing and Urban Development make loans for sewer and water supply systems, and the U. S. Public Health Service continues to be responsible for drinking water health standards. Water as a factor in industrial growth is of concern to the Department of Commerce, and a vital component in regional development activities. The Soil Conservation Service, particularly in its small watershed projects, is also concerned with water programs. And still other agencies too numerous to list have a variety of lesser interests in water programs, policies and projects.

In the field of energy, federal responsibilities are of three types: *first,* the actual production of energy, involving TVA, the Bureau of Reclamation and the Corps of Engineers; *second,* the promotion of energy production technology involving primarily the Atomic Energy Commission; and *third,* the regulation of large segments of the energy industry (natural gas, electricity, coal mine safety) involving primarily the Federal Power Commission which sets rates and determines accounting procedures, and the Securities and Exchange Commission which supervises certain financial aspects of the energy industry. Some federal agencies also have transmission and distribution responsibilities. Not so readily classified on an organizational basis are federal policies, such as the depletion allowance and oil import policies. Also important are policies concerning public lands, including oil and gas production from the intercontinental shelf, and recently, policies seeking to protect the environment as in the case of the Alaska pipeline.

State and local responsibilities in these three program areas are more limited or at least less complex. With respect to transportation, the major state role, as measured by dollar amount involved, is highway and road construction. Involving significantly fewer dollars are state activities regulating certain transportation rates, controlling safety standards, and policing transportation particularly on the highways.

The state role with respect to water has until quite recently been very modest, or nonexistent. California is a major exception to this statement. Under the stimulus of federal grants, however, the states have become involved in a variety of water-related programs, including research. Relatively more important has been the role of local governments in water supply and sewage management. Until quite recently, water supply systems were exclusively a local responsibility, and remain primarily a local matter even today, although financial aid comes from state and federal governments. Water pollution control activities, such as they were, were also primarily a local government task, involving thousands of miles of sewers and some sewage treatment plants. Drainage and storm water runoff was similarly primarily a local responsibility, except where federal flood control expenditures may have been involved. But since 1956 federal grants for certain water pollution control activities have become very important, and with the grants varying types of federal control. Sweeping reforms were enacted with respect to water pollution control, over the President's veto, in October 1972.

The patterns of intergovernmental relationships in these three fields vary considerably, both within a particular field and among the three fields. Although local and state governments are consulted by the Corps of Engineers, in planning flood control and other projects, their roles are largely advisory. Once Congress authorizes a project and appropriates money for it, the Corps handles the construction under its normal bid and contract procedures. In contrast, the Bureau of Public Roads, spending approximately 5 billion dollars a year since 1956, maintains a relatively small staff of federal employees (about 5,000) and spends most of its money through grants to the states, and through the states to local governments. Actual construction, as well as detailed design, is done by private contractors.

The programs of the federal government involving the production of electric energy are exclusively federal except that congressional enactments and agency policies have kept a major part of the federal role at the generation and transmission levels, leaving distribution to local governments, cooperatives, public utility districts, or private power companies. And while state regulatory agencies have been important in determining the rates of private utilities, federal power wholesale distribution is considered to be outside of the responsibility of the state regulatory agencies. And TVA uses its supply contracts to control local resale rates.

Since the 1930's the federal government has in various ways tried to involve the states in planning of water development projects, emphasizing particularly planning on a river basin basis, but the final responsibility for decisions on water development projects has remained almost exclusively with the federal agencies, subject only to the normal political constraints of the system.

Whereas water is almost exclusively a governmental responsibility, with the federal government playing a major role, publicly produced energy is a small portion of the total, although private sector companies come under a variety of governmental controls.

MANAGEMENT CONTROLS AND FORMAL COORDINATIVE DEVICES

No account of the project evolvement process, especially at the federal level, would be complete without reference to the complex structure of control agencies and coordinative organizations. These agencies and organizations influence policy and apply their power and influence at many points in the evolvement process, and it is not unusual for them to be regarded with hostility by those seeking project approval.

Perhaps most important is the Office of Management and Budget (OMB) the major control unit in the Office of the President. Its day-to-day responsibilities include the formulation of the President's budget program (and in this connection OMB supervises agency budgeting and accounting), as well as control over important relationships between the executive establishment and the Congress, particularly as related to legislative proposals. Thus, the phrase "not in accordance with the program of the President" serves as a more-or-less effective constraint on agency legislative initiative.

OMB determines annually the amount of funding for a particular program, and has power to keep these amounts lower than congressional authorizations. It thus influences project schedules. A classic illustration of budget powers developed early in the Nixon administration when congressional authorizations for water pollution control programs were one billion dollars annually, and the Administration appropriation proposal was less than 300 million dollars. In this case, conservation groups were able to arouse sufficient public concern to have the Congress increase the amount to 700 million dollars. In fiscal year 1973, a major controversy was developing as a result of President Nixon's withholding appropriated funds.

Another illustration of the control power of this office in the field of water is evident in the establishment of the interest or discount rates for public projects. Since the interest or discount rate is an important element in cost-benefit analyses, reflecting costs over time, an increase in this rate may have the effect of precluding authorization of particular projects because costs then may exceed benefits. Professor Arther Maass has argued that the Office of Management and Budget has preferred to use indirect control, as illustrated in the interest or discount rate, as a means of limiting construction of water projects. He suggests that a direct confrontation of the program issues involved in particular projects would be a more forthright and politically responsible course of action.[18]

Another agency in the Office of the President with some influence on project evolvement is the Council of Economic Advisors (CEA). Created by the Employment Act of 1946, the Council has the responsibility of reporting both to the President and to the Congress on the state of the economy. In this connection CEA is empowered to recommend public works construction or curtailment as in its judgment may seem wise. The Council, under Presidents Kennedy and Johnson, took a considerable interest in these policies, but President Nixon's Council has seemed to be relatively passive with respect to public works projects, being more concerned with issues of general inflation and control of the economy.

Still another agency in the Executive Office of the President is the Council on Environmental Quality created by the National Environmental Policy Act of 1969 (NEPA). Its responsibilities include the formulation and recommendation of national policies to promote the improvement of the quality of the human environment, but its small staff precludes effective direct control over the line or program agencies. It should be noted, however, that the provisions of its authorizing statute (Section 102) requiring the preparation of "environmental impact statements" by all federal agencies, provides a potential basis for control. At present, the Council is in fact overwhelmed by the number and length of these statements and as a result is known to give all but the most important only a cursory review.

It is probably accurate to say that coordination of water resources programs has received more emphasis and concern than any other program area involving the federal government. Since the 1930's, when water programs were expanded to provide a stimulus to the economy, a substantial number of national and regional arrangements have been established seeking more effective coordination. Out of these many efforts ultimately emerged the Water Resources Council (WRC) established by the Water Resources Planning Act of 1965. WRC has a continuing responsibility to report on the relationship of water supply and demand throughout the nation. The Water Resources Council, possessing few explicit control powers, relies primarily on persuasion, and on formal and informal discussions among water agencies. In the process, problem reports and procedural and policy recommendations are made which influence agency action. There has yet been no careful appraisal of WRC accomplishments. Perhaps one of its more significant efforts has been its concern for project evaluation policies and procedures (including a critical review of cost/benefit analysis).

Its most recent efforts in this area began with a specialized task group which developed a draft report on the subject in 1969. After an elaborate coordination process, the Council in 1972 issued new evaluation criteria which are to be applicable to Federal water projects.[19] Essentially, these new criteria seek to broaden the traditional basis for economic assessment to include not only the importance of a project for national economic growth, but also its contribution to regional development and to environmental enhancement.

The Water Resources Council (WRC) also conducted studies in connection with the change in the interest or discount rate by the Office of Management and Budget referred to above.

The WRC emphasizes river basin regions as appropriate planning units, and in this connection, supervises a variety of regional planning programs, including the activities of the

river basin commissions. It also administers a small program
of federal grants to states to aid them in comprehensive water
and related land resources planning. But like the river basin
commissions which it oversees, WRC tends to operate on a consensus basis, relying most heavily on reports and studies to
accomplish policy changes.

A control agency of considerable importance to all federal
activities is the General Accounting Office, (GAO) headed by
the Controller General, of the United States.   The GAO is an
agency of Congress rather than of the executive.   Although it
maintains a continuing surveillance of public expenditures, its
influence on project evolvement is most apparent in special
studies which it conducts from time-to-time at the request of
Congress or of particular congressional committees. (Legislation is pending to expand the program audit functions of the
General Accounting Office, so as to provide Congress with systematic program performance reports).   Over the years, the
studies and analyses of the General Accounting Office have covered a wide range of topics. Thus, for example, in the fall of
1969, the GAO released a critical evaluation of the results of
the water pollution control grants program, pointing out that
despite expenditures of 1.2 billion dollars, there seemed little evidence that water quality in the streams had been significantly altered.

In addition to its concern for program effectiveness the
General Accounting Office also investigates a wide range of
administrative procedures and practices.   Thus, the effect of
GAO activities on public projects has been and will continue to
be far reaching.

A number of other federal agencies exercise control responsibilities with an impact upon public projects.   These are
primarily the regulatory agencies, such as the Federal Power
Commission, the Inter-State Commerce Commission, and the Atomic
Energy Commission.   Space does not permit a review of the effects which such agencies may have, but it is important to note
that they can be of considerable importance to public projects
(e.g. FAA Airport safety and noise control requirements).

In another category are the standard setting agencies,
such as the Public Health Service and the Environmental Protection Agency.

Still another type of agency, often overlooked, is that
involved in the collection, compilation, and dissemination of
statistics and other data and information.   This would include
the U. S. Geological Survey, the Coast and Geodetic Survey, the
Bureau of the Census, as well as many other agencies.

At the state and local level, there are equivalents for the control and management agencies found at the federal level. However, the states have placed major emphasis on the regulatory function, sometimes granting authority to administrative agencies and sometimes establishing regulatory policies by statute (e.g. the setting of taxing and borrowing rates) which limit such project developments as sewer and water works. Although some states have established planning offices, and the federal government has through a variety of programs been seeking to expand state planning responsibilities, the overall impression is that the states have in fact performed very ineffectively in this regard. State plans tend to be of an *ad hoc* character and seldom reflect fundamental policy decisions. Even in the field of water, where collaboration with the federal government has been most extensive, the role of the states is minimal, leaving most of the initiative to the federal water agencies.

THE LEGISLATURE AND PUBLIC PROJECT DECISIONS

In the 19th century public projects were usually initiated by individual legislators. Quite understandably their motives were often simply to "do something for the home folks." It is in this context that the so called "pork barrel" and "horse trading" traditions developed. Today the role of legislators in initiating public projects has shrunk, except perhaps in the case of the U.S. Army Corps of Engineers (as reflected in the so called "18 steps" referred to earlier in this chapter). The Corps continues to regard itself as the construction agency of Congress, and thus more-or-less overtly takes its cues directly from Congress rather than from the President and the executive branch. As a result, its approach continues to place emphasis on particular projects, as authorized specifically by Congress, although the definition of "project" has often been broadened to include an entire river basin or region.

But although less direct, the role of the legislature in the initiation of public projects continues to be dominant. Even though proposals are initiated by executive agencies, no public project is undertaken without legislative approval. The first step in congressional procedure is authorization, which may be general or very specific. The next step is the appropriation which usually is quite specific and may be for only a part of a project. In some few cases, particularly at the state and local level, referendum procedures may involve the public directly in project decisions, but even in these cases, implementing legislative action is often required. This important role of the legislature and legislative process reflects in part the idea of a government by law, by which is usually meant legislative enactment, rather than administrative order.

In part, it reflects centuries of struggle over the power of the purse, as a result of which it is still tradition in the Congress for appropriations bills and tax measures to be initiated in the lower house.

Yet the 20th century has seen a marked change in the relative roles played by the executive and the legislative branches in project initiation. It was almost literally true, particularly in the period from the Civil War to the administration of Theodore Roosevelt, that project proposals, as virtually all expenditure proposals, were initiated in Congress. Moreover, the control over expenditures was indeed very detailed. There was no central budgeting; the agencies dealt directly with Committees of Congress; and Congress controlled even the salaries of particular employees. This continued to be the case even after civil service reform began to insulate public employees from political pressures and to stress merit in selection. It is only in the context of 20th century administration and 20th century relationships between the executive and the legislature, that procedures for planning, authorizing and implementing public projects have been revised.

The shifting relationships which resulted in the executive having increasingly greater initiative with respect to the public projects did not occur without conflict and struggle. Although the size of the tasks made the dominance of executive agencies inevitable, Congress did not give up its prerogatives without a struggle. To illustrate: when President Theodore Roosevelt "dared" to submit an outline of a proposed bill in a message urging legislative action, a number of senators criticized the President for invading the legislative sphere of responsibility. By the time of the New Deal, however, the presence of the floor of an Administration Attorney to assist House Majority Leader, Pat Harrison, in explaining and defending some complex financial control legislation was accepted.

Two factors have undoubtedly contributed to the change in the relationships between Congress and the executive with respect to project proposals and legislation in general. One of these is the increased complexity of the subject matter. A corollary is the tremendous increase in the size of the federal bureaucracy, as well as in its specialization. The second major factor is the sheer volume of legislation which the Congress must now consider. Following from this is the virtual impossibility for any single member of Congress to attempt to master all of the material on which he must act. Congressmen and senators also find it necessary to specialize, and in this context a certain fragmentation in decision making inevitably occurs. Some semblance of coordination and integration may thus be provided by the executive--through the control agencies, through various organizational arrangements, and through program proposals designed to deal with problems in a broad context, and delegating implementation to the Executive, often through administrative rules and orders.

The evolvement of public projects and the relationships of Congress and the executive cannot be understood separate and apart from the relationships of Congress and the executive to the public (or better the many publics which characterize American society), to particular constituencies, and to interest and pressure groups.

Equally important in the decisional system are the political parties. Earlier in this chapter the lack of program focus in the parties was discussed. Here comment is introduced on the party process, procedure and organization, which are important both for what they do and for what they do not do. The role of political parties in the United States is confusing, and frequently misunderstood. First, although we speak of a two-party system in the United States, this is inaccurate. In fact, American political parties are highly decentralized with the primary focus of power and influence being at the local level. It should be noted that the term "local" in this context may refer to the precinct, to the ward, to the municipality, to the county, to combinations of these several units, or to the state. Patterns vary region-to-region, state-to-state, and time-to-time. The important point, however, is that such national parties as we may have are built out of coalitions of local entities. The national party structure is highly fragmented and highly pluralistic. About the only time when we have something that appears to be a national political party is at the time of the presidential election, and even then the national focus is evident for only a brief period--really from the time of the nomination of the candidates until the election.

Prior to the nomination of the candidates, their efforts are usually devoted to building coalitions of local political organizations. The primary system tends to emphasize states as the proper unit, but in many of the states, the importance of locality continues to be crucial in the selection of convention delegates, so that state delegations also are coalitions. At the same time the national political party has little or no control over which local candidates are to bear its banners. Party membership is a matter of individual choice, involving no commitments to ideology or program. Similarly, the decision by candidates to run for office under a particular party label is their own to make, the primary system of nominating candidates simply confirming this pluralistic and uncoordinated approach.

National or state parties, and even local parties, have no sanctions for holding elective officials accountable in the absence of party discipline. Elected officials, particularly legislators, go to their jobs as free agents. Their major concern tends to be their own re-election, and thus they are constantly assessing how their activities and actions may affect their chances for re-election. At the same time, not being committed to party programs or positions, their actions are determined by alliances and alignments developed over the years as these contribute to this goal of election and re-election. As previously indicated, it is frequently emphasized that the major function of American political parties is the selection of candidates. This means, simply, that parties provide the machinery by which we select candidates; they do not determine the programs and policies to which the candidates are expected to adhere. In this connection, the British system is often suggested as offering a sharp contrast in party responsibility, and in the control of the national parties over the behavior of members, particularly those in Parliament, primarily through control over nominations. In our system, patronage, privilege, log rolling and horse trading provide only a semblence of discipline.

To restate, the American legislator, although he bears a party label, has himself chosen that label, and does not regard it as limiting his behavior as a legislator. Such controls as influence the behavior of legislators stem not from the fact of party existence, but from power relationships to constituents, to colleagues in the legislative body itself, and to the executive and administrative agencies. Thus, conflict, bargaining, and personal interaction is generally much more important than party organization. Party votes do on occasion occur, but these are more often related to power relations in the legislative body itself, rather than to basic program or policy issues. It is significant that the rules of both parties in the U.S. Congress permit a congressman to depart from party caucus positions (which are seldom taken) when his conscience is involved or when the interests of his own constituency dictate. So what remains?

Thus, it is not surprising that local benefits and pressure group support are often politically more important in determining whether public projects should go forward, than are considerations of national interests, whether economic or environmental.

GOAL SETTING

To state the matter differently, the American political system is weak when it comes to determining public goals.

Since political parties are not programmatic or issue oriented, they play little or no role in debating or shaping national goals. Such leadership in this area that does occur tends to come from the president, key officials in the executive branch, from various organizations (e.g. U.S. Chamber of Commerce, the AFL-CIO, the Farm Bureau) and from a few individuals, often members of Congress, who have for one reason or another assumed a responsibility for leadership in a particular field.

The extend to which major legislation often bears the name of its sponsor or sponsors is evidence of this situation; e.g. the Sherman Antitrust Act, the Taft-Hartley Act. The Tennessee Valley Authority would never have been created without the persistent leadership of Senator George W. Norris of Nebraska. And a careful analysis of any single Congress will reveal how senators and representatives become associated with certain program areas. Presently, Senator Edmund S. Muskie has chosen water pollution as a field of primary concern, and in the House, Congressman John A. Blatnik has been a major figure in this field. Similarly Senator Henry Jackson is associated with energy policy; and Senator William Proxmire with economic policy.

But the point to emphasize is that our system lacks the capability to establish long range, concrete goals and to formulate policies and programs for their achievement. Presidential leadership at times fills the gap, but the interests of presidents are not constant, and the crisis nature of their focus of attention prevents continuous leadership. The result (which has already been referred to) is that the functional bureaucratic agencies take over much of the responsibility for program and policy formulation, as well as for its implementation. But the bureaucratic agencies are specialized and narrow in focus. Hence a perennial problem of American government, particularly in Washington, is coordination and integration. With no national planning machinery it is difficult, if not impossible, for agencies to relate their concrete program proposals even to some kind of inferred national goals. (It is in this context, incidentally, that the development of the National Environmental Policy Act may become significant, because it places on all federal agencies a new kind of responsibility for coordination and integration, while at the same time recognizing primary agency responsibility for action.) Thus while the complexity of the system necessitates a great deal of attention to coordination and integration, coordinative techniques and procedures easily tend to become substitutes for goal decisions and choices, cause lengthy delays, and often prevent the examination of alternatives and consequences flowing from particular program activities.

Against this general background of the American governmental process, it is now possible to turn to particular subject areas of project evolvement with which this volume is concerned, namely transportation, water development, and energy.

PROJECT EVOLVEMENT IN TRANSPORTATION

The United States does not have a single unified transportation policy or program. Although one of the reasons for creating the Department of Transportation was to bring about more effective program coordination among the several modes of transportation, this remains an unrealized objective, as suggested for instance in the current political struggle over the reallocation of highway trust fund monies. So long as the private automobile remains a major factor in the transportation picture, planning and coordination are necessarily limited, since in many respects the automobile epitomizes the market economy and individual free choice.

In a country so large as the United States, transportation has always been important. From colonial times, government has had a major interest in improving transportation, particularly in the development of a variety of public projects to make transportation easier. The colonies and later the states would probably not have evolved into a nation if less adequate transportation had been available. Fortunately, in the first two centuries the tidal rivers with which the country abounded provided much needed transportation routes, particularly since the ships of those days were of light draft. Well into the 19th century water transportation was an important means for keeping the Nation united.

It is curious to note, however, that despite the obvious importance of transportation to the nation, one of the major early constitutional controversies concerned the question of whether the national government had authority to finance so-called "internal improvements" by which was meant river improvements and highways. It is equally curious, that President Andrew Jackson, representing the new west which depended on transportation, vetoed an internal improvements measure on constitutional grounds. Although he did not intend it to be so, in a sense his veto marked the end of the river improvement and canal building era under public auspices, and the beginning of the railroad era under private, corporate auspices, which would dominate transportation from Jackson's time to the end of the World War I, when automobile and airplane took the place of railroads in public thinking on the subject.

Initially, railroad development was a matter of individual initiative, the capital for railroad development coming from private investors, local governments on occasion, and in a few instances from states. The sophistication of our financial

system today makes it difficult to appreciate how hard it was
to accumulate the capital necessary for railroad extensions.
A good deal of foreign capital was attracted, and in the period
before the Civil War, involvement of the federal government in
railroad policy was minimal, a major exception being land
grants *through the states* to what became the Illinois Central
Railroad.

After the Civil War, with the issue of internal improvements finally laid to rest, the nation entered a short period
of massive land grants for railroad development west of the
Mississippi. In just a few years the federal government made
land grants of thirty-seven million acres to a variety of railroad companies (an acreage equivalent to about 1/9th of the
area of the forty-eight contiguous states). The land grant era
lasted less than a decade, its end being in no small part due
to corruption associated with the generous distribution of public domain lands to railroad developers.   In retrospect it
seems clear that in many cases, the major impetus for railroad
development was not transportation but the desire on the part
of promoters and speculators to secure large quantities of the
public domain.   It is not at all clear that the land-grants
were necessary for railroad development nor that such development would not have occurred in the absence of the land grants.
Although intended as a source of capital, a considerable portion of the land grants made no measurable contribution to
meeting capital requirements.

The experience with railroad development has not been much
different from many public projects where major political support comes, not from those concerned with the primary or direct
program objectives, but from those concerned with secondary or
indirect objectives (benefits).

Economists have properly pointed out that secondary benefits should not be counted as a measure of *national* economic
growth and development, resulting from public projects.   One
result has been that secondary benefits have often been ignored
in the literature on public project policy.   Yet, from a political point of view (in terms of gaining support for particular programs and projects) secondary benefits may be major
motivating factors.   Major political support for the Reclamation Act of 1902 came from railroads, which hoped to increase
their eastward bound freight, from real estate developers and
land speculators, who hoped to gain better prices for land in
irrigated areas, and from local businessmen, who saw in irrigation agriculture a way to increase their trade.   Reinforcing
these secondary interests in irrigation were generally held beliefs in the "manifest destiny" of the United States to settle
the continent from shore-to-shore.   There was no widespread
concern for producing more food through irrigation, although a

romantic image of "making the desert bloom" was frequently referred to. In short, the development of irrigation was land and settlement policy, not agricultural policy, and perhaps this is one reason why the reclamation program was located in the Department of the Interior, rather than in the Department of Agriculture.

But to return to the project evolvement processes with respect to rail transportation. By the 1890's the basic railroad system was in place--but there had been very little transportation planning, and no assessment of transportation needs. On April 6, 1808, Albert Gallatin submitted to the Senate his report on building roads and canals with federal funds, but this proposal received little support. Rather, decisions were made on the basis of faith, hope, and charity (charity particularly for the land speculators). The hope was that transportation development would encourage economic growth, and the faith was that such growth was inevitable, a part of the "manifest destiny" of the American nation.

Somewhat later (1870's and 1880's), political interests in rail transportation took the form of a demand for rate and safety regulation. Stimulated by Populist demands in the 1870's, many states sought to regulate, first, rail rates and, ultimately, rail safety. By 1887, under the interstate commerce power of the Constitution, political pressures were sufficient to result in the creation of the Interstate Commerce Commission which remains an important instrument for federal railroad regulation and control.

From an engineering point of view, the railroad system of the United States was impressive. From an economic point of view, it was a catastrophy, and the strains of World War I made this fact all too evident. At the same time, the development of automobile transportation by that time began to have a growing impact on the economics of the railroad system, initially on interurban rapid transit and ultimately on all traffic. Interurban rapid transit systems reached their peak in about 1920, when it was possible to travel from Madison, Wisconsin to New York City or Boston by electric street cars paying nickel and dime fares along the way. Ironically by the end of World War II most of this rapid transit system had been dismantled so now we must seek to reestablish mass transit systems in and between metro areas through massive federal subsidies!

The problems of urban and metropolitan transportation pose important public policy and program issues, but the solution, except in the largest metropolitan areas, are not conceived of as involving direct federal projects. In cities like Denver, mass transit is being developed (by a local transit authority) through reliance on bus transportation and regular streets. In cities like San Francisco, Chicago, and a number of others, projects expanding subway and related systems are underway. In both situations, the federal government plays an important role through grants or subsidies, leaving planning and operating decisions in the hands of local governments subject to federal standards and guidance.

With the exceptions noted, the development of rail transportation has been primarily a private sector responsibility, government involvement being limited to grants, subsidies, loans, and regulation. Amtrak, a public corporation operating the National passenger rail system is too new for comment.

The public role is considerably different in the field of highway transportation, where the entire system has been financed with public funds. In the early days of the republic, Congress authorized the construction of the National Highway (which became U.S. 40 in later years) which was planned to cross the country from Philadelphia to St. Louis. Actual construction was spasmodic, partly because of the constitutional debate over internal improvements, partly because of lack of funding. Through most of the 19th century, once steamboats and railroads became operative, there was very little concern for national or even regional highways, emphasis being on construction of local roads radiating from towns and cities. In the cities this was accomplished by contract with road builders, while in rural areas many of the roads were built under township auspices, often by abutting land owners. The survey of the public domain into townships and sections made it natural where terrain permitted to follow section lines for the road system, and land grants usually reserved an easement along section lines for local roads. In an era in which horse drawn plows and other equipment were the primary means for road construction, engineering problems were generally either simple or impossible, the route of roads following the natural terrain. In towns and cities, before the automobile, pavement of roads was rare, and maintenance minimal. The first major impetus to paving city and suburban roads followed the invention of the bicycle! Local roads were financed from local property taxes. In rural areas, roads were even less frequently surfaced, construction being limited to grading. Thus, the highway system at the end of the 19th century was decentralized with many of the decisions being made at the most local level, often by the citizens assembled in town meetings. The change agent in this situation was, of course, the automobile.

THE IMPACT OF THE AUTOMOBILE: PHASE I, 1916-1956

In the automobile era, federal involvement in highway construction began with the Highway Act of 1916, which set the

pattern for intergovernmental relations still being followed
today. Under this statute, grants were made to the states to
aid in highway construction. Within the first decade of this
intergovernmental relations program, concepts of a highway network developed, and at the same time the Bureau of Public Roads
(now the Federal Highway Administration) began to set minimal
construction standards. From the start, federal involvement
was based upon grants to the states, with the states taking
major responsibility for design, location, and construction of
the highways under contract with private construction companies.
Initially, federal monies were a small part of total state and
local expenditures for road and highway purposes, even though
road and highway building represented a major field of public
works, reflecting both population growth and change in location,
as well as increased use of automobiles. The constitutional
basis for federal highway grants was originally the commerce
power, and the postal power to which later the spending power
was added. The Bureau of Public Roads was located in the U.S.
Department of Agriculture since rural roads were the primary
federal concern and the political power of the Farm Block important. The Great Depression, beginning in 1929, put a halt
to much of the road and highway construction, but several New
Deal programs designed to stimulate the economy included highway construction, an important emphasis being on farm-to-market
roads. At the same time, organizationally, rural highway construction became more-and-more centralized at the county and
even state levels, all states ultimately using gasoline taxes
for financing non urban highways. Cities continued to have almost sole responsibility for local roads, while state agencies
undertook the development of state highway systems. In most
states, however, counties and even townships continued to have
some responsibility for the lower quality rural roads.

THE IMPACT OF THE AUTOMOBILE: PHASE II, 1956-PRESENT

In the first phase of national highway development (i.e.
1916-1956) an extensive national system had been developed including approximately 40 thousand miles of highway. Over the
years, federal monies became available for highways other than
those designated as part of the United States highway system.
And after World War II, the backlog of need, together with post-
war prosperity, contributed to a demand for improving the national highway network substantially. And this interest resulted in 1956 in the enactment of a new national highway program which authorized the development of the "interstate system" of limited access highways totaling forty-four thousand
miles.[20] It should not be inferred that the demand for this new
super highway system came directly from the citizen public.
Rather, it represented an alliance of truckers, contractors,
and similar interest groups, and state and federal agencies.

Again, the emphasis was on federal matching grants to the states, with substantial portions of the system receiving 90% grants from the federal government. Although the system that existed to that date had in many respects been financed through dedicated funds, raised primarily through gasoline taxes, the 1956 Highway Act carried this manner of financing considerably farther, establishing the Federal Highway Trust Fund into which were paid federal gasoline taxes as well as a number of other special taxes such as those on tires and auto accessories. The Highway Trust Fund, supplemented by some general revenues, tended to produce 4-5 billion dollars a year, and this income, together with state matching funds, has been used since 1956 for the construction of the "inter-state" highway system.

This construction program is undoubtedly the largest construction program in the history of the world. When completed, it will make it possible to travel throughout the land without leaving the system except for food and automobile service.

Several comments on the present system need to be made. First, although primary responsibility remains with state highway departments, the Federal Highway Administration is exercising far greater surveillance of location, specifications, construction quality and most recently of environmental impact. The present system is unquestionably dominated by the federal government. The original 1956 act provided for public hearings related particularly to highway location, but many states were inclined to treat such hearings as a *pro forma* requirements. More recently pressures from the federal government, as well as more broadly based citizen concerns, have resulted in statutory revisions looking to a system more fully responsive to citizen participation. In part, the demand for participation resulted from resistance to massive highway development in some larger cities, where, there was evidence that location of expressways could be used to bring about economic or racial segregation. Construction of urban expressways required the destruction of many housing units, in some cases without adequate compensation or provision for their replacement. Thus, hostility, sometimes in the form of demonstrations, was encountered. More recently, concern for adverse environmental effects of highway construction has occurred, and revisions in federal statutes as well as enactment of the National Environmental Policy Act of 1969 has forced highway agencies to pay more attention to a wide range of environmental factors and effects.

During both phases of highway development, but particularly in the second phase since 1956, much of the most effective political support for highway projects has come, not from highway users (who often tend to be disorganized and inarticulate), but rather from those who seek to gain secondary benefits from the expenditure of public funds. These secondary

beneficiaries include engineering firms who design and construct highways,   industries that supply earth moving and other
construction equipment,   industries,   that provide   cement and
asphalt for highways,   and so forth.   Here again is the phenomenon,   noted early in this chapter,   of how support for public
projects occurs.

One aspect of highway construction (particularly of the
interstate system) that has frequently been noted but not too
well explained is the   fact that most highways generate more
traffic than original estimates predicted.   Thus, in the cities, expressways seem to increase congestion (and air pollution), and between cities, the interstate system, by increasing
the mobility of the American people, has undoubtedly contributed to the accessibility of outdoor recreation facilities and
opportunities, and encourage commercial and private travel. At
the same time, this magnificent system has been a great boon to
the auto.   Thus unanticipated secondary consequences may often
be most important.

To a substantial degree local roads continue to be financed by real property taxes.   At the next level, state and
local gasoline taxes and federal grants are used to finance
main arteries both in cities and in rural areas. As indicated,
the interstate system is financed through the Federal Highway
Trust fund with state matching funds ranging from 10% to 50%.
An important point, politically, is that most of our highways
and roads are financed by the users through user taxes, although only approximately in proportion to the extent of use.
A current political struggle, therefore, concerns the attempt
to divert some of the highway tax monies to the correction of
highway-induced problems, such as air pollution and congestion,
particularly through subsidization of mass transit systems.

There is an interesting anomoly here.   In many respects,
the automobile industry is the model of the free enterprise system. Yet it is clear that the automobile industry would probably have developed much differently had American citizens
been unwilling to tax themselves to provide the unexcelled system of roads and highways on which the automobiles travel.   It
seems almost impossible to separate out the interests of the
individual traveler (or trucker) from the interests of those
who construct and maintain the highways and those who manufacture the vehicles which travel on the highways. Even more difficult is the identification of those who feel the effect or
the impact of the highway system or of any particular part of
it. It is easy to be carried away by a concern for those directly affected by construction of a limited access highway,
while at the same time failing to assess the significance of
the highway to the level of living and quality of life of people many miles away.   For example, there is no doubt that the

tremendous variety of fruits and vegetables available to American consumers all through the year is a direct result of the unexcelled highway system which this country has constructed. That highway system has literally freed us from the seasonality of fruit and vegetable production and has brought to even small communities a variety of food which clearly has become an important aspect of the quality of life in the United States. Similar illustrations can be made with respect to other products carried on the highway system, but perhaps more important than such economic consequences, is the freedom of movement which the highways (and the automobiles) have made possible--a freedom which is taken for granted, but which will be severely affected by any program to limit auto use.

BARGE TRANSPORTATION

Public rivers and harbors projects have been important almost from the first days of settlement, because rivers and canals provided early routes by which the frontier was opened, and by which it changed rapidly from a wilderness. The reason for this acceptance of public responsibility is simply that rivers and harbors were regarded as public areas, and hence it seemed natural that their development should be at public expense. Symbolic of early development is the Erie Barge Canal which, among other things, made New York City the major port of the United States, and opened the mid-West to settlement. The steamboat era, of course, made the change even more rapid, but this era was short-lived, for railroad competition brought about a decline in the importance of most waterways. It was at this point in time perhaps that river development projects shifted from simply providing improved transportation to patronage and "pork barrel" expenditures. And while rail development was largely a matter of private sector investment, river and canal development continued to involve public sector expenditures. Ultimately river development for transportation was coupled with flood control, water supply, power, and where appropriate, with irrigation and most recently recreation development. Thus the Corps of Engineers has been the major river development agency, and as such was well-protected by congressmen and senators with potential river development projects in their districts or states. As a result, the inland waterways system, including the Mississippi River, the Ohio River, and Tennessee River, the Arkansas River and many others, has continued to be improved for barge traffic. Harbor development, closely related to river development, has also continued, a current concern being increasing the depth of harbors to permit access by some of the largest ocean-going tankers. Since improvement of rivers and harbors has traditionally been a responsibility of the Corps of Engineers, the evolvement of public projects in this field was historically a matter of so-called "pork barrel" politics, bargains and trade-offs, which meant simply that an emphasis was given to local interests, rather than to regional or national interests, in deciding on the scope of public projects. And while cost-benefit analysis has been applied to major water storage projects in the last twenty-five years, rivers and harbors improvements have suffered less restrictions from this analytical technique. Perhaps a recent classic example of river development for local political advantage is that of the Arkansas River, initiated under the forceful leadership of the late Senator Robert S. Kerr, who was known for his constant efforts on behalf of his home state, Oklahoma.

AIR TRANSPORTATION AND AIRPORTS

The air transportation industry is made up of the air carriers and so-called "general aviation" (the private individual plane sector). The government's role with respect to the air carriers was initially that of regulation and subsidy, particularly in the early days of air transportation. Subsidy continues for feeder airlines, and for airports, and all air transportation is regulated by the Federal Aviation Administration and the Civil Aeronautics Board. Federal regulatory responsibility includes rates, quality of service, safety, and airport location (if federal funds are used). The public project aspects of air transportation involve primarily the location and construction of airports, including installation of a variety of electronic safety devices. Here again a pattern of intergovernmental relations has developed, with funding coming from both federal and local sources. Public goals have included: speed of movement for people, for merchandise, and for mail; technological leadership, including publicly supported research, particularly in connection with military aircraft; the development of a national air transportation network; and safety, including both vehicle and system safety, the latter involving a network of radio, radar, and other navigational aids.

Public regulation of commercial air traffic has included a concern for frequency and quality of service. It has also been concerned with the possibility of overdevelopment of airport facilities, and in this connection federal policy has emphasized the desirability of regional airports serving a number of cities.

Because the technology of air traffic has developed so rapidly, and because of the substantial growth and relocation of urban populations, airport construction has been a continuing task. In most respects, the national system is in place, but what remains is its further expansion, development, and modernization. Because of increases in traffic and increases in the size of planes (which in turn require longer runways and more elaborate servicing facilities) many communities have had to relocate their airports, and in the process have been confronted with serious conflicts, as illustrated most dramatically in the case of the Miami jetport halted because of feared adverse environmental effects on the Everglades. In many other communities, problems of jet noise and air pollution have been arousing increased citizen attention. In terms of system operation the issue of the distances from the cities to the aiport locations has also become important.

PROJECT EVOLVEMENT IN WATER DEVELOPMENT

Water projects are of many kinds. The responsibility for them rests with many different agencies at federal, state, and local levels. Hence, project evolvement will differ considerably depending upon the purpose of the project and upon the agency and level or levels of government which may be involved. Federal government concern with water development has included navigation, discussed above in connection with barge transportation, flood control, irrigation, power generation, soil conservation, and most recently water pollution control. Local responsibility for water projects (and to a limited degree state responsibility) has included water supply, sanitary sewage and storm water management, and drainage, and in collaboration with the federal government, flood control. Today, in all of these areas, where traditionally the local interest has been dominant, the federal government has in one way or another developed a companion interest. That interest is still modest in the case of water supply and in storm water management and drainage. It is massive in the case of sanitary sewage management (water pollution control) and flood control.

In terms of the agencies involved at the federal level, the Corps of Engineers has the longest history, having been concerned with navigation projects since early in the 19th century and after the Civil War with flood control. As the approaches to navigation and flood control have broadened to include entire river basins, the scope of the Corps' programs has similarly broadened. In the context of river basin planning, the Corps devotes attention to a wide spectrum of water related developments, including recreation, urban water supply, and power generation.

The Bureau of Reclamation, created by the Reclamation Act of 1902, is the second oldest federal agency with a responsibility for water development, while the Tennessee Valley Authority and the Soil Conservation Service were products of the New Deal. The most recent federal interest is that expressed in the programs of the Department of Housing and Urban Development and of the Environmental Protection Agency.

In addition, the federal government has for many years sponsored what might be called "supporting programs" for water activities: including the activities of the U. S. Geological Survey, the Soil Survey of the U. S. Department of Agriculture, the research sponsored by the Office of Water Resources Research in the Department of the Interior, and the research and data gathering efforts of the National Oceanic and Atmospheric Agency and of the Environmental Protection Agency, as well as the more general data collection of the Census Bureau.

Not only are the programs and agencies concerned with water very diverse--the stimuli to action, as well as the constraints to action are worthy of note. Reference has already been made to how the program for water navigation grew out of the transportation needs associated with the early settlement of the country. Similarly, the development of urban water supply systems reflected the demand for water in growing cities. And as knowledge of bacteria developed, cities became concerned about water quality in public health terms. The federal interest in flood control was a direct response to the catastrophies associated, first, with the floods in the Mississippi River, and then more recently with floods in other major watersheds.

Two factors essentially of a technological or scientific character shaped the federal water programs; these were the recognition in river basin planning of the importance of upstream storage reservoirs, and a more recent concern for *comprehensive* planning which would take account of land use, and social and economic factors. Both of these have been important in the federal approach to water[21] development, at least since the late 1930's. This concern with the geography of water and with the relationship of water to other human activities is, in fact the distinctive mark of federal involvement in water development. Thus the federal approach is distinguished from that of most local governments, where the concern continues to be much more narrowly focused upon the problems of adequate water supply, functioning sewer systems, and, under state and federal pressures, sewage treatment facilities.

But in addition to the responses to particular *water* needs, it is clear that projects in this field of activity have often been motivated by or stimulated by other than a direct concern for water. As in the case of highway development, so in the case of water projects, those most articulate and concerned have not infrequently been the secondary beneficiaries, that is, those who construct the projects, those who sell supplies or services, and those who will benefit from employment or secondary economic effects. In the 1930's for example, many water development projects were justified simply as providing employment and stimulating the economy. More recently this objective was an important element in the Appalachia Study, which sought ways of channeling public expenditures to improve the economy of that region. In a different category is the more recent stimulus to water development expenditure and projects based upon a concern for pollution control. In this field, the effort is designed, through grants to local governments, to provide higher levels of sewage treatment. Considerable confusion exists as to the nature and extent of the problem and how it can be measured. It is not at all clear, for example, what "quality" water is and whether the emphasis should be on stream quality or on effluent characteristics. Congress opted for effluent standards in the law of October 1972. In many respects, the action approach to water pollution control is still largely at the individual project level, not unlike that which characterized water planning fifty years ago. There are those who feel, moreover, that the emphasis is not so much on solving pollution problems, as on spending available funds.

PROJECT EVOLVEMENT: ELECTRIC POWER

The generation of electric power is still largely a private sector responsibility, subject to state and some federal regulation. Only in the Tennessee Valley Authority Service area and in Nebraska is power provided exclusively by public agencies. But in all parts of the country expansion of service is based on market demand studies, and investment in generating capacity. Historically, the public interest in hydro-power projects goes back to the era of Theodore Roosevelt when his forceful vetoes put an end to giving away power sites to private electric utilities. At the same time, under the leadership of Gifford Pinchot, the Roosevelt administration began to use its control of the public domain as a means by which to regulate those power companies which sought the privilege of building power lines over public lands. The controversies thus aroused eventually resulted in the enactment of the Federal Power Act of 1920, and provided the basis for federal regulation of hydropower development, and ultimately of interstate power activities generally. In the New Deal period a number of federal hydropower developments were initiated, including TVA and the Columbia Valley development in the northwest. Also during this period, the Rural Electrification Administration was set up to bring electric power to rural areas. More recently, federal interest in electric power has included peaceful uses of atomic energy.

Government concern for electric power is dominantly a matter of traditional regulatory functions performed by federal and state governments. More recently the federal interest has included a concern for air and water pollution. In the case of TVA and certain water projects constructed by the Corps of

Engineers and the Bureau of Reclamation, the government interest includes power marketing. But the federal role is minor, partly because most major economic projects have been built. There is reason to believe, however, that the general interest of the national government may expand, if predictions that we are on the threshold of an energy supply crisis prove correct.

Environmental groups have emphasized the need to develop a national energy policy which should seek to relate various sources of energy, and various uses of energy in a rational way. It is conceivable, therefore, that such a policy, if it were to be developed, might include a larger role for government in project evolvement, possibly including construction of power production facilities (atomic), financing, planning and regulating. Insofar as the present situation is concerned, the responsibilities of the Atomic Energy Commission for atomic power plant siting is of some significance. And national policies with respect to oil, gas, and coal might also become important. Even though the Tennessee Valley Authority is the world's largest integrated producer of power, the amount of electric energy produced by government nationwide is relatively small. Power production at federally constructed hydroplants tends to be regarded as a byproduct of other programs, hence, except in the case of TVA, the question of project evolvement for power tends to be subsidiary to the other objectives.

AUTHOR NOTES

This chapter is authored by Dr. Norman Wengert, Professor of Political Science at Colorado State University. Dr. Wengert has been involved in the field of water resources development and natural resources development for over twenty-five years. He has written many articles dealing with various facets of natural resources development. He is author also of a well known book, *Natural Resources and the Political Struggle*, which in philosophy is the precurser of much of our thinking today with respect to political decision-making in the area of natural resources development. Dr. Wengert has held a variety of positions, in several major universities and in the federal government over the past twenty-five years. Approximately ten of those years have been spent in positions with the United States Corps of Engineers Institute for Water Resources, The United States Department of the Interior, The United States Outdoor Recreation Resources Review Commission, and the Tennessee Valley Authority. Dr. Wengert Holds a B.A., J.D. and Ph.D. from the University of Wisconsin, and an M. A. from the Fletcher School of Law and Diplomacy, Tufts University. He is a member of the Wisconsin Bar.

REFERENCES

1. John Wesley Powell, *Report on the Lands of the Arid Region of the United States, with a more detailed account of the Lands of Utah*, Washington, 1878.

2. *U.S. Conference of Governors, Proceedings*, Washington, GPO, 1909.

3. *Report*, National Conservation Commission (Gifford Pinchot, Chairman), 3 vols., Senate Doc. 676, 60th Congress, 2nd Sess., 1909.

4. *Village of Euclid v. Ambler Realty Co.*, 272 U. S. 365 (1926).

5. Tennessee Valley Authority Act, 48 Stat 58.(1933); Act establishing the Soil Conservation Service in the U. S. Department of Agriculture, 49 Stat 163. (1935).

6. Anthony Downs, *Inside Bureaucracy*, Boston 1967, p. 75.

7. Charles Lindblom, *The Intelligence of Democracy: Decision Making Through Mutual Adjustment*, New York 1965.

8. Paul Diesing, *Reason in Society*, Urbana, Illinois, 1962. See also Carl J. Friedric, *Rational Decision*, Nomos VII, Yearbook of the American Society for Political and Legal Philosophy, New York 1967.

9. The most complete statement of group theory is David Truman, *The Governmental Process*, New York 1955.

10. The Federalist Papers are so readily available that no citation is given.

11. Kenneth Crawford, *The Pressure Boys*, New York 1939.

12. Stuart Chase, *Democracy Under Pressure*, New York 1945.

13. Kenneth Galbraith, *American Capitalism: The Concept of Countervailing Power*, London 1952, especially Chapter 10.

14. David Truman, The Governmental Process, Alfred Knopf, New York 1956.

15. Mancur Olson, Jr., *The Logic of Collective Action*, Cambridge, Massachusetts, 1965.

16. Harold Seidman, *Politics, Position, and Power*, New York 1970, p. 4.

17. Clinton Rossiter, *The American Presidency*, New York 1956, p. 149.

18. Arthur Maass, "Public Investment Planning in the United States," in *Public Policy*, Vol. 18, No. 2 (Winter 1969),pp. 211-243.

19. *Federal Register*, Tuesday, December 21, 1971, Washington, D. C., Vol. 36, No. 245, Part II: Water Resources Council, Proposed Principles and Standards for Planning Water and Related Land Resources.

20. Federal Highway Act of 1956, Stat

21. Gilbert White, *Strategies of American Water Management*, Ann Arbor, Michigan, 1971.

3

ENGINEERING THE PHYSICAL INFRASTRUCTURE:

TRANSPORTATION BY ROBERT F. BAKER

WATER BY FRANK C. BOERGER

POWER BY LAWRENCE M. ROBERTSON

Engineering is a profession. The common thread in all engineering is to manage men, money, and materials to bring about some intended structure or device which has human utility. Examples are diverse and might include rockets, radios, bridges, a petroleum processing plant, etc. The branch of engineering which plans, designs, and constructs dams, canals, pipelines, highways, sewers, waste treatment plants, etc. - the physical infrastructure - is called *civil engineering*. It is the oldest major branch of engineering.

The charter of the engineer is extremely broad. The civil engineer bears full responsibility, as a professional, for performance. Thus he is expected to realistically project demand so that the facility he designs is not outdated too soon; nor should there be a waste of capacity. The facility must also be designed technically sound. For example a bridge or a building must withstand the loads imposed, a pipeline must deliver the flow of water anticipated, and a water treatment plant must provide safe palatable water. In addition the project must be financially feasible and conform with relevant laws. Because of this total responsibility the engineer has an extremely broad charter. The dictionary definitions, then - "to contrive," "to manage," "to plan" - are all consistent with this

87

professional charter. So the engineer is a planner, a designer, and a manager.

Synthesis is also a tradition of engineering. While the engineer is not a scientist he must understand the relevant sciences - physical, biological, and social - in sufficient depth to bring to bear the important principle from those sciences in the solution of practical problems.

These overall concepts are important in understanding the three divisions - transportation, water, power - of this chapter. The function of the engineer has been to provide certain levels of transportation, water, and power for the *human utility* afforded by their easy access. To be operational as a professional in these subject areas, as in others covered in this treatise, would require formal education coupled with years of professional practice. Involved is more than information. Professional organizations, legislation of various sorts, modes of practice, much operational sensitivity, *and* technical knowledge are involved.

Each of the respective authors attempt to give some insight into the engineering process as related to the three major problem areas: transportation, water, and power. The treatments are neither comprehensive nor inclusive - nor are they intended to be this. But through the expositions of three experienced professionals it is possible to glean some degree of operational insight into the engineering process for each of these three problem areas.

A. TRANSPORTATION

Transportation is the system for transferring persons and goods to meet a variety of needs by a wide spectrum of users who have varied criteria for fulfilling the transfer. Transport administration and management are modal problems. At the policy level where public projects are approved or rejected, however, major consideration should be the ultimate ends (quality of life) of individuals and groups of individuals and how these ends will be influenced by a change in transport quantity or quality. By considering the function of transportation, it is possible to isolate the ultimate end from the consideration of the needs of a series of modes.

To the extent that an individual's need for transportation is not essential, the *quality* of the transport system will influence his decision. If an activity in society depends upon individual decisions to participate (recreation, cultural activities, etc.), success may well hinge upon the quality characteristics (cost, time, accessibility, reliability, etc.) of transport which is available.

Modes

Modes of transportation have developed in accordance with the resources and technology that are available. New systems cannot be developed, nor can existing systems be improved, without both. Furthermore, modes have become more or less successful (or are replaced) in direct proportion to how shippers of goods or materials or how many persons have found the system output useful for a desired end. As noted in the preceding section, the desirability or need for transportation varies from optional movements to those that are essential for human survival.

The development and growth of a new transport mode has always paralleled a major advance in system output or performance (quality of service). System output for transportation can be classified as follows:

1. Direct Effects (on the user)
   a. Cost to user
   b. Time relations

c. Safety
d. Comfort and convenience (user satisfaction)

2. Indirect Effects (on the environs)
   a. Physical
   b. Psychologic

The direct effects are those system outputs that define the quality of the service to the user; that is, how well does the system perform the transfer relative to the desires and needs for transportation. The indirect effects are the system outputs that influence the environs and are peripheral and unrelated to the purposes of the transport function. Indirect effects may be either beneficial or adverse in character.

The difference that exists between modes is the difference in system output or performance particularly in the direct effects. One mode will be more costly to the user, but will provide faster service. Some are more hazardous than others. Highway transportation, because of its flexibility, has become the primary mode for personal movements. Within each mode, sub-groupings can be identified (such as bus transit and taxis for highways).

Over long periods of time (20 to 50 years) the success of a mode has been directly related to the quality of service (output) that is provided; that is, the characteristics of cost, time, safety and user satisfaction. If there is insufficient need for a given type of service (horse-and-buggy), or if some mode of sub-grouping of a mode has superior off-setting characteristics (automobile vs. passenger train), the relative cost of the service with the lower demand may become too great to be borne by those who find the mode most satisfactory.

The existence, or life, of a transport mode, therefore, can be greatly influenced by how much the user must pay for the service. To some extent, it is possible to increase the use of a mode by the provision of public support (subsidy) which provides a reduction in direct costs to the user. New forms of transport have frequently been subsidized as a means of accelerating their acceptance or usefulness. It should be noted that such public support has generally *not* been for the purpose of "better transportation"; rather, a need has been felt for the activities (or functions) of society which the subsidized transport will make possible.

One of the modal problems which is most difficult to evaluate is the extension of the life of a mode through public support, and the problem is further confused when there are conflicting laws and different operating conditions. The best example of such a conflict is rail vs. truck transport. Thus,

if the government operates part of one mode, but not of another,
it is difficult to obtain a competitive economic situation.
The problem is further compounded when there are greatly different indirect effects (social costs).

If public support (subsidy) is to be used as a means of
furthering a given mode, the system output, the functions which
are best served by that system output, and the public resources
required are the dominant political factors. For example, low-
cost transport of bulk materials serves one important social
and economic need. Fast dependable transport of persons and
goods for trips of over 500 miles serves a completely different
set of social and economic needs. The relevant issue with regard to public financial support is not the mode, but the functions which the system output makes feasible; the relative
importance of that function to the survival, advancement, and
fulfillment of life for individuals; and the environmental
impact.

The major modes of transportation are: highway, railroad,
air, waterway, and pipeline. In subsequent sections, additional detail is provided for the various modes, including
their extent, system output or performance, public policy, and
processes for providing the service. As to its individual role
in society, each of the modes provide contributions to many
phases of the life of individuals because the need for movement
of persons and goods is quite pervasive.

Some modes, of course, are more effective than others in
fulfilling specific needs. The effectiveness of a mode in fulfilling human needs and desires is dependent upon the output or
performance (quality of service) of the mode. For freight
movement, the dominant characteristics are cost to the user,
time in transit, and reliability. For movement of individuals,
the problem is more complex, but control of one's actions is an
important consideration in a large percentage of the transport
for personal mobility. Cost to the user is a significant factor where public transportation is an alternative. With private automobiles, however, highway users appear to be willing
to pay a premium price for the flexibility or control of their
movement. The extent to which the total transport costs are
perceived by highway users is questionable, and the social
costs (or adverse indirect effects) are rarely considered. The
separation of an activity from its indirect effects is not
unique to highway users, of course, or to transportation in
general.

Until the 1960's, transport modes evolved under a user-pay
philosophy, with and without substantial subsidies. Indirect
effects such as neighborhood and rural area disruptions, air
and noise pollution and other social impacts were considered

secondary to the needs for transport. Attempts to achieve
greater economic efficiency through mass transportation were
not successful in urban areas, nor with rail passenger service
for most intercity movements.   The need for a more balanced
transport system became apparent.   Beyond a reduction in the
percentage of the transport served by the use of personal automobiles, however, general acceptance was not possible for a
quantitative basis for defining "balanced."

The fundamental problem in developing a balanced transport
system is that the flexibility of the highway mode, particularly for individual movement, provides very positive contributions to an individual's way of life. Replacing this flexibility means either a loss of individual control, a system that
has competitive flexibility, or a reduction in desire or need
for transport.

The choice of transport modes, then, is controlled by the
character of service that is provided to the user. Restrictions that must be imposed in the public interest will need to
consider the impacts on individuals, both positive and negative,
in deciding on the best compromist for advancing the quality of
life of a community.

Requirements

To view transportation (or a mode) as an individual system
is useful for transportation management purposes. However,
from the preceding, or from the view of groups of individuals,
industries, or government, however, it is a sub-system; a part
of a larger system.

As such, there are quantity and quality requirements which
should be specified by the system of which it is a component.
If transportation develops its own quantity and quality requirements, sub-optimization is probable. Sub-optimization results in less than optimum performance of the larger system.

The past practice of developing transportation needs has
grown out of the view that society was a set of independent
activities or "boxes."   As a result, transportation needs were
developed by extrapolating the growth curves of the various
modes, as if they were operating in a vacuum. One of the major
contributions of systems engineering in recent years has been
its emphasis on the interaction and interdependence of seemingly disassociated activities. Certainly individual modes and
transportation itself can be most effectively developed by considering the ultimate impact on the life of individuals and the
community.

The Parkinson Law principle as applied to market-place
economics will lead to the continued growth of successful activities and difficulty in accelerating the development of new
and better concepts. For transportation, the extrapolation of
demand curves is a useful technique, particularly over the
short-term (three to five years). But unless one is to be
overly controlled by history, if one is to escape historical
trends, then a more effective technique for obtaining quantity
and quality requirements must be developed.

The first step in the systematic development of transport
requirements should be the preparation of estimates of the need
for transportation by the various activities of the private
sector and of governments. The medical, educational, public
safety, recreational and similar programs should not operate as
if their need for transportation did not exist, nor should the
administration of these programs neglect to estimate the extent
to which their degree of success is dependent upon the quantity
and quality of transportation. The assumption is consistently
made in developing these program needs that necessary transportation will be provided without cost to the program, and the
planning is directed to the maximization of their own output.
It is apparent that if the transport assumption is not valid,
their optimization plans are no longer valid.

Given the transport requirements of various programs, the
quantity and quality of transportation needs can be developed
so as to meet the needs of those programs to which highest priority is assigned by the political process. Equally important,
the effect of fiscal, environmental and other constraints on
transportation can be translated into an ultimate impact, rather than solely upon the narrow view of the user or special
interests.

The requirements for transportation, then, must be stated
by the system that transportation serves. In reality, this is
the way that controversies on transport requirements have ultimately been resolved in the past. While extrapolation of demand curves has been the basis for estimating the needs, Congress and state legislatures have looked to the total needs of
the nation and the states, particularly for majority groups.
To produce the most effective transport system, policy requirements, cleared by the political process, should include the resources available, the quantity and quality characteristics
that are to be provided, and the constraints within which the
indirect effects must fall. After the requirements have been
set by the policymakers the provision of the system should become primarily a technological problem.

EXISTING TRANSPORTATION SYSTEM

Mobility is viewed by some as an inherent quality of life; that is, to have greater mobility is to have a better life. On the other hand, the approach taken herein is that it is the things which mobility makes possible that is important.

In the world today, the degree of development of a nation has been equated to the characteristics of its transport system, particularly its economic development. While such a surrogate or index of development cannot be a fundamental characteristic, it nonetheless reflects the not uncommon view of the great importance of mobility of persons and goods.

Massiveness

Table 3-1 contains data on expenditures that suggest the massiveness of the transportation system. With $161 billion, twenty percent of the GNP, going for transportation-related activities, the magnitude in comparison to other activities is evident.

Table 3-1. Total transportation expenditures in 1970* (public and private - $ millions in 1965 dollars).

	Passenger	Freight	Total $	Total %
Highways - Total	65,670.2	65,346.2		
Auto	63,355.3		131,016	81.0
Bus	2,134.9	205.2		
		65,141.0		
Air	10,307.2	1,275.0	11,582	7.2
Rail - Total	687.7	10,381.0	11,069	6.9
Rail Transit	329.7			
Commuter	151.7			
Intercity	206.3			
Water - Total	2,335.6	4,203.0	6,539	4.1
Freight		4,008.0		
Commercial				
Fishing		195.0		
Pipeline		1,256.0	1,256	0.8
TOTAL	79,007.0	82,461.9	161,462	100.0

*U.S. Department of Transportation, September, 1971.

Tables 3-1 and 3-2 show the relative importance of freight
transportation.   It will be noted   that more   was expended for
transferring freight than persons. Even on the highway system,
the expenditures are practically equal.

Table 3-2.   Intercity freight movement - 1970*.

	Ton - Miles in Billion		% of Total	
	1960	1970	1960	1970
Truck	285.5	412.0	21.7	21.4
Rail	579.1	773.0	44.1	40.1
Inland waterways	220.3	306.0	16.8	15.9
Pipelines	228.6	431.0	17.4	22.4
Domestic airways	0.8	3.4	0.1	0.2
TOTAL	1,314.3	1,925.4	100	100

*1971 Motor Truck Facts, Automobile Manufacturers Association

The rate of growth of freight transportation   should   also
be noted.   In ten years,   there was nearly a fifty percent increase in intercity ton-miles. Of some significance is the relatively large increase in pipeline transportation.

There are many consequences   that result from the existing
size and   diffusion of   transport services.   Not the   least of
these is the difficulty in changing the existing system or its
pattern of growth.   Over time periods   as short as five years,
it is next to impossible to make *any* substantive change without
more serious adverse effects than any benefit can offset.

This is particularly true of highway transportation, with
half of the population with driver licenses and one vehicle for
every two people in the country. Not being permitted to drive,
or as an owner, being required to modify a vehicle can set off
a significant chain of events, that can influence all activities that are based upon personal mobility. Unfortunately, insofar as producing a change is concerned, the life of an automobile averages ten years; that is, to completely replace the
existing fleet will require ten years.

The highway itself poses problems.   The 3.9 million miles
of roads and streets have evolved over centuries, particularly
the rights of way.   The investment is large. Furthermore, the
progress that can be made in modifying such a large system is
also troublesome.   Abandoning large segments is not easy because of the immediate consequences to all individuals and desired interests who are at present dependent upon that segment.

95

Since highways account for nearly 80 percent of transportation expenditures, other modes are not so massive, nor quite so pervasive. Nonetheless, they have all evolved over many years, and rapid change is not a reasonable anticipation without consequences that are related to the goals of individuals and of society.

The adverse effects of transportation are most clearly evident in urban areas. Congestion, pollution, and neighborhoods that are disrupted or threatened by new facilities are the most common examples. In rural areas, degradation of many forms are inevitable if transportation projects are required.

There is a wide range of alternatives, of course, for solutions to both the urban and rural problems related to transportation. Some of the possibilities deal with reducing the need for transport through greater accessibility (land use control), or better communication (television conferences). Some deal with improved transport efficiency through electronic control of vehicles or new transit systems. Some of these alternatives should and could have been in the process of implementation over the past ten to twenty years.

The most important reason for the failure to change is related to the size of the system, and the time required to produce a significant effect. To elective officials, a change that cannot be achieved in two to four years has very little attractiveness. In addition, the views of the voting citizenry can change in a very erratic manner.

Institutional Structures

The institutional structures related to the existing transportation system are as unbelievably large, complex and cumbersome as its massiveness might suggest. There are over 33,000 separate government agencies with jurisdiction over some part of the highway system mileage, and over 12,000 airports under separate operating authorities.

Prior to 1966, transportation activities were conducted in individual modal agencies. There was no unified transportation activities within government, outside of the military establishment and some few state or local governments. With the establishment of the U. S. Department of Transportation, it was possible for the interest of the federal government to address itself to total rather than modal needs. Many states and regional transportation authorities were established following the federal government action, and there is now a greater potential for a unified transport capability.

The institutional problems were broader than just government. For years, regulation by the United States Interstate
Commerce Commission had interceded to prevent constraint of
trade and monopolistic practices. But these activities had to
be addressed to a diverse operation which involved government
interest and private sector activities in totally different
ways in one mode than another. For example:

1. *Highways*, were built and maintained by one government
agency, operated (or controlled) by another (state police) and
used by the private vehicle owners and truckers.

2. *Air* transport ground facilities were provided in part
by the federal government, airway safety and operations were
controlled and operated by the federal government, industry and
private individuals owned and operated the planes, and municipal authorities built, maintained and operated the terminals.

3. *Railroads* were owned and operated by the private sector until Amtrack was established in 1971 to operate passenger
service.

4. *Waterways* were policed, controlled and maintained by
government, as were part of the shore facilities. The private
sector owned and operated the ships.

5. *Pipelines* were owned and operated by the private sector with limited government activity related to safety.

A major institutional problem today is this division of
responsibility between government and the private sector. The
division creates confusion relative to subsidization. For
example, under present policy, the government provides the
highway and airport facilities, and operates the airways. The
private sector provides the facilities for the other modes of
transportation. Accordingly, if the private sector provided
highway and airport facilities and collected tolls (or taxes),
the amount of funds required of government would be substantially lower. This additional funding, from government would
constitute a government subsidy. Under existing policy, there
is some private sector funding of highways (real estate developments) and airport facilities. The extent to which the governments subsidize transportation can be estimated by the
amount of general funds required in excess of the taxes paid by
users for the express purpose of providing transport facilities.
In the 1960's, approximately twenty percent of highway funds
came from other than highway use taxation.

Furthermore, the Federal Highway Safety Act provided that
the Executive Branch of the federal government specify the

safety requirements for highway transportation. (This condition has always existed in air transportation). Safety is one of the four criteria by which system performance is measured, and control of safety can and has influenced all other system outputs.

Institutionalized patterns also produce problems insofar as the planning, design and construction of facilities is concerned. This is more true for highways and airports than for other modes of transport. Generally, planning, design and construction are conducted (1) by different, unrelated groups in the government, (2) by different, unrelated groups within one government agency, and (3) over a period of five to twenty years. In addition, the feedback system from citizens to policy-makers and to the executive branch has been inadequate, and this has led to further complications in the orderly development of responsive transport systems.

Transportation planning is now integrated, in part, with other planning. This is not true to a significant extent in the federal and most of the state governmental departments, however, where transportation agencies are still responsible only for transportation planning and funding. Integration of plans is now required for highways, and to some extent for airports and other modes. With the control of funds still on a modal basis, however, there is a strong tendency for the planning to be fragmented.

At the design level for highways and airports, more and more attention is being given to benefits which are broader than the economic efficiency of transportation. This is an improvement over earlier procedures that either neglected or over-powered other benefits. Problems are still prevalent, due largely to the fact that there is no sound basis for resolving the problem of conflicting interests or benefits. This will be discussed in more detail in the next section.

Existing lobbyists, trade associations, and private interests constitute influential institutional structures. Typical of these is the American Association of State Officials (AASHO), a group of state employees responsible for planning, designing, constructing, maintaining, and operating (in some phases) the highway facilities. The group was formed in 1914, under encouragement from the federal highway organization who wished to improve engineering competency at the state level. At the moment, AASHO has great influence on national standards of design, construction and maintenance.

Highway transportation policy has been influenced by the "transportation establishment," particularly the highway lobby, but the influence appeared to be less effective in the early

1970's than during the preceding twenty years. In effect, this
lobby is similar to that which exists for any other programs.
It is probable that the increase in highway funding has been
due largely to the public desire for highway travel which has,
and still is, increasing at the rate of six percent per year--
doubling in less than twenty years.

At the national level, many lobbies are as strong or
stronger than the highway lobby, as events in the early 1970's
will indicate. Recent legislation has limited and constrained
the development of transportation in the area of vehicle safety; environmental pollution; compensation for displacement produced by highway and airport projects; protection of parks,
wilderness areas, historic landmarks, and urban neighborhoods
and communities.

Institutional structures pose serious problems for changing the transport systems. The endless interweaving of governments and the private sector preclude any simple or rapid technique for improvement of the methods for providing transportation or for improving transport concepts.

Fiscal and Legal Policies

Fiscal and legal policies for transportation vary widely
from mode to mode, both because of the independent conditions
under which modes evolved, and the responsibility that the government has assumed relative to the mode.

The most straightforward conditions exist with freight
transport by railroads and pipelines because the government has
maintained primarily a policing action with little or no operational responsibility. To the extent possible, these have
operated under the private sector as a user pay, profit-making
activity. Subsidies have been provided, particularly when it
was concluded that the public interest warranted such support.

Legal policies throughout transportation evolved primarily
to restrain unfair competition and to prevent the development
of monopolies. Thus, railroad companies were not permitted to
operate trucks for fear of their not developing trucking that
would reduce rail shipments. Freight rates were also controlled, again with a view to protecting the public against an
improper restraint of one mode.

For *highways* fiscal and legal policies grew out of a climate that said as late as 1880 (in Ohio) that roads were a local affair and of no interest to the state. By 1920, the federal government because of interest in getting agriculture products to the market, assumed an interest in upgrading highway
engineering.

Prior to the arrival of the federal government on the scene, accusations of corruption and misuse of funds by state officials were a common election-time occurrence. The practice of taxing gasoline was initiated and proved a good source of revenue. At the urging of federal highway officials, most states passed a constitutional amendment that earmarked highway user resources for highway use. Most of this legislation was passed in the 1920's and early 1930's with the argument that this was the means whereby the costs could be recovered by the governments for building and maintaining roads and streets for automobiles, trucks, and buses. A public utility analogy was put forth, wherein the gas tax was represented as the means whereby one paid his highway bill.

Up until late in the 1960's, it bordered on the heretical to suggest to highway engineers that governments should use gas tax revenue for what had been the private sector responsibility; that is, the funding of vehicle operation (buses) over the highways. The idea grew out of the continued decrease in survival of urban transit systems under a user pay fiscal policy.

With reference to fiscal policy, the government faces the responsibility of constructing, maintaining and operating highway facilities over which all classes of highway vehicles must operate. A fiscal policy that would include the financial support of transit systems would be new, and would be in addition to the existing policy for funding the other activities related to providing the facilities.

There is no "right" or "wrong" fiscal policies, of course, in any absolute sense. Earmarking tax revenues is a policy of government and as a policy is subject to change in the public interest. So also is their policy on the quantity and quality of the facilities that will be provided.

There are pervasive arguments for reallocating resources, within or without highways. The lack of enthusiasm from political administrations to change past policies is related to uncertainty as to whether *not* rebuilding worn out roads and streets, loss of maintenance to existing facilities, and failure to remove ice and snow would be acceptable to the voters.

Legal policies are also unique to highways. Federal funding reaches the state only after expenditures have been made by the state for construction on projects that were previously approved by federal officials. No maintenance or operation funding is provided. Also, all of the funding must go through state highway officials. The Governor of a state can control the official, but he does not intercede directly. Furthermore, assistance to cities, counties, and local governments can only be given through the same state highway officials. Again, the reasons for this go back to the federal concern for local political corruption.

Federal funding for highways has been primarily for interstate highway construction ($4 billion per year) since 1956. Other highways have received approximately one billion dollars per year since the late 1960's with most of the funds going for primary highways. It is interesting to note that even though federal officials convinced the states to earmark highway revenue, the Highway Trust Fund of the federal government was not established until 1956, and only then as a means of assuring the funding of the Interstate Highway System.

Until late in the 1960's, two other legal policies were related to federal highway activities. One was that compensation for right-of-way could not exceed "the fair market value." This led to inequities, particularly for low-income urban residents. Furthermore, costs of moving or relocating had not been permitted under earlier interpretations of federal law. These practices were also a product of the "game-playing" in the early days of road building where misuse of public funds were often traced to over-payment for right-of-way.

A second new policy that came about with the establishment of the United States Department of Transportation was that a total interest in highways was assumed. Thus, the need, desirability and interests of truck and bus transportation were relevant, not just the provision of facilities. Until the early 1970's, however, the historical policy of "hands-off" any responsibility for aiding bus and truck operators was being followed except at the highest policy levels. While substantive changes in support of bus operators is now a factor, no significant change has been introduced relative to the support of truck interests.

A final highway legal policy that is relevant has to do with vehicle size. From the earliest days of federal funding a restriction against increasing the size of highway vehicles has been one of the requirements for receiving federal aid. Weight, width, length and height were of concern because of the probable adverse effects on the existing pavements and bridges. The policy has been to "freeze" the size in order not to produce serious damage to the 3.9 million miles of facilities. As a result, productivity by trucking has been seriously constrained. This policy is still in effect even though there is evidence that significant increases (up to ten times the present limits) would be a sound economic investment. Major restraints against increasing the size of trucks come from highway engineers and from those concerned with safety and user satisfaction of automobile users operating on the same facility with such "behemoths." The highway officials are concerned that funds for pavement and bridge replacement would be needed and would not be forthcoming.

For air transportation, fiscal and legal policies are more recent and less influenced by historical practices. Until 1969, federal aid for airports was decided by Congress and varied as the demand for funding could be met. The passing of the Airport Development Act of 1969 produced and Airport Trust Fund to assure a continued level of funding for upgrading and building aviation facilities and for operating the airways.

The division of air transport activities consist of (1) aviation facilities, (2) airways and (3) terminal facilities. Each of the three have different fiscal and legal responsibilities. Aviation facilities include all of the taxiways, runways, drainage and other airport facilities not related to the terminal or to flight operations. The airway system is the taxing, take-off and landing, and flight between airports of aircraft. Within specified distances of an airport the federal government controls all aircraft operation. Once clear of these areas, the pilot controls the operation within established guidelines, flight plans, etc., that are set by the federal government. The terminal facilities are owned and operated by local authorities. This includes parking aprons. On major hubs, airlines can build, maintain and operate their own terminal under a leasing arrangement. There is no federal aid provided for the terminal facilities. Some subsidies are provided to airlines. Most of the federal aid to aviation (outside the military) goes for the airway system (approximately $2 billion per year in the early 1970's.

For public projects, approximately $300 to $400 million of federal aid was available in the early 1970's for aviation facilities. The normal matching ratio is fifty percent, but large airports will frequently pay for facilities if the federal aid is not adequate. At the other extreme, federal aid may go as high as eighty percent if the federal government deems the facilities are of sufficient importance. Only construction or reconstruction can qualify for federal aid; that is, maintenance is a local responsibility.

The airport authority has a number of ways to obtain funding. Generally, tax funds are not used. The airport is expected to be self-sustaining. Landing fees are charged, and rental of space for many uses, parking lots, and ground transportation franchise are major sources of revenue.

The management of transportation facilities for which public projects are needed varies widely from mode to mode. Most rail and pipeline requirements have been and are being defined under private sector practices that have been governed primarily by economic criteria. While this is still true to a large extent, urban transit can now be supported by the federal government through the Urban Mass Transportation Administration (UMTA).

In the following sections, primary attention is directed to the management of highway facilities. Air and water transport are local government or local authority activities under partial federal control due to the retention by the Federal government of the safety and use of airways and waterways.

Policy Development

The conduct of national policy studies for transportation has been largely in the domain of economists over the past fifty years, at particularly the international analyses. Viewed as an "economic activity," the generation of economic resources has been a dominant factor in the studies.

Policy has also been highly modal oriented, and controlled by a fiscal policy that called for a user-pay, self-sustaining activity to the extent that this was feasible. In no case have subsidies been entirely missing, an indication from the government that some public funding was in the public interest or was required to generate business activity. There is and has been considerable argument as to which mode was "more subsidized" than another, and in the end these analyses depend upon the definition that is used for subsidy and user charges.

Policy decisions by the legislative branch of governments, on the other hand, have been greatly influenced by many factors, and do not appear to have been controlled, or particularly heavily influenced, by the results of economic analyses. This is certainly true for highways wherein the dominant interest of the decision-makers was the continuing popularity for highway travel. By any standards, from the early days of the automobile through the 1950's, there was practically no resistance to the provision of highway facilities and very strong, popular support to the initiation and construction of major urban and rural highways. The continuing exceptions were those who were displaced by the highway, but these voices were ignored as purely self-interest expressions that were opposed to the "good for the majority." To a considerable extent, the popularity of transportation extended to all modes, with airport facilities, in particular, becoming a focus for sight-seeing and recreation.

In the mid-1960's the picture changed dramatically and
relatively suddenly. Strong and persistent groups resisted
what were termed impersonal and over-bearing extension of
transportation facilities. Past policies were seriously questioned, particularly those that encouraged the continuing
growth of highway and air transportation.

It is in the area of policy-making that the role of transportation (and the facilities that are required) is ultimately
defined. Much of the difficulties in transportation in the
early 1970's can be traced to the absence of the relatively
precise policies which existed through the 1950's. No longer
is there a basis upon which transport facilities can be provided. Each project which is advanced can be challenged due
to adverse environment or social impacts, and forced into a
process requiring a policy decision by the highest court, legislative body, or government executive.

If vague policies relative to transportation were to be
replaced with more precise statements, subsequent analyses
would be facilitated. Otherwise the analyses will need to consider the entire range of impacts, the entire range of type of
facilities, and an infinite number of compromises. Furthermore,
none of the "decisions" by any one group will have any authority, since in the absence of precise policy, the decision can
and will be appealed to a higher authority.

Policy development, more than anything else, is a process
for producing results upon which priorities can be based. In
turn, the policy decision provide a basis for the necessary
compromises in the planning and design phases. Due to the national character of transportation (the difficulty in restricting vehicles and drives to smaller political sub-divisions),
the most critical policies--those concerned with the quantity,
quality, resources and constraints--must be stated by Congress.
These policies need to be stated precisely, but with sufficient
flexibility for local authorities to apply local priorities.
For example, the maximum total emission of air pollutants that
would be permitted from transportation sources would be set for
a given sized area. Local authorities would then have the option of making the constraint more severe through restricting
of vehicle movements if the conditions so warranted. The relative effects of aircraft vs. various other transport vehicles
could be dealt with on the basis of local effects. If maximum
emissions from a given type of vehicle (automobile) is not set,
the local governments would have various options ranging from
special actions in times of emergency to some form of rationing.

Policy development for transportation at the local level
showed clarity rather than set national policy. Local policies

that would tend to set national policies are those that restrict interstate or inter-community transportation through
more restrictive standards than imposed under national policies.
These have the effect of imposing local priorities on other
communities. Standards on the *quantity* of transportation, on
the other hand, can provide an opportunity for the application
of a community's own priorities in order that the community
needs can be fulfilled.

Planning

Transportation planning is the development of the means
whereby a policy decision is implemented. The planning for
highways and airports proceeds under broad policy guidelines on
funding and requirements for facilities. These guidelines are
derived from policy-approved growth patterns.

Early in the 1960's, a policy was established that required that the planning for urban roads and streets be conducted in concert with broad, comprehensive urban planning.
This recognition of the interactions between urban activities
led to the combining of urban planning activities and to the
establishment of urban planning groups where non had previously existed. The new planning approach did not call for
any change in priorities, requirements or constraints relative
to the need for urban transportation. As a result, many of the
planning groups that did not agree with existing guidelines
attempted to elicit policy decisions on an *ad hoc* basis, and
much of the merit and effectiveness of the planning process was
lost.

Planning will be most effective if it proceeds under given
policy directives which include the quantity and quality requirements, resources available, and constraints on the indirect effects. Without these policy-cleared directives, there
is no firm basis upon which to plan, nor is there a policy-
approved basis upon which compromises can be made. The absence
of clear policy required planners to make assumptions or to
combine planning and policy development. This procedure tends
to produce advocacy planning, and incomplete results on community impacts. Everything being equal, policy studies will be
more acceptable if not conducted by an advocate or by those who
are responsible for implementing one phase of the policy.

Highway planning and programming can be separated into
rural and urban activities. Rural problems require the identification of the corridor through which the highway is to be
placed, as well as analyses of the adequacy of an existing road,
and how much of an existing facility should be re-located or
re-constructed. Urban planning is the same in principle, but
is more complex when new, limited access facilities are being
considered.

Under procedures that existed through the 1960's, the primary basis for evaluating the adequacy of a given road, street or highway was a benefit/cost analysis, based upon benefits to the user. Thus, if the cost to the user to accomplish an improvement was substantially less than he would receive in economic benefits, then the project was economically sound. Highway projects generally show the greatest benefits from time savings, and these are converted to dollars. No highway project is considered for federal funding that does not show benefits to the users that are three to five times as large as the costs.

The planning analyses proceed through a comparison of existing conditions with several alternate improvements for an anticipated growth in traffic; that is, the dollar benefits which are estimated are really a *savings* that the improvement will make possible. A benefit/cost ratio in absolute terms is not conducted: rather, it is assumed that the original facility had been justified. In this sense, the new facility reduces the adverse effects that will result if the existing conditions are not modified.

Two types of problems must be solved in planning studies. One has to do with whether or not the users will benefit sufficiently if a facility is constructed or improved. This involves a benefit/cost analysis that considers only the impact on the user. The second is the analysis of the impact on the community that the facility will serve. This requires both an economic and a social impact study. Conventional highway economic impact studies have considered only the material well-being of the total community or region that is influenced. Differential impacts have also been made on occasion (analysis of impacts on groups of individuals). Studies of social impact have been largely concerned with identifying the adverse effects (costs) of the improvement; that is, the social improvements (benefits) are usually considered as being included in the benefits to the user.

Benefit/cost analyses of intangible values would be most easily comprehended under cost effectiveness terminology. Thus, benefits (some level of clean air, or not destroying a neighborhood) per dollar expended to achieve the improvement. The inclusion of social costs should be paralleled by consideration of social benefits: for example, benefits (lives saved with better facility through a reduction in highway facilities, better emergency services, etc., or number of low-income families who could reach a job within thirty minutes) per dollar expended for the improvement. These analyses provide a spectrum of opportunities for influencing the welfare of a community. In most cases, the decision will call for a compromise of one benefit in order to gain another.

Present highway planning is necessarily controlled by the existing fiscal and legal policies. Accordingly, the funds available from highway users and the distribution of funds between classes of highways are used to program the improvements, wherein programming involves setting priorities on projects within the anticipated revenues. The first consideration in establishing priorities is the benefit/cost ratio; that is, the higher the ratio the higher the priority. This is subject to an administrative review which accounts for other factors, such as the relative importance of the facility, the volume of traffic, and the political attractiveness of the project. Such reviews will often result in a shift of priorities.

The planning for the interstate system is conducted under a different set of conditions. The policy of the federal government is that the system is to be built. The resources, quality characteristics and time table have been specified. The states were given practically a free-hand in selecting the sections of the system to be provided first. Here again, however, the state has to initiate the request, the plans, and the construction must be accomplished and paid for before the federal-aid is received. In the case of the Interstate, this is ninety percent of the engineering, right of way, and construction costs. For other facilities, sixty percent is the federal-aid matching ratio.

Most of the problems in the 1970's arose from the results of the planning efforts. Resistance to new facilities in both urban and rural areas registered due to the elimination or degradation of some set of values. Since improved facilities generate traffic, continuation of policies that encourage highway (and air transport) growth were challenged.

Arguments between transportation efficiency vs. social efficiency will continue at the planning level as long as policies and criteria are indefinite, and as long as there are differences of priorities. Under these conditions, planning is a misnomer. It is more properly, a policy definition stage wherein the responsibility for policy making is both obscure and ill-defined.

Most of the present transportation problems are related to large urban areas and for a limited number of hours per day. Unfortunately, these problems are of such political importance that they overshadow many local problems. Furthermore, there is a tendency for a community to accept the relatively severe conditions in large urban areas as being comparable to the "severity" of their own adverse conditions. Unfortunately, this leads to "severe congestion"--a relative term--being diagnosed for a community where the importance of the traffic problem has been magnified.

The role of transportation planning in the urban planning process involves among other activities, the combining of estimates of transport needs for various alternate land use plans. Everything else being equal (that is, the same amount of education, law enforcement, health features, etc.), the decision on land use defines the ultimate requirements for transportation, and vice versa. For a given level of technology, and for a given attitude of individuals for mobility, the fiscal demand for transport facilities cannot be separated from the effects of how land is used. The reverse is also true; that is, by providing a certain type of transport facility, land use will inevitably grow in response to it, under the existing political-legal system for controlling land use.

Design

The design of a transportation facility is controlled by engineering economic principles. Other factors are considered, but the dominant consideration is the economic benefit to the user. This is, of course, an extension of the principle of private sector economy which requires that expenditures be offset by sufficient benefits to the purchaser of the product or service.

For the past thirty years, benefit/cost ratios have been used as the primary criteria for comparing new, alternative facilities with existing ones. If the benefits to the user were sufficiently great, and if funds were available, the project was scheduled for construction. The concentration on the interest of the user, with little regard for interactions or interference with other programs, was paralleled in other activities and programs. Thus, facilities for communication, energy, education, health, manufacturing, housing, etc. were provided with primary regard for the users of the product or service.

While modern concerns for the adverse effects of transportation are reasonable, there is certainly sound reason for arguing for an improvement when dollar and other benefits to the users can be realized. The relevance of the adverse effects has to do with estimates of the *total* impact on individuals and the community; that is, society may get better transport service, but be worse off than it was before.

By the time a transport facility improvement reaches the design stage, the *total* societal needs and the impacts of the modifications should have been comprehended. Constraints should have been set on any changes that would be made in a facility. The bases for justifying an improvement in service should also have been stated explicitly. At the design stage,

the problem is to provide for the change that has been specified by the planning process and within the stated constraints.
The trade-offs between benefots from improved transport services or those retained by not making a change cannot be fur-
thur debated without challenging the work of the policy and
planning levels. Question at the design stage, however, is
whether any change in transport facilities can be justified,
within the constraints that have been imposed.

For transportation facilities, user interests other than
dollar benefits are also considered, such as safety, time-
savings (and related characteristics), esthetics, (and other
comfort and convenience considerations). Generally, these factors are assigned a dollar value and introduced into the bene-
fit/cost analysis. Indirect effects on the environs are included, but primarily through consideration of right-of-way
costs. Other impacts on people and programs are not, except as
compromises become necessary due to resistance to the proposed
improvement of transport services.

The problems of design fall into two broad categories:
traffic operations and structural strength. Within traffic
operations, considerations are primarily those of geometry;
that is, the extent of vertical and horizontal curvature, the
number and width of lands and shoulders, the width of the right-
of-way, esthetics (to the user), signing and signaling, intersections, etc. The major criteria that are used are economic.
Safety considerations are incorporated on a semi-economic basis
--the minimum standards are based, in part, on the resources
that are available. This is not as cold-blooded as it sounds,
because comparisons are being made with existing conditions,
and a reduction in risks does constitute an improvement.

Prior to 1940, esthetics to the user was given very low
priority. Emphases were on changing winding, narrow, muddy
roads into straighter, wider, all-surface facilities. Dollars
spent on esthetics were not available to get people out of the
mud, or to save lives through safer facilities. To a major extent, the same priorities exist today, but more wisdom can now
be incorporated into the engineers' view of esthetics.

Highway geometry can only be partially influenced on any
specific project. Acceptable design standards for curvature
and grade have been established by AASHO and the Federal Highway Administration (FHWA) for various types of facilities. The
standards are under constant review and constantly upgraded.
On federal aid projects; it is all but impossible to obtain
changes in the standards. The basis for these standards are
the aforementioned criteria for economy and safety for a predicted volume of traffic, and a given design speed.

Structural problems are concerned with stable slopes, drainage structures that must resist traffic and embankment loads, pavements and bridges that can carry the legal loads, and tunnels that are safe from collapse and falling rock. Erosion is generally a threat to the stability of slopes and drainage structures, in addition to its esthetic characteristics.

The location of bridges and pavements is decided under geometric analyses, combined with economic considerations. Length of span and height above the ground are the major economic factors in bridges. Swamps, areas with a high-water table, and other soft ground pose economic problems for pavements.

The legal vehicle loads are set by state legislatures, and only temporary pavements (or bridges) and old bridges can be restricted to loads less than the legal limits. The legal loads vary from state-to-state, with larger loads permitted in the East and far West than in the Midwest. To obtain federal aid, the legal limit cannot be changed so as to exceed the federal limit set for the Interstate Highway System. Since limiting the use of heavy vehicles is not considered politically feasible, the restriction for Interstate Highways has led to the same legal limits throughout a state.

Construction, Maintenance and Operation

The construction, maintenance and operation of transportation facilities are directed to producing the greatest benefits to the user of the facilities. Similarly as for design, the focus is on the improvement of conditions for the user, and not for other activities and programs.

These three processes are controlled in an initial way by the policy, planning and design procedures. Thus, the design of new facilities can and should account for adverse effects that are related to construction, maintenance and operations.

For existing facilities, the problem is more complex. For example, the use of chemicals to control snow and ice on highways and airports will have certain adverse ecological impacts. To eliminate their use will require a trade-off between reduced transport service, more hazardous conditions, and improved ecological conditions. Two of these involve the quality of transport service, whereas the third has to do with the quality of the environment.

The construction of most transportation facilities is accomplished by the private sector under contract to a government or industry. Maintenance and operations, on the other hand, are usually conducted by the owners' personnel.

Construction control is through plans, specifications and inspection by the owner. To influence the character of construction, either the plans or specifications should be changed prior to the letting of the contract. Changes after construction has started are costly.

Maintenance and operations do not receive federal aid (the notable exception being the operation of the airway system by FAA). To influence these activities, top policy makers must normally be involved, particularly if significant resources are involved.

SYSTEM OUTPUTS

Transport systems are for the sole purpose of transferring persons and goods from one place to another, and their output can be classified into the aforementioned categories of direct and indirect effects. Transportation's contributions to and adverse effects on society *cannot* be changed except by changing the output; that is, one can improve its impact on social and economic programs by improving one or more of its direct effects; or by reducing or improving its indirect effects. As can be seen, improving its indirect effects may very well require a reduction in its transport effectiveness.

In considering changes in transport systems, direct and indirect effects can be a most useful tool for estimating the social and economic benefits to the various activities for which transportation is the means to an end. In principle, the technique involves estimating the impact of a change in direct or indirect effects on various programs (education, recreation, public safety, economic development). This technique is discussed in greater detail in a subsequent section. The most effective solution--through consideration of the greatest number of alternatives--will be obtained if the change in output is considered before a system change is contemplated.

Another important characteristic of system output is its variability from moment to moment, and from individual to individual. Arithmetic averages, the basis for economic analyses, fail to deal with extreme values. For example, in emergency transport, the major variable is time. While "average time" is useful for estimating total effectiveness, the "longest time" may represent an intolerable condition, and a system to reduce such time-delays may be deemed necessary. Needless to say, this may well be a totally different system than one that would reduce the average time.

The variability of system output should be developed in statistically significant terms. Thus, the variability of individual time, cost and safety of personal transportation is quite large, and different groups of individuals will be assisted by different changes in the transport system. For example, if a reduction in poverty is a major community goal, then the *average* home-to-work travel time is of less importance than the home-to-work travel time of the low-income group. Statistical reliable data will be needed if effective transport changes are to be produced.

Competition Between Values

In any transportation project, one is certain to encounter competing "goods;" that is, to obtain one benefit another must be sacrificed. Safety of the highway traveler, for example, will be reduced with (1) longer routes to avoid parks, historic shrines, etc., (2) retention of trees near the roadway for esthetic purposes, (3) failure to replace antiquated highways in urban and rural areas, and (4) placing airports in isolated areas thereby increasing access travel mileage. The problem of choice between two or more "goods" is present in every public project. The specificity with which requirements (quantity, quality, resources, constraints) are stated will facilitate the resolution of conflicting values. Such policies are not yet being issued, and therefore, a continuation of debates can be anticipated.

National vs. Local Problems

The problem of influencing national policy will not be encountered as regularly as are those for local problems. Much more frequent will be projects which require a specific solution. The location of a road, street, or airport is typical.

Occasionally, the problem has well-defined national policy guidelines. Thus, the interstate system must be located within a relatively narrow corridor. Public policy states that it *must* be built. Even this policy has been successfully challenged, but generally, the need and desirability of *not* providing a section of the interstate system is not likely to succeed outside an urban area.

In large metropolitan areas, local transport problems can be solved in a number of ways, because mass transportation is feasible; that is, there would be a large number of persons who would find a variety of services which would meet their needs.

In smaller areas, however, the alternatives are much more restricted. Fortunately, the magnitude of the transportation problem is not so large and existing conditions are less intolerable than in major metropolitan areas.

Within a given area, the requirements for transportation
should be developed, including quantity, quality and constraints. Frequently, quality limits will be set by government
as a prerequisite for financial assistance. Such limits are
based upon providing a relatively safe facility that will not
become congested for at least twenty years. Not infrequently,
the requirements from the state or federal governments are for
a facility that is in excess of that deemed necessary by local
officials. Once again, competing values enter, since part of
the quality requirements are for safer facilities.

A major decision at both the national and the local level
is whether or not new facilities are required. Past bases for
justifying new facilities are no longer adequate. Past assumptions that congestion, available funds, or "status" requires a
facility are being replaced by more rigorous analyses which
lead to more definitive bases for concluding that new facilities are essential. Once there is a clear understanding of
what social and economic activities of a community would be
enhanced by a new facility, the debate can center on what can
best be sacrificed in order to achieve the desired end.

INTERDISCIPLINARY STUDIES

The provision of facilities for transportation has required the use of a wide range of physical sciences. In recent
years, behavioral sciences have become increasingly more important as the relation between the operation of a facility and
its geometry has been organized.

The evolution of a total system concept has led to greater
recognition of the need for relating total community needs to
the benefits and adverse effects of operating a transport
facility. As a result, interdisciplinary studies in transportation, utilizing any and all theories and organized bodies of
knowledge, can be effective.

In other chapters, further details on various disciplines
are provided. In the following section, a summary discussion
is provided on the manner in which interdisciplinary efforts
can be incorporated into transportation studies.

Transport Efficiency

Improving transport efficiency will be an alternative
objective as long as there are individuals and communities;
that is, transport systems will never be perfect. When and how
transport should be improved is derived through studies of
social efficiency.

Transportation efficiency involves many modes and many elements within the quantity and quality characteristics. It is beyond the scope of this discussion to do more than describe the principles of how such activities can be conducted by interdisciplinary study. As noted in a preceding section, the approach is greatly simplified if the requirements for output have been specified. Also noted earlier, if these requirements have not been stated, the first order of business in a transport efficiency study is to develop or obtain politically acceptable requirements.

Given the requirements, the development of alternative solutions becomes the objective of the transport efficiency studies. The objectives will vary from reducing costs, reducing time in transit, increasing safety, reducing air and noise pollution, etc., for some quantity levels. It is axiomatic that any change will involve some trade-off of benefits, that is, reduced costs will generally mean sacrificing some other characteristics, or reducing pollution will require an increase in cost, reduction in safety, or reduction in quantity.

One of the most frequently encountered transport problems at the local level is the construction of new highway facilities. Such problems are examined in the following paragraphs to demonstrate how interdisciplinary studies apply to transport efficiency.

The policy guidelines. New facilities are generally proposed in urban areas because of congestion, and in rural areas because an existing facility needs a new pavement or the curvature is too severe for modern traffic. In the first case, the transport benefits that are most frequently sought are savings of time and operation costs and increased capacity (greater quantity). In rural areas, the same benefits plus safety are normally desired.

Increased capacity, time and operating cost savings, and safety can be obtained in a number of ways other than the construction of new facilities. Some of these would require longer implementation time and greater investments (for example, electronic controlled vehicles would permit higher speeds, less headways and significantly higher capacities). Reducing the quantity of traffic (particularly at peak hours) could produce the results for a lower quantity of transport and for a change in some of the quality levels (user satisfaction). Since the greater capacity is a long-range characteristic, policy guidelines are essential relative to whether greater transport quantities will be required for the long term.

The provision of new facilities also has a number of alternatives which will be influenced by social efficiency

questions. For example, assuming that new facilities are a requirement to meet community needs, then what types of structures, heighborhoods, industries, parks, rivers, etc. can be
disrupted (or cannot be disrupted). Also, preferential treatment could be noted for mass transit, import access, hospital
access, etc. Since there is no surface location that will not
be objectionable to some individual group, the social efficiency process will need to delineate the type of adverse effect that will be tolerated in order to obtain the additional
transport capability. A tunnel (sub-surface location) may be
the most acceptable solution. As noted earlier, these questions that involve conflicting benefits are political and cannot be resolved by interdisciplinary analysis.

Interdisciplinary studies of indirect effects will be most
productive. Intrusions on neighborhoods, parks and wilderness
areas; disruption of aquatic life through relocation of watercourses; air and water pollution; and other impacts on the environs can be identified. Corrective measures for those adverse effects require analyses of the consequences--both the
benefits and adverse effects. For example, a more circuitous
route to avoid a park may increase the length of the highway by
a small amount. If this added length is 0.1 miles on an important highway carrying 100,000 vehicles per day, then one fatality can be expected every five years because of the added
length. (In 1970, the fatality rate was approximately one
death per 20 million vehicle-miles of travel.) While other
actions can be taken to reduce the probability of fatalities,
accidents are a function of exposure or miles of travel.

In subsequent chapters, the contributions that can be made
by various disciplines will be discussed. The major requirements for successful interdisciplinary studies are: (1) the
objective and the constraints must be stated in operational
terms, (2) the contribution or objective of each individual
must be established, (3) no individual should be expected to
provide expert opinions in other than his own area of specialization, and (4) two or more solutions with incompatible benefits should be listed as alternatives if each falls within the
policy guidelines.

B. WATER

No nation or society has been able to develop beyond the most simple agrarian level without a high degree of water resources development. The kitchen middens accumulated in the prehistory of man mark the progress of our tribal ancestors from stream to stream in Britain and Scandinavia; Israel is rebuilding irrigation systems first used over 2,000 years ago; Roman aqueducts are still in use; and the ancient civilizations of Egypt, Greece, Troy and Carthage sought rivers of the sea for supply and transportation.

For millennia the rivers have given food and water for man and his domestic animals and crops; they have often given him the land itself. Their winding courses marked the routes for roads and rails while serving as arteries for water-borne commerce. From the water wheels that powered the looms of the early eighteenth century to the hydro-electric power of the twentieth, they have been essential to industrial development. They have diluted and cleansed our wastes; absorbed heat in thermal power generation; furnished tranquility to the troubled spirit; and provided recreation for the gregarious swimmer, the solitary boatman, the hopeful fisherman, and the adventurous naturalist.

They have also stimulated, through these means, some of the most magnificent engineering achievements of modern man. From the Aswan Dam in Egypt to the greatest earthmoving project in history--the taming of the Mississippi--their threat of disaster and promise of plenty have provoked the effort and aroused the concern of both primitive and civilized societies.

In the past, however, water resources planning has been used to accommodate, rather than shape, the future. The magnificence and plentitude of this resource in the United States, until the present, has engendered the attitude that it is on the periphery, rather than at the center, of means to achieve our goals, ill defined though the latter are.

These resources have been so essential to the progress of history of the United States that the eye and mind sweep past them in search of less familiar territory.

To help place the functions that are performed by water management projects in the perspective of historical development, a brief discussion illustrates the types of projects that have evolved for man's use.

River Basins

Over the past two centuries, the river systems and the coastal areas of this nation have been utilized increasingly to support the activities of man. The first developments, in addition to those for water supply, sprang from the fact that early settlements in the flat river bottoms close to water transportation expanded as years went by. The need for levees or flood walls to protect these communities against periodic high outflows of the river basins grew increasingly urgent. Locks and dams to further channelize the flows and control them for larger tugs and tows created the shipping routes to sources of major agricultural and industrial products while providing the means for moving raw materials to the factories. Figure 3-1 illustrates a third use for the streams followed naturally from the first two; they served as large conduits to disperse and dilute human wastes and the effluents from industrial processes.

Because precipitation was seasonal, rivers flooded during parts of the year but suffered low flows at other times. Large dams were built to modify these cyclical variations and therefore to provide more efficient utilization of the waters. The dams served a variety of purposes: Retaining the water in the tributary areas for use by individual farmers; preventing erosion of topsoil into the major river systems; and providing a year around water supply for irrigating lands that served as additional sources of agricultural products. The reservoirs also guaranteed that both municipal and industrial demands for water supply could be met. Protection from floods and maintenance of summer flows in the downstream areas furnished many benefits. In addition to providing releases for navigation, fish and wildlife, flows maintained an adequate dispersion or assimilative capability for the wastes that were being discharged into the streams. In many areas, due to the hydraulic head that could be developed, it was found appropriate to install hydroelectric systems for the generation of power to meet the needs of the people and their residential, commercial and industrial activities. The impoundments behind the dams in the river channels and in the upstream areas created lakes that were used for recreation.

Figure 3-1. The San Joaquin River in California supports
a deep water port at Stockton, recreation and water supply in
the delta, shown here, and further water supplies for the San
Joaquin Valley. Stripping of vegetation from levees was an
unpopular maintenance technique.

Ports and Harbors

At the points where these river systems entered the oceans
and at other points along the coasts, ports and harbors were
developed to provide for the transfer of people and materials
from land modes of transportation to seagoing vessels. Port
and harbor development includes the construction of breakwaters, the dredging of channels and anchorage areas, construction of piers and wharfs to facilitate the transfer of
materials from the vessels to the shoreline. The ability of

the nation to move materials freely along the coastal areas
and to engage in international trade was essential to the
United States' development as a major participant in world
trade.

FUNCTIONS OF PROJECTS

The utilization of this great natural resource has come
about by construction of many types of water management projects. Their functions can be characterized in general as conservation and use of the water itself, use of water for pollution control, and use of water for navigation. Concomitant
uses generally developed in connection with some of these other
major uses include use for recreation, improving fish and other
wildlife habitats, and for esthetic purposes. Table 3-3 gives
an order of magnitude insight regarding the amount of water
used for some of the withdrawal purposes.

Table 3-3. Water withdrawals for selected years and purposes in
the United States, including Puerto Rico.

| | | | (Billion gallons per day) | | | |
Year	Total Water With- drawals	Irriga- tion	Purpose of Public Water Utilities	With- drawals Rural Domestic	Indus- trial and Misc.	Steam Electric Utilities
1900	40	20	3	2.0	10	5
1910	66	39	5	2.2	14	6
1920	92	56	6	2.4	18	9
1930	110	60	8	2.9	21	18
1940	136	71	10	3.1	29	23
1950	200	110	14	3.6	37	40
1960	270	110	21	3.6	38	100
1970	370	130	27	4.5	47	170

SOURCES: Withdrawals reported for 1900 to 1940 are taken from
PICTON, Walter L. (March 1960). Water Use in the United States,
1900-1980, prepared for U. S. Department of Commerce, Business,
and Defense Services Administration. U. S. Government Printing
Office, Washington, D. C. Withdrawals reported for 1950 to
1970 are taken from MURRAY, C. Richard & Reeves, E. Bodette
(1972). Estimated Use of Water in the United States in 1970,
Geological Survey (From National Water Commission Review Draft.)

Water supply projects are developed at all levels of government. Local municipalities and private water companies develop supplies as close to the area of need as possible. Wells to draw from groundwater aquifers are developed where possible. In cases of larger communities and in areas where groundwater is insufficient to meet the demand, other sources of water are developed, usually by projects to conserve surface water in reservoirs along streams. The dams with the reservoirs behind them, together with appropriate arrangements for overflow control such as spillways and with aqueducts or canals for transportation to areas of use, are constructed by utility districts or municipalities. In other cases they are constructed by state or federal agencies that sell the water as wholesalers to the distributing districts. Facilities of this type range from small dams on relatively small streams to very large dams in downstream locations.

Federal agencies were specifically authorized by Congress to enter into developments of projects of this type by the Water Supply Act of 1958. Under the provisions of that act, the federal agency will construct the project; it will enter into a contract with local water supply over a fifty-year period. The California Water Project is another example of an overall project consisting of reservoirs in the upstream area, conveyance facilities then pick up the water from the delta area in central California and transport it southward into the dry upper San Joaquin Valley and over the Tehachapi Mountains into the southern California area; there a series of regulating reservoirs store the water for sale to water vending agencies who build the treatment plants, operate and maintain the facilities that constitute the distribution system.

Irrigation

Water supply projects are also developed for agricultural use. Local districts, formed under a variety of state laws or by the cooperation of adjacent landowners, divert water from rivers or streams, from the groundwater, or from conveyance facilities that are constructed specifically for that purpose in order to irrigate otherwise arid land. Since 1902, the Bureau of Reclamation, an agency in the Department of the Interior, has been developing specific water management projects to provide a supply of water for these local irrigation districts. They have constructed some 270 dams and reservoirs throughout the western states to serve this basic purpose. In general, their reservoirs are major downstream reservoirs of considerable size.

Another federal agency within the Department of Agriculture, the Soil Conservation Service, has the authority to use federal monies to construct dams to make reservoirs in the upstream areas to provide for conservation of the water supply for use in agriculture. The size of reservoirs that Congress gives the Soil Conservation Service the authority to construct is limited.

Industrial Supply

Another important use of water provides the needs of industries. Water for processes and, more important, water for cooling of industrial processes are developed both in individual projects by industries themselves where they draw out of natural streams or by these industries drawing water from water districts. Most industries requiring cooling water, for example in power plants and refineries, locate themselves on natural water courses so that this important asset is readily available.

Pollution Control

Another of the important uses of the water courses of the nation has been for pollution control. That is the streams, rivers, lakes and ocean areas have been used to assimilate the wastes developed not only from the industries mentioned before but from the commercial and residential areas, the so-called sanitary wastes. The majority of these developments has been undertaken by municipalities or local sanitary or sanitation districts established under various state laws. These local agencies customarily construct facilities to meet the needs of the community based on local bonding or taxing programs. In more recent years under the Federal Water Quality Administration and now under the Environmental Protection Agency, federal assistance for programs of this type has increased greatly.

Flood Control

Another basic water management program is the flood control program. Since many of our communities were established where periodic heavy rainfalls or large snowmelts have caused an excessive amount of water to come down through the valleys, flood control measures have been found necessary. This program in some small areas is administered by local flood control districts who built levees or other retaining structures to keep the rivers within their banks. To provide for broader approaches in larger basins requiring major projects, Congress in 1936 authorized the Corps of Engineers to study water basins to see what flood control needs should be met by structural means. This program of federal assistance provides for funding of projects specifically authorized by Congress with local contributions to provide for lands, easements, and rights of way for

the project and for relocations of utilities. The types of
projects include levees, flood walls and other channel improvements to redirect the stream or river. They also include dams
in upstream areas that impound the water for later release
during low flow periods. In many basins, a series of projects
constructed over a long period of time has gradually brought
major river systems under control. Along the lower Mississippi
Valley, for example, a system of levees and dikes with some by-
pass channels, with flood walls protecting portions of cities,
has allowed control of most of the Mississippi River floods.
In the Missouri River under a long-term plan developed in the
early late 30's and early 40's, the so-called Pick-Sloan Plan,
a cooperative plan between the Corps of Engineers and the
Bureau of Reclamation, a series of dams in Montana, North
Dakota and South Dakota was constructed to provide for flood
control on the Missouri system, a major tributary of the
Mississippi.

Inland Navigation

The developments on these two river systems also has provided for another important use of our water resource, that is,
to support navigation. The inland waterway system of the
country acts both as a means of taking agricultural products
from the farm to market areas and for moving raw materials from
their natural locations to their manufacturing areas, but also
offers the possibility of distributing agricultural and industrial products to the consumers. Some of the rivers, like
the upper Mississippi and the Ohio, are controlled by a series
of locks and dams that provide for slack water navigation, a
stair-step effect down the river valley. In other rivers like
the Missouri, the open-river type navigation is used where the
depth of water is maintained by the releases from the lowest
dam in the chain of dams in the Missouri to keep an adequate
depth of water from southern South Dakota to the mouth of the
Missouri without utilizing locks. In the open water channel
groins and dikes are used to direct the water so that the current maintains adequate velocities to scour out the sediments
to help provide for minimum maintenance of the channels and to
reduce erosion of the banks. In some of the river systems
canals have been built to interconnect natural channels and
therefore provide a transportation network. The program for
development of this type of navigation system has been a federal program that has been under the direction of the Corps of
Engineers since its inception in 1824. The overall development of the navigation channels themselves and the dikes, dams,
locks and other facilities then are constructed at federal expense. Local projects to utilize these channels include the
wharfs, piers, docks, loading and unloading facilities and the
other means required for transfer from water transportation to
land transportation at origin and destination. These facilities

are constructed in some cases by local agencies, more often by private organizations.

Deep Water Navigation

Another use of our waters for commerce involves facilities required for deep water navigation. Ports and harbors along the coasts of our oceans are supplemented by facilities in the Great Lakes as well as our estuary areas. Here again the federal government has played an important role in the development of the major channels by providing the dredging that is required to provide adequate water depths for the ocean-going vessels. The combination of natural waterways, dredged channels, locks and dams that comprise the St. Lawrence Seaway provide an example of this type of improvement with an international favor.

Breakwaters provide for the development of sheltered harbor facilities in most of our ports. In general, the landside development has been the responsibility of local governments as well as private entrepreneurs.

Recreation

Another parallel use requiring water management projects is the use of our waterways, both natural and manmade, for recreation. A variety of facilities provided by all levels of government as well as by private clubs includes breakwaters for protecting marinas, the marinas themselves, boat launching ramps, and service facilities, so that these waters can be used by the growing number of people that take the opportunity for leisure time recreation.

Hydroelectric Power

The energy available in falling water was captured in the early days by the waterwheel; it has been used since to develop hydroelectric power. Dams hold the water in upstream areas; penstocks or pipe lines divert the water downhill so that the head (difference in elevation) can be used to drive the blades of turbines to generate electric power. The advantage of this type of power is its flexibility; that is, it can be started up and shut down again comparatively easily when compared to thermal type plants. It does not in itself cause pollution of the air or the waters. It can be constructed to provide for a reversible process known as pumped storage so that the water can be pumped back up the hill during low-power demand periods so that it can be available for peaking power at later times.

Facilities of this type have been built by the private utilities in many parts of the country. They have also been included in major projects built for other purposes such as flood control dams and dams built for irrigation. In projects such as the Central Valley Project of the Bureau of Reclamation in California, the sale of power from dams has helped to reduce the cost of the water that is delivered to the farmers for agriculture.

Multi-purpose Projects

It is apparent from the preceding enumeration of the various functions that individual water management projects can be designed to provide for several of these uses; they are then called multipurpose projects. Likewise the assets of an entire basin can be coordinated so that several of these functions can be accommodated in a series of structures to obtain a broad range of benefits. The dams on the upper Missouri River for example, as stated before, provide for releases of water to maintain a navigation channel downstream from Sioux City, Iowa during the navigation season from late spring to late fall. The same releases from the dams in the series are used to develop hydroelectric power that is delivered to Bureau of Reclamation power centers for use throughout the Mid-West. The water that is stored in the reservoirs is used to provide for municipal supplies for towns in the area and is being developed for irrigation for many of the relatively dry farm lands of North and South Dakota. The reservoirs are filled each year by the runoff during heavy rains and by the snowmelt and therefore provide for flood control in the downstream areas by retaining the water during peak storms. The water that is released for the navigation during the summertime also provides low flow augmentation of the Missouri River thus providing a help in assimilation of the sewage and industrial wastes that are discharged into the river. Maintenance of these steady flows has also improved the usefulness during the summertime for recreational purposes including boating and fishing. Projections of these multi-uses are made in advance and are balanced by an interagency council representing the agencies with direct interests; a balance is maintained between the various needs of the basin in order to insure optimum operation of the entire project. The reservoirs themselves, of course, have provided a vast new still-water recreation resource in the entire area.

EFFECTS OF WATER MANAGEMENT PROJECTS

Many changes in the natural environment have been wrought by the water management projects enumerated above. In most cases, the positive "intended" effects had been preplanned; the projects were designed to accomplish specific purposes and that they did. Examples of the magnitude of the accomplishments of some of these typical projects are shown in Table 3-4.

Table 3-4. Water Management Projects

Navigation (1970)	1,531,696,507 Tons/Year
Foreign Imports and Exports	580,969,133 Tons
Coastal Ports and Inland Waterways	950,727,374 Tons
Hydroelectric Power Capacity (January 1, 1972)	55,898,039 Kilowatts
Recreation Use (Corps of Engineers, 1969)	250,000,000 Visitor-Days/Year
Reclamation Development	
Storage Capacity (1970)	133,000,000 Acre Feet
Annual Delvieries	600 Billion Gallons/Year

The major effects have been to adapt the countryside to the needs of an evergrowing population, to provide for facilities for commercial and industrial growth and for the transportation facilities needed to provide mobility within the country and the ability to carry on commerce with the other nations of the world. Other of these projects modified the regimen of nature by retaining waters that would have otherwise caused flooding in lowland areas by storing waters during time of heavy precipitation and allowing them to be delivered at a different time to another place when the water could be used more advantageously.

The projects were conceived and constructed to· meet future needs that were projected based on growth trends that have been reflected in the most recent past. The growth ethic therefore is frequently urged though it allows more people to settle in a flood plain. Additional water supply provides part of the basic needs so that more people can move into an area that might otherwise not have become overcrowded. Only very infrequently has the planner looked upon a development as a way to control and direct growth, more frequently it is a means to accommodate it. Thus, the development of water management projects has not always led to optimum land use and population planning.

FEDERAL AGENCY INVOLVEMENT

In describing the various functions of water management projects, various Federal agencies were mentioned. There were brief discussions of the interrelationships between them and the local agencies in the planning and construction of these projects. A recapitulation of the various areas of operation of the agencies is shown in Table 3-5.

The method by which the various agencies participate in projects varies. For example, individual projects proposed by the Corps of Engineers, the Bureau of Reclamation and the Soil Conservation Service are authorized as specific items by Congress. Subsequently Congress appropriates money for the advance planning and eventually the construction of those projects.

In the field of water pollution control, the Environmental Protection Agency and its predecessors have made grants to local agencies through a state clearinghouse for the construction of the projects, instead of actually acting as a construction agency itself.

In some cases, however, projects are constructed by local agencies with federal money contributed to the project for specific purposes through the construction agencies mentioned before. As an example, in the Oroville Dam project in the State of California, a part of the state water project, a portion of the storage in the reservoir is designated for flood control purposes. Therefore, Congress appropriated money that was released through the Corps of Engineers budget to the State of California to pay for the flood control storage portion of the cost of the project.

Local Participation

The participation of local sponsors in projects is undertaken in two general ways. The first, as mentioned before in flood control projects, requires local agencies to participate in the original construction of the project by taking care of certain specific parts of the project. An example is the buying of the land, easements, and rights of way for a flood control project and paying for the relocations of utilities that might be necessitated by the construction of the project. On the other hand, in a water pollution control project, current laws provide that the federal government may pay a certain percentage of the project cost. The remainder therefore, must come from the local sponsor. In some states, for example in the State of California, where the state participates as well, the federal and state share will provide a grant for eighty percent of the project cost on approved projects. The local people will be required to fund the rest.

The second method of participation is by repayment to the federal government for a portion of the project. As mentioned above in water supply projects, the local entity is required to repay over a fifty-year period the portion of the project costs that are allocated to water supply under the provisions of the Water Supply Act of 1958. In the case of recreational developments under the Water Projects Recreation Act of 1965, local sponsors are required to repay to the federal government

Table 3-5. Federal Agency Activities In Water Projects

Department	Agency	Federal Aid Program
Agriculture	Farmers' Home Administration	Development of Rural Community Drainage Projects Development of Rural Community Irrigation Systems Loans to Soil and Water Conservation Districts Association Grazing Facilities Rural Water and Waste Disposal Systems Loans to Develop Rural Community Recreational Facilities
	Soil Conservation Service	Watershed Protection and Flood Prevention River Basin Surveys
Commerce	Economic Development Administration	Public Works and Development Facilities Including Sewerage and Waste Treatment Plants
Environmental Protection Agency		Waste Treatment Works Construction Water Pollution Control - Research Projects Water Pollution Control - Methods Water Pollution Control Training Water Quality Planning
Housing and Urban	Community Resources Development Administration	Public Facility Loans Water and Sewer Facilities Grant Program Public Works Planning
Interior	Bureau of Reclamation	River Basin Surveys Irrigation Projects Water Supply Reservoirs
	Office of Water Research and Technology	Water Resources Research Saline Water Research and Development
	Bureau of Outdoor Recreation	Recreation Facilities

Defense	Army Corps of Engineers	River Basin Studies
		Flood Control
		Water Supply Reservoirs
		Navigation
		Beach Erosion
		Recreation Facilities
		Wastewater Management

fifty percent of the costs of the facilities for recreation over a fifty year period. In most cases, local sponsors are also required to pay for maintenance and operation of facilities if built by a Federal agency; this would naturally be the case with water pollution control facilities that are constructed and owned by the local entity.

In the case of each of the federal agencies and of the state agencies that have been developed in particular areas, and of the regional agencies that have been specifically started to perform particular functions in the field of water resource management, the principles applied are those that established by the laws of Congress and by the precedents that derive from those laws. Regulations and executive proclamations have been promulgated to lay out the interrelationships for each one of the departments for each type of project that it designs or constructs. These regulations are constantly under review and revision and modernization. Planning and design at a particular point in time, then, requires that one familiarize himself with the current regulations and with the potential developments toward changes that are underway in Congress or in the federal establishment.

DEVELOPMENT PROCESS

Some of the problems in planning water management projects have been described and it is perhaps understandable that the process takes a considerable period of time. Some of the externalities in the evolvement process were discussed in Chapter 2. Projects sponsored by local districts can reach realization in two or three years. After preliminary project planning, revenue programs become the critical control. Bond issues for general obligation or revenue bonds generally require voter approval. Tax increases to provide for capital improvements are also used to support public agency investment.

Many water pollution control facilities are eligible for federal assistance under the Clean Water Program of the Environmental Protection Agency. Annual programs within appropriations established by Congress are available based on priorities established by the states for regional projects. Proposals being considered by Congress in 1972 could provide eighty percent of the project funds with state cooperation.

Federal construction agencies, in the past, considered seven years as a reasonable goal for bringing projects to fruition but seldom achieved it. The time elapsed between the idea and its realization is now approaching twenty-five years. Figure 3-2 illustrates the many steps which must be taken, using the Corps of Engineers as an example.

After local people ask for assistance, study is authorized and funded. The agency conducting the study holds a public hearing to receive testimony on the character, scope, importance and urgency of the problem and to receive suggestions for its resolution. Notices of this meeting are sent to all federal, regional and local agencies, conservation and development associations, and all citizens who express an interest. The inputs of this public meeting provide the direction for the planning effort. They also establish problem areas that require solution; in some cases they bring out implications that arouse public sentiment both for and against project development. In any case the public meeting establishes a preliminary basis for further study of the potential problems so that the analysis can proceed.

Formation of an advisory committee, to insure public participation in planning, is frequently considered for major projects. This approach adds yet another dimension to the development processes as illustrated in Chapter 2.

The immediate objectives of many who join in planning the project are antithetical. The most difficult job of the planner is to attach relative weights to each of these objectives and balance the solution among them. What evolves is either a good or a bad compromise. A good compromise, however, may be politically unacceptable while a bad compromise may achieve consensus. In any event, the manner in which the decision is reached is becoming as important as its content. It is for this reason, as well as for reasons of efficiency and logic, that the interdisciplinary approach is essential.

PLANNING OF WATER MANAGEMENT DEVELOPMENTS

The enumeration of the agencies involved points to the fact that most of them are fundamentally user-oriented; they conceive the projects, sponsor them, plan them and construct them to provide for a particular use that can be characterized as single purpose development. Each organization, therefore,

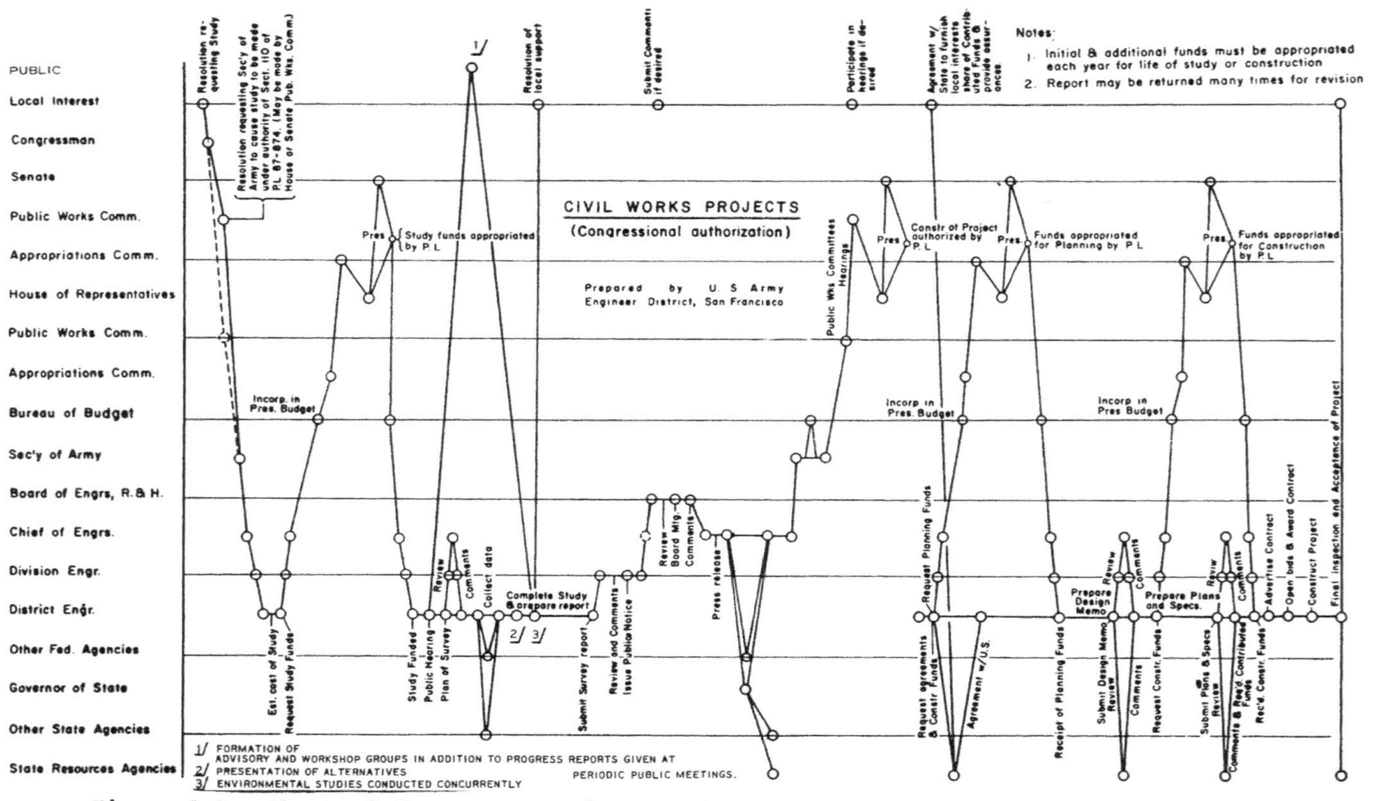

Figure 3-2. Chart of the sequence of events between a projects inception and completion.

bases its approach to planning for its structures and facilities on its own particular needs and requirements for the future. For example, as long as there are adequate supplies of fresh water available, there is no need for the water supply district to concern itself with the function of reclamation of wastewaters or of recreational use. The navigator on a waterway concerns himself with adequate depths and widths of channel and does not particularly heed the need for water quality nor environmental preservation. As the growth toward utilization of our natural resources has continued, each of these single purpose entities gradually has tended to encroach on some other single purpose agency or single use; the need for coordinated planning for a multi-purpose approach became apparent.

Multi-Purpose Approach

Multi-purpose planning has been established, then, through an evolutionary process based both on mandates from Congress in specific actions, and in precedents set in establishing certain projects. In order to overcome the relatively limited single agency approach in projects at the federal level, the Senate established new guidelines in 1962 in Senate Document 97.

In addition, executive directives and mandates have been established that have contributed toward the evolution of multi-purpose planning by federal agencies. As a result of these authorities, planning agencies now have the ability to consider the multi-purpose uses of the waters of the country in formulating plans for public works projects. Thus a variety of the uses and structures mentioned above now are, as a matter course, planned into a particular project and the relative effect of the various uses are weighed in analyzing the best course of action and the best plan to be developed.

Landmarks in this progress toward comprehensive planning to encompass multi-purpose objectives have been augmented by explanatory and augmenting legislation between the major laws passed by Congress. The metamorphosis as applied to the Corps of Engineers is traced to provide an example.

As indicated, the initial legislation in 1824 extended the Corps of Engineers responsibility for internal improvements to rivers, waterways, and harbors to develop internal transportation and promote foreign trade. Congress, by the Refuse Act of 1899, also gave the Corps authority to monitor water quality. A recent interpretation of this act requires the Corps to grant or deny permits for industrial discharges on the basis of water quality. This activity was initiated in 1971 in a cooperative program with the Environmental Protection Agency.

Another landmark was the River and Harbor Act of January 1, 1927, in which Congress authorized the Corps to study the water resources potential of the principal rivers of the United States. In carrying out this assignment, the Corps prepared some 200 comprehensive reports on individual river basins outlining the possibilities for their development for flood control, hydroelectric power generation, irrigation, navigation, and other purposes. These are known as the "308 Reports," the name corresponding to the number of the legislative document. Figure 3-3 illustrates a typical type of double benefit from these types of projects.

On the basis of these reports, Congress initiated the general flood control program in 1936. Such reports provided the foundation for basin-wide plans of development subsequently undertaken in great river basins such as the Ohio, the Missouri, and the Columbia, as well as the early projects of the Tennessee Valley Authority.

In the early 1970's the Corps engaged in a new series of studies and comprehensive planning undertakings which, by the time they are completed, will encompass every major region and river basin in the country. These basin studies are being carried out as part of a larger effort in cooperation with other federal and state agencies. The Water Resources Council has established major regions of the country for this planning; some are organized into formal agencies, e.g., the Columbia River Commission. Others are controlled by a coordinating group, e.g., the Pacific Southwest Inter-Agency Committee. In any case, the responsibility for providing leadership and review is formally vested in a group that has the authorities to require broad consideration of overall requirements in the basin to insure proper balance in proposed incremental solutions.

An extension of this system might be used to provide a basis for examining the relationship of water management projects to other public projects and to other types of national needs. A similar pattern of broadening programs has been authrized for the two other federal construction agencies. These authorities have permitted both the Bureau of Reclamation and the Soil Conservation Service to take a basin approach to water management projects. They too then participate in the Basin Commission or Committee programs mentioned before. Thus, coordinated and cooperative systems are developed with each entity--federal, state, and local--providing its share in the planning, construction, maintenance and operation of certain projects.

Figure 3-3. Flooded farms on the Salinas River in California suffered serious losses in 1969. The imbalance between winter flows and the need for irrigation supplies in summer are corrected by water resources projects.

Multi-Agency Development

The evolution that has been developing then has been changing the requirements for multi-agency planning. In the earlier years in project planning, it was normal for one agency to make the plans and to move its proposals through the chain of reviews to action by the appropriate agency with only minimum participation by other agencies and a minimum of review by other interested parties. A gradual change in this procedure now has involved many agencies at the local, regional, state and federal levels in continuing review during the planning and preconstruction period. Among laws which have promoted this

involvement are those governing fish and wildlife protection,
water quality, and consideration of recreational facilities as
a project purpose. Encouragement to local agencies to participate has been given by Congress which has, in effect, made the
federal government the banker, or mortage holder, for construction of facilities in federal projects for water supply
and recreation. Local agencies repay the cost of water supply,
and fifty percent of the cost of recreation, over a fifty-year
period with no initial investment required.

Furthermore, the Water Resources Council has, in the "yel-
low book," a new outline of procedures to be used in the analysis of water resources projects. Comments on this proposal have
been received by the Council and are being analyzed for action
prior to the adoption of this new federal directive from the
Executive Branch for water resources planning.   The guidelines
adopted will be yet another step in the direction of insuring
multi-agency coordination in planning; therefore they should
add to the ability to engage in truly interdisciplinary planning for projects of this type.

PLANNING PROCESS

The general process for the accomplishment of water management projects has been described above.   Although the details of the actual engineering planning and design for the
various types of water projects varies, it is possible to discuss the general approach that has been used and cite examples
of specifics that are applied to some of the particular projects.

Requirements

First, the establishment of the need for a project must be
determined.   The project might be required to meet present
needs.   For example, in a water pollution control facility, it
is possible that the effluent from the facility does not meet
current water quality standards and therefore the project is
needed immediately to upgrade the processes to meet those standards.

On the other hand, a project can be foreseen as necessary
to meet an upcoming need.   For example, a water supply project
may be needed in an area where the population has still not
reached the level at which it is using all the capacity of existing water treatment plants but, with growth envisioned in
the near future, an expansion of the water treatment facility
is required.   A similar projection could be made for a region
that would demand an entirely new impoundment to provide for
additional water supply to take care of the overall projected

growth of the region that would demand an entirely new impoundment to provide for additional water supply to take care of the
overall projected growth of the region. Even if a facility is
needed to meet existing problems, it is generally considered to
be good practice to plan for future expansions while you are
planning to meet the present needs and therefore projections
are needed in that case also. Projections of future needs follow the pattern of trying to determine what future population
growth is expected, what the land use constraint may be, what
per capita consumption or demand will be required, what changes
in processes are possible, taking advantage of new technology.
The engineer, working with the planner, the economist, and the
government entities involved, must make these projections in
order to determine what level of needs must be met. Here again,
the multipurpose approach, that is considering a combination of
needs in laying out the alternative projects, is usually in the
best interest of the entire community.

Herein also enters the question of the philosophy of the
particular region that is to be served. In general in the past,
it was presumed that continuous growth was important and necessary and that it had to be supported. That premise is being
questioned at the present time and must therefore lead to an
evaluation of the desire to have a project as well as an evaluation of the need to have it. In some types of public projects, the client that makes this determination tends to be
amorphous. Therefore, the process for finding what the desires
and needs are has become more of a public process than it had
been in the past. Public participation through advisory groups,
public hearings, use of news media and published reports will
be discussed in more detail in Chapter 12.

Example

An illustrative example may demonstrate the various approaches. In Figure 3-4 three projections of population growth
from the 1970 census are shown for a community; they show a
range of growth rates that seem probable based on analysis of
demographic trends. For the sake of simplicity in this example,
it is *assumed* that water requirements will increase in direct
proportion to population; that is, that total demand for domestic consumption and for industrial and agricultural needs will
maintain a constant per capita requirement.

Existing water supplies are shown on solid horizontal line
at the left. Based on earlier projections of need, Reservoir A
was planned and is under construction. It will be completed
and filled by 1975 adding the increment shown to the available
firm supply. Depending on which rate of growth actually occurs, this additional water may be needed in 1975 or may not be
required until 1979 (under minimum growth rate).

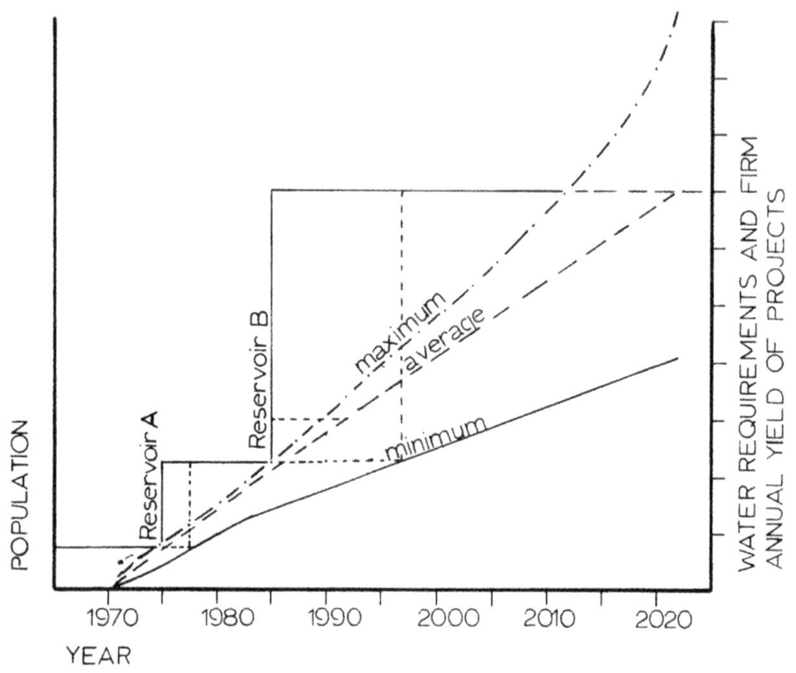

Figure 3-4. Typical water requirements projections.

Since the community recognizes that another increment of
storage may be necessary and that federal help will be needed
to develop it, a further projection has been made. Under maximum growth, it will be required by 1985; under minimum growth
it will not be needed until 1998.

Assuming that the vertical line Reservoir B is the optimum
project that could be developed at the most logical site, it is
apparent that that project could provide for average needs past
2015. It also can be seen that a project half as large could
provide for average demands past 2005 and for more than needed
to meet 2020 needs on the minimum growth curve. To put it
another way, even the half-sized Reservoir B could tend to encourage growth and usage since the supply would be available.

On the other hand, if no Reservoir B project were built at
all, it would tend to inhibit growth past the level of minimum
growth in 1998 that Reservoir A can support. Thus, a utility
or community service could be used to influence growth rate.
The other variables such as per capita domestic demand and the
influence of availability of water supply on new industries
were omitted here as stated. These factors would have to be
considered in a full-scale analysis.

136

Feasibility

But presuming that there is a real need to be filled and a desire on the part of the client to fill that need, the engineer can proceed with his feasibility studies. A great deal of physical data needs to be gathered and analyzed. For example, in a water supply or flood control development in a particular basin, an analysis of rainfall records, of stream flow and of current diversions of water have to be made. This inventory amounts to a measurement of the resource that is available. At the same time, preliminary geologic and soil investigation need to be made to determine whether foundations are adequate for dam structures, what construction materials are available, and what the record of seismic activity is. A survey of existing facilities, including highways, railroads, utilities and communities, has to be considered in determining the potential effects of a new project or these types of man-made structures.

Environmental Considerations

At the same time an environmental inventory and assessment should be made. General observation of the fauna and flora of the area with particular emphasis on rare and endangered species is appropriate. Effects of new structures on feeding grounds of wildlife or their living areas should be evaluated. The effect of a structure in the stream upon fish life should be checked, including consideration of the spawning grounds and the other types of water life that constitute the food chain of the fish. The effect of reducing some heavy flows to lower flows and the effects of maintaining higher minimum flows during dry periods of the biota of the area should be projected.

Analysis of Alternatives

Based on preliminary analysis of the factors described, some alternative courses of action can be laid out. The courses of action in general are concerned with types of structures, alternative locations of structures, and phasing of construction of structures to meet projected needs. Alternative ways of meeting a need should also be considered: for example, in a water supply project, consideration should be given to reclamation of wastewaters, to groundwaters as well as surface waters, to transport schemes from adjacent basins as well as to inbasin development, in laying out these alternatives. The alternative of constructing no project at all would have been ruled out if the need and desire were established carefully as described above.

Then preliminary plans and cost estimates of the alternative plan proposals should be made to determine their relative merit from the economic point of view. Environmental assessment should be made to find their relative merit from the environmental, ecological and aesthetic point of view. Preliminary judgments can then be made on the advisability of proceeding further with the planning process towards refining estimates and plans for particular alternatives. These plans then need to be developed through more extensive field explorations and data gathering if necessary. During this period also, the possibilities for concurrent development to meet other needs of the community should be carefully considered. The whole range of purposes that a water management project could serve should be considered. The ways and means of accomplishing these project purposes from the structural point of view, the environmental point of view and the financial management point of view should be investigated.

Project Design

After selection of the best alternative following the evolvement process including public participation previously described, the design of the project can be undertaken. The design will be based on detailed information concerning all facets of the project, for example complete geologic and soil exploration based on test boring of the site, plus an inventory of the amounts and types of the construction materials available, will determine construction methods necessary in order to provide an adequate structures. Planning and design of relocations necessary and staging of construction to provide for diversions of the stream must be considered carefully. The overall setting of the structures into the landscape must be considered. Plans for appropriate mitigation for fish and wildlife must be carefully laid out.

DISCIPLINES APPLIED TO PLANNING

The planning process described above has demanded the application of a wider range of disciplines to the planning for these projects. As technology has advanced and our knowledge of nature and its processes have proceeded, greater specialization in each of the disciplines has also been required.

Engineering

Civil engineering, the fundamental engineering discipline for most public works projects, has engendered many facets that needed to be studied in detail; therefore, there developed the hydrologist, the earth sciences mechanic, the power specialist and the expert on materials. Hydraulics required specialists in open channel and conduit flow plus studies of reservoir currents and water strata, wave action, and littoral movements.

Structures require detailed analysis not only in their static loadings but under dynamic loadings as well. Figure 3-5 illustrates one facet of hydraulic engineering.

Figure 3-5. Hydraulic engineers study proposals for water supply, navigation and water quality on a model of the San Francisco Bay and Delta Region.

Needed also are the surveyor and mapmaker to establish existing topography; the geologist determines the subsurface formations that must be known for foundation design and availability of construction materials. The sanitary engineer examines the water quality problems and designs systems for treatment facilities.

The other branches of engineering also expanded their detailed knowledge. Electrical engineering expanded its field from power to electronics and communications. Mechanical and chemical engineers were required for specific aspects of planning and design. Systems analysis and programming became important as multipurpose projects were conceived in order to make analyses of the multitude of variables that impinged upon the design of the structures. Applications of computer techniques were utilized as needed.

As the development from single purpose to multipurpose proceeded, a more sophisticated need for economic evaluation required that economists be added to the planning team. Concern for visual harmony and appearance led to the employment of architects and landscape architects to include their approaches to design of the projects.

Environmental Considerations

In accordance with multipurpose development recreation planners were also required. Biologists and other representatives of the ecological sciences were needed to analyze the effect of the public works project on the flora and fauna, including the fish and wildlife and natural vegetation. Meteorology and climatology have been important factors and now assume added importance in light of the need to analyze air pollution potentials. Archaeologists and historians were needed to interpret the effect of the project on the remnants of previous civilizations and cultures in the area, to acquire and interpret data and to rescue significant artifacts.

Social Sciences

The impact of the projects on the development in an area was presumed in early work. Today ever increasing importance to the socioeconomic effects must be given and thus the social scientists are drawn into the planning process. Greater public review has inevitably resulted and thus a greater recognition of the political processes involved and the legal remedies and constraints must be considered as the planning for any large public project proceeds. Demographic projections, including land use, the distribution of economic classes, and effects on citizens' health and welfare, have brought into the planning process disciplines from almost every federal cabinet level department.

Coordination

The additions of members of other disciplines to the planning team and the organization of the disciplinary planning therefore has grown in many ways, depending on the agency having the responsibility for carrying out the planning. The ultimate authority, that is, the person or persons who make the final decisions in matters where there are conflicts, has not always been clear. Varying interpretations of how best to measure the needs and how to meet them continue to cause problems throughout the planning process. At present, in each agency there are checks and balances that are reflected in the consolidated report that the agency develops.   The consensus, if there is one, comes after considerable cooperation with outside agencies that are involved in the formal review process after the planning has proceeded to the point where a document can be published.

APPLICATIONS IN WATER MANAGEMENT PROJECTS

Though no examples can provide a complete understanding of the planning for water management projects, some samples may establish the requirement for detailed analysis of all facets of the problem.

The typical water resource development project consisting of a dam and reservoir is developed after a careful analysis of the hydrologic history of a drainage basin. The sizing of the reservoir is based on providing adequate storage to meet dry year needs by carrying storage over from wet years; the scheduling of use and releases for a planned reservoir is based on historic runoff records.

Reservoir Allocations

The storage in the reservoir would be allocated to specific purposes as shown in Figure 3-6. In that period prior to expected precipitation and flood runoff, the reservoir would be drawn down to the top of the conservation pool to permit maximum storage during any flood. The conservation pool would be used for low flow augmentation to provide for adequate water for fish life and also to provide for consumptive uses for municipal water supply or for irrigation.

A minimum pool would be maintained even in the driest cycle to provide for recreation use and to allow for siltation of the reservoir due to upland surface erosion.

Referring back to the discussion of cost sharing, local water supply sponsors would repay the federal government for the prorated costs for the storage for water supply, including a share of the siltation pool.

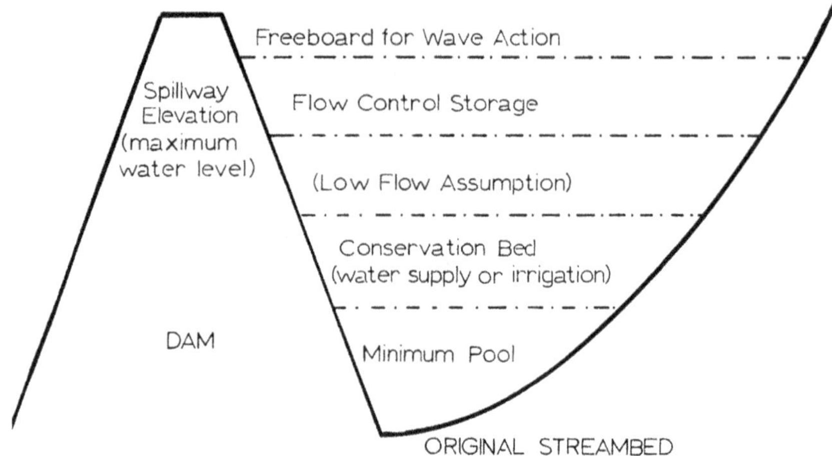

Figure 3-6. Allocation of reservoir storage.

Flood Control

The amount of storage set aside to control large floods is based on projections of maximum floods that might occur. A hydrograph showing the reduction in flows for an historical flood is shown in Figure 3-7. The analysis of the standard project flood--the presumed maximum that could occur--is done similarly. Flood control storage is determined and spillway sizing is established by analyzing these large floods; the downstream area is offered maximum protection by lowering the peak flow and the dam is protected by avoiding overtopping.

Topography and Geology

Other factors affecting the size and location of the dam and the reservoir are the topography and geology of the site. Figure 3-8 shows a layout of the important structures appurtenant to the dam itself and the surface features that determine the design. A geologic map showing subsurface features is constructed also.

Foundation conditions determine the type of structures used: concrete gravity section, thin arch, or rock or earth dam. The cross-section of the structure also is based on materials available. An example of a zoned rockfill structure is shown as Figure 3-9. Detailed analyses by soils engineers provided assurance that the dam would be safe under various pool levels; seismic evaluations were considered in proportioning the component parts.

Adequate diversion structures are necessary to carry the stream flows during construction. In many cases they are

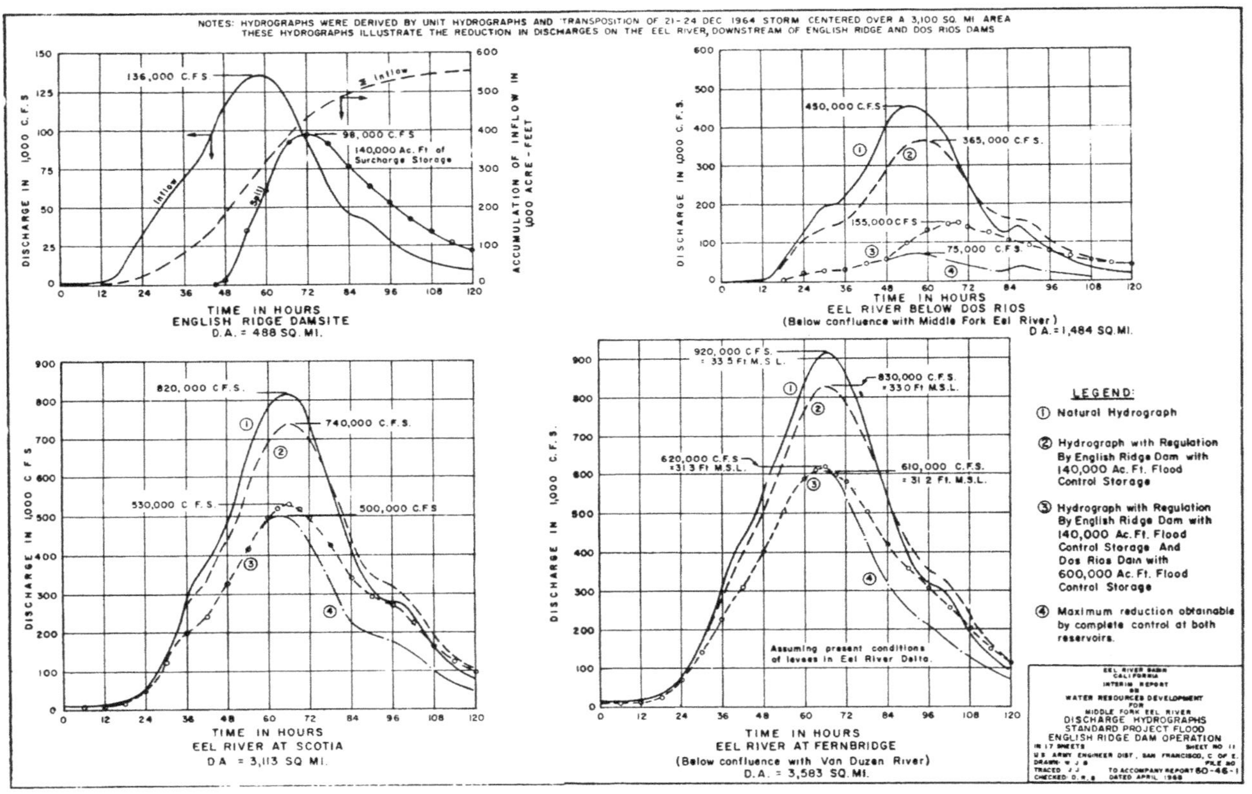

Figure 3-7. Reduction of flows for an historical flood.

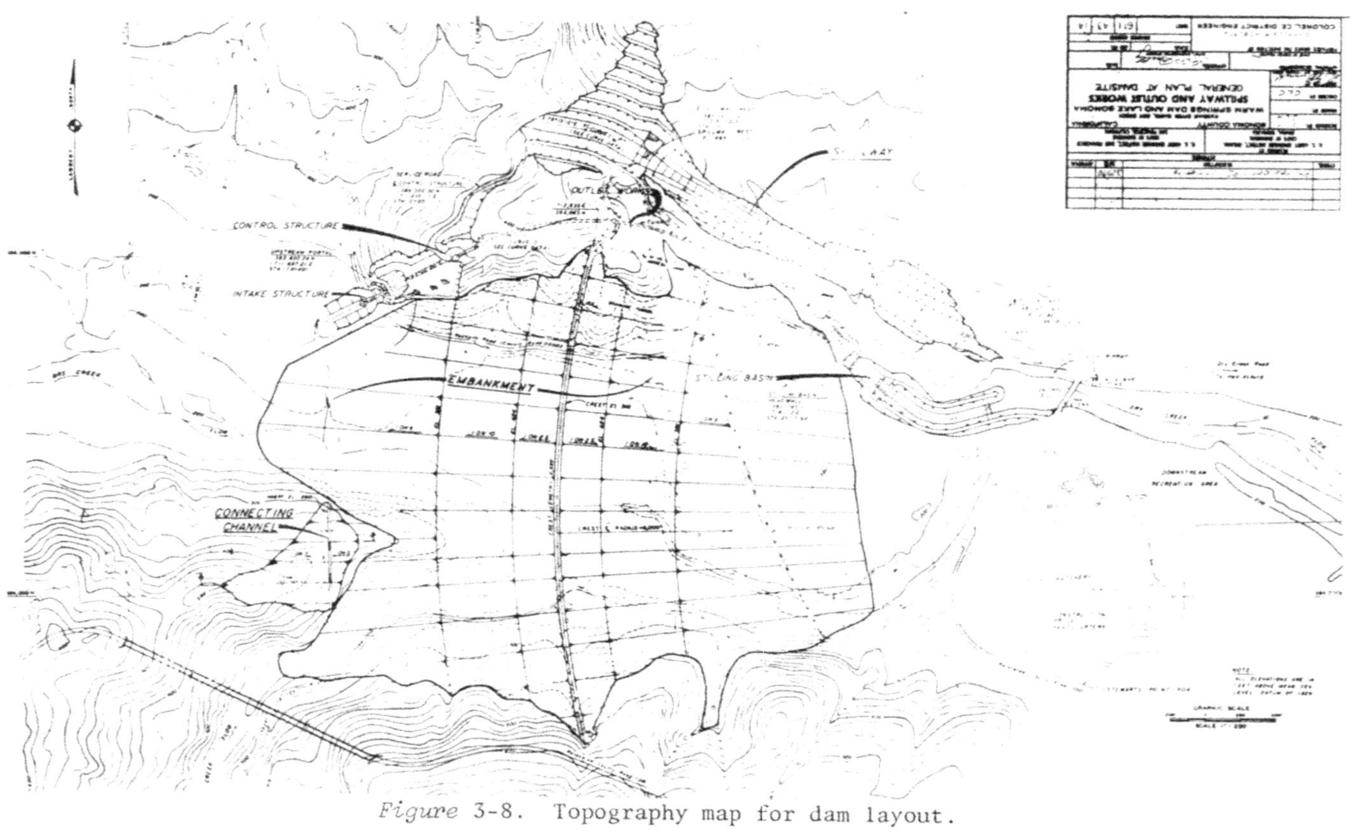

Figure 3-8. Topography map for dam layout.

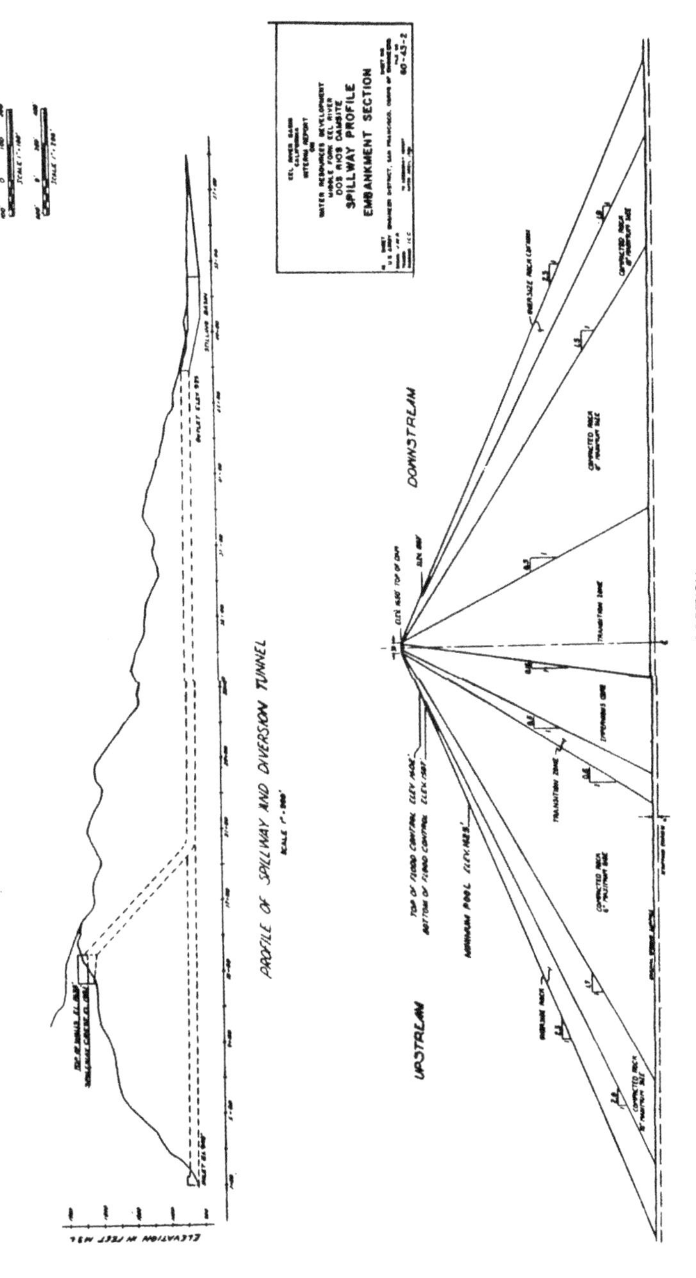

Figure 3-9. Example of zoned rockfill structure.

incorporated into the final scheme to provide outflow from the
reservoir to the downstream areas.

Works of Man

Solutions are required to replace roads, railroads, and
utilities that must be relocated to accommodate the new facilities.   In some cases these relocations involve establishing
new communities and entire towns; they also may include moving
cemeteries and finding new areas for statues and other memori-
als.

Environmental Considerations

The variety of possible solutions, to provide for consideration of environmental constraints, is many.   Some are discussed in detail for a specific project in the following paragraphs.

Watershed Management

The Soil Conservation Service of the Department of Agriculture developed a land use, or conservation treatment, plan
for the watershed area to be acquired for the project.   The
total estimated for acquisition is 17,650 acres, consisting of
13,600 acres for the dam and reservoir; 1,420 acres for recreation; 1,340 acres for a game management preserve; 290 acres
for relocations; and 1,000 acres for borrow sites. The net result of the property acquisition is that the actual areas that
can be set aside for recreation and game management are 4,000
acres and 2,400 acres, respectively.

There are four objectives for this land conservation plan:
to prevent erosion and the production of sediment in order to
reduce siltation in the reservoir; to promote recreational development; to enhance the natural beauty of the area; and, finally, to improve the environment as a wildlife and human habitat.   Varying treatments are proposed for the different areas
shown in Figure 3-10; rather than list these, a compilation
of treatment and cost has been prepared as Table 3-6. Area E
is the 2,400 acre wildlife management area; areas A, B, C. G
and H will contain most of the recreational development.

There are vast acreages that will require fertilizing;
this can be done by aircraft at an estimated total cost, including fertilizer, of $18 an acre.   Unit costs for fencing
were estimated as $2,000 a mile; for seeding, $10 per acre; for
tree planting, $80 per acre; for preparation of road embank-
ments for seeding, fertilizing and mulching, $500 an acre for
approximately 160 acres.

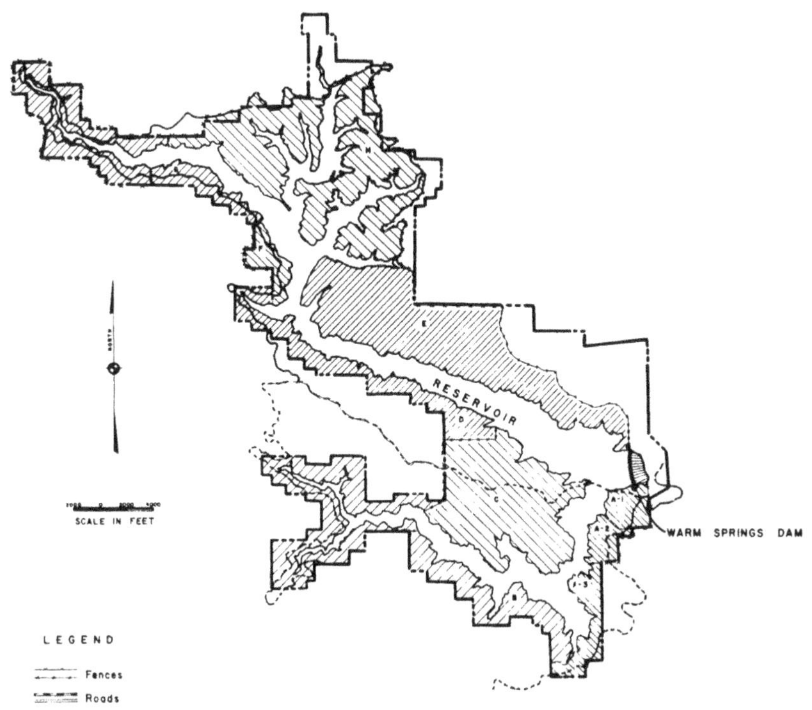

Figure 3-10.   Location of conservation treatment   are as related to warm Springs Dam.

The total plan cost,   therefore,   is $355,000. The effect of this expenditure,   however, will be remarkable;   some of the land is now badly used, little is developed, and the potential both for reclamation and environmental improvement is tremendous.   It can be hoped that the local   soil conservation districts will attempt,   as part of their educational program, to organize the local residents to institute a   similar voluntary watershed protection program over the whole drainage basin.

Water Quality, Including Sanitation and Pollution Control

Water quality includes, of course, sanitation and pollution control,   but,   as these are important in different areas both geographically and in responsibility,   it is possible to obtain guidance from two organizations.   A report on water quality needs was prepared by the U. S. Department of Health, Education and Welfare.   Recommendations on the scope and type of sanitation facilities needed were submitted by the Regional office of the Federal Water Pollution Control Administration,   Department of the Interior [now a Region of the Environmental Protection Agency (EPA)].

Table 3-6. Land treatment measures related to Warm Springs Dam and Reservoir, Sonoma County, California.

Conservation Treatment Area	Measures	Unit	Amount	Unit Cost	Total Cost
A(1)	Fertilizing	Acre	200	$ 18	$ 4,000[1]
A(1) & (2)	Fencing	Mile	2	2,000	4,000
A(3)	(No treatment)				
B	Fertilizing	Acre	1,900	18	34,000
B	Fencing	Mile	0.5	2,000	1,000
C	Fertilizing	Acre	1,900	18	34,000
C	Fencing	Mile	2.5	2,000	5,000
D	(No treatment)				
E	Fertilizing	Acre	1,200	18	22,000
	Seeding	Acre	500	10	5,000
	Fencing	Mile	5	2,000	10,000
	Water Development	Number	4	1,000	4,000
	Internal Improvements (Misc. trails, fencing, seedings, etc.)				15,000
F	Fencing	Mile	1	2,000	2,000
	Tree Planting	Acre	30	80	2,000
G	Fertilizing	Acre	1,300	18	23,000
	Fencing	Mile	6	2,000	12,000
H	Fertilizing	Acre	2,700	18	49,000
	Fencing	Mile	7.5	2,000	15,000
	Tree Planting	Acre	270	80	22,000
Roads[2]	(Seedbed, seeding, fertilizing & mulch)	Acre	160	500	80,000
Construction Cost					$343,000
Engineering & Design					4,000
Supervision & Administration					8,000
TOTAL					$355,000

[1]Rounded to the nearest $1000   (1967 price levels)

[2]40 miles of road requiring 4 acres of treatment per mile

The study by the former agency indicated that proposed
releases from the lake, for water supply and the fish hatchery,
will meet water quality objectives. These objectives were not
met without cost, however, as will be explained more fully under the fish and wildlife section. The intakes for water supply and hatchery use will be located at four separate levels
and, for purposes of flexibility, can be used separately, interchangeably, or together for each purpose.

Fish and Wildlife

Wildlife: As mitigation for submerging a considerable
area now inhabited by deer and their smaller companions, a
2,400 acre wildlife management reserve is established. As has
been noted in the watershed management program, this area is
fenced to avoid the depredations of domestic animals; seeded
and fertilized to afford cover; planted to provide browsing
areas; and located to include four springs. The height and
type of fencing will not deter deer and smaller animals from
entering and leaving the reserve.

The area will offer intriguing possibilities to humans as
well; however, a sanctuary for bird-watchers, as well as birds;
a setting for naturalists to conduct serious studies; an experiment in living for wildlife of which the lessons learned may
be transposed to other areas.

Fish: Having deprived the animals initially of their natural habitat, no credit is received for creating a new home for
fish; instead, the project is encumbered by exercising the
right of eminent domain over their ancestral home. The matter
is adjudicated, and just recompense determined; that this en-
visions a piscatorial population explosion is a matter for determined anglers to resolve.

The State Department of Fish and Game will operate a
hatchery, which will be built to their specifications; i.e., to
take 1,500,000 King Salmon, 200,000 Silver Salmon, and 7,200,000
steelhead trout eggs each year. From these eggs they will rear
about 1,000,000 King Salmon fingerlings, 110,000 Silver Salmon
yearlings, and 300,000 steelhead trout yearlings. They will
also make incidental plantings of about 50,000 Silver Salmon
fingerlings and 500,000 steelhead trout fingerlings, plus perhaps four or five million steelhead trout swimup fry.

In order to provide water at the correct temperature for
the fish, a multilevel intake structure is being designed. At
the base of the dam are the outlet works through which flood
flows and excess water can be released at a rate of 7,000 cfs,
at elevation 221 feet above mean sea level. Above is a tunnel
containing the lowest of four additional intakes at elevations

270, 350, 390 and 430 feet m.s.l. Each of these has a capacity of 300 c.f.s. for municipal and industrial water supply and can also supply the 75 c.f.s. for the fish hatchery. There are separate conduits for the fish hatchery flows. The structure is designed to make separate releases, through separate transport systems, for municipal and industrial water and for the fish hatchery. Through single or multiple intakes, the temperature and turbidity of releases can be regulated for both purposes; the possible combinations are many.

Recreation

Recreation uses must be planned for in conjunction with other uses of the reservoir. The questions governing the expenditure of money for recreation facilities should be: are the facilities wanted, needed, and can they be economically justified? By every yardstick, it can be calculated that the annual visits to this site will be an average of 1,500,000 annually over the next 100 years. A smaller reservoir reached the 1,000,000 mark in 1966, eight years after completion. Large land investments made in the region by private interests tend to demonstrate that the estimates are conservative.

In planning for the recreation area, these same objectives were sought: the blending of recreation uses with nature; the use of nature for man's entertainment, pleasure and knowledge; and the protection of nature from human abuse.

The terrain, topography, and access routes in the area determined many of the decisions that were made. The two principal areas to the north and west, while semi-enclosed in arms of the reservoir, therefore giving a sense of wilderness and isolation, are on the two main access roads: the Dry Creek Road plus its tributaries to the east area, and the Hot Springs Road in the north.

The relocation of these two roads and their branches was governed largely by recreational demands. The Dry Creek Road runs into the relocated Stewarts Point-Skaggs Spring road which winds past the fish hatchery, then along one of the most spectacular view routes in the area to the first lookout point, at the end of the spur, which affords a magnificent view of both arms of the lake. This view may be enjoyed by picnickers as well.

To avoid indiscriminate use of all areas of the lake, zoning was used in its different parts. Recreational development is oriented, broadly, to accommodate day use and overnight use. The first provides for picnicking, fishing, boating, swimming and sightseeing; the second is designed for the camper and will

give him access to the water for boating and fishing.   In addition, there are primitive areas whcih can be reached only by
boat, including much of the shoreline shown undeveloped on the
east side.

The zoning for land areas will be, then, day use, for
which most of the below dam and eastern area is reserved; overnight use, most generally in the northern area; and primitive
use, in those areas difficult to reach.

Another unusual feature will be zoning of the water areas.
The Warm Springs arm of the lake is reserved for primitive water activities--no waterskiing, no power boats--only canoes,
rowboats, and sore backs. From the dam to a point opposite the
wildlife preserve on the Dry Creek arm are areas for waterskiing and motorboating, and in the upper area, with its many
inlets, we hope to provide tranquility for the fisherman and
silence for the afflicted ear of the urban dweller.

The design and architecture of all these facilities, including the homes for those who will live and work there will
be unobtrusive and, in the sense of blending with the natural
environment, organic.

The visual environment can be improved not only by the
measures listed in the section on watershed treatment, but by
restoring this somewhat neglected terrain to full health.
Studies involving an experimental planting area at a nearby
area will determine what trees and plants thrive best under
various conditions not only of nurture but of degrees of use by
visitors. The experiment includes planting in areas periodically flooded as the level of the lake varies.

CONSIDERATIONS IN CONSTRUCTION

During construction of a water management project, careful
consideration of engineering factors and of environmental implications is important.

In the case of a large dam, early stages include relocations of roads, utilities, and inhabitants. Sometimes entire
communities have to be moved from a valley location to a site
adjacent to the newly formed lake. Cemetery relocation is a
particularly difficult part of this phase of the project; next
of kin of all remains have to be contacted to determine the
choice for a new resting place.

Diversion of the flow of the stream must be accomplished
in order to protect the dam and appurtenances while under construction.   Provision must be made for adequate flows for fish

and other marine life; controls to prevent turbidity that would adversely affect the fish also are provided.

Construction operations should be scheduled in a way to avoid disruption of fish and wildlife during the spawning and mating seasons. Protection of the young of the species and of the food chain that supports them is important.

Earth moving operations should be laid out to provide minimum disruption of life processes, including consideration of air and water pollution and noise levels. Borrow areas should be restored to natural slopes and ground cover insofar as possible.

The influx of workers into an area must be planned for by insuring that adequate community facilities, including school and housing, are available. The social and economic patterns can also be adversely affected by the rapid dispersal of the work force at completion of the project.

CONSIDERATIONS IN OPERATION

The method of operation of a facility must take all of the multipurposes into consideration and judge the relative importance of each in an effort to try to find the optimum way of handling each. An example would be a reservoir that contains storage for flood control, water supply, hydroelectric power, recreation, water quality and fishery enhancement. The latter two could control a dual purpose allocation from the pool: releases for water quality minimum flows could also provide that quantity needed for the river fishery. Similarly, there could be a dual-purpose, overlapping storage area used for both flood control and water supply. Just prior to the rainy season, the flood control pool would be empty and the reservoir drawn down even into the water supply pool. As the dry period approaches and the threat of flood lessens, the reservoir would be allowed to fill partly into the flood control pool. At the same time, maximum releases could be made during periods of high power demand, with minimum releases during non-critical periods. Water for the recreation pool would provide some stability for the lake level, avoiding extremes in drawdown; recreational facilities would be designed to accommodate fluctuations in the water level.

Environmental factors should be considered also in determining methods of operation of the facilities. For example, drawdown of the reservoir during the spawning season in the spring can leave all the eggs high and dry out of the water and thus cause serious losses to fish population. In many cases the multi-purpose uses are in competition. For example, use of

a reservoir or lake for power boating,   including water skiing,
is not always compatible with its use for fishing due to the
noise and wave action.   Zoning of water areas can be used to
help find reasonable solutions regarding these competing uses.
In streams downstream from reservoirs, problems of water temperature and turbidity have been created by the flow from reservoirs.   The effects on fish and other underwater creatures
as well as on the recreation potential for fishermen has demonstrated that improvement is required. Multi-level outlet works
that permit the selection of water temperature and control of
turbidity, depending on the season of the year, provide partial
solutions to these difficult problems.

An example of coordination of several facilities to provide for multiple uses occurs in the Central Valley Project of
the Bureau of Reclamation.   Trinity Dam forms a reservoir in
steep sloped valleys, therefore during summer releases for irrigation flows the lake is drawn down leaving bare slopes between the upper shore line and the water level.   These areas
tend to limit recreation use; they also are cause for aesthetic
consternation.

The waters from Trinity that are diverted into the Central
Valley flow through Whiskeytown Lake; this reservoir is regulated so that in general inflow equals outflow. Therefore with
a relatively stable water surface elevation, it is highly desirable for recreational use.

C. POWER

A review of the nature and history of the Electric Power Industry may be helpful in understanding the operations and objectives of the industry and its plans and efforts to preserve and enhance the quality of the environment.

Historical Background

The industry got its start in 1882 when the Pearl Street Generating Station began supplying power to an area of about one square mile in New York City. The power was fed into a direct current network to which customers were connected. The concept of a network had been patented by Edison in 1881. Subsequently, the idea of alternating current distribution was evolved.

In the case of the direct current network, the voltage of the customer was the same as that of the network, which placed a low limit on the system voltage. In the case of the alternating current distribution it had the advantage of having a higher voltage for distribution and could use transformers to provide the required voltage for the customers. This made it possible to transmit larger amounts of power over longer distances and to larger areas, resulting in the large systems that we have today consisting of larger generating stations feeding transmission and distribution networks throughout the nation.

Electric Load Demand

The electric peak load demand for the nation has incr ,ased from a few kilowatts (KW) in 1882 in New York in one system, to a maximum of 276,000 megawatts (MW, thousand KW) in 1970 on about 3,500 systems. The forecast by the Federal Power Commission in its 1970 National Power Survey is that the peak demand in 1980 will be 589,000 MW, about 1,000,000 MW in 1990, and about 1,950,000 MW in 2,000.[4] The energy required was 1.53 trillion kilowatt hours (KWH) for 1970 and is forecast to be 3.07 trillion KWH for 1980, 5.8 trillion KWH in 1990, and 10.36 trillion KWH in 2000.

The Electric Utility Industry is made up of about 3,500 systems, some owned by private companies (investor-owned utilities), some owned by the federal government or by other public bodies, such as municipalities, states, or public utility districts, and some owned by rural electric cooperatives.

The investor-owned segment is the largest.   It has about 250 generating systems with 77% of the total nation's capacity and 78% of the ultimate consumers as of December 31, 1970. Nearly all of the 200 major investor-owned utilities operate integrated generation, transmission and distribution systems. There are about 150 smaller investor-owned utilities, most of which are engaged solely in electric distribution.

There are some 40 systems in the federally owned segment which is the main supply power in bulk for local distribution and resale by others.   With the exception of the Tennessee Valley Authority, the major systems are operated by agencies of the Department of the Interior.   This group accounts for 11.5% of the United States total generating capacity.

Utility systems owned by the public bodies other than the federal government or by consumers they serve, account for 10.5% of the U. S. generating capacity.   Of some 2100 systems, about two-thirds are solely engaged in distribution and resale of electricity from bulk power suppliers, such as municipally owned and most rural electric cooperatives.   The other one-third operate generating facilities either as part of an integrated generation-transmission-distribution system or to supply power for distribution by others.

About 1,000 electric cooperatives supply power in many rural areas.   Most are relatively small and over 90% are engaged in only distribution of electricity.   They account for less than 2% of the U. S. generating capacity but serve over 8% of the ultimate customers.

Major Demand Components

In spite of demands by some individuals that power growth be curtailed, the best judgment of those whose business it is to study demand trends is that the U. S. electrical requirements will nearly quadruple between 1970 and 1980 and by 1990 electricity will have increased its share of total energy consumption to about 41% from 25% in 1970 (see Table 3-7).

The outstanding thing here is that for the next two decades, the use pattern will remain about the same except for the increase in commercial use.   The use in industry is much

larger than in the home and it would take a 10% reduction in
today's home consumption to achieve only a 2.5% reduction in
today's total consumption.

Table 3-7. Categories of electric power use.

Category of Use	Percent of Total Use		
	1965	1970	1990
1. Industrial	41	40	41
2. Residential	24	25	24
3. Commercial	18	18	20
Sub-Total	83	83	85
4. Miscellaneous, inc. losses	17	17	15
Total	100	100	100

There are many reasons why the rate of growth in use of
electricity will not decrease appreciably in the two decades
ahead. The projected growth for the next twenty years is based
on events which are already in motion. The children who are
here will form new families in the 70's and 80's. New houses,
apartments, shopping centers, office buildings and industrial
plants will require electrical service and more are being designed.

This is seen also by the projected U. S. Bureau of Census
population projections, summarized in Table 3-8.

Table 3-8. Population growth in the United States.

	1970	1980	1990	2000
Population (Thousands)	205,167	227,510	254,720	280,740
Present Increase		10.90	11.96	10.20

Also it is clear that measures needed to upgrade the nation's environment are creating substantial new requirements
for electric power. Examples are recycling scrap paper and
metal to reduce the accumulation of solid wastes; sewage treatment plants to reduce water pollution; prospects for electrical
rapid transit to reduce automotive congestion and air pollution; and power consuming devices to control pollution by power
plants themselves.

The installed generating capacity in the U. S. at the end of 1970 was about 340,000 MW and was based on a 20% reserve margin. The net generating capacity required at the end of 1980 is projected to be 665,000 MW and at the end of 1990 is 1,260,000 MW. In 1963 it was 211,000 MW.

In 1963, eighty percent of electricity generated was by fossil fuel plants with coal as the major fuel--76%, natural gas, 26% and residual fuel oil was 7%. Nuclear power supplied only 0.1%.

The largest unit placed in service up to 1960 was a 335 MW capacity, at which time the first 450 MW unit was added. In 1970 a large part of the new base load capacity added was 500 MW or larger. Many units 1,000 MW and larger are under construction and sizes may reach 2,000 MW by 1980 and 2700 MW by 1990 (see Table 3-9).

Table 3-9. Projected growth of utility generating capacity.

	Installed Capacity End-1970	Projected Capacity End-1980	Projected Capacity End-1990
	(Millions of Kilowatts)		
Hydroelectric			
Conventional	52	68	82
Pumped Storage	4	27	71
Thermal			
Steam-Electric			
Fossil-Fuel-Fired	260	393	557
Nuclear	6	147	500
Gas-Turbine-Diesel	19	31	51

The Conventional hydro units would be of seasonal operation. The pumped storage and the gas-turbine and diesel units would be peaking. The nuclear and newer fossil-fuel-fired steam electric units would be on base load operation and the older fossil-fuel-fired units would operate in an intermediate role.

The nuclear plants accounted for only about 2% of the U. S. 1970 base load capacity, but they are expected to account

for about 50% of new thermal base load capacity additions during the 1970's and about 70% during the 1980's.

Pumped storage additions are expected to increase in capacity beyond those of gas turbines and diesels for peak load
service, as they offer the cheapest and most dependable sources.

Plants in the intermediate service operate at capacity
factors between 20 and 60%. This includes older, smaller,
fossil-fuel-fired units which were built for base load service.
Also, some diesels and some gas turbines fit this use, as well
as cycling steam units are being built for this operation.

It is interesting that of the total amount of base load
capacity projected to be in use by end of 1990 that about 83%
will be of the post-1970 vintage and more than half of the
total will be less than ten years old at that time.

Total Energy of the United States

Mr. Leon P. Gaucher, before his retirement from Texaco in
1969, headed Research Project Evaluation and was responsible
for keeping Texaco up to date on all facets of energy. Writing
in the Chemical Technology magazine of March, 1971, he stated,
"What we think of as big business today--580 million tons of
coal per year, 14 million barrels per day of oil and over 2
billion cubic feet of natural gas per hour--will be dwarfed by
the corresponding business of the future." Further points of
interest are summarized as follows:

> Comparatively, the coal business is small. The production
> of the coal industry in 1970 was worth only about 2.5 bil-
> lion dollars, which is less than half of the gross sales
> of any *one* of several large oil companies.

> Today's economy is profoundly dependent upon oil and gas.
> Petroleum provides the fuels for practically all of the
> nation's transportation, and oil and gas together, satisfy
> over 90% of the nation's requirement for heat, three-
> fourths of the energy used in commerce and industry, and
> nearly a third of the fuel used for power generation.

> In addition, almost all of the organic chemicals that we
> use today, including lubricants, surface coating, plastics,
> and synthetic rubber, are derived from natural gas and
> petroleum.

Energy Consumption Patterns of the Future

> *Hydroelectric Power*--The total hydroelectric power in the
> year 2200 amounts to only 1.4% of the total energy that is
> expected to be required by that time.

Fuel Cells--We do not expect fuel cells to make any significant impression on the energy picture of the future.

Coal--The consumption of coal--will continue to increase because of the phenomenal anticipated increase in electric demand. However, retention of this market by coal will not be easy because, in addition to the problem of avoiding air pollution, coal will continue to be pressed on all sides by nuclear and other fuels. As a result, it seems reasonable to expect that the use of coal for power will be confined to those locations where good coal is most abundant.

Natural Gas--Even though there is an abundance of it now, natural gas is likely to be the first energy source to encounter shortages.

Coal and Oil Gasification--Even before 1990, however, one can expect that a small amount of high BTU gas--i.e., "natural gas equivalent"--will be made by gasification of coal and oil.

The technology that is involved is well along and has been studied intensively both here and abroad for a great many years. This work is being continued by the Office of Coal Research. Two pilot plants have been announced: a lignite gasifier in Rapid City, South Dakota--using the CO_2 Acceptor Process developed by the Consolidation Coal Company, and a large coal hydrogasification pilot plant, operated by the Institute of Gas Technology in Illinois.

Nuclear Power--The demand for electric power will increase more rapidly than any other form of energy. We estimate that by 2200, half of all energy consumed will be in the form of electricity, and from now on, most of the large base load plants will be nuclear.

This presumes that economical *nuclear breeding* will be developed by 1990. Otherwise, the availability of uranium will not be great enough to support the rate of growth shown.

Fuel Supplies for Use in Thermal Power Generation--1970-1990

Four basic fuels are used in thermal power generation. The three fossil fuels are coal, natural gas and oil. The fourth, uranium, is today the basic fuel of nuclear power. Table 3-10 shows the projected distribution in use.

The nation's aggregate resources of coal are vast. However, coal is not uniform, varying greatly in physical and chemical properties, and in costs to mine and to transport. The bulk of the low sulphur and inexpensive coals which are found in the West are located far from utility demand centers, which are predominantly in the East. Utilities are finding it increasingly difficult, if not impossible, to locate deposits that are large enough to supply the requirements of a large power plant that can be mined and transported at a reasonable cost, and that have low enough sulphur content to meet the increasingly stringent limits of environmental regulations.

Table 3-10. Projected distribution of fuel use for thermal power generation 1970-1990 (percent of heat input).

| | Year | | |
Fuel	1970	1980	1990
Coal	54%	41%	30%
Natural Gas	29	14	8
Residual fuel oil	15	14	9
Nuclear	2	31	53
TOTAL	100%	100%	100%

Many utilities will be required to install equipment to remove sulphur from the furnace gasses before discharging them from the plant stack. Stack gas sulphur removal processes have been under development for a number of years and some are being demonstrated, but not yet satisfactorily at a commercial scale. Removal of organically found sulphur in mined coal requires full chemical treatment. Indications are that limitations on sulphur dioxide emissions drastically increase cost in coal-burning plants.

There are other factors that affect the coal cost, such as tightening mine safety standards which will act to increase the cost of coal production in underground mines, and tightening environmental regulations will increase the cost of and sometimes prevent strip mining. The doubling of output of utility coal makes a difficult problem for the coal industry and will entail significant cost increases at the mine, in shipment or at the plant.

The natural gas supply has tightened and wholesale prices are rising. The utilities which have depended on this energy form are uncertain as to future availability of gas and the price. It is expected that supplies will be marginal during the 1970's and inadequate during the 1980's unless new supplies are developed on some techniques found for producing gas from coal.

Residual fuel oil, especially of low sulphur content, is obtained largely from foreign sources since the economics of U.S. petroleum refining do not favor production of this fuel form. With U.S. reliance of foreign sources and world require- ments for oil increasing rapidly, difficult problems could develop.

The world and U. S. resources of uranium are vast--many times larger than those of fossil fuens--but realization of more than a small fraction of the potential awaits development and use of breeder reactors, which are under intensive develop- ment in the U.S. and several other countries with the objective of having full-scale plants on the utility systems during 1980's. They will probably not account for a major portion of total nuclear capacity until the 1990's. The proven reserves and stockpiles of low cost uranium ore are the result of ex- ploration programs carried out in the mid-1950's in connection with defense requirements; they are more than adequate to meet the power industry's projected needs over the next decade. Large scale exploration efforts have resulted in locating new deposits in the last several years here and abroad. If the ex- ploration is continued and there are no serious delays in the breeder reactor development, the outlook is for adequate ore sources. Major buildup of U.S. uranium mining and milling capa- cities will be required and the prices will increase from the present depressed level as the nuclear power buildup takes place.

Construction of Breeder Reactor

In August, 1972, steps were taken to provide the United States with its first demonstration breeder reactor plant, which is to be in operation in 1980 at an estimated cost, in- cluding escalation of $700 million by a memorandum of under- standing between the two principal utilities -- Commonwealth Edison and TVA -- and two special corporations formed to carry out the project -- Project Management Corp. and Breeder Reactor Corp.

President Nixon in 1971 called for the successful demon- stration of the LMFBR (Liquid Metal Fast Breeder Reactor) by 1980, and Congress appropriated additional funds for the pro- gram. The AEC had reported in 1962 to the President that the development of the breeder reactor was needed to effectively utilize our uranium resources in the long term. Congress in 1969 provided statutory authority for AEC to initiate project definition phase of an LMFBR demonstration program, and in 1970 it passed legislation for AEC to engage in a definite, cooper- ative arrangement for the development and construction of an LMFBR demonstration plan on a utility system.

In November, 1972, Westinghouse Electric Corp. was selected by Project Management Corp. and approved by AEC as the
lead reactor manufacturer. Project Management Corp., a non-
profit organization, was established by Commonwealth Edison Co.
and TVA and has the responsibility for design, development and
construction of the LMFBR on the TVA system on the Clinch River
near Oak Ridge, Tennessee. General Electric Co., and Atomics
International also made proposals, and the work will be divided
so that all three companies will participate "in a meaningful
way" and a competitive industry will be developed.

Dr. Edward E. David, Jr., Director of the Office of Science and Technology, in September, 1972, states:

> Most, if not all of the breeder technology has been suc-
> cessfully demonstrated at least on a reduced scale. What
> remains is to demonstrate quality control and engineering
> components on a commercial scale, and their integration
> into a full-scale operation plant (Table 3-11).

Table 3-11. Financing for breeder reactor program.

Project Resources	Millions
AEC Direct Assistance	$ 92
Applied Base Program (AEC)	320
Furnished Special Nuclear Material	10
Utility Contributions	254
Expected Contributions from Reactor Mfg.	20- 40
Total Estimated--Range	$696-716
Total Estimated Costs	$ 696
Escalation calculated at 30% of cost.	

Power from Fusion

Existing nuclear power plants produce electricity from
energy released when a heavy element splits apart by fission.
Fusion is somewhat the opposite. Energy is released when nu-
clei of light elements join together, or fuse, to form a new
heavier element. The sun releases energy through fusion of hydrogen. The same thing is proposed for a power plant using a
form of one deuterium atom to 6000 normal hydrogen atoms. It
is relatively easy to extract and is cheap--$150 per pound.
The energy in a gallon of sea water is equal to that of 300
gallons of gasoline. It is estimated that there is enough in
all the oceans to last 10 billion years at the present rates of
use.

A major problem in developing fusion is controlling the reaction at very high temperatures required. Deuterium fusion occurs at five hundred million degrees F., which is hotter than the interior of the sun. Investigations are being made of fusion of deuterium (H^2) with a nucleus of tritium (H^3), heavy hydrogen isotopes. This reaction occurs at a lower temperature of 100 million degrees. No containment vessel can be built.

Investigations have been made on controlling the plasma at these temperatures in a magnetic field of circular cross-section or the kidney-shaped cross-section of the toroidal containment vessel of Gulf General Atomic.

Recently, a new way to control fusion is being investigated using high energy, powerful laser beam focused on frozen pellets to produce fusion temperatures without the confining magnetic field. The pellets are frozen, solid deuterium-tritium. A laser is a device for generating or amplifying a beam of light whose waves are both monochromatic (all the same wave length) and coherent (all in step).

The controlled fusion reactions described above represent a potentially inexhaustible source of inexpensive, efficient, "clean" energy for supplying all mankind's future power requirements and a bomblike explosion is impossible.

Mr. R. A. Hause of Public Service Electric and Gas Co. of New Jersey foresees units as large as 10,000 MW of fusion power plants in his system by the year 2000. Capital costs would be about the same as for a fission plant and cost of electricity produced would be less than conventional power plants.

Transmission of Electric Power

Transmission is steadily growing in importance to electric supply, as more generation sites are located distant from the load centers and more interconnections are used to enhance reliability and economy. With growth there has been an increase in capacity of individual lines by use of higher voltages and by use of direct current in a few cases. In some instances new transmission lines have been placed underground.

The transmission voltages have been increased successively from only 13 kilovolts (KV) to 23 (KV), 46, 69, 115, 138, 230, 345, 500 KV and in the fall of 1969, the first 765 KV line was placed in service. A high voltage ±400 KV (800 KV) direct current line has been placed in service forming a northwest-southwest intertie on the west coast. The overhead lines have bare conductors carried on wood, steel, or aluminum structures and are located on various types of rights-of-way. They might have storm damage or lightning faults, but this is rare and such conditions can be readily located and repaired. Outages of overhead lines are usually of short duration--on the order of a fraction of a second.

Results of an Ultra High Voltage (UHV) research project have been given in an Interior Department report in 1972 on the American Electric Power (AEP) and Allmanna Svenska Electrika Aktiebolaget (ASEA) in cooperation with Ohio Brass Co. The project was:

> to determine the economic feasibility of AC transmission at 1000 kv and above and to establish the ceiling of system voltages and producing equipment that will make transmission at those voltages a practical reality.

Tests on impulse and switching surge voltage and interference have shown that the dielectric strength of air will allow, from a technological point of view at least, a system voltage of at least 1500 kv and a higher level, up to 2000 kv, may be possible. A switching surge level of 1.5 p.u. is now possible. UHV apparatus can be built. (A 1500 kv transformer has been built). Corona and induced effects can be controlled. These conclusions have been reached after three years of intensive study.

Power transmission can also be accomplished by using insulated conductors which are buried in the ground or placed in ducts, pipes or tunnels, but at a much greater cost than overhead construction for the same voltage. The cost of underground transmission lines is about nine times as much as for 138 KV overhead lines and 16 times as much as for 345 KV in suburban areas. In rural areas the difference averages 14 to 26 times and in places may go to 40 times due to special conditions. Underground transmission lines are free of weather problems but are subject to some failures which are difficult to locate and interruption of service may last a few days to several weeks to locate and make the necessary repairs under the exacting conditions required. Research work is being conducted on 500 KV and 750 KV cables.

Reliability of Electric Power Supply

One of the major objectives of the electric power industry throughout its history is to make power available whenever needed and minimize service interruptions. The loads are increasing rapidly and there have been expansion of transmission networks throughout most of the nation. This, together with the extensive power failure in the northeast in November, 1965, has brought new attention to the reliability of interconnected systems. In July, 1965, the Federal Power Commission published

a three volume report "Prevention of Power Failures," which recommended as a key step in bulk power supply reliability, the
establishment of regional joint planning and operation of electric power facilities.

In response to the evident need, the industry has greatly
expanded its coordination arrangements. By the end of 1970, 21
contractual agreement power pools were in operation, representing some 60% of installed capacity, as compared to only 9
pools in 1960. In addition, there are at least 13 informal organizations of utilities in the contiguous U. S. which engage
in more limited aspects of inter-system coordination. Beginning in 1967 the utilities established a number of regional
reliability councils and in 1969 formed the National Electric
Reliability Council (NERC) to encourage improvement of coordination at both the regional and national levels. At the end of
1970 the nine regional councils included virtually all major
electric utilities in the 48 contiguous states, comprising
about 90% of the total installed capacity. Provision is made
for participation by the smaller utilities.

*General Conditions Relating To and Objectives of the Electric
Power Industry*

The Electric Utility Industry policy is that the industry
has the responsibility to supply electric energy in adequate
amounts, reliably, economically, in an environmentally acceptable fashion, and with due regard for the conservation of our
natural resources. The industry has some special characteristics which are unique to its activities and which are not present in other businesses. Electricity cannot be stored like
other commodities and be ready for use as desired. It must be
instantaneously generated in a power plant by conversion to
electricity from some other source of energy such as coal and
deliveries through transmission and distribution facilities to
the consumers at any place, at any time and in any amount that
they may require. The sum of the requirements of all the customers at any instant is the instantaneous demand on the whole
system which must be supplied. This is continually varying and
depends upon, such things as, the time of day, day of the week,
season of the year and weather. The lowest demand is in the
early morning hours and is a maximum during the day or evening
hours. Each utility must anticipate the demands upon its system and plan for and build additions to facilities to meet
these needs, plus sufficient reserves to take care of scheduled
outages for maintenance and unscheduled outages due to equipment failures. "Pooling" or sharing of reserve capacity can be
accomplished by interconnecting systems and also provide a
means of interchange of economy power when a system may purchase low cost, surplus power from another.

In rare situations in which the utility does not have sufficient operable plant capacity to meet the demand and cannot obtain sufficient reserve capacity from other systems, it can call on voluntary curtailment by customers or it can reduce voltage by a small amount for a short time. The reduction cannot be for too long a time or below a certain voltage. If there is still insufficient capacity to carry the load, it must "shed load" or temporarily interrupt load in a predetermined emergency plan which takes into account priorities of community and human needs. This situation may be brought about by delays in additions to facilities due to delays in construction or granting of permits or licenses.

A problem affecting utilities supplying power is that projects require 5 to 8 years from inception to completion. To this must be added several years of hearings and procedures before licences, permits and approvals are obtained. In the meantime, the load continues to increase at a rapid rate, and at inflated costs. The service is essential to the nation but there is great concern today as to this problem and its solution. The delay in completion of major new generating units rapidly erodes the reserve margin and thereby weakens the supply capability and degrades the quality of service because of more frequent reduction of voltage or "load shedding." The delays are resulting in greatly increased costs due to inflationary pressures, interest on investment, operation costs of less efficient units.

Concern of the public and the industry about environmental questions is affecting power costs in other more direct and even more substantial ways. For one, utilities must spend progressively larger sums on environmental protection and appearance-improving features of power installations. For another, the fossil-fuel sources on which they can draw are being progressively narrowed by regulations requiring the use of low-sulphur fuels, which are presently in short supply and have been commanding stiff premiums. It may be true, as some contend, that the public is prepared to pay substantially more for electricity in the interests of reducing the environmental effects of power operations, and, if so, the price effects of these new costs may be accepted by the rate payer without serious challenge. As matters now stand, however, not enough time has passed for the true impact to be felt and so the question is still at issue. It is already clear that the consequences flowing from the Calvert Cliffs Decision (requiring Atomic Energy Commission to review all environmental aspects of nuclear power plants) of July, 1971, cannot yet be fully measured in terms of the added time for analyses and procedures and, perhaps added environmental controls needed to conform fully with the National Environmental Policy Act of 1969. Furthermore, the nation's electric power program of the next two decades is

critically dependent on successful introduction on schedule of
tremendous increments of nuclear power.   As a matter of fact,
very serious delays are taking place now in 1972 in getting a
large number of capacity units in service even when construction is completed.   This includes fossil-fuel-fired units as
well as nuclear plants.

The cost pressures from inflation trends, higher environmental cost, higher fuel costs, unusually high interest rates
of recent years and still continuing rapid escalation of construction costs have had heavy impact on the power industry
(the most-capital intensive and one of the fastest growing of
all major U.S. industries).

There have been various estimates of price of electricity.
The 1964 Survey of the Federal Power Commission pointed out
that the ·price trend had been downward.   The average was 2.7
cents per KWH in 1926 to 2.2 cents in 1940, 1.7 cents in 1962
and 1.54 cents in 1968, while the price of almost every other
item was increasing.   In the period 1940 to 1962, the price of
electricity was reduced by nearly 25% while the average price
of consumer goods (consumer price index) rose more than 200%.
With the increase in costs mentioned in the previous paragraph
the present estimates are that the price of electricity will be
1.83 cents per KWH in 1990 based on 1968 equivalent dollars,
plus whatever inflation the dollar experiences up to 1990 (see
Table 3-12).

Table 3-12.   Cost of electricity to ultimate consumers (cents
per kilowatt hour).

	1990 Projected Cost				
	1968 Actual Cost	1968 Equivalent	With 1%/yr Inflation	With 3%/yr Inflation	With 5.7%/yr Inflation
Production	0.77	1.09	1.36	2.10	3.18
Transmission	0.20	0.30	0.37	0.57	.89
Distribution	0.57	0.44	0.55	0.84	1.28
Total	1.54	1.83	2.28	3.51	5.35

RESEARCH AND DEVELOPMENT GOALS

As stated previously:

The electric utility industry has the responsibility to
supply electric energy in adequate amounts, reliably,

economically, in an environmentally acceptable fashion, and with due regard for the conservation of our natural resources. This statement was reiterated in the report of the R & D Goals Task Force to the Electric Research Council. The report goes on to state that, "to fulfill this obligation will require further improvements and advancements in our overall operations. The rate of progress is largely dependent on a meaningful, well-planned and coordinated research and development program, which must be fully supported by the electric utility industry, the manufacturers of utility equipment, and the government.

The report goes on to state,

At the present stage of technological development, electrical operations--all their contributions to man's progress notwithstanding--intrude on the natural environment more than either environmentalists or the industry itself would like. But within our reach is the capability to produce and deliver electricity in ways that will further reduce the environmental consequences and permit man to use electricity increasingly to improve his life.

The problems in the power field hinge on technological progress and this depends on research and development. The industry is examining its research and development needs. It realized that the power system in the nation is the most advanced in the world and has a very high degree of reliability and provides service at a low price which has permitted us to our high standard of living. It is feld that we can progress further by research and development.

The Electric Power Industry formed the Electric Research Council in 1965. It consisted of 12 industry executives--eight from the investor-owned companies and one each from the U. S. Department of Interior, TVA, the American Public Power Association and the National Rural Electric Cooperative Association. The purpose of the Council was to encourage cooperative sponsorship of reserach that had industry-wide importance. As time passed, the Electric Research Council assumed increasing control over the industry's research effort; all with the help and strong support of Edison Electric Institute staff and member companies.

In the fall of 1970, the Council set up a Research and Development Goals Task Force to survey R & D needs in depth, to set priorities and schedules to the year 2000 and to develop an estimate of probable cost.

The Task Force published its report in June 1971 entitled: "Electric Utilities Industry Research and Development Goals

Through the Year 2000." It is the most complete study the Electric Power Industry has ever made on the subject. The report has been accepted by Electric Research Council. A suggested organization to carry on the acceleration of research contemplated in the report was suggested by the Task Force. The name of the organization was suggested Electric Power Research Institute. The suggestion has been adopted and the Electric Power Research Institute was incorporated in 1972 in the District of Columbia as a non-profit corporation.

Industry, government, universities and other interested groups will be represented on advisory committees. Research will be contracted on a competitive basis with existing laboratories and research organizations. For the present, it is not planned to set up central laboratory facilities.

Mr. W. G. Meese, President of ERC, started the funding by presenting a plan to the EEI Chief Executive Officers' Meeting in December, 1971, of increasing industry's cooperative sponsorship of R & D over a three year period to a level of about 0.1 mill per KWH in 1974. This plan was submitted to the EEI Board of Directors at the annual meeting in July. This level of contribution would be about four times the former contributions of about one-fourth of one percent of gross electrical revenue. This was about $40 million. An additional amount of about $110 million was spent by utility equipment manufacturers. The estimate for utilities would be substantially higher if there were included the "incremental" costs the industry incurred in building and operating pioneer nuclear power stations. The costs incurred over and above the normal for conventional power installations of comparable size.

The estimate of cost of the goals for R & D will average, over the next 29 years, about double the current level of combined expenditures of government, manufacturers and utilities of about $600 million, about half of which is government activity in nuclear research. In terms of 1971 dollars, the program will cost an average of $1,120 million for each of the next 29 years starting with $667 million in 1972 and peaking at about $1.2 billion in 1977. There are five areas of R & D which have been given top priority and Table 3-13 gives the projected costs of each one from 1972 to 2000. The research program areas are reviewed in their entirety under the headings 1-6 in the following pages. The program not only suggests the present state of the art in various facets of the industry, but creates through the program goals the state of the art for the future. Both the availability of resources, increasing demands, and environmental sensitivity combine to form the basic themes for the program.

Table 3-13.   Summary of total R&D costs to utilities, manufacturers, and government (millions of 1971 dollars).

	Energy Conversion	Transmission and Distribution	Environ- ment	Utili- zation	Growth and Systems	Total
1972	500	110	47	8	2	
1973	619	123	91	8	3	
1974	701	147	108	8	4	
1975	799	157	103	8	5	1,072
1976	841	156	104	15	5	1,121
1977	922	160	107	15	5	1,209
1978	873	160	109	15	5	1,162
1979	857	167	107	15	5	1,131
1980	806	168	105	15	5	1,099
1981 1985	3,875	913	503	67	22	5,380
1986 1990	4,098	1,036	503	64	21	5,722
1991 2000	8,430	2,474	1,006	128	51	12,088
Total	23,319	5,772	2,893	369	133	32,486

R & D Goals Task Force Report to the Electric Research Council, June 1971

1. *Energy Conversion*

   a. Establish Nuclear Breeder Reactors as being commercially available for purchase of mid-1980's for central station baseload applications.

   b. Improve present methods of generation in efficiency, reliability and environmental impact. Continue development of gas-steam combined cycle.

   c. Establish scientific feasibility of Nuclear Fushion within 5 to 8 years and make it commercially available for purchase by the mid-1990's for central station baseload applications.

d. Establish gasified coal fuel as economically available for gas turbines, MHD and conventional boilers by 1975. Continue research on other methods of fuel preparation such as hydrogen production and solvent processing.

e. Establish open cycle MHD as being commercially available for purchase by the mid-1980's using gas, oil, coal or coal derived fuel for central station baseload applications topping either steam or gas turbines. Establish the MHD position of these combined cycle plants for peaking and emergency power requirements.

f. Establish Fuel Cells in the 10-20 MW size range as being commercially available for purchase by the late 1970's for substation application, fueled by natural gas, hydrogen or fuels derived from coal or oil.

g. Continue research on high energy bulk storage batteries for peaking purposes.

h. Continue R&D for unconventional cycles such as potassium-steam binary cycle and Feher CO_2 cycle. Continue research on thermionics for topping nuclear and fossil generating plants.

i. Proceed with additional research on solar energy at a moderate funding level.

j. Continue basic research for new methods of energy conversion.

2. *Transmission and Distribution*

a. Bulk Power Transmission

(1) Develop new and improve existing *underground transmission* systems.

(2) Increase the capacity of existing and new *overhead transmission* rights-of-way by improving design and equipment, both ac and dc.

(3) Improve designs and equipment for *substations and dc terminals*.

b. Tools and Techniques

(1) Shorten time and reduce the costs of splicing present day cables.

(2) Decrease installation costs by developing improved
techniques and equipment applicable to distribution
systems and underground and underwater transmission.

(3) Develop new and improved tools, techniques, hardware
and equipment for installation and maintenance of EHV
and UHV transmission lines.

c. Aesthetics and Compactness

(1) Design increasingly attractive transmission structures capable of carrying the structural load at reasonable cost.

(2) Develop reliable and economical underground equipment
in particular transformers and switching devices, in
order that more distribution components can be placed
underground.

d. Systems and Control

(1) Establish quantitative terms for the security required, continuous quantitative assessment of the
state of security by on-line computers, guides for
bringing the two into agreement and eventually automatic implementation of corrective measures.

(2) Improve present media and exploit new media of intelligence transmission to attain the high quality, capacity and reliability needed for computerized system
security and control.

(3) Exploit all exitation control for stability; fast
prime mover control; fast switching of lines, series
or shunt capacitors, shunt reactors and braking re-
sistors for system stability both transient and dynamic; modulation of dc transmission; development of a
high-speed phase shifter or equivalent for modulation
of ac transmission; and in conjunction with development of energy storage systems, the exploitation of
fast control of energy flow for benefit of system
stability.

(4) Organize system control into an hierarchy for efficient allocation of functions and expand system control to include the analysis of system state, at
first on an operator guidance basis leading to an ultimate direct and comprehensive control effort.

(5) Advance the science of power system instrumentation for higher reliability, higher resolution, compatibility with telemeter and computer systems, and capability to sense system state variables for computer control. This includes the development of in-service monitoring techniques, particularly for transformers, and new methods of fault detection with their response speeds.

e. Test Facilities

(1) Build a high power, high voltage industry test facility starting immediately.

(2) Expand test facilities for HV and EHV cables and provide for testing dc as well as ac cables.

(3) Provide additional test facilities for testing EHV and UHV lines, both ac and dc.

3. *Environment*

a. Air Quality

(1) Determine basic physical and chemical data related to removal processes for particulates and oxides of sulphur nad nitrogen from stack emissions.

(2) Further improve systems to restrict SO_x emissions in present and future power plants.

(3) Further improve systems to restrict NO_x emissions in present and future power plants.

(4) Further improve systems to restrict particulate emissions for new and retrofit operation.

(5) Develop a feasible process for desulfurizing coal and oil fuels.

(6) Determine health effects of short-term, high-concentration exposures to oxides of sulphur and nitrogen particulates on sulfuric acid.

(7) Determine the effect of major constituents of power plant effluents on vegetation.

(8) Study the synergistic effects of the major stack gas constituents in the atmosphere.

(9) Develop standardized portable measuring instruments
for the measurement of low concentrations of oxides
of nitrogen and sulphur and particulates.

(10) Determine the distribution and effects of minor constituents of stack emissions.

(11) Improve dispersion technology.

(12) Study effects of alternate cooling systems.

b. Water Quality

(1) Investigate effects arising from disposal of solid
and liquid wastes from power plant operations.

(2) Determine the effects of once-through cooling systems
on aquatic biota.

(3) Develop viable alternate methods of cooling heated
water discharges.

(4) Evaluate the effects of heated water discharges on
the health and aging of aquatic ecosystems.

c. Nuclear

(1) Further reduce radiation release from the overall nuclear cycle.

(2) Determine cumulative effects of small amounts of ra-
dioisotopes on biological systems.

d. General Environment

(1) Determine the health effects of exposure to electro-
magnetic and electrostatic fields.

(2) Develop siting criteria on a regional basis for fossil fuel and nuclear power plants.

(3) Develop methods of noise control and noise specifications for power plant equipment.

(4) Develop ecosystem and climatological model to predict
long-term effects caused by power generation.

4. *Energy Utilization*

a. To increase the efficiency of consumer devices in home
and commerce, thus reducing the consumption of energy per

unit of light, heat or cooling and contributing to wise
use and conservation of energy.

b. To develop concepts and equipment (batteries foremost) to
make possible widespread use of electric energy to power
mass transportation, personal and specialty transportation, thereby reducing air pollution.

c. To develop beneficial uses--such as space heating, aqua-
culture and agriculture applications--for the presently
man-usable low grade heat now rejected in the production
of electricity, and thus reduce thermal problems.

d. To develop high efficiency air pollution controls for industries other than the electric industry, including more
efficient filters, scrubbers, collectors, electrostatic
and ultrasonic precipitators. (Air pollution controls for
the electric industry are goals of other chapters.)

e. To improve and develop new equipment to clean up our rivers and lakes and to improve treatment of sewage and solid waste.

f. To study, test and develop possible further application
of electric energy to improve manufacturing technology,
improve product, increase productivity, reduce pollution
and conserve fossil fuels.

g. To study and develop new methods of food production, processing, storage, preservation, and distribution that
utilize electricity, thus reducing man's dependence on
weather and improving his diet.

h. To improve communications technology, including gathering
processing, transmitting and utilization of information.

i. To improve and develop new equipment to reduce noise from
operation of electro-mechanical equipment of all kinds.

j. To improve and develop new electric residential task saving and leisure time equipment.

5. *Industry Growth and System Development*

a. Develop and continually update a National Fuel Model in
order to gain knowledge of fuels and their availability
and thus to help determine the nature and fuel limitations of future energy conversion systems.

The National Fuel Model would include determination of:

(1) Long range availability of "non-renewable" fossil and nuclear fuels.

(2) Long range cost of recovering and transporting these fuels.

(3) Long range requirements of various fuels based on energy conversion cost.

b. Develop and continually update, through use of modeling techniques, the capital and operating cost of transporting energy in the best method of ultimate delivery to the customer and to help determine optimum plant location. This would include:

(1) Electric Power, overhead and underground ac and dc, at various voltages and frequencies.

(2) Coal by train, ship or in pipe (with water, oil or other fluid).

(3) Oil by train, pipe or ship.

(4) Gas (natural, synthetic or hydrogen) by pipe, train or ship in gaseous or liquid form.

(5) Nuclear fuels, such as uranium, thorium, deuterium, lithium in natural form or refined.

c. Develop an overall power system model from the fuel to the consumer during the next 30 years, in order to gain a balanced view of the economics of the total system as well as the physical shape of the system. This would include:

(1) Forecast of electric energy requirements.

(2) Fuels.

(3) Modes of conversion.

(4) Energy storage.

(5) Mode of energy transport.

(6) Transmission systems.

(7) Reliability.

(8) Costs.

d. Develop overall design concepts and the effect on reliability of distributed generation units and their integration into a power system, in order to reduce siting problems, waste heat dissipation problems and transmission requirements. These would include:

   (1) Fuel cells using natural gas, liquid fuel or hydrogen piped to or stored at the site.

   (2) MHD.

   (3) Small rotating generating equipment using fuel listed above.

e. Develop the overall concept and effect on reliability of large generation parks or off-shore installations and their integration into a power system, in order to make more efficient use of sites including use of by-products and spent fuel from the production of electricity. These considerations would include:

   (1) A group of large fossil, fission or fusion generating units.

   (2) Fuel delivery and processing problems.

   (3) Waste processing provisions.

   (4) Environmental problems.

   (5) Electric energy transport problems.

f. Study the interrelation between the availability of electric energy and the community as well as its effect on social and financial structures, including:

   (1) Availability of electricity as a factor in social advancement.

   (2) Use of electricity as a function of cost.

   (3) Financing problems of government and investor-owned utilities.

   (4) Interrelations of government and investor-owned utilities.

g. Provide for physical protection of the power system, including:

(1) Protection against subversive activity.

(2) Protection against natural and man-made catastrophies.

(3) Methods to minimize damage and duration of a power outage.

h. Study and advance methods for recycling waste and restoring land, including:

(1) Use of trash and garbage as a fuel.

(2) Use of sewage as a fuel.

(3) Restoring areas used for mining, generation or transmission when they are no longer useful.

i. Continually update studies of population and load characteristics, including:

(1) Population growth.

(2) Load growth.

(3) Load densities.

(4) Power utilization.

j. Study and develop methods for direct utilization of dc by consumers with the aim to eliminate the cost of conversion to ac when direct current is generated by MHD. fuel cells or other dc generators.

6. *Fundamental Research*

a. To learn more about the environmental effects of our activities and how they affect the health and well-being of the populace.

b. To learn more about the physical and chemical processes that take place in our materials and equipment so that we can reduce the cost and further improve the reliability of our electric power supply.

c. To learn more about the basic processes involved in alternative schemes for generating, transmitting, distributing and utilizing electric energy.

d. To learn more about the control and operation of large, expensive and complex systems for the purpose of enhancing the economy and reliability of our service to customers.

THE ENVIRONMENT AND THE POWER INDUSTRY

The introduction of the report to the FPC by the National
Power Survey Task Force on Environment states:

> In our time, we have seen commanding new social values
> arise, and among the most important of these is a new re-
> spect for the conservation of the environment, and the
> need to adopt our energy sources and supply to the re-
> strictions this imposes upon us.

The problem is to protect and where possible, to upgrade
the outdoor environment without losing the gains that have been
made in indoor and outdoor environment, both now and in the future. It is agreed that almost any development may have some
effect on the environment.

The electric power may have effects on the outdoor environment such as:

1. Air quality from discharge into the air of products of com-
    bustion from fossil-fuel-fired power plants or trace
    amounts of radioactivity from nuclear plants.

2. Effects on water quality from discharge of waste heat from
    thermal power plants, or discharge of chemicals from power
    plants, or trace amounts of radioactivity from nuclear
    plants.

3. Land-use effects, including ecological effects of hydro-
    electric power projects, accumulations of fly ash from
    coal-fired power plants, use of large tracts of land for
    power plant sites or for transmission rights-of-ways, and
    areas for storage of radioactive wastes from nuclear plants.

4. Esthetic effects such as noise and appearance.

Table 3-14 shows the electric power industry is far from
being the major source of air pollutants except for sulphur di-
oxides. However, it is a significant and concentrated source.

The fossil-fuel-fired plants use various kinds and grades
of fuel which produce various amounts of wastes. Natural gas
is the cleanest and coal of low heat value and high sulphur
content is the most troublesome.

The Clean Air Quality Act Amendments of 1970 have greatly
strengthened the regulatory powers over air pollution and ordered the establishment by 1971 of national air ambient quality
standards and individual states must present implementation
plans designed to control emissions from stationary and other

Table 3-14. Estimated nationwide discharges of airborne pollutants, 1968 (million tons per year).

	Carbon Monoxide	Particulants Matter	Sulfur Oxides	Hydrocarbons	Nitrogen Oxides	TOTAL
Power plants	0.1	5.6	16.8	Neg.	40.0	26.5
Other fuel combustion in stationary sources	1.8	3.3	7.6	0.7	6.0	19.4
Transportation	63.8	1.2	0.8	16.6	8.1	98.5
Industrial processes	9.7	7.5	7.3	4.6	0.2	29.3
Solid waste disposal	7.8	1.1	0.1	1.6	0.6	11.2
Miscellaneous	16.9	9.6	0.6	8.5	1.7	37.3
TOTAL	100.0	28.3	33.2	32.0	20.6	214.2

Note: Sulfur oxides expressed as tons of sulfur dioxide and nitrogen oxide as tons of nitrogen dioxide.

emission sources in such a manner that the ambient air quality standards will be met by 1975.

Air Quality--Fossil Fuels

Utilities have taken major steps to minimize emissions of air pollutants from existing and future generating plants and more work will be done. Steps already taken include removal by treatly improved mechanical and electrostatic precipitators; discharge of gases through high stacks at high velocities, so that they enter upper atmosphere, thus, having minimal effect; use of low sulphur fuels as much as possible; reduction of nitrogen oxide formation through modification of combustion techniques and flue gas recirculation; and rapid and massive introduction of nuclear power into base load operation.

The bulk of coal burned by utilities has a sulphur content of 2 to 4% by weight. Increasingly utilities are being required to burn fuel with a sulphur content of 1% and in some instances, the allowable limit has been set below 0.5%. The available supplies of coal or other fossil fuels as low as 1%, much less 0.5%, are very limited in relation to the massive needs of the power industry. Unless practical sulphur-removal systems can be developed quickly and put to use, the fossil-fuel supply outlook will be very bleak and those areas that have established low sulphur restrictions may either have to relax them or face prospects of power shortages.

A number of different sulphur removal processes have been studied and several piloted in small medium-sized installations. However, present indications are that it will be several years before full-scale systems are ready for routine commercial service.

Air Quality--Nuclear Fuel

The release of radioactivity from nuclear power plants is restricted by regulations of the U. S. Atomic Energy Commission which has statutory responsibility for licensing all nuclear facilities.

Since December 1970, responsibility for setting radiation standards has been vested in the Environmental Protection Agency which has made no change in the limits set by the Federal Radiation Council.

AEC has consistently sought to keep radioactivity releases from nuclear power plants as low as possible. In mid-1970, the AEC proposed an amendment to its regulations to limit nuclear plant releases to 1% of the exposure established by the Federal Radiation Protection guides for the general public. This action reflects two important nuclear power trends: first, the actual experience with radioactivity releases from nuclear power plants to do an even better job of radioactivity confinement.

Water Quality

The electric power industry uses but does not consume about four-fifths of the total water used by all industry for cooling purposes. It also accounts for nearly one-third of the total water uses for all purposes.

The principal ways in which electric power operations can affect water quality are through discharge from thermal power installations, of heat, chemicals and, in case of nuclear installations, trace amounts of radioactivity. Of these, heat is the most significant and has attracted the greatest amount of attention from conservationists, ecologists and the public.

It is a basic law of thermodynamics that whenever heat energy is converted to another form, there is a net loss that becomes "waste heat." In modern fossil-fuel-fired power plants, up to 40% of the heat released in the boiler is converted into electrical energy. The unused heat energy is discharged to the plant environs, about 90% going into the water that is used for condenser cooling and the balance being carried into the atmosphere by the stack gases. Most of today's nuclear plants operate at a lower thermal conversion efficiency, about 31%, and thus discharge more waste heat per KWH of electrical output. HTGR (High Temperature Gas Cooled Reactor) nuclear plants and the proposed breeder reactor plants will have thermal conversion efficiencies comparable to modern fossil-fuel-fired power plants. There is no loss through a stack from combustion process, essentially all the waste heat from a nuclear plant goes into the condenser cooling water.

Usually the condenser cooling water has been handled on a "once through" basis, that is taken from a body of water such as a river or ocean, pumped through the condenser and then discharged directly back to its source. Large flows are involved—about 1590 cubic feet per second (cfs) in a modern 1000 MW fossil fueled plant. Typically the discharge stream is from $10°$ to $30°F$ warmer than the inlet stream. However, as the warmed water rejoins and mixes with the receiving body of water, the temperature difference quickly reduced and, depending on local conditions, is often undetectable a few hundred yards from the discharge point.

There is concern that the local temperature differences and their effect on oxygen content of the water and other conditions important to aquatic ecosystems together with the movement of so much water, may upset the natural balance of aquatic life.

Much research is being done and the indications are that when properly controlled, little, if any, adverse effect on water quality is present and in some localities there may be a beneficial effect to fish life.

There has been public concern and government agencies have established or proposed strict regulations of warm water discharges. The Water Quality Act of 1965 was enacted and in 1968, comprehensive Water Quality Criteria were issued by the Federal Water Quality Administration, then a branch of the Department of Interior and now incorporated into the Environmental Protection Agency (EPA). In most states, no thermal power plant can be built until the responsible regulatory agency is satisfied that applicable water quality standards will be met.

Alternatives to "once through" cooling or cooling ponds, cooling towers either "wet", releasing heat to the atmosphere by water evaporation, or "dry" where heat is transferred by air conduction as in an automobile radiator. This involved considerable added cost, as compared to "once through" cooling.

There might be some discharge of chemicals, such as those used in control of algae growth in turbine condensers and trace amounts of radioactivity to be considered in water quality control. Mechanical tube cleaning can be employed.

Land Use

Several hundred acres are required for a site for a large generating plant and EHV transmission right-of-way requires 20 acres or more per mile of line.

The quadrupling of electric power operations in the next two decades will require some 300 new major thermal generating plant sites and construction of some 90,000 miles of additional 230 KV and higher voltage transmission lines. There may be less usable existing sites that may be abandoned and combined rights-of-way can be of advantage.

Electric Power Transmission and the Environment

The Federal Power Commission has produced "Guidelines for the Protection of Natural, Historic, Scenic and Recreational Values in the Design and Location of Rights-of-Way and Transmission Facilities." (Commission Order No. 414, dated November 27, 1970.)

The guidelines cover selection and clearing of rights-of-way routes; the location of transmission towers and overhead lines; the design of transmission towers; the maintenance of transmission line rights-of-way; and the location of appurtenant above-ground facilities. In addition to the written guidelines, the report includes drawings showing the "preferred" method for planning and designing the facilities and in selecting and clearing rights-of-way. Some of these guidelines are reviewed in Chapter 7.

Esthetics

There have been cases in the past of construction of unsightly transmission and distribution lines, plants and other facilities. However, in recent years this has been given very careful thought and commercial artists have been engaged to assist in improving appearances. Nuclear plants and substations are given careful thought for attractive appearance by architecture treatment and landscaping. Much progress has been made on underground construction of distribution facilities.

WESTERN SYSTEMS COORDINATING COUNCIL ENVIRONMENTAL GUIDELINES

The Western Systems Coordinating Council, one of the nine regional Councils in 1971, had 40 member systems and 13 affiliate member systems--all interconnected. The Council set up the Environmental Committee in December 1970 to develop and publish environmental guidelines. They are designed to enhance and maintain a high quality environment while continuing to provide reliable electric service in the areas served by the utilities of the Western Systems Coordinating Council. The guidelines are offered to member companies for assistance in their case-by-case considerations of environmental matters.

The Guidelines were completed and submitted to the Council Chairman, on September 27, 1971 and have been published in booklet form.

The statement of the basic underlying principles of the guidelines is:

The basic underlying principle of the guidelines is to establish a timely, positive and effective program of long-range planning for the siting, design, timely construction and operation of electric generating and transmission facilities, with full recognition of environmental impacts and constraints for preserving, protecting and enhancing environmental quality.

This is an excellent publication and is recommended for use and study of the details of the subject as indicated above.

REFERENCES

1. The 1970 National Power Survey - Federal Power Commission.

2. The Electric Utility and the Environment - A report to the citizens advisory committee on recreation and natural beauty by the electric utility industry task force on environment by President Johnson's Citizens Advisory Committee on Recreation and Beauty. Laurence S. Rockefeller, Chairman - November 1966-68. Library of Congress Catalogue Card Number 68-57661.

3. Houston H. Poor - Steam Generation Research in the Future - Institute of Electrical and Electronics Engineers - Environmental Research in the Power Industry - The Manufacturers' Role - 1972 - 72 CHO 695-7-PWR.

4. Electric Utilities Industry Research and Development Goals through the Year 2000 - Report of the R&D Goals Task Force to the Electric Research Council - June, 1971 - ER & Pub. No. 1-71.

5. Power Engineering Society of the Institute of Electrical and Electronics Engineers - Newsletter - July, 1972.

6. *Power Engineering Magazine* - January, 1973.

7. *Electric Light and Power Magazine,* E/G Edition - September, 1972. "Fusion - Can We Simulate the Sun?"

8. *Scientific American* - June, 1971. "Fusion by Laser."

9. *Transmission and Distribution Magazine* - January, 1973. "UHV Research Project Interim Report."

10. *Science Magazine* - December, 1972. "New Means of Transmitting Electricity: A Three-Way Race."

AUTHOR NOTES

Transportation

Robert F. Baker is a transportation consultant in Bethesda, Maryland. He has over 34 years of experience in the field of transportation engineering. This has included a tour of duty with the Corps of Engineers in World War II when he was involved in construction of airfields. From 1954 to 1962, Mr. Baker was on the faculty of the Civil Engineering Department of Ohio State University where he also established and directed their Transportation Engineering Center. From 1962 to 1967, he was Director of Research and Development for the U.S. Bureau of Public Roads. As a private consultant since 1967 Mr. Baker has conducted a wide range of studies dealing with such diverse subjects as research management and engineering analyses of slope stability. Mr. Baker is a registered engineer in four states, and has more than 100 publications related to soil mechanics, transportation, research management, and analyses of policy. He received the 1967 Roy W. Crum Award from the Highway Research Board.

Water

The section on water was authored by Colonel Frank C. Boerger, USA (Retired). Col. Boerger was District Engineer, U. S. Army Corps of Engineers, in San Francisco from 1966-1969. In 1969 he retired from the Corps to enter private consulting practice in the civil works field.

While with the Corps of Engineers, Col. Boerger served in Europe, the Far East, and the United States and was involved with a large variety of military and civil works construction projects. In California he supervised the planning for the Warm Springs (Dry Creek) Dam, the operation and maintenance of Lake Mendocino, and the planning and construction of other water resources development and navigation projects. As a private consultant he has continued his involvements with major civil engineering projects at policy, planning, and design levels.

Power

The section on power was authored by Dr. Lawrence M. Robertson, Denver, Colorado. Dr. Robertson is former Vice President for Engineering, Public Service Company of Colorado, Denver, Colorado. He was with Public Service Company from 1922 until his retirement in 1968. During this period he saw the industry evolve in technology and grow with the population.

His contributions to the industry have been numerous and in many areas: technical, legal, planning and others. He has served in numerous public service roles and was a member of the 1964 National Power Survey of the Federal Power Commission. Since his retirement he has been active as a private consultant on electric power problems for numerous private organizations.

Mr. Victor Koelzer, Professor of Civil Engineering at Colorado State University and formerly Chief Engineer for the National Water Commission has reviewed the chapter section on water and made a number of helpful comments.

Mr. Richard F. Walker, Vice President of Engineering and Planning, Public Service Company of Colorado, Denver, Colorado has reviewed the chapter section on power and has made a number of suggestions which have been incorporated into the text.

PART II

SUBSTANTIVE ELEMENTS

The core argument here is one of explicating substantive areas and technical inputs in order to provide both a sensitivity to the discipline areas involved, as well as practical considerations for multi-disciplinary project design. The analysis of project consequences to the physical world and associated biological systems, is followed by in-depth consideration of perturbations to human communities and the assaults to aesthetic integration.

4

THE PHYSICAL SUBSTRATE

BY THE EDITORS

The earth's crust is the physical basis for a metabolizing society. From it resources are derived, it is a structural component of all building, and it is a sink for residuals (i.e. wastes). The air, the water and the land are both passive and active mediums, permitting the physical metabolism of an industrial society. Furthermore, these mediums are also the bases for various ecological systems.

Thus, since it does indeed comprise our physical world, the earth's crust must be understood in the context of its structural properties, its resources, its esthetic qualities and its ecological significance. Further the natural continuing processes that cause constant change to the earth's crust must be understood. Public projects are impositions on the earth's crust which will definitely cause new equilibrium conditions to emerge as natural processes are disrupted. To provide some capacity to evaluate, judge, and merely be aware of these changes and their extent - and then to know how to mitigate them - is the purpose of this chapter.

PRINCIPLES OF CHANGE

The earth's crust is a dynamic system from both physical and biological points of view. Fueled largely by radiant energy from the sun, photosynthesis by green plants captures and stores a portion of this energy, giving the energy basis for life. In other words, green plants comprise the base of the food chain. In a natural system the flow of biological energy is essentially steady state - that is on a long term basis, replenishment is about the same as consumption through successive tropic levels of consumers.

189

For the physical system, weather contains the energy causing degradation of the earth's crust. More commonly this degradation is called "weathering." Weather is created through
vaporization and precipitation of water, coupled with large
scale movement of air masses caused by differential heating and
cooling of the earth's surface, and the Coriolis force. The sun
is a "heat pump," giving a continuing supply of new energy on a
global scale.

All of this means that the earth's crust is continually
changing - over geologic time. Thus water moves as a part of
the hydrologic cycle over land, and through soil and rock formations, as groundwater, and finally into streams and lakes.
In the process it leaches minerals from the soil, and erodes a
part, carrying some of both to the sea. The system strives for
static equilibrium; but the equilibrium state can never be
reached as long as solar energy is continuously added.   It is,
however, in a state of "dynamic equilibrium" in that the processes are stochastically unchanging over the millenia.

Table 4-1 summarizes some of these ideas using energy as a
basis for implying a taxonomic structure for understanding
changes occurring in the earth's crust. From this structure the
relationship between specific phenomena and subsystem functions
can be seen.   Conceptually such a structure may be useful in
discerning the set of consequences likely from an intervention
by man. A dam across a river for example will cause a chain of
undesired consequences at the most independent level of related
consequences (that is far to the right in Table 4-1). However,
if one tampers with the global insolation from the sun, say by
causing carbon dioxide levels to increase, then the aggregate
effects are beyond comprehension (in this instance the intervention would be far to the left in Table 4-1).   So discerning
the point of invervention is essential, from this the sequence
of possible impacts and consequences to both natural systems
and man.

Therefore, when man intervenes in this natural system, the
changes occurring over geologic time may be rapidly accelerated,
causing a chain of unstable situations to occur.   The dynamic
equilibrium condition is upset. Thus a dam across the Nile River
stops the flow of sediment in the river, doing several things.
It causes more erosion in the river channel, flooding of agricultural lands ceases, and with it new deposits of rich soil,
and then the sardine fishery in the Mediterranean is lost due
to cut off of essential nutrients in the river. Although the
intervention of the dam is at the far right in Table 4-1, it is
clear from this example that the sequence of events stops with
the last entry in that table.   Indeed a chain of primary, secondary, and tertiary consequences could be seen by further un-
folding the table with additional entries.

Table 4-1. Taxonomic structure of earth processes using solar energy as a basis.

Energy Base	Major Subsystems	Functions	Manifestations	Specific Phenomena	Related Consequences
Weather	Hydrologic cycle	Lifts water in elevation	Mechanical energy available to erode earth's surface and transport it Convection of water masses Mixing of water masses	Streams carrying silt Removal of pollution from site of generation Pollution is attenuated	Replenishment of beach sands Supplies estuarine nutrients Stream channel either builds up or cuts
		Distills water	Chemical energy available to dissolve mineral substances, and gases	Natural water has chemical composition indicative of past geologic exposures	
	Air movement	Departs kinetic energy to air	Mechanical energy erodes and transports earth's surface. Air masses are convected providing different air environments to a given spatial location. Mixes air, attenuating any characteristics	Acceleration of chemical reactions by dispersion of concentrated suspended and gaseous constituents.	
Photosynthesis	Synthesis	Provides the chemical energy to run the food chain through successive trophic levels	Carbon and other minerals and nutrients are fixed from diffuse sources in the soil and atmosphere and are recycled by: carbon cycle nitrogen cycle	Summary themes of all life processes in aggregate: Carbon cycle, Nitrogen cycle, Phosphrous cycle, Sulfer cycle.	Maintenance of carbon balance in air, nutrient balance in soils, etc.
	Degradation	Uses chemical energy stored by producers to synthesize new cell materials	Metabolism of animals (bacteria and higher forms)		
Man developed	Resource extraction	Develops mechanical energy to cause structural and chemical changes to materials handled.	Industrial activities, building activities	Waste production	Upsets to ecosystems by perturbations and reduction habitat

BASELINE CONDITIONS

In any operational assessment of the effect of a man made
or man caused intervention it is first necessary to establish
"baseline conditions." This must be done in a very quantitative sense for the system receiving the perturbation.

Thus to assess the effect of a dam on a river one must
first determine natural conditions. Implied here are measurements of chemical constituents, suspended sediment load, flows,
etc. over the annual cycle. It is important that the natural
system be thoroughly understood not only as one in dynamic
equilibrium, but that any indicator will exhibit a stochastic
variability over the annual cycle, and as annual means are compared from year to year. The baseline description must have
measurements, however, over a long enough time period to understand the system being intervened.

EFFECTS OF INTERVENTION

As previously outlined, intervention with the earth's
crust or with the processes occurring on the earth's crust is
understood in the context of understanding the overall natural
systems and the specific subsystems in question, documented by
measurements. Types of interventions are characterized best in
the context of the metabolism concept. For illustrative purposes some brief excursions into a whole multitude of intervention possibilities are outlined in the following.

Mining

Although mining and pollution are not the central thrusts
of this book, they are activities connected with many public
projects. The physical infrastructure is constructed of materials obtained either by mining or by harvesting lumber and
other renewable resources. Then the fossil fuels used to ener-
gize the society are products of mining. These activities result in tremendous quantities of residuals, ranging from tailings from processed ore to brine water associated with oil production. The brine is of course a salt hazard to adjacent
ground or surface water bodies. Mine tailings are less obvious
in their environmental hazard. Any kind of disturbance of the
earth - mining or otherwise - will open new opportunities, or
exposures, for weathering. The history of the west is replete
with case examples of water and air pollution by mining - ranging from leaching of radio-nuclides into the Colorado River to
release of iron oxides into Blackbird Creek in Idaho, causing a
film on all rocks and over the gravel bottom and resulting in
an abiotic stream.

A further example is open pit and strip mining, which contain the same hazards of exposures to leaching of minerals and hazardous chemicals--an *intensity* effect - but in addition has a large *extensive* effect just by the sheer number of acres of land overturned. The land has been changed, and if rehabilitated it will require a readjustment to a new ecological equilibrium condition. This may or may not be the same as the old one. Mine to mouth power plants, built and planned in the Western United States are the most prominent examples. However, a modern highway and other building construction requires vast quantities of rock aggregate which must be obtained by open pit mining in alluvial formations. Recently these activities have come under intensive scrutiny.

In handling these situations, several problems are of concern. These include: pollution hazards, visual disharmony, ecological perturbations and possible waste of otherwise good land. Pollution can be controlled in such instances by minimizing the surface area exposed to leaching. This of course is not easy - nor is it inexpensive. Earth cover and land restoration by proper grading and then planting is necessary. In the case of oil shale residues, with highly saline leachate, it is important to keep the buried materials from becoming exposed to the hydrologic cycle.

Aggregate borrow pits may become eye sores to an area, and due to the increasing prevalence of aggregate mining, land rehabilitation is advisable.

Construction

The demand for basic materials of construction requires prodigous quantities of aggregate, Portland cement, asphalt, steel, copper, wood, ad infinitum. Thus construction activity cannot be separated from the demand for raw resources. But beyond this aspect lies the recognition that the earth's crust is a structural component of the structure being built. An arch dam for example depends, in addition to the strength of the arch, upon the capability of the canyon walls to withstand the thrust. Along another vein, geologic mapping will aid in planning such that fragile or valuable land areas might be avoided. Then an analysis of the slip hazard caused by a highway cut could lessen damage to scenic values if perhaps a steeper cut could be allowed.

The potential for underground construction in minimizing environmental problems has scarcely been examined. Buildings, power plants, utility tunnels, rapid transit, etc. could often be sited underground to minimize esthetic impact and some practical difficulties. The Eisenhower tunnel bypassing Loveland Pass in Colorado certainly minimizes visual and ecological impacts of I-70; also the reliability of the highway in terms of that link in the interstate highway system is increased. Similarly, the twin tubes under San Francisco Bay for the BART rapid transit system eliminated totally any esthetic impacts which could be caused by another span.

Residual Production

Again waste disposal problems are not the thrust of this discussion. But the problems of waste disposal in a metabolizing society does relate to the dynamic equilibrium theme of the chapter.

Solid waste generation in the United States is about six pounds per capita daily. Reuse although sporadically implemented cannot presently cope with the massiveness of the problem, and landfill is still the only real alternative. A well placed landfill could allow land rehabilitation. But the materials disposed of could relieve some of the huge demand for raw materials. Economic incentives and institutional capabilities must be allowed to evolve first however.

The traditional approach to problems of air and water pollution has been to assume that sufficient dilution, coupled with stream degradation, and then convective removal from the point of generation - with subsequent dispersion - will take care of associated problems. However, the chemical and heat additions to air and water does cause perturbations to the normal chemical and physical processes in the respective subsystems.

Thus in examining the effects of any large structure - or community for that matter - the demand for resources implied, and the production of wastes must be considered along with the structure per se. Though the focus of this effort is toward the latter concern, holistic planning cannot ignore the other concerns either.

5

ECOLOGY

BY WILLIAM T. HELM AND JOHN C. PETERS

Public projects do cause ecological changes. They are physical intrusions on biological communities, which will - to greater or lesser extent - upset existing patterns of equilibrium. Some of these changes are direct and immediately evident while others are indirect and perhaps more slowly perceived. Some are beneficial and others are harmful - the judgment depending upon the respective value position of the various interest sectors in our society. Today it is important that all effects of public projects be considered and evaluated as a part of the planning process. Ecological changes are a prominent category of effects to be delineated with respect to their relationship to public projects.

Public projects, after all, relate to man, and according to ecologic principles man is part of the ecosystem and is regulated and influenced by the same set of principles which regulate all other living organisms. Man's failure to recognize this in the past has often resulted in an endless series of problems, frequently compounded one upon the other, until the resulting megaloproblem seems to defy solution.

The first step in solving a problem is to recognize that one exists. The successful administration of public projects and programs can depend a great deal on how responsive the decisionmaker is toward ecological principles before deciding upon, designing, and constructing a public works project. This chapter presents the interrelationships between ecology and public projects, emphasizing the interactions between the ecologist and the engineer. The focus is on defining ecological problems and pointing out opportunities for solving them using the ecologist and engineer as members of a team.

TERMINOLOGY AND PRINCIPLES

The word ecology is derived from the Greek "oikos," meaning "home" or "place to live."   Technically ecology is usually defined as the study of the interrelations of plants and animals among themselves and with their environment. A less technical description is that ecology is the study of how the household of nature operates.   Ecology and economics come from the same root word, and a sense of the relationship between ecology, economics and sociology can be gained from the following description from Clarke (1966).

> Man, being egocentric, began this type of study in his immediate surroundings.   Not until long afterwards did he realize that man's economics is but a special case of the broader subject.   In the words fo Wells, Huxley and Wells (1939), "Ecology is really an extension of economics to the whole world of life." Economics and sociology might be thought of as the "ecology of man" in a broad sense. The realization that the relations of man to his environment, both physical and social, form a distinct and most important study is reflected in the increasing use of the term human ecology in sociology and in other fields.

Living organisms, plant and animal, are interrelated. Certain species of plants live together in mutual harmony because their requirements and adaptations are suitable for a particular situation. A different set of plants would be found in some other location with a different slope, exposure, density of trees, etc.   The species of animals associated with the plants would differ widely according to the nature of the plant community. because it is the assemblage of suitable individuals which make up the community.

Community ecology however leaves out the important physical-chemical components of the functional system. The entire complex, plants, animals and the physico-chemical substrate is an ecological system, or *ecosystem*.   Both living organisms and their nonliving environment influence the properties of the other and both are necessary for the maintenance of stable life systems.   An ecosystem is the largest functional unit in ecology.

Ecosystems can be divided into three major categories: terrestrial, marine and freshwater.   Each of these categories contains several ecosystems.

> Terrestrial - deserts, grassland, forests, tundra.
> Marine - seas, estuaries, seashores.
> Freshwater - rivers and streams, lakes and ponds, marshes.

Each of these ecosystems can be further subdivided on the basis
of various physical and chemical characteristics. Deserts can
be classed as hot or cold, and lakes as oligotrophic (few nutrients) or eutrophic (well nourished or abundant nutrients).

Within any ecosystem are to be found many species and
countless numbers of organisms. The place, or type of place,
an organism inhabits is called its *habitat*. Habitat may also
refer to the place occupied by an entire community. Thus the
habitat of a *Trillium* plant is a moist, shaded situation in a
mature deciduous forest, while the habitat of the "sand sage
grassland community" is the series of ridges of sandy soil occurring along the north sides of rivers in the southern Great
Plains region of the United States. A description of the habitat of an organism (*Trillium*) or group of organisms (a population) includes other organisms as well as the physical or abiotic (nonliving) environment, while a description of the habitat
of a community (many organisms of many different kinds) includes
only the latter.

Ecological niche differs from habitat by requiring a
broader and more inclusive set of descriptions. Niche refers
not only to the organisms habitat, but also to its position or
role within a community and ecosystem. This role is determined
by the organisms structural adaptations, physiological responses
and behavior. To determine this role would require an investigation of its activities, foods, range of movements, predators,
its effects upon other organisms and the extent to which it
modifies important processes in the ecosystem.

All life in every ecosystem is based upon the ability of
one type of organism to utilize energy derived from the sun,
plus basic raw materials, to manufacture protoplasm. These organisms, the chlorophyll bearing or green plants, form the base
of a *trophic pyramid*. All other organisms depend, directly or
indirectly, upon these primary producers. Within a trophic
pyramid there are a series of nutrient or *trophic levels*. Plant
producers (algae, grass, shrubs) form the first or base level,
herbivores (which feed directly on plants) constitute the second
or primary consumer trophic level, and the primary carnivores
(which feed on herbivores) form the secondary consumer or third
trophic level. Additional levels are composed of secondary and
tertiary carnivores.

The trophic level concept is often illustrated with a triangle in which the primary producers constitute the broad base
of the triangle, the herbivores perch on top of them and at the
apex of the triangle are the carnivores; Figure 5-1 illustrates.
A similar triangular figure can also be used to illustrate the
numbers or biomass of some particular ecosystem or even the biomass of an individual species within the community. The latter
is illustrated by the tremendous numbers of young fish at the

bottom of the pyramid and the relatively few very old fish at
the top. An illustration of the relationship between the number
of grazing animals and the vegetation upon which they depend
would depict the production of vegetation at the base of the
pyramid and the production of the animals toward the top. In
this case the production of animal flesh is considerably less
than the production of plant tissue. There is a loss of energy
at each step of this food pyramid because the processes of assimilation and growth are not 100% efficient. The greater the
loss between trophic levels the less the height of the triangle
in relation to the width of the base.

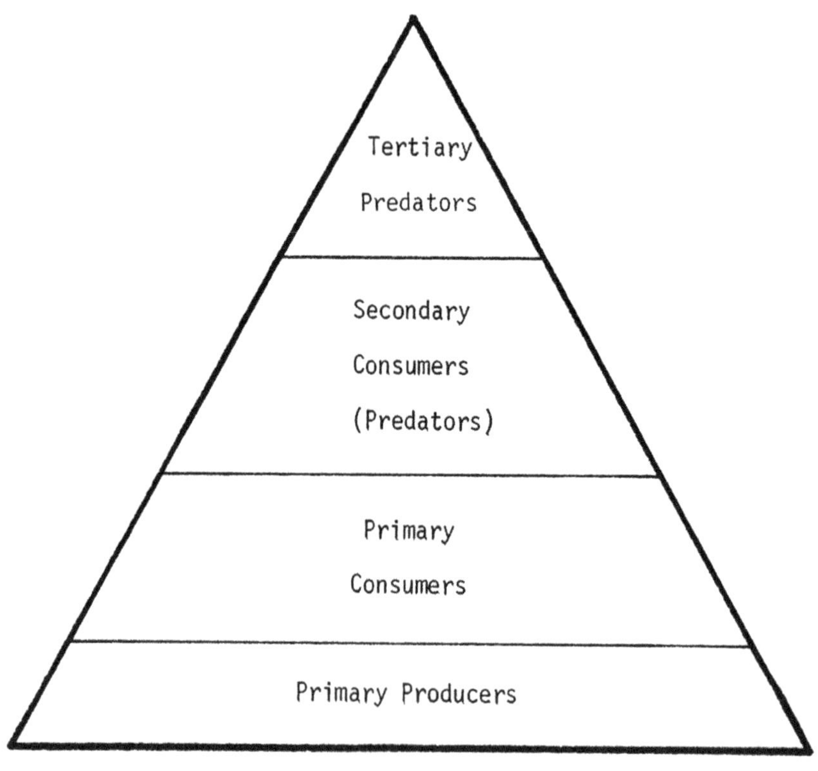

Figure 5-1. Trophic level pyramid.

In some ecosystems the base of the triangle may not be
composed of the primary production within the ecosystem. Many
relatively small rapid flowing streams do not have sufficient
algae and aquatic plants growing in the stream to form the necessary supply of material for the upper trophic levels. The
energy for the base of this triangle is provided by organic

matter produced on the land in the vicinity of the stream. The
residue of leaves, twigs, branches and various animals from the
land mass around the stream forms the base of this particular
triangle. The second level, the consumers, utilize this supply
of organic matter in the same fashion as if it had been produced within the stream.

It should be apparent that for the entire pyramid, all life
is dependent upon the abundance of primary producers (plants)
in the first trophic level. This abundance is influenced by
such factors as nutrients, heat, light and water, but not necessarily in an easily described fashion. Complete absence of a
necessary nutritive element would prevent growth. In some cases
a plant will utilize a substitute (strontium instead of calcium)
if there is not quite enough of the required nutrient present.
Requirements vary from one species to another, but in this variation is the key to understanding at least part of this complex problem. Plants do not require equal amounts of all nutrients, and all plants do not have the same requirements. The
nutrient in short supply cannot be identified merely by determining which nutrient is least abundant. Calcium, although one
of the more abundant materials, could nevertheless be short of
what was needed by the plants. One or more essential substances
may be present in concentrations below that required for growth
and reproduction. This limiting of plant growth is identified
as Liebig's *"law"* of the minimum.

Too much of some factor such as heat, light or water may
be just as limiting as insufficient quantities of an essential
nutrient. The concept of the limiting effect of both maximum
and minimum is referred to as Shelford's *"law"* of tolerance.
Thus the limits of tolerance are the upper and lower levels beyond which growth would be limited, and between which are the
acceptable growth and reproductive conditions. These limits
vary from one species to another and from one season of the
year to another. Odum (1953) lists the following subsidiary
principles to the law of tolerance:

1. Organisms may have a wide range of tolerance for one factor
    and a narrow range for another.

2. Organisms with wide ranges of tolerance for all factors are
    likely to be most widely distributed.

3. When conditions are not optimum for a species with respect
    to one ecological factor, the limits of tolerance may be
    reduced with respect to other ecological factors.

4. The period of reproduction is usually a critical period
    when environmental factors are most likely to be limiting.
    The limits of tolerance for reproductive individuals, seeds,
    eggs, embryos, seedlings, larvae, etc. are usually narrower
    than for nonreproducing adult plants or animals.

Neither of the above concepts can be considered alone, because limiting factors and tolerance are inextricably interwoven in their action. Thus soil structure, pH, moisture and temperature may be suitable but an insufficient quantity of Boron may limit plant growth, or conversely there may be ample Boron and suitable soil structure, pH and temperature but too much moisture for the plants to grow. Another type of interaction occurs when requirements differ at various stages in development of an organism. As hatching time approaches, the oxygen requirement of fish and aquatic insect embryos increases. The amount of oxygen which will dissolve in water varies inversely with temperature, being greater at low temperatures than high. Eggs which are close to hatching may perish if the water temperature rises, not as a result of the higher temperature being a limiting factor, but rather because of interaction between the effects of increased water temperature and the changing requirements of the embryos. At higher temperatures the metabolic rate of the embryos becomes higher and thus the quantity of oxygen required increases. Concurrently warm water contains less dissolved oxygen than cool water, so that even at saturation there may be insufficient available oxygen for survival of the embryos.

The chemical elements required for growth circulate in the biosphere, from the environment into the bodies of plants and animals as they grow, back to the environment when they die and decompose, and again into the bodies or organisms. This generally circular pattern is variously called an *inorganic-organic cycle, nutrient cycle* or *biogeochemical cycle* as indicated in Figure 5-2. Recharge rate or the rapidity with which elements are recycled varies from one situation to another. Materials are not equally plentiful in the various phases of these cycles, and the various loops are not completed in equal periods of time. Some essential substances may be sidetracked by deposition in places where regeneration is excessively slow, such as in very cold regions, or in poorly aerated soils or mud. Natural biotic communities depend upon, and indeed are attuned to the recycling of nutrients, since new nutrients are not derived from the weathering of rocks and parent material fast enough. In the case of phosphorus steady depletion of the total supply by soil erosion is a critically serious matter. Over-abundance in some lakes and streams contributes to excessive enrichment, while rapid depletion in the soil requires expensive replacement. Continued erosion losses then render such replacement temporary, a doubly critical situation since phosphorus is a relatively rare material.

Nutrient cycling is readily seen in lakes and ponds where the supply of dissolved nutrients ebbs and flows with the different seasons of the year. After the ice breaks up in the spring in the norther latitudes the supply of nutrients is high as a result of the circulation of all of the water in the

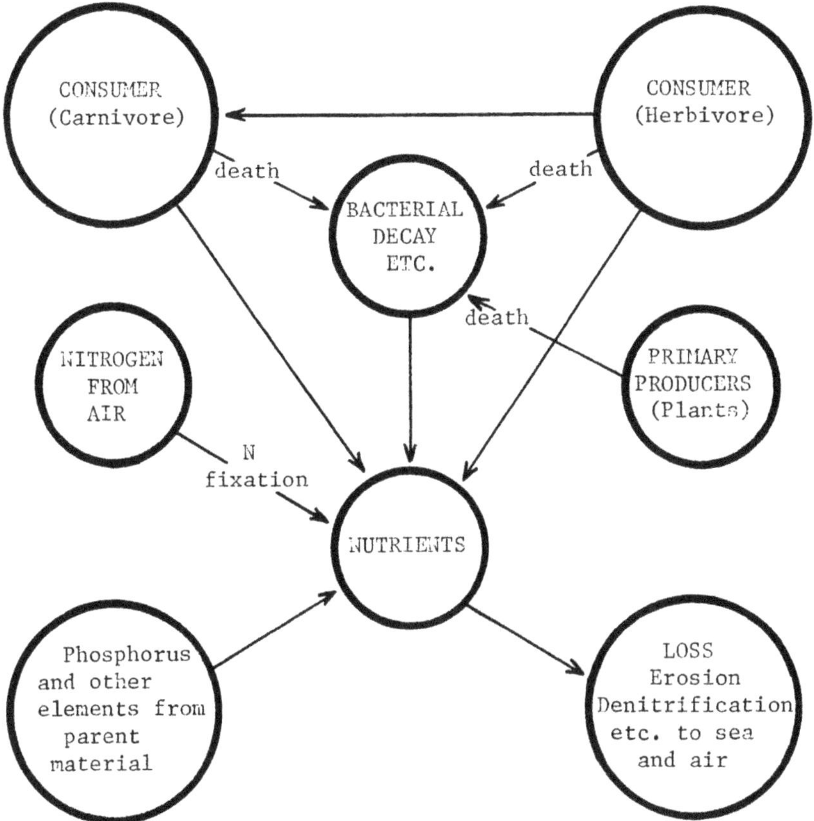

Figure 5-2. The Nutrient Cycle. A simplified diagram illustrating the basic components and paths by which nutrients circulate throughout an ecosystem.

lake basin. A large variety of different species of algae develop and take up much of the nutrient supply in the growth of new protoplasm during the spring and early summer. Measurements of the amount of nutrients dissolved in the water show a great decrease at this time, demonstrating the uptake by plants. Death and decay of algae will then return nutrients to the water for use by other algae at a later time. The decay and return to availability, of course, takes quite a period of time, but this is a continuing process and there are algal cells constantly being added to the supply of decomposing material.

An example of the balance between a terrestrial plant community and the nutrient cycle can be observed in the eastern deciduous forest community. Nutrients are extracted from the soil and incorporated into vegetation. Leaves, branches, old trees and animal bodies contribute to the organic matter supply

on the ground, and are decomposed to supply nutrients for new
growth. As long as this growth cycle continues very little of
the nutrient supply is washed into streams. When the forest is
cut the cycle is interrupted, and relatively large quantities
of nutrients are translocated into the streams. Once in the
stream much of this supply moves rapidly away downhill. New
forest growth must then be initiated with a diminished nutrient
supply, and the plants must depend upon the lower soil regions
and the weathering of parent material for a larger portion of
their nutrient requirements than did the previous stand of vegetation.

Changes in the physical structure and species composition
of a stand of vegetation can be expected to influence the population of animals which inhabit the area. Invertebrates, birds,
mammals, and even fish may be affected. Some effects may be
direct, the organisms habitat is destroyed and it can no longer
survive there. In other cases, the effect may be indirect, and
although the organism itself could survive in the new environment its food supply cannot. A thorough understanding of the
foods and feeding habits of the organisms in a population is
necessary before the ecological relationships within that population can be explained.

A typical example of the food interrelationships of a
small fish, Figure 5-3, illustrates the reliance of that individual upon a variety of other organisms and also of some other

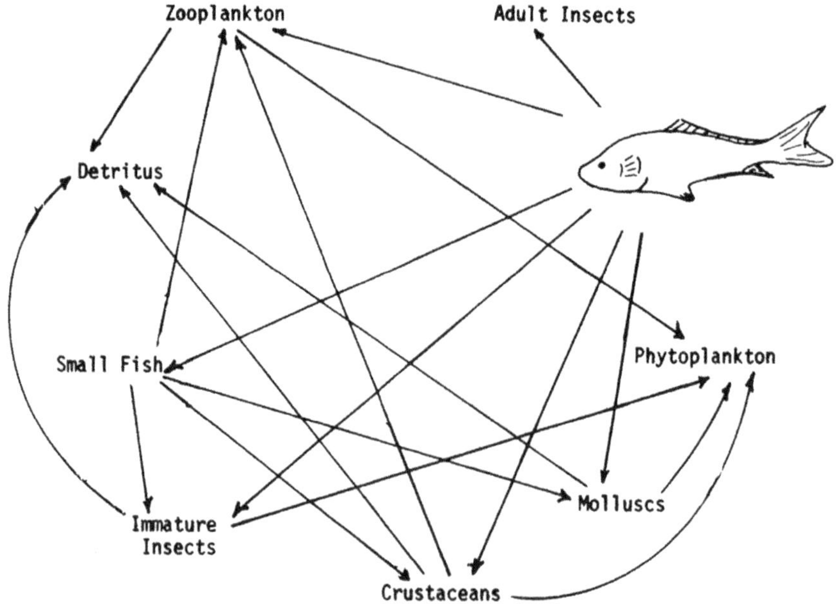

Figure 5-3. Food web.

organisms upon that small fish. The flow of energy is much branched and has many directions, and has been called a *food chain* but is more properly a *food web*. As in nutrient cycling food webs are not unidirectional and simple but rather multidirectional and quite complex. Eliminating one source of food from a web simply shifts consumption in other directions. It should also be apparent, however, that the fish when very young and small will require very small food organisms, and as it increases in size it may require larger and larger food items. In natural communities there will be a wide variety of species and sizes. Late springs, very cold or very warm periods will of course change the time of appearnace and development of many organisms, and we note the results in very poor survival of one year's production of animals. By and large, however, the wide variety of organisms present serve as a buffer and the species involved survive.

The term diversity refers to the species composition of communities. A community with a large number of species is said to have more diversity than one with few species. In tropical or subtropical areas there are many species of plants, insects, birds and animals compared to boreal forest areas (Figure 5-4), which means the resources are divided among a greater number of species in the former. A change in weather or climate, an outbreak of disease or change of any sort which would result in unfavorable conditions for organisms with certain habitat requirements could result in severe reduction in numbers for the affected species. With few species in the boreal region, loss or severe depletion in number of a relatively few species would have a greater effect on the population than loss or depletion of a similar number of species in a tropical region. The more trophic links there are (Figure 5-4), the greater the likelihood of compensatory mechanisms operating if one species becomes rare or abundant. A change in the environment may cause one species to become very abundant and its competitors to become less abundant. Predators may then shift to the more abundant species. Predators may also shift from the most abundant species to alternate prey if the abundant species declines markedly in number. Note also that euryphagous species would then have an advantage over competitors which consume a limited variety of species.

Elton (1958) has presented a six-point discussion illustrating that population stability is greater with highly complex food webs than with simple food webs. He points out that invasions and outbreaks of species occur in the simplified cultivated or planted communities, not in natural communities. There are also more problems with insect pests in the boreal forest regions dominated by only a few tree species than in the tropics where there is a tremendous variety of species.

	Number of species	Number of species lost	% Lost
Fish	300	3	1
Mammals	150	2	1+
Birds	250	2	1
Insects	4,000	20	<1
Plants	2,500	25	1
Fish	25	3	12
Mammals	40	2	5
Birds	70	2	3
Insects	170	20	12
Plants	400	25	6

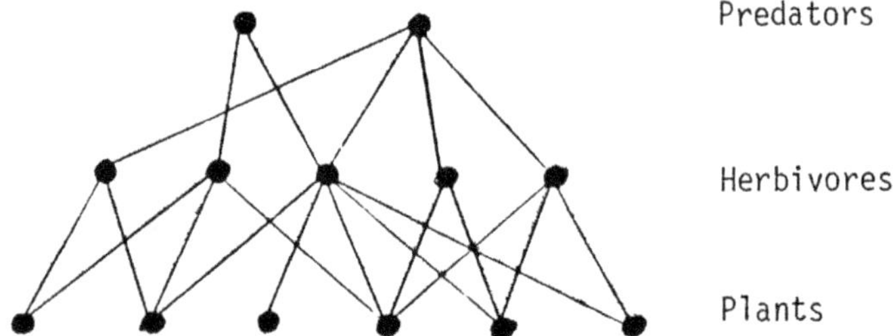

Figure 5-4. Diversity-stability and trophic web pattern. Number of species is based roughly on subtropical and boreal areas. In the trophic web each dot represents a species, and lines between dots indicates the species in the upper level eats the species on the level below. (Adapted in part from Watt, 1968).

Great diversity in floral species results in great diversity in habitats, with the resultant great diversity in fauna. The probability of a highly complex trophic web in such a community is much greater than in a simpler community, thus species diversity, which is equated with community stability, can also be said to be directly related to habitat diversity.

Populations of various organisms do not necessarily enjoy uncontrolled growth and expansion. There are a number of *control mechanisms* which regulate the growth and ultimate size of various species populations. Limitation on the amount of sunlight penetrating through the forest canopy would regulate the germination and growth of shade in tolerant vegetation. Competition for food and space or for breeding areas would also regulate numbers of certain insects and fish. Populations of algae in geologically young lakes may be described as having great diversity but few of the species produce sufficient numbers of cells that they become very abundant. An increased nutrient

base, from agricultural runoff or domestic effluents, would
support an increased density of algae. Some of the original
species would continue to exist but others would be replaced by
species better adapted to utilize the increased nutrient supply
and to develop denser algal populations. With the increased
density of population, however, the toxic materials excreted by
some of the algae may serve as a control mechanism to limit the
increase in cells and thus limit the population density to a
certain level.

Immediately after the hatching of eggs laid by some stream
invertebrates the density of the population will be quite high.
Since the average size of the individuals is very small, the
total biomass present does not exceed the capacity of the
stream system. With increased size of individuals as a result
of growth, the biomass of invertebrates increases to the point
where the combination of social pressure and competition for
food exceeds the capacity of the system. Some of the invertebrates then begin to drift downstream, thus decreasing the density of the population that remains behind. With continued
growth and development of the invertebrates additional drifting
is necessary. This phenomenon of drifting is called stream
invertebrate drift or organic drift and is believed to be directly related to the productivity of the stream. There always
remains a quantity of invertebrates roughly equal to the carrying capacity of the stream.

Fish are often used as an example of how control mechanisms
operate and of the resultant effects upon the controlled organisms. A population of pan fish under adequate control either by
natural predation or by man managing the population will exhibit good growth rates and adequate replacement through reproduction. Populations which are not adequately controlled will
develop too many young fish for the habitat to support and the
increased competition among individuals will result in a decreased growth rate. If this situation persists long enough the
entire population will be composed of slow growing, stunted and
thus small individuals.

ECOSYSTEMS

Any discourse on impacted ecosystems must be related to
the norm, or natural ecosystems. Upsets and adjustments due to
the man-induced perturbations can then be examined with an eye
toward modifying those attitudes and procedures which result in
unsatisfactory ecological consequences. All of man's activities
cannot be conducted in such fashion that some undesirable ecological ramifications will not occur, but a knowledge of how
different systems function, and how various types of alterations affect them, would permit more intelligent planning and
execution of projects. As should be evident already, ecological

systems are intricate, with many links and processes at one
level which are dependent upon links and processes at other
levels. The system responds to a change in one location with
an adjustment, or a series of adjustments in one or more subsystems, a process varying in magnitude with the immensity of
the original change. Following are descriptions of some common
systems, how they function normally and how certain alterations
affect them.

Lakes

Ponds and lakes are described as lentic or quiet waters in
contrast to lotic or flowing waters as in streams. Lack of
constant unidirectional currents in lake waters establishes
some physical restraints on the way various lake inhabiting
communities develop. The physical substrate, water, exerts a
profound influence upon the form and structure of aquatic organisms. Plants in lakes can be of two types, those that drift
in the water and those that are rooted to the bottom. Drifters,
primarily the algae, are moved about by the wind induced currents but manage to maintain a vertical position in the lake by
a variety of flotation schemes. These include deposition of
fats and oils, inclusion of gas bubbles and the form and size
of individuals. Algae are primary producers, that is, they
utilize sunlight and nutrients to form protoplasm and thus produce the basic materials to support the variety of animal life
which cannot utilize the sun as a primary energy source.

Rooted plants, as do the algae, utilize the sun's energy
and nutrients to form protoplasm. Unlike the algae, the rooted
plants are restricted to life in the littoral or shallow water
zone around lake shores. The depth of this zone depends upon
the transparency of the lake water and thus the depth to which
sunlight can penetrate. Primary producers, the algae and rooted
aquatic plants, depend upon the proper combination of sunlight
and nutrients for their existence. Increasing the turbidity of
lake water through the addition of finely divided soil particles from erosion of cultivated or forested lands, industrial
wastes such as stains or finely divided particulate matter, or
adding large amounts of such particulate matter as dust or
smoke to the air from power plants, smelters and industrial
plants would limit the depth to which the available sunlight
could penetrate. This of course would limit the depth to which
rooted aquatics and algae could grow.

A similar decrease in light penetration can occur as a result of the addition of large quantities of nutrients from sewage disposal plants, industrial facilities or drainage from agricultural fields. The response within the aquatic community is
a decrease in number of algal species, replacement of some
species by others better adapted to the new conditions and a
much increased density of algal cells. This dense population of

algae then effectively shades the lower regions of the lake producing the same end result as increased turbidity; the depth of the euphotic zone is restricted. These two changes, increased algal density and restricted euphotic zone, combine to induce further changes in both the habitat and biota. The extra organic matter produced would settle to the bottom of the lake upon the death of the algae. Decomposition and transformation of the organic matter by bacteria would utilize oxygen, and this increased utilization of oxygen in deep water would diminish the oxygen supply. In shallow lakes where wind action is apt to cause nearly continuous circulation of water this increased oxygen consumption would have a minimum effect. In deeper lakes, however, stratification occurs as a result of a thermal barrier. This barrier is established in early summer when the surface waters are warmed by the sun and the deeper waters remain cool. The interface between the warm and cool waters is known as a thermocline or metalimnion. This layer proves to be an effective barrier to water circulation, preventing a vertical exchange between the epilimnion (the upper warm layer) and hypolimnion (the lower cold layer). In eutrophic lakes the supply of organic matter is sufficient to severely deplete or even exhaust the dissolved oxygen in much of the hypolimnion. Concurrently the vertical restriction of the euphotic zone has eliminated phytoplankton photosynthetic activity in the hypolimnion, terminating partial replenishment of the oxygen supply via that source.

Some species of fish are physiologically adapted to life in cool, well oxygenated water. Such fish could no longer survive in the anoxic hypolimnion, thus we have an additional change in the aquatic community. In parts of the northern tier of states this group of fish, the trout and whitefish, have gradually disappeared from many of the lakes receiving domestic effluent. At the time of lake overturn, when the surface temperature of the lake is altered until it approximates the thermocline and lower regions of the lake, the entire lake will circulate. This mixes the upper and lower strata of water and rejuvenates the oxygen supply in deep water. In lakes in which the oxygen demand is not quite sufficient to eliminate all of the habitat below the thermocline this rejuvenation may permit the continued marginal existence of some of the coldwater species. Even the fish population in the upper zone is not unaffected by the change in the algal community. Fish that can directly utilize the large quantities of readily available algae will tend to have a competitive advantage over those which must depend upon consumers or predators for their energy supplies, and thus the fish community in this portion of the lake may also be altered.

From original pristine conditions to man-altered conditions there are a number of other changes which occur. Originally the lake contained fewer total numbers of algae, aquatic plants, invertebrates and fish than in the intermediate and final stages of alteration. Thus, the number of individual algal cells, invertebrates and fish increased while the number of species decreased. This change in the lake community has thus decreased the species diversity and in turn decreased the stability of this aquatic system. The altered lake is now subject to greater fluctuations of algal and invertebrate populations than was the original lake. Blooms of algae and tremendous hatches of various aquatic insects prove to be annoying to those living near the lake. The dead algae wash up on the shores, form deposits on the bottom of the lane and even cause toxicity and oxygen depletion problems in effluent streams.

Most invertebrates, both the drifter-swimmers or zooplankton and those generally restricted to the bottom or benthos are primary consumers depending upon plant material for their food supply. In most cases this food supply or organic matter, is produced to a large extent within the lake (Figure 5-5) by a combination of phytoplankton and rooted aquatic plants. When the proportion contributed by sources outside the lake, either as organic matter or plant nutrients, greatly exceeds that produced within the lake, responses noted above can be expected to occur.

Man, as part of the ecosystem, is not unaffected by the changes noted above. The green fetid water is no longer desirable for swimming, the odors from the dead and dying algae make lakeshore living an ordeal and the change in species composition of the fish population lessens the enjoyment of fishing. Real estate values decline. The welfare of nearby communities suffers and some people begin to understand that man is indeed part of his environment.

Frequently, portions of lake shorelines are marshy. Some marsh characteristics are very similar to lakes in that the algae, aquatic plants and zooplankton are the same genera and in some cases the same species as found in lakes. Water temperatures in the marsh tend to be somewhat higher in the summer than in open lake water and the composition of the algal and zooplankton populations may be somewhat different than in open lakes. Just as with terrestrial forest vegetation, the marsh has an upper and a lower story. In the marsh the upper story is the emergent vegetation, extending from one to several meters above the surface, while the under story is comprised of submerged aquatics and various types of algae. Like all vegetation, marsh vegetation utilizes carbon dioxide in the conversion of sunlight and nutrients into protoplasm. A byproduct of this process is oxygen, a necessary ingredient in the air for man. The expanse of shallow water and muddy bottom, housing a vast population of bacteria, serves as a reducing area for such undesirables as sulfur dioxide in the atmosphere, and in this way prove to be valuable ecosystems in the vicinity of industrialized cities.

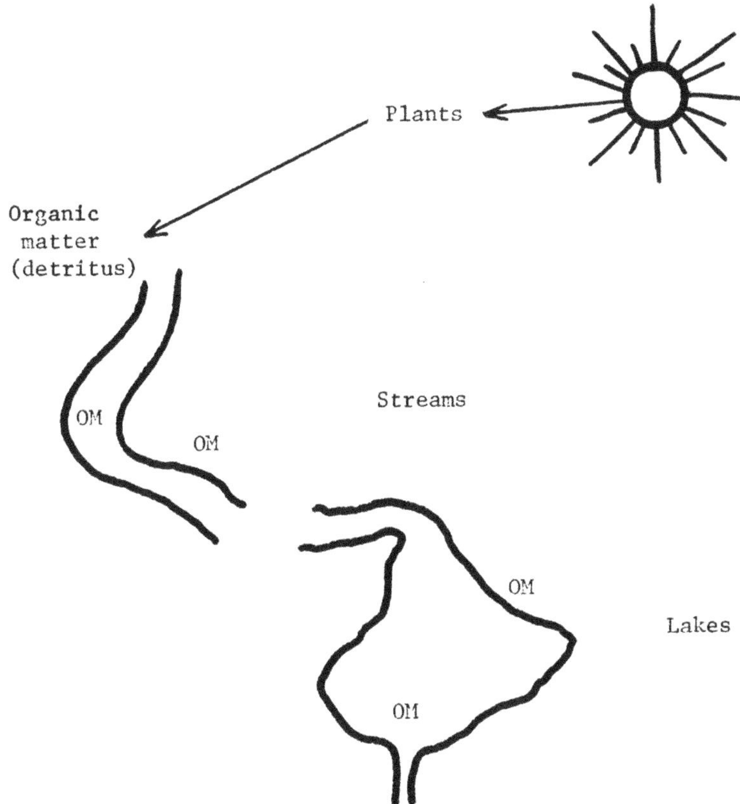

Figure 5-5. Sources or organic matter, the food base for invertebrates in lakes and streams. The proportion supplied from within and outside of the aquatic systems is proportional to the size of the latters.

Marshes exert an influence on the contiguous streams and lakes by providing a spawning habitat for particular types of fish. Northern pike and sometimes walleyed pike utilize marshy areas for spawning in the spring. Where the edge of the marsh is sufficiently deep the large adults inhabit this area throughout the year. A number of game and pan fish also utilize marshes for breeding during spring and early summer. In most cases the young utilize the rich plankton population in the marsh early in life, during a period when they are small and susceptible to predation by other fish, birds and even some of the insects in the lake. The value of a marsh near a lake can sometimes be appreciated most when it has been eliminated. Real estate developments on the shores of lakes in southern Wisconsin have filled marshes to gain additional developable land and in the process have eliminated essentially all of the breeding

areas for such species as the northern pike. The result has
been almost complete elimination of this fish, one of the prime
attractants to the people buying homes near the lake.

Alterations to marshes involve draining and/or filling to
eliminate the standing surface water or to provide a solid
foundation upon which to construct buildings. Elimination of
the low-lying region thus deprives the lake of a safety valve
for the temporary storage of excess water and also affects the
supply of water for the recharge of the ground water table.
The very intimate relationship between the water in a lake or
marsh and the elevation of the ground water table can be altered considerably by such activities as the paving of streets,
covering of a considerable portion of the landscape with houses,
driveways and sidewalks, and the elimination of much of the
surface water. This combination of activities, building and
draining, has often depressed the water table and decreased the
rate of recharge, thus influencing the extent and use of water
from that supply.

Streams

Streams, in contrast to lakes, are flowing water systems
with unidirectional current and characterized by inhabitants
adapted to water currents. Velocity of flow varies from tor-
rential in the steep mountainous regions to relatively slow and
quiet in the lower elevation river valleys. Streams, because
of the movement of water, do not stratify thermally as do lakes.
The nearly uniform temperature distribution thus imposes a
somewhat different set of conditions upon the inhabitants than
those described for lakes. In addition, the development of
free-floating plankton communities would be a self-defeating
process since the flow of water in the stream would tend to remove them. Invertebrate communities in streams are closely al-
lied with the bottom substrate. Since the substrate can vary
all the way from large boulder-sized particles to fine silt and
clay the variety of invertebrates, in terms of shape, size and
conformation, is equally broad. In fast water the invertebrates
are adapted for clinging to stones or to creeping down between
the stones and living in the semiprotected areas beneath and
between them. Adaptation to swift currents then takes the form
of a combination of shape, and agility for those organisms exposed to the current, and shape and size for those avoiding the
current by staying below the surface of the substrate. Invertebrates adapted to burrowing in soft substrates would be found
in protected areas of the swift streams, where fine sediments
have been deposited. In slower streams with substrates composed
of finely divided particles, the invertebrate population would
be adapted to burrowing and thus avoiding suffocation by shifting sediments. Diversity in terms of species would be highest
in the streams providing the greater variety of habitats and
lowest in those with a monotonous habitat. The slowest streams

begin to take on some of the characteristics of lakes and in some cases, where large rivers have broadened into a basin and the current speed is very low, there can indeed be a number of similarities between rivers and lakes.

Primary producer communities in streams have some similarities but many dissimilarities with those found in lakes. Again, the very nature of the medium, a unidirectional flow of water, determines the types of algae found in streams. Attached or sessile algae retain their position on the bottom and thus avoid being washed continuously downstream. Primary production then is limited to those regions where the water is sufficiently clear and shallow that the algae can obtain adequate sunlight. There may also be rooted aquatic plants in regions where the bottom sediments are sufficiently fine for root penetration. On unstable bottoms, those in which the stones tend to roll and shift as a result of current action, the supply of algae is quite limited. In many streams the population of primary producers is inadequate to supply the needs of the consumer population. Energy must then be supplied from an outside source, the detritis from the terrestrial environment (Figure 5-5).

Fish inhabiting streams are also adapted to the constant current, although some of the same species may also be found in lakes. The breeding characteristics of stream fishes are such that the eggs will be deposited in locations insuring their survival and hatching without being tumbled along the bottom where abrasion would destroy them or being washed into quiet areas where fine sediments would cause death due to oxygen depletion. Salmon and trout bury their eggs in nests called redds in gravel bottoms. Such burying retains the eggs beneath the surface where they are incubated and hatched. To survive the eggs require a constant supply of oxygen delivered by water moving through the gravel substrate of the stream. Interruption of such water flow would decrease the supply of oxygen and induce mortalities among the eggs.

The fish themselves have varying requirements for feeding and resting. There can be very intricate divisions of the available space among the various sized individuals of different species of fish. One species may be dominant in most situations yet smaller individuals of other species may be found in close proximity but in positions less desirable than those occupied by the dominant species. Depth, velocity, turbulence and proximity to certain types of cover or obstructions appear to be criteria for selection of location. Brown trout in small to medium sized streams appear to have a preference for pools located near the bank of the stream with brush overhanging the water. Such bank pools, with or without an overhang of the bank itself, appear to be far more desirable to the fish than pools located in the center of the stream. Some deep, slow flowing water and either an overhanging bank or overhanging brush

appear to be requirements for good populations of this species.
Other trout appear to have some of the same requirements although in somewhat differing degrees.   Diversity of habitat is
again equated with diversity of populations.   Great diversity
of habitat types in streams insures diversity in the invertebrate and fish populations.

Turbid water during the spring is a characteristic in most
streams.   The runoff of snow melt in the colder regions of the
U. S. and increased flow as a result of spring rain in others
causes the transport of erosion products into streams, resulting in increased turbidities. At the same time the increase in
flow results in higher velocity currents and increased capacity
for the river to transport such material.   Since the size of
particles which can be transported by flowing water is proportional to the velocity of the water, not only can the fine erosion products be transported downstream but particles large
enough to be stable during normal flow periods can also be moved
at this time. With much of the stream bottom being moved about,
most of the fine particles will be transported through regions
of adequate flow and deposited in quiet areas like large pools,
reservoirs or lakes.   During low flow periods, however, with the
attendant slower currents the addition of silt can be very destructive.   Current velocities at this time may not be sufficient to transport this material for any great distance and the
interstices in gravel beds may be filled, eliminating the living space for invertebrates and the breeding grounds for trout.

Marshes along a stream system tend to discourage land development in the near vicinity and thus preserve the river flood
plain, insuring that during periods of high water there is ample
relief for floodwaters to flow up onto the land and thus lessen
the flow in the channel.   Such relief is a valuable part of the
natural system since it reduces the severity of flooding in
nearby developed areas.   It is after marshes have been filled
and eliminated that their role in damping floods is appreciated.
Recent events in southeastern United States have demonstrated
this relationship.   A major flood control project resulted in
the straightening and diking of a meandering, marsh surrounded
river.   Flood waters then rushed down the new channel and into
an area incapable of accepting water at the accelerated rate.
The agency promoting the flood control project ultimately identified the value of the meanders and marshlands along the natural channel and recommended returning the river to its original
course.

Marshes along the edges of streams provide spawning areas
for some fish and nursery grounds for the young of many others.
Small fish often require protection for their survival, and
since in many cases there is a limited amount of protected water
in the stream channel the young, to survive, must find an area
out of the channel.   Small areas of marsh, which do not result

in excessive warming of the stream, also serve as valuable re-
fugia for small trout.   Social and predator pressure, as well
as an inability to maintain position in spring high flow forces
very young trout to seek the protection of vegetated stream
banks or marshes.

Terrestrial

The following example and description of a terrestrial
ecosystem was chosen not so much because it applies to a significant or insignificant proportion of this continent, but because increasing numbers of large public projects have been undertaken here in recent years. In general the construction and
operation of a project affects a limited land area, unless gaseous or particulate wastes are released into the environment.
Such releases, through direct toxic effects or chronic alterations in light or soil conditions could affect vast areas.
Various ancillary developments, like alterations in population
distribution with concomitant highway system expansion, opportunistic urban sprawl, etc., would extend the sphere of terrestrial effects.   Although virtually all projects affect aquatic
systems, the following discussion will be limited to the terrestrial ecosystem.

Areas which receive limited precipitation and have high
evaporation rates and/or rainfall concentrated in winter with
dry growing seasons may be classified as desert or desert-like.
Cold deserts, characterized by extensive stands of sagebrush,
are found in seven western states.   Vegetation is intermediate
in density with the tallest stratum, shrubs, generally less than
two meters in height. A sparse understory is comprised of various short grasses and forbs capable of rapid maturation after
the onset of the normally dry summer.   Plant spacing is not as
wide as in the well known hot deserts, yet the soil is not entirely covered.   Consumers range from ants, conspicuous not in
themselves but because of the circular denuded areas around
their central mound, to various small rodents, jackrabbits and
occasionally antelope.   Secondary consumers include badgers,
coyotes, hawks and golden eagles.   Because of limited precipitation the soils have not been leached of nutrients, and in
fact are often strongly alkaline.

This ecosystem is attuned to but limited by the amount of
water.   Nutrients are recycled efficiently, with the deeply
rooted sagebrush translocating nutrients to the surface. There
is less diversity of habitat, and thus less diversity and less
stability throughout the system than in the eastern deciduous
forest for instance.   Populations of producers and consumers
exhibit wide fluctuations in numbers.

Interjection of a common type of public project, a dam,
could produce extensive systematic changes in this cold desert

ecosystem. Scenic degradation occurs when gravel and fill
sites, construction roads, and power lines are expediently located. These effects can readily be anticipated, but they may
be minor compared to other effects. It is often necessary to
include several types of benefits to insure project authorization, thus along with power generation, irrigation and recreation may be cited as benefits. Irrigation of desert soils is
not always a beneficial and profitable land use. Irrigation
return flow water carries a heavy load of dissolved salts to
create problems in receiving streams or with previously vested
irrigation rights downstream. Desert lands may be abandoned
after the initial good crop years, to go through the very slow
plant succession characteristic of arid regions. Soil erosion
from snow melt and strong winds could damage receiving streams
and deplete the upper soil horizon.

Concurrently with the conversion of desert lands to agriculture, the consumer-predator trophic levels would undergo adjustments. Burrowing forms would not persist in frequently
disturbed fields, while those consumers (now pests) which are
adapted to utilizing the irrigated crops would flourish. Jackrabbits would decline while some gamebirds might increase.
Predatory birds, previously dependent upon the assemblage of
first level consumers, would not decline in number. Abandonment
of the farmland would not result in rapid restoration of the
raptors because the very slow plant succession would not support the necessary consumers for a considerable period of time.
This would be doubly tragic, since these birds as a group have
already absorbed a double dose of man's largess. Widespread
animal control programs and depredations have been superimposed
upon population declines generated by pesticide accumulation.

Preservation of a breeding area for predatory birds could
also be cited to broaden the list of project benefits. An adequate reservation for breeding would not necessarily provide
sufficient hunting area for these wide ranging birds, and thus
conversion of the nearby desert communities to agriculture
would defeat part of the purpose of the breeding reservation.
Construction of roads, observation posts and a visitor center
would additionally stress the remaining fragments of the original raptor population, since these birds will not tolerate such
disturbance.

In summary, although the project involves the construction
of a dam, a wide variety of terrestrial ecosystem alteration
could occur. While some of the changes might be acceptable as
a result of valid needs for additional agricultural production
and power generation, to have them occur as unplanned and uncontrolled consequences in spite of their pernicious effects
upon ecosystem components supposedly benefited by the project
would not reflect well on the planners.

INTERFACE AREAS

Ecological investigations include not only a determination of the various plants and animals present in an area but also an assessment of the various physical and chemical factors influencing these populations. An assessment of a terrestrial community would of necessity include the compilation of information on the soil types, slope, direction in which the slope faces, moisture, wind direction and velocity, humidity, amount of solar radiation, daily and seasonal variations in temperature as well as average daily temperatures and a list of the plants and animals in the area. Specialists in various disciplines are often consulted to obtain information on many of the items mentioned above. Examples of these interfacing disciplines would be geology, soils, hydrology, and meteorology. In some cases it is possible to anticipate the requirements for assistance from these other professions prior to initiation of ecological studies. In others it becomes apparent at various stages during the ecological investigations that additional information beyond the capabilities of the ecologist would be required. In any event the inclusion of some or all of the interfacing disciplines should be anticipated for any extensive evaluation.

CONDITIONS REQUIRING ECOLOGICAL PLANNING

A relatively few years ago the concept of the land ethic, or the proper treatment of land and water to retain natural beauty and productivity, was not particularly widespread. Today there is increased awareness among the American public of ecological problems and the damages that can occur as a result of not utilizing adequate planning. Ecological planning must be incorporated in public projects. The foregoing descriptions are intended to provide a basis for nonecologists to understand some of the ecological principles and to increase their awareness of how ecological planning can be utilized. Ecologists should be viewed as an integral part of a planning team and not as cosmetic technicians to be utilized after the damaging decisions have been made. Additionally, the team must be interdisciplinary with the members interacting at all levels, from posing basic questions to exploring the interfaces of disciplines. Assembling a group of experts to have each do his own thing without interaction would be chaotic and nonproductive.

At the time of initial consideration of a project, when economics and engineering feasibility are being discussed, ecological input could be most valuable. To the untrained eye one site may be so similar to another there appears to be no reason to suspect that what could be done in one location could not be done in another. A knowledgeable ecologist might well identify serious problems at the initial discussion stage. There seems to be few if any large public projects which should not include ecological planning at every step in the process. The National Environmental Policy Act in fact states interdisciplinary involvement as a basic goal. In view of the trend in recent years there appears to be no reason to expect major changes away from the philosophy of this act. Structural engineers, hydrologists, geologists and other specialists are commonly utilized at all stages in the planning of large projects because their talents are recognized as being essential for the proper planning and implementation of the project. Ecological expertise should be included as one of the essential components in any planning group.

Decisions on when to call in members of a particular discipline at various stages in the planning and construction of a project are not necessarily easily made. Past experience in many cases dictates which professions are consulted and the stage at which this consultation occurs. Since the inclusion of ecologists is a relatively new occurrence, and since many organizations have had little experience, there is a tendency to impose their own opinions and conclusions as to whether ecologists should be consulted, and if so, at what stage in planning the consultation should occur. It would be far safer to assume that virtually all projects in all conditions require some type of ecological planning. Inclusion of such planning at an early stage in program development permits more valuable input of ecological thinking than inclusion at too late a time.

As an example of how early ecological input could be utilized, an exploration of ecological alternatives in the siting of nuclear power plants could be instructive. In many cases the once only use of cooling water would result in a heated water discharge of sufficient magnitude to create problems in the receiving lake or stream. Existing uses may already have resulted in a serious increase in critical summer temperatures in the river in question. Any further increase could cause extensive losses to commercial and sport fisheries. In the absence of a comprehensive design for the utilization and protection of the river basin, there would be no method to evaluate the various commercial uses which result in water quality changes, and thus no way of assigning priorities. Unless the proponents of the power plant could purchase the heat pollution "rights" of an existing user this project, from an ecological standpoint, would not be feasible.

Assuming some arrangement could be made to release waste heat, or a self-contained recycling cooling system was used, there may still be some additional problems. Sophistic claims nonwithstanding, there have been enough incidents to indicate there is a possibility of an accidental release of radioactive materials. Contamination of a river system could result in radionuclide concentration in algae, further concentration in consumers, and yet additional concentration in secondary consumers resulting in unacceptably high levels in fish. Shellfish in the river, estuary and along the sea coast could also be contaminated.

Use of river water for cooling, or even siting the power plant close to the river would not be advisable from an ecological standpoint. The river ecosystem is more readily damaged than some nearby terrestrial ecosystems. Considerations of safety, dispersal of population and recreation reinforce the ecological reasoning in this case. Construction and operation of the plant in an accessible yet sparsely settled inland area would provide desired economic benefits, avoid concentrating more of the work force close to an existing city and not use valuable river front land when there was no need for access to the river. Radioactive materials, in the event of an accident, could be retained on the property instead of being distributed by a river.

Opposition to such projects from the general public and from special interest groups can be expected to continue and perhaps increase because to them it appears the planning has been poor and/or information released has been misleading if not false. Much of the past opposition would have been avoided if ecological considerations had been given sufficient weight. Trends during the past several decades indicate more, not less consideration must be given to environmental protection. Appending ecological expertise after site decisions have been made negates much of the value to be derived from this discipline.

Additional considerations which will become increasingly important involve cultural and moral aspects of public projects. To date many projects have been initiated without properly assessing the impact of the project on the public, either in the United States as a whole or in the project area. Sophistic statements of project initiators have not set forth the types and magnitude of adverse effects to be expected. Providing increased supplies of a potentially limiting resource to an area already overcrowded and suffering increasingly from air and visual pollution should be subject to much public debate and honest analysis. Delicate ecological, cultural and behavioral patterns can be upset by the brutal intrusion of a massive project, and the subsequent opportunistic exploitation of neighboring lands. Completing plans for an irrigation project today without including candid planning discussions with ecologists, economists, sociologists and environmental planners may well spell failure for the project. It makes little sense to dam and divert rivers to generate a few acre-feet of water when the losses, in the form of increased siltation and salinity of receiving streams, decreases in the already scarce trout streams, disfigurement of scenic areas and foreclosure of alternate

future development far exceed the actual benefits to the general
public.   The arguments that the new reservoir will provide increased recreation, and that economic benefits will accrue have
usually been specious. From planning discussions through public
hearings all facets, good and bad, should be included in the
total process.

METHODOLOGY

Fundamental to any ecological evaluation of a project is a
knowledge of the various plant and animal systems in the vicinity of the proposed project.   To obtain such information it is
necessary to conduct an inventory.   This inventory can be
structured so the initial efforts are directed toward the rapid
assessment of the major plant and animal communities without
expending a great deal of time in detailed study.   Depending
upon the size and nature of the project this type of inventory
can be accomplished by inspection of existing aerial photos or
by overflying the region of concern, noting the various major
vegetational types and then examining selected areas on the
ground. Purpose of the reconnaissance inventory is identification of the various plant and animal communities in the region
so some preliminary judgments can be made ' about the probable
effects of the project upon these communities.   An additional
value of such an inventory would be the possible inclusion of
information obtained from the inventory in the planning of the
project. There are no general rules on the amount of land that
can be examined over a specific period of time to produce a reconnaissance inventory because this depends upon the terrain,
the type of plant communities and the size of the project involved. The reconnaissance inventory is a qualitative inventory
in that most of the information relates to the type of plant
and animal communities and not to any quantitative aspects of
those communities.

If the project now appears to be sufficiently sound that
feasibility studies should be undertaken, the ecological inventory then moves to the comprehensive stage.   At this point the
information derived from the reconnaissance inventory auter-
mines the direction to be taken and the amount of effort to be
expended to obtain comprehensive information on those plant and
animal communities which may be affected by the project.   Here
again the experience and knowledge of the ecologist is necessary to identify those areas which might be affected. There are
examples of communities being altered or essentially eliminated
as a result of large projects when these communities were not
contiguous or even close to the project.   In other cases, although they were close, incorrect assumptions were made as to
the potential effects of the project upon them.   Comprehensive
inventories include a series of quantitative measurements of
the various communities present and of selected abiotic factors.

Such measurements can be expected to identify rare and endangered species, species particularly sensitive to a change in the environment which might occur as a result of completion of the project, an assessment of the relative importance of various communities in the project area and those abiotic factors which, if altered, could cause changes in the biotic communities.

Inventories in the vicinity of a proposed dam may yield information on the general vegetative types and an inventory of the various animals, fish and insects with a proposal for the amount of time and financing necessary to complete a comprehensive inventory. The comprehensive inventory would specify the types and amounts of vegetative communities which would be inundated by water if the dam is constructed to various elevations, the loss of habitat for various birds and animals, the probable effects on surviving vegetation of changing water levels, changing ground water tables and changing land use patterns. Animal populations in the region would also be evaluated and the probable effects of a dam on such things as the migration of big game animals, change or loss of habitat for other animals and the net result of these changes could be shown. A comparison of the species diversity of the fish population in the river to be impounded and in reservoirs on that same river system, or reservoirs on some nearby and similar river system could be made. From this an evaluation of the resulting fish populations can be compared to public desired and the resulting information utilized in evaluating the feasibility of the project. An example of a very important consideration would be whether the construction of a particular facility would adversely effect a population of fish of considerable economic or aesthetic value.

PLANNING PROCESS

There are several noteworthy steps or stages between the time a public project is conceived and the time it becomes a functional, operating facility benefiting John Q. Public. The basic stages are: preconstruction, construction, and operation, and collectively they can be considered the primary elements of the process dealing with public projects. Each of these stages can be broken down further into smaller units based on time sequence. For example, the earliest stage of preconstruction is often called "project planning" and involves reconnaissance and feasibility levels of investigation. Also, it is convenient for budgetary and other administrative purposes to consider the operation stage in a broader sense and hence it is often called "operation and maintenance." Nevertheless, the commitment of resources for a public project involves a process based on time sequences. An understanding of the process and the role of the ecologist in this process should be clearly understood by the decisionmakers. While Chapter 2 contains a more general elaboration on project evolvement, some more specific detail here is useful.

As previously stated, the earlier the ecologist has direct input into a project the more valuable his contribution. It does little good to bring in the ecologist at the construction or operation phase of the project after most of the ecological mistakes have been made. Entry of the ecologist should occur at about the same time as entry of other planning specialists and he should be a full partner to a team effort. Any discussion of the process of utilizing ecological talents in planning and executing public projects must include information the ecologist requires from others, and the relationship between ecologists, engineers and other disciplines in the work-flow pattern of the project.

An ecologist can no more work in the absence of other disciplines than can certain components of an ecosystem exist without the other components of that system. To be most valuable to the process the ecologist will require input from other members of the planning group. From the engineer he will require details on the location of the project, the approximate operating conditions, the various types of inputs and outputs from the system during the construction and operating phases, and, in general, information on how, when, where and why a multitude of various things are to be done. A work-flow diagram indicating the relationship between the ecologist and engineer would perhaps explain some of these interrelationships. The inclusion of only the two professions is not meant to imply that there is not a flow of information between ecologists and other professions on the one hand or engineers and other professions on the other. A presentation of the work-flow interactions between these two disciplines follows and points out in more detail some of the specific responsibilities of each professional, and the structure of their mutual relationship.

PRECONSTRUCTION STAGE

Reconnaissance Investigation

ECOLOGY: Examine wide areas in vicinity of each site--compile extensive inventory of plants and animals present--identify areas for systems which could be affected by project.

ENGINEERING: Explore for sites in general area. Identify potential sites, routes, etc. and auxiliary support areas.

Input: Engineering input should include basic characteristics of project, including sufficient detail so

ecologists can estimate potential effects in various situations, identification of all possible alternatives as soon as possible and memos on changes
and additions which might alter the environmental
impact of the project. Alternatives should provide
general information on sites, structures, economics, and environmental effects.

Output: Interaction at this level should include reports
on any situations in which the proposed project
would require special designing and/or operation
to meet NEPA standards, identify situations in
which cancellation of the project may be advisable,
but not any detail on estimated effects of proceeding.

Move to next step when alternatives are located and/or delineated, even though some detrimental effects are predicted. Alternatives beyond agencies' field of expertise must be considered. Determine role of public involvement.

Feasibility Investigation

ECOLOGY: Conduct more intensive inventory, to es-
tablish precisely what
plants and animals are present, their distribution
and numbers, and any pertinent related information
(wet or dry areas, stable
or fluctuating water levels,
etc.) which would influence
the stability or longevity
of the ecosystem.

ENGINEERING: Evaluation of
sites, routes, etc.

Evaluation of methods,
structures, etc. in relation
to sites.

Input: Continual communication on changes in sites,
routes, etc. as well as information on possible
construction methods, location of material sources
for construction, and any processes during construction or operation of the project which will
result in solid, liquid or gaseous wastes or aesthetic impacts. Analyze potential problems associated with project. For example, with a dam we
must evaluate supersaturation of dissolved gases,
channelization, water temperatures below dam, timing and quantity of flow releases, entrapment of
nutrients, etc.

Output: Reports on the various ecosystems in which the alternatives would be located with more detailed
predictions of probable effects based on studies

and further evaluation. At this point it should be possible to present a preliminary critique of the strengths and weaknesses and the feasibility of continuing with the project from the standpoint of ecological insight into probable alterations in and effects of proximate ecosystems, local cultural patterns, visual harmony, land uses likely to be adversely influenced by decisions which may have to be made to offset disbenefits created by the project, etc. Present alternatives to public and to decisionmakers.

Design of Project

ECOLOGY: Identify sites, routes, etc., which would have the least effect on the ecosystems involved, or which would take advantage of the most stable characteristics of each system--evaluate effects of proceeding or not proceeding with project, on the basis of predicting the effects of the project on the organisms identified during the inventory phase-- predict wherever possible, the magnitude and duration of effects, and the rate and extent of recovery.

ENGINEERING: Determine construction standards, design structures, etc. in corridor or location determined during previous steps. Deal with selective withdrawal and multilevel outlets, supersaturation of gases and design of spillways, etc. to take into account environmental protection. Prepare specifications dealing with air and water pollution and aesthetic values.

Input: Detail on probable design features; size of project; operating criteria; daily, weekly or seasonal changes in construction or operation procedures; amounts, types and sources of fill and construction materials which might cause problems because of dust, toxicity, changes in surface water supply, alteration of animal movements, pollution of streams or lakes, etc.

Output: Reports on probable or predicted effects on those ecosystems involved, of various designs or sizes of project under consideration. Identification of which ecosystems (if any) appear to be most impacted by the particular designs involved, with analysis of methods of reducing or mitigating damage. Identification of potential problems during operation of project, such as waste discharge safeguards, containment of toxic or other material which should not be permitted outside of the project area, etc., which should be considered at the design stage. Presentation of details to public and decisionmakers. Complete the specifications to meet air and water quality standards and aesthetic values.

CONSTRUCTION STAGE

ECOLOGY: Observe construction to modify processes, materials or methods which are damaging to the ecosystems involved.

ENGINEERING: Supervise construction, modify or adapt methods and materials to facilitate desired project progress.

Input: Notice of changes in design, materials, methods of wastes from those originally proposed. Information on materials and methods to be utilized, and wastes to be discharged. Access to all construction areas, records of chemicals used and their disposition, and in general to all information pertinent to an understanding of how the construction could affect the surrounding area.

Output: Identification of those methods, materials, processes or wastes which would have an adverse effect on the ecosystems involved, with suggestions for minimizing or eliminating the adverse effects. Identification of previously unforeseen adverse effects from the location or operation of the project with suggestions for minimizing such effects. Conference(s) with contractor to meet environmental needs in compliance with contract specification.

OPERATION AND MAINTENANCE STAGE

ECOLOGY: Observe normal operation and maintenance procedures for abnormal or emergency situations. Change procedures to comply with environmental needs.

ENGINEERING: Modify project features to improve operation and/or reduce maintenance. Change contracts to meet current legal requirements.

Input: Access to operations and maintenance manuals, and to the facility to observe actual procedures.

Output: Identification of procedures, processes or methods which would cancel or subvert the protective measures designed into the project, with provision for modifications which would restore the enviornmental protection originally envisioned.

ECOLOGICAL R$_X$

An ecological prescription can be interpreted in various ways, from bitter medicine to be taken to relieve project ills to preventative procedures to avoid the appearance of such afflictions. It is in the latter vein the following material is presented. Examples of types of public works activities are analyzed to show not only that some procedures have an adverse effect upon proximate ecosystems, but also why and how these effects occur and finally, presentation of possible ways to avoid environmental mistakes of the past.

Public projects are often conceived, designed, and operated in terms of specific project boundaries. Too often little concern is directed toward impacts outside project boundaries or in "secondary impact zones," which result in degradation over wide areas not taken into account by project planners and administrators. Deposition of silt in watercourses, releases of water considerably warmer or colder than receiving waters, releases of toxic or radioactive wastes or large quantities of particulate matter (even though the actual concentration in the effluent is low), releases of gases and particulate matter into the atmosphere, releasing large quantities or organic matter or plant nutrients, and filling in coastal marshes to create real estate all tend to exert permanent environmental destruction over extensive areas beyond the boundaries of the project. Those people responsible for public projects must define primary and secondary impact zones, and consider direct and indirect impacts to the environment. In this way they can improve the overall quality of the project and could conceivably receive more general support for it.

The following procedural summaries for several selected problem areas reflect on some of the environmental problems. Problem areas discussed include power plants, highways, dams, and channelization. The discussion points out the types of problems associated with each general category of public works activity. The dialog is by no means complete in its presentation of all possible problems, but rather limits itself to a general discussion of key issues currently before the public.

Powerplants

Powerplants consume large amounts of fuel resources, which must be transported to the plant site; they produce large quantities of waste materials, some of which may be controlled; they employ many people; and finally, they are a massive visual intrusion upon a given site. Figure 5-6 is a photograph of the

Figure 5-6.   Four Corners Powerplant.   Wet scrubbers are operating on units 1 and 2 (stack on far left). Bureau of Reclamation photograph, reproduced by permission of the *Southwest Energy Study*.

Four Corners powerplant at Farmington, New Mexico.   Associated with this plant is a nearby strip mine, shown in Figure 5-7, and transmission lines, Figures 5-8 and 5-9. Thus each individual site must be carefully examined to identify its accompanying environmental problems.   For instance powerplants, whether fueled by fossil fuel or nuclear energy, by their nature produce heat, which is not completely utilized.   This excess or waste heat must be dissipated in the vicinity of the powerplant. In the past, powerplants were constructed near streams or lakes with a sufficient volume so that this water could be used for cooling.   There are numerous situations today where the necessary volume of cooling water exceeds the capacity of the receiving water to accept heat without drastically changing the aquatic environment. An increase in the temperature of a north temperature zone lake could result in later formation of ice cover in the fall, or perhaps lack of ice cover for part or all of the winter, with the resulting loss of ice fishing, skating, and ice boating. Fish moving rapidly from the heated discharge area into adjoining cold water may die of thermal shock, and the normal aquatic invertebrate population may undergo a drastic readjustment of species and numbers. In some cases, alterations may be beneficial, depending upon the segment of society

Figure 5-7. Navaho strip mine near Four Corners. National Park Service photograph (reproduced by permission of the *Southwest Energy Study*).

Figure 5-8. McCullough switching station. Bureau of Land Management photograph (reproduced by permission of the *Southwest Energy Study*).

Figure 5-9.   Typical   transmission   line   and maintenance
road. Bureau of Land Management photograph (reproduced by permission of the *Southwest Energy Study*).

making the evaluation. Heated discharges may increase the temperature of a lake sufficiently to prolong the growing season,
produce greater annual growth of those fish adapted to that
particular situation,   promote a longer open water boating season, or even provide open water where waterfowl can overwinter.

There are other environmental problems besides waste heat
associated with powerplants and their siting.   These include,
for nuclear plants,   radiation hazards and public safety, and,
for fossil fuel plants, air pollution.   There is presently no
institutional mechanism at the state or Federal level to guide
the siting of powerplants or to provide for enlarged public
participation in relationship to their social and environmental
impacts. Such a mechanism is needed and is currently under
consideration by Congress and many state legislatures. The lack
of such a mechanism hampers effective environmental programs
not only for power generating plants, but also for accompanying
features such as transmission lines, mines, haul roads, switch-
yards, water supply, etc.

Rapid progress has been made to develop technology to minimize air pollution of coal-fired generating plants and meet
the requirements of the Clean Air Act of 1970.   Scrubbers and

precipitators are being improved and used to remove SO_2, NO_x, and particulate matter.   However, we do not know the long-term effects of SO_2 on the environment and more field level investigations are needed, especially in the arid and semiarid west to determine the impacts of SO_2 on biota.

Powerplants require transmission lines to transport their power to market as indicated in Figure 5-9 and 5-8. Significant progress has been made in locating major transmission lines with a minimum of environmental impact.   Guidelines have been developed by the Western Systems Coordinating Council (1971) that consider aesthetic and environmental impacts. These guidelines have been field tested and have received public acceptance thus far.   The guidelines recommend that utility representatives must be fully appraised of the specific environmental concerns of the people in the areas affected and have an awareness of the need to evaluate the visual impact and physical infringement on the landscape.

The Western Systems Coordinating Council report further recommends that to minimize the environmental impact of a transmission line, the following criteria should be considered in the selection of the route:

1. Locate the route in an area of minimum conflict with present and future planned land uses.

2. Replace or upgrade existing lines.

3. Add new lines parallel to existing lines provided that reliability criteria can be maintained.

4. Develop joint use lines with other utilities.

5. Avoid:

    a. Line locations which create unusable spaces between lines.
    b. Heavily timbered areas, steep slopes, proximity to main highways and scenic areas.
    c. Crossing main highways in the vicinity of interchanges or bridges.
    d. Crossing highways at points where the structures can be seen from a long distance or silhouetting the structures against the sky, if possible.
    e. Wildlife concentration areas, such as nesting areas and flight corridors.
    f. Parks, monuments, scenic, natural, historic, archeological, and recreational areas.

6. Cross canyons up slope from roads which traverse the length of the canyon.

7. Use the topography or natural cover to screen the line from
    highways and other areas of public view.

8. Consider access and construction roads and locate the route
    for minimum damage from road building activities and for
    minimum use by unauthorized personnel. Permanent roads
    should be minimized.

9. During field reconnaissance, guard against the natural ten-
    dency to favor the locations which are easily observed.
    Tangled underbrush and other areas which must be observed
    by rough foot travel may hide the best route. Find all
    practicable routes joining the termini.

Perhaps the most environmentally acute problem associated
with powerplants today is the strip mining of coal on public
and private land. (Figure 5-7 illustrates.) Millions of acres
of public land will be mined and problems associated with aesthetics, ecological disruptions, reclamation of mined lands,
and water pollution need to be dealt with before full public
acceptance of strip mining is forthcoming. The problem of re-
claiming lands following stripping needs research and field
testing in pilot areas located throughout the country, in all
climates - humid, subhumid, semiarid, arid, etc. Few effective
techniques for reclamation of lands affected by strip-mining
are available. Tests have been limited primarily to small plots
and effective large-scale land reclamation procedures are lacking.

Highways

Examples of incidents which have caused environmental
problems are widespread throughout the country. Areas of concern that have received most public attention in regard to
highway location and construction are impacts dealing with aesthetics, parks, and streams. In the highway construction process, it is difficult to separate aesthetic considerations from
erosion, they often are closely related. A stripped hillside
with steep cutslopes is not only judged aesthetically unpleasant by many, but is also a potential source of sediment pollution as a result of erosion.

Typically, the construction of a highway involves the
movement of vast quantities of earth. The area surrounding the
new highway grade is often stripped throughout the entire
length prior to earth moving, and extensive areas are denuded
of protective cover. Topsoil is not always stockpiled to be
reused later but often ends up in the fill. Rain falling on
unprotected soil causes far more soil erosion than rain falling
on protective vegetative cover.

The impacts of erosion can result in serious secondary impacts. King and Ball (1964) found that the amount of silt eroded from the area of highway construction was sufficient to fill many of the pools in the Red Cedar River in Michigan and reduce the depth of most others. They concluded that the result was permanent disruption of the fish-producing capabilities of this stream. They did not attempt to assess aesthetic impacts on the land resulting from erosion, but land impacts must have been serious in order to permanently harm the river to such a degree. This illustrates the type of damage which can occur during construction and presents a situation in which something else could have been done. Additional planning could have indicated protective measures necessary to temporarily impound or impede rapid runoff of rainstorms and would have greatly diminished the amount of silt reaching this stream. An assessment of the environmental impacts associated with this construction could have changed the location or design of the roadway if the problems had been clearly identified.

Extensive denudation of soil, construction of bridges and other structures, and channelization has resulted in the muddying of water for long periods of time. It is the public's concern for trout and salmon streams that has created direct confrontations between conservationists and roadbuilders. Stream trout populations depend to a major extent upon insect food which is produced on or in the stream bottom. The greatest production of insects occurs in gravel or rubble bottoms with sufficient space between the bottom particles for insects to move. When silt is added to such areas it tends to settle out in the interstices between the bottom stones and fill in the living areas. If sufficient silt is added to completely fill in these areas and to coat the bottom of the stream, the living space for insects is virtually eliminated and the trout population can be drastically affected.

Typically, trout and salmon bury their eggs in gravel streambed materials by scooping out a depression (called a redd) and depositing their eggs in this depression. The redd is subsequently covered by gravel and protected from light and predators. With large quantities of silt, the intra-gravel spaces within the redd can be clogged by silt dropping out of suspension and the movement of water through the gravel is impeded. Fish eggs require oxygen to carry out metabolic processes and a steady supply of water to wash away waste products given of by developing eggs. Impeding the flow of water through a redd prevents the eggs from obtaining sufficient oxygen supply to maintain life and/or sufficient intra-gravel flow to get rid of the accumulated wastes and the result is death to the eggs. The effects of silt on trout egg survival and fish numbers in a Montana stream were reported by Peters (1965; 1967). He pointed out the importance of keeping soil on the

land where it is useful and out of the water where it is harmful to a trout stream. A summary of the results of his studies
are presented in Table 5-1.

Table 5-1.   Numbers of fish and fish egg mortality in relation
to stream silt load.*

Stream section	Average suspended sediment concentration over two yr. period (ppm)	Highest sediment concentration (ppm)	Number of fish captured		Average trout egg mortality (%)
			Brown trout	Rough fish	
I	21	39	216	7	3
II	81	160	197	17	22
III	176	378	55	225	54
IV	240	754	8	1201	70
V	464	2280	6	378	47

*Eggs were eyed rainbow trout eggs introduced into gravel bot-
tom to determine survival rate.

The location and/or design of a highway can drastically
affect big game animals. Big game animals, especially mule deer
and elk, will move out of the high country into foothill areas
during the late fall and early winter to spend the winter in a
less severe environment.   In the spring, they migrate and disperse back into the high country.   Blocking or disruption of
migration routes by highways has drastically impacted big game
animals.   The location of a highway adjacent to or in a key
winter game range area is often catastrophic and could severely
deplete a herd.

The possible location of highways through parks or scenic
landscapes has resulted in a great deal of controversy.   Problems associated with noise, right-of-way requirements, adequate
protection of aesthetic values, etc., are especially difficult
to solve for parks and scenic areas. Local public input during
the earliest stage of planning of highway locations in or near
park and scenic areas can prevent rather serious confrontations
with the public.   Unfortunately, once a road has been located
and a design completed, changes are costly and not easy to make.

Dams

Typically, people think of water resource development projects as being dams and reservoirs.   But there are other features involved in water resource development besides dams and
their environmental impacts must be taken into account too.

Features such as pumping plants and aqueducts for irrigation projects and transmission lines and substations for hydropower dams have singificant impacts on the environment and must be considered. In some cases roads must be built into roadless areas for access to the damsite and/or reservoir area. Often rather extensive preliminary drilling and other exploratory work must take place to determine feasibility. Therefore, all features or elements of a water resource development project must be taken into account and analyzed for their environmental impact.

Impact assessment must be made not only for the freshwater aquatic environment, but also for the marine and terrestrial environments as well. The area of primary impact is often the freshwater environment, but significant secondary impacts occur to marine and terrestrial environments with water resource development. Water placed on the land for irrigation can "make the desert blossom." The change from a desert landscape to irrigated agriculture with crops is accompanied by other significant changes in the terrestrial environment. These changes affect the population of animals and often result in substantial shifts in abundance of certain animal species. The cactus wren is replaced by the ubiquitous sparrow; jack rabbit populations are decreased, cottontails become more abundant; etc.

As a result of water resource development in a coastal drainage, changes in inflow quantities or timing of inflows can seriously impact an estuary. Many species of shellfish and finfish have rather specific requirements tuned to the natural rhythms of the estuary, especially during the early stages of their life. Changes in natural regimes or rhythms in an estuary are often accompanied by changes in water quality. The trapping of organic matter in an upstream reservoir, the lowering or raising of water temperatures, the increase or decrease in the suspended or dissolved load can result in a decrease in productivity of the estuary. Ultimately, valuable natural resources in the estuary such as commercial and/or sport fisheries are seriously modified by upstream water resource development.

The direct impacts of the construction of a dam and the subsequent impacts of the inundation are well understood. A stretch of river (length in miles) is replaced by a reservoir covering an area of a certain dimension (number of acres). The size of the impoundment has a capacity of a given number of acre-feet. Certain types of lands will be inundated and are enumerated by size in acres. The quantity of material required for the dam is presented, e.g., the number of cubic yards of concrete or the volume of the earth embankment. The amount of shoreline that will remain in public ownership and the extent of development of public facilities is shown. The physical setting of the dam and reservoir is illustrated by using photographs or artistic drawings.

The effects on the native flora and fauna are not as well understood as the physical impacts, but certainly can be determined. The expected general effects of a new reservoir on biota are both direct and indirect. The former occur in the inundated area while the latter occur in the area adjacent to the reservoir. For example, if a reservoir is constructed on a river, the direct impact is fairly obvious. A certain amount of land and stream is inundated and a variety of preimpoundment habitats is replaced by a monotypic aquatic habitat. The net result of this environmental change is the preimpoundment biota must move, adjust, or perish. Those organisms that are able to move out of the impoundment area are forced to encroach on adjacent habitats. Survival becomes a question of competition and/or finding a suitable replacement habitat.

In the impoundment area, the resulting environmental change may offer increased habitat to certain species which find it suitable. However, a new species complex or community becomes established often quite different from the old. Over time, a new reservoir should become populated by representatives of the original fauna which adjusted to the new environment and a lentic fauna which invaded the reservoir in the post-impoundment period.

Outside the impoundment area, the changes in the adjacent ecosystem are not as dramatic, but are predictable. They are dependent upon land use management and downstream water operation. The downstream river ecosystem is modified by changes in waterflow, temperature, and dissolved solids concentrations, and silt load regimes. Once the balance is altered, the downstream flora and fauna will be impacted and new equilibrium states will emerge. Optimum flow patterns similar to natural regimes should be approximated to maintain beneficial existing populations.

In addition to the displacement of animals from the impoundment area of new reservoirs, there are migrational adjustments, decrease in range use and availability, and disturbance to herds or populations that must be considered. Change in use of adjacent land areas does result in impacts and must be analyzed.

The relative significance of the impacts of a water resource development project must be shown. A rather complete listing of the expected impacts associated with water resource development is found in Battelle's "Environmental Evaluation System for Water Resource Planning" (N. Dee, et al., 1972) and a paper on the "Environmental Impacts of Water Projects in California" (R.M. Hagen and E.B. Roberts, 1972). The former classification scheme is outlined in Chapter 11. However, no evaluation system is currently accepted by the majority of the environmental community that adequately assesses environmental impacts of water resource development projects.

Channelization is an outgrowth of public works activities which has come under attack by the environmental community in recent years. Streams and rivers are channelized by Government agencies in conjunction with highway construction, dam construction, etc., generally for flood control, erosion control, and navigational needs. Chapter 14-B contains examples of several stream reaches channelized by a highway.

Physical activities related to channelization that have harmful impacts on natural stream channels are: dredging, clearing and snagging, riprapping, diking, sand and gravel removal, sluicing, and dumping. Often certain of these activities are carried on simultaneously with channelization. For example, a channel may be first cleared and the snags removed, then dredged and rerouted, and finally its banks are armored with riprap.

In the channelization process, meander loops are cut off and the channel is rerouted. Often there is a significant loss of channel length. Peters and Alvord (1964) surveyed 13 Montana streams and rivers and reported a loss of 68 miles of channel in the 768 miles inventoried. The results of channelization are the stream gradient becomes steeper and average water velocity increases - and this in turn causes a drastic change in the flow regime.

Also, there is significant change in the shape of the channel with channelization. In cross section view, natural channels have tilted bottoms with deep water areas toward the concave bank and shallow water areas toward the convex bank. Following channelization, a channel has flatly sloped banks and a flat bottom with water at a uniform grade, which eliminates pool and riffle sequences associated with natural channels. Bars and islands are also removed, along with other natural materials occurring in the channel.

Typically, streambed and floodplain vegetation are removed to facilitate the channelization operation. The rate of revegetation depends on the stability of the new untreated channel banks or the extent of riprap used to armor new banks.

Major changes occur to reaches of natural channel above and below channelization work. In the process of adjusting to the increased gradient, there is a tendency for channels to scour upstream and deposit downstream from channelization projects. Often this process will continue over a number of years before a new equilibrium is reached.

Cutoff meanders are often left high and dry or are filled in and no longer are capable of sustaining aquatic life. Those cutoff meander loops remaining with water typically support invertebrate aquatic fauna found in lentic environments and are too shallow to support game fish species.

The composition of fish populations can change significantly in the new flumelike channels resulting from channelization. Game fish species are usually replaced by nongame species. Significant changes in fish population structure occur with fewer fish two years or older found in altered channels. Without exception, channelized streams show a reduction in both invertebrate and fish populations.

From an aesthetic viewpoint, a straight reach of altered channel with riprap on both banks represents an unnatural scene. Even though this channel may pass a high flow without erosion and function well transporting debris, it may be judged to be unacceptable by environmentalists. When channelization is necessary, the design of the new channel should consider fisheries and aesthetics. Figures 5-10 and 5-11 illustrate the design and planning which has gone into retaining stream meanders in conjunction with highway construction in the state of Montana.

THE PRESCRIPTION

No attempt has been made previously to include a neat prescription to prevent or minimize environmental degradation as a result of a particular type of public project. Rather, each activity was presented to point out the more sensitive.or key environmental problems associated with a type of public project. Nevertheless, certain general conclusions can be reached regarding an ecological prescription as follows:

1. The ecologist must be involved throughout the entire sequential process of a public project. He must be a member of the first team during preconstruction, construction, and operational stages in order to be most valuable.

2. Just as data are required to plan and design an engineering feature, so too are data required to provide an adequate environmental assessment of a public project. The cost of obtaining environmental data must be an integral part of all public projects.

3. The impacts of a project must be examined beyond specific project boundaries. Secondary impacts should be assessed as well as the primary or direct impacts.

4. Environmental guidelines should be used to prevent or minimize harm resulting from a public project. However, each ecosystem is unique as is each public project. Therefore, guidelines should be used as a means to an end not the end itself.

Figure 5-10.   Morris Weaver Meander, Clark Fork River,
Montana.   Two meanders were built by the road builders in the
Clark Fork River adjacent to Interstate 90 near Drummond in
western Montana.   They were built so that the river channel
would be as long after construction as before construction.
Each meander is a little over one-quarter mile in length. They
are named after the local landowners from whom the land was
purchased and are called the "Hazel Marsh" and the "Morris
Weaver" meander.   These loops are designed to approximate the
shape that nature itself builds. They have deep water areas on
the concave or cutting side of the meander. Note that the riprap has been limited to the cutting side and there is none on
the deposition or convex bank.   The old channels have been
filled in, topsoiled, and reseeded.

5. Monitoring systems are essential on projects with public
    controversy.   Regardless of how well project features are
    planned and designed, with the environment in mind, there
    is a difference between the conceptual and real world.   A
    construction specification or an operating criterion must
    be made to fit to a real situation. This can best be
    achieved through the monitoring and feedback mechanism. It
    gives credibility to expectations of the concerned public.

6. A public project is a highly complex entity, most often
    made up of many separate features.   Each feature must be
    examined as a unit for its environmental impact and then

Figure 5-11. Scale Model of Interstate 90. A scale model
was utilized to locate and design a stretch of Interstate 90
near St. Regis, Montana.   The model was used to place the road
in the narrow St. Regis Canyon in such a way as to minimize
disturbance to the river and the surrounding canyon walls.   The
existing primary highway will be left in place and provide easy
access for those desiring to use the river valley for recreation. An interdisciplinary mix of biologists, engineers, landscape architects, land managers, fishery biologists, etc., provided input.   Agencies represented included the Montana State
Highway Department, Montana Fish and Game Department, U.S. Forest Service, Federal Highway Administration, and Bureau of Sport
Fisheries and Wildlife.   The scale model was a useful planning
tool in maintaining the integrity of a highly scenic canyon and
at the same time providing a transportation facility.

the cumulative impact of all project features analyzed for
direct and indirect impacts. In other words, the new equilibrium states must be projected and compared to the old
for each alternative anticipated.

In conclusion, today's public projects must satisfy the
needs and values of all segments of the public. Public projects
must be planned, designed, built, and operated to minimize environmental damage or wherever possible enhance the environment.
As a member of a team, the ecologist can make a good public
project better.   Properly used he can provide another level of

input into the public project process that will help the decisionmaker make the best decision.

GLOSSARY

Anoxic   - insufficient oxygen for animal respiration.

Biosphere   - that portion of the earth which contains liv-
                      ing organisms, thus in which ecosystems oper-
                      ate.

Epilimnion   - uppermost of three strata in those lakes which
                      stratify thermally.

Euphotic Zone   - portion of lake receiving sufficient light for
                      photosynthesis.

Euryphagous   - broad diet, organism consumes a wide variety
                      of species.

Hypolimnion   - lowermost of three strata in those lakes which
                      stratify thermally.

Invertebrates   - animals without internal skeletons or back-
                      bones, hence without vertebrae. Includes such
                      forms as molluscs, worms and crustraceans as
                      well as insects.

Monophagous   - one-food species, organism's food is limited
                      to a single species.

Organic Matter - derived from living tissue, plant or animal -
                      detritus refers to fragments or organic matter.

Raptors   - birds of prey, eagles, hawks, owls, etc.

Stenophagous   - limited diet, organism consumes a limited va-
                      riety of species.

Thermocline   - also Metalimnion, middle layer or stratum in a
                      stratified lake, where the temperature change
                      per unit of depth is greater than in either of
                      the other two strata.

REFERENCES

G. L. Clark, *Elements of Ecology* (New York: John Wiley and
Sons, Inc., 1966), 560 p.

N. Dee, J. K. Baker, N. L. Drobny, K. M. Dulce, and D. L. Fahringer, *Environmental Evaluation System for Water Resource Planning* (Battelle-Columbus Laboratories, Columbus, Ohio, 1972),

C. A. Elton, *The Ecology of Invasions by Animals and Plants* (London: Methuen and Co., Ltd., 1958), pp. 145-150.

R. M. Hagan and E. B. Roberts, "Ecological impacts of water projects in California," *Journal of the Irrigation and Drainage Division*, ASCE, Vol. 98, March 1972, pp. 25-48.

D. L. King and R. C. Ball, "The influence of highway construction on a stream," *Research Report No. 19*, Michigan State University Agricultural Experiment Station, 1964, p. 1-4.

R. MacArthur, "Fluctuations of animal populations, and a measure of community stability," *Ecology*, Vol. 36, 1955, pp. 533-536.

E. P. Odum, Fundamentals of Ecology, (Philadelphia: Saunders, 1953).

J. C. Peters, "The effects of stream sedimentation on trout embryo survival," *in Biological Problems in Water Pollution*, Third Seminar, 1962, U.S. Dept. Health, Education and Welfare, Public Health Service Pub. 999-WP-25, pp. 275-279.

J. C. Peters, "Effects on a trout stream of sediments from agricultural practices," *Journal of Wildlife Mgt.*, Vol. 31, No. 4, Oct. 1967, pp. 805-812.

J. C. Peters and W. Alvord, "Man-made channel alterations in thirteen Montana streams and rivers," *Trans. N. Amer. Wildl. Conf.*, Vol. 29, 1964, pp. 93-102.

K. E. F. Watt, *Ecology and Resource Management*, (New York: McGraw-Hill, 1968) 450 p.

Western Systems Coordinating Council, *Environmental Guidelines* (Chairman Environmental Committee, So. Calif. Edison Co., P. O. Box 351, Los Angeles, 1971), 96 p.

AUTHOR NOTES

This chapter was a collaborative effort between Dr. William T. Helm, Associate Professor of Wildlife Science, Utah State University, and Dr. John C. Peters, Environmental Specialist, U.S. Department of the Interior, Bureau of Reclamation. Both are fisheries biologists by discipline background.

At Utah State, Dr. Helm has taught courses in fishery science and aquatic and general ecology, and has been Director of the Bear Lake Biological Laboratory. Recent projects that he has been actively involved in include studies on energy flow and habitat and the effects on populations of stream insects and fish, effects of shoreline developments on certain aquatic populations in lakes, and an evaluation of the impact of the nuclear testing program on the freshwater ecosystem of Amchitka Island, Alaska. Also he has worked together with a variety of other professions in consulting work, and in short courses and seminars related to public project developments.

Dr. Peters has been with the Bureau of Reclamation at the Engineering and Research Center in Denver since 1971 where he administers an environmental clearinghouse program and acts as an environmental advisor for the Bureau's planning, research, operation and maintenance, design, and construction activities. In these roles he has been involved in ecological aspects of design decisions, court cases, review and preparation of environmental impact statements, and in formulating environmental policies for all Bureau activities.

Before joining the Bureau, Dr. Peters was with the Montana Fish and Game Department for 12 years. He was instrumental in 1963 in the passage of Montana's Stream Preservation Law. This law, the first of its kind in the United States, provided for the submission of plans for construction and hydraulic projects affecting streams. He administered this law from 1963 to 1971 for the Fish and Game Department, working closely with the Montana Highway Department.

The development of this chapter was influenced also by the insightful commentary of Mr. Jack W. Ward, Applied Construction Research Associates, Phoenix, Arizona, during the Vail Conference in 1972. Mr. Ward, formerly Chief Engineer, Alabama State Highway Department, critiqued the chapter in both overall direction and on specific content from the point of view of a construction engineer faced with the reality of building.

6

THE HUMAN COMMUNITY

BY EVAN VLACHOS

Many parts of the present treatise have explored in depth various aspects of the surrounding environment within which public projects take place. Throughout all discussions and analyses continuous reference is made to the "human factor" or "social aspects" invariably associated with the design, construction, and maintenance of major projects. Yet, despite the general concern and continuous preoccupation with this general category of the surrounding social environment, the topic lacks specificity and concreteness usually found in analyses of physical dimensions of the environment.

The purpose of the following presentation is not to provide further general discussion of the environmental ethic or to offer social criticism concerning the potential negative effects of technological interventions on society. Rather, it draws attention to efforts of describing in more concrete terms the dimensions of "human community," the component parts of the "social system" and the strategies and tactics involved in comprehensive planning and evaluation of the effects and consequences of public projects.

Given the public project as the stimulus for an analysis of the surrounding human community, the present chapter attempts to synthesize existing knowledge around three major themes and concerns:

1. Provide a description and analysis of the social system and of the "human community." This implies the development of a vocabulary of major concepts, basic definitions, and generalizations concerning the essential parameters of an integrated social system.

241

2. Explicate in a more specific manner the social domain of real or assumed effects and consequences of public projects. Specifically, we need to classify the impacts on various environments characterizing the total physical and social world, and delineate some general principles of social planning.

3. Develop working procedures and synthesize methodological considerations in the analysis and "measurement" of the effects (impacts and consequences) of public projects on social systems.

The connecting argument in the above areas is essentially the understanding of how various social or human dimensions of any community act as constraints or facilitators in absorbing, accommodating, or even rejecting technological change.

HUMAN COMMUNITY AND SOCIAL SYSTEM

Basic Vocabulary and Essential Distinctions

Both *community* and *society*, despite their centrality and importance, are among the more ambiguous concepts in sociology. Sometimes they are used interchangeably with a reference to a group of people living together and sharing a common culture. It is important, however, to keep them distinct by reserving the term *society* for the most inclusive, complex, and self-sufficient type of social grouping; *community* on the other hand, places particular emphasis on a concrete collectivity, usually in relation to a spatial dimension, (with its concreteness being more or less a semantic convention). In addition, the term "human" underlines primarily man's adjustment to other men living in a common habitat, while the term "social" would indicate the more abstract relations and mutual awareness and expectations in collectivities.

Sociologists have been using the term *community* in essentially three ways. First, it is a synonym to reflect various institutional establishments and organized groups. Secondly, *community* has been often used to refer to a cultural or spiritual concept of belongingness, as an illusive philosophical concept to reflect a sense of identify and unity with one's group and a feeling of involvement and wholeness on the part of the individual. Thirdly, *community* may be understood as a composite of social and territorial organization whose population shares common ties.

Given these terminological variations (and the underlying theoretical differences), there is no widespread agreement as how to describe community as a sociological entity. Furthermore, although such multiple uses of the term community may be

unavoidable and part of the elusive character of social phenomena, nonetheless they make things difficult for those who seek
to study communities as a distinct form of social organization.
The problem is accentuated by the overlapping between such concepts as "community," "social structure," "social system," "society," and "social organization." We may avoid for the time
being further semantic confusion, by indicating some agreed
upon elements of community:

> *People:* a demographic base
>
> *Place:* a given geographic area
>
> *Identity:* feeling of belongingness
>
> *Common culture:* sharing of knowledge, beliefs, customs,
> laws, etc.

As a result of these four major elements there emerge recurrent
patterns of behavior which provide the basis for the delineation of a social system. In adopting throughout the present
chapter the term "social environment" we may be closer to some
widely accepted definitions of human community as involving
three key variables: a territorial variable (physical environment); a sociological variable (social interaction and organizational and institutional networks), and a cultural variable
(common ties and normative system). If we also introduce as an
additional dimension in the understanding of human community
its temporal character (the rhythm, timing, and diachronic
character) we can have a more dynamic expression of a given
social system.

A particular social system implies a collection of people,
devices, and procedures intended to perform some function. A
social system model is a working model of a social organization
which is capable of achieving a goal and involves the systematic exploration, analysis and evaluation of all the possible
consequences of proposed alternatives to an on-going set of interdependent role components surrounded by a boundary of social
norms.

Rather than using such a formidable vocabulary, we may
simplify things when we observe that social system models can
be analyzed as to the following elements:

1. The *level* of the system, ranging from a micro to a macro
    scale.

2. The *structure* of the model, i.e., how are the elements of
    the system arranged relative to each other.

3. The *integrative* mechanisms peculiar to the model or the
    "glue" that holds the system together, such as shared goals,
    shared values, common resources and reciprocity.

We need to elaborate first of all the point about the
level of the system, since this notion will guide the presentation of the component variables of a social system or a community.   In defining level we refer not only to size but also to
the quality of interdependency and interaction.   Basic as size
and structure may be, there are other important variations in
social organization, such as the degree of formality, the extent of control, the span or endurance of the system, which
will permit us to distinguish in the continuum of increasing
level among three major types of environments (or levels of the
system):

1. *Micro-environments* which have as basic units the individ-
    ual and the family as well as the immediate work place.
    Such micro-environments are the basic units of the society
    within which personal contacts are maximized, where infor-
    mal control prevails, and individual problems of values are
    faced.

2. *Meso-environments*. Such environments involve larger social
    units, such as a given spatial community (including varying
    sizes of residential units), social aggregates, and a number
    of institutions which vary from small units to larger local
    organizations.   These meso-environments, which indicate an
    intermediate scale of personal contact and larger social
    considerations, refer to groups of individuals who have
    common interests or needs and who are mostly bound together
    by common rules of conduct. Particularly important to such
    a discussion is the term "community" which, as it has been
    said previously, need not be restricted to a physical en-
    vironment or a territory but it may be expanded to incor-
    porate also aspatial aggregations.   If a "community" in-
    volves very "large" segments of the population it may be
    part of the next wider category of "macro-environment."

3. *Macro-environments*. These involve ultra-large systems, such
    as metropolis, or larger urban agglomerations, or the total
    society within national boundaries.   In other words, the
    entire interaction of spatial, cultural, and institutional
    configurations. Such ultra-large systems encompass also
    regional constellations, international relationships and in
    a final analysis a global orientation towards the entire
    human community on the face of earth.

Implicit in such differentiations is also a recognition of
a scale of requirements for planning and evaluation. There are
different design and methodological requirements when we plan
for a small by-pass versus a trans-mountain highway system, or

when we evaluate impact of a small irrigation project as contrasted to an inter-basin exchange.

A general systems approach may be useful not so much for reasons of providing a specific methodology, but more as an organizing scheme of core ideas and key variables and as an illustration of the potentialities involved in improving the study of complex social systems. If nothing else, systems theory is a vehicle for making meaning out of chaos and a means for showing the significance of various subsystems to the whole. Since there is a tendency in systems analysis to concentrate on modeling and problem solving with less emphasis to problems of conceptualization and verification, there is an urgent need for a clarification of the assumptions characterizing aspects of social life and the limitations of various qualitative dimensions of social phenomena.

In trying to organize our data and review in some meaningful way the parameters involved in any social system we need to answer the following major questions involved in any system design.

1. What are the major variables involved in an examination of a social system?

2. Can these variables be classified in any meaningful way?

3. What are the major types of relationships that ought to be investigated?

A systematic way of describing human or social environments is through the utilization of a dynamic approach exemplified in systems analysis. Such an approach has to be considered in the context of the following requirements:

1. Demarcation of the relevant objectives of goals as well as alternatives to be achieved in a given system (system outputs).

2. Description of the "system," that is delineation of the boundaries of the particular environment or "community," as well as system interfaces.

3. Delineation of constraints of the system or the inputs which tend to characterize the dynamic aspects of a particular social environment (resources).

4. Consideration of time constraints, another expression of the dynamic aspects of the system especially as they refer to short- versus long-range objectives.

5. Elaboration of feedback mechanisms, or the arrangements
    found in all stable systems for controlling the input rate
    as a function of the output. Feedback loops provide a
    "community" with the ability to survive within larger sys-
    tems than themselves, with respect to both input resources
    and output demands.

6. Specification of techniques for systems analysis and of the
    proper methodology for being able not only to describe but
    also explicate both technological and non-technological as-
    sumptions of our system.

7. Evaluation of the performance of the system, especially in
    the juxtaposition between efficient and effective alterna-
    tives, i.e., towards a search for economically viable and/or
    socially desirable alternatives.

The analysis of social systems in such a manner has major
advantages in terms of the clarification of concepts, the recognition of interdependencies, and, therefore, in a clearer
characterization of inputs, system boundary and output. The
last one seems to be one of the major difficulties which comes
forward when we discuss the impact of public projects in a given
society, i.e., the definition of goals and the assessment and
allocation of resources within a framework of priorities of the
overall system. We are looking at a modern society, whose
overall goals most often are not well defined. It is also a
society, whose various closely interrelated sub-systems provide
a vast array of multiple and often conflicting subgoals. To cap
this situation we also have a technology, the by-products of
which many times precipitate major social and political crises.

The analysis of social systems and the attempt for a systematic synthesis of perceiving the impacts or spillovers of
technological perturbations give rise to quite a number of pitfalls. These include the gap between real system and theoretical formulation; the confusion between scientific knowledge and
value convictions and/or assumptions; and, especially, the
omission of important input and output components as well as
system's parameters. All of these, especially "hard" scientists, consider research into social systems as an "art form"
rather than legitimate scientific inquire. Males and Gates, for
example, have written cogently on the nature of the beast when
they raise a number of points concerning the non-linear, "organic" approach of the social sciences, the ethical stance demanded of the researcher, and the general differences in modes
of thinking and communicating. They conclude with both a per-
ceptive distinction in environmental management thinking and
with a hopeful note when they write:

    In general, research in social systems relating to en-
    vironmental management falls into two main schools of

thought, the normative (prescriptive) and the descriptive.
The normative school of thought concerns itself with the
manner in which things *should* function, and tends to accept the formal statements as real, e.g., an organization
chart describes the organization. The descriptive school,
on the other hand, is more concerned with defining the
system as it *actually* functions; the normative aspects are
of concern insofar as they represent constraints, goals,
or hypocrisies of the real system. It is the opinion of
the researchers that a normative approach has little value
in examining social systems relating to environmental management; it is, however, quite common... . If the problems
of social systems in environmental management are pursued
sufficiently long and diligently, it is inevitable that
social systems research will cease to be an "art form";
however, to act as though this is true at present is not
only misleading but erroneous.*

Without adding to the cogent presentation of Chapter 10
and the thorough discussion of systems analysis, we can only
emphasize here that a social system can be part of the same
methodological considerations which delineate the interaction
and interdependence of all environmental systems and their components. The ultimate goal of such operations and analysis is
to minimize, if not control, environmental "crises," technological perturbations, and to upgrade the quality of life. Perhaps, rather than providing here another long discussion and
obtuse definitions as to what each part of the social system
entails, we may use as an example the construction of a dam and
its relation to a surrounding community.

The external environment, i.e., the ecological configurations and the conditions of an existing social structure provide the constraints (and the facilitators) or the *inputs* characterizing our overall system. Such considerations include
natural resources (available water, land configurations, etc.),
demographic characteristics (total population, age, sex composition, vital rates, etc.), normative and legal constraints
(values of the community, prior appropriation doctrine, etc.),
the economic viability and potentialities of the area, the administrative apparatus (sources of local government, degree of
autonomy, etc.), state of technology (degree of automation,
technological innovation, etc.).

In building a dam and with consideration to existing inputs
(as well as additional inputs from other environments, such as
state, regional and federal policies), we need then to consider

*Richard M. Males and William E. Gates, *Decision Processes in
Water Quality Management.* Oakland, California, Engineering-
Science, Inc., 1971, p. VI-2.

the processing mechanisms, or the concrete organizational *system* devised for maximizing desired goals. System or thruput involves physical structures (buildings, canals, etc.), and organizational infrastructure, such as rules of operation, patterns of leadership and command, efforts for control, integration, information, and communication and ways of interacting with other organizational environments.

The desired goals or objectives exemplified in the construction of a dam denote the *outputs* of the system and revolve around a variety of goods and services for individuals, communities, and regions. Recently, in addition to such obvious objectives as increased productivity, water supply, wealth, land improvements, local benefit, security, etc., increased mention is made of such broad goals as enhancement of the quality of life, social well-being, aesthetic satisfaction, regional development environmental quality, and others, which provide us with infinitely more difficult problems of evaluating the performance and effectiveness of a new public project.

Finally, the construction of the dam of our example may generate through *feedback*, additional inputs. Achieved objectives may provide, then, the impetus for new capital investments, increased population, and new attitudes or values vis-a-vis the changing environment (such as let us say renew confidence on the ability of the surrounding community for survival, influx of new citizens and cosmopolitanism, effect of tourism on the region, etc.).

The crucial point in the discussion and example provided above is that the construction of a dam is closely associated with all major components of a given social system: with its inputs (as expressions of the conditions of the external environment); with the organizational arrangements devised to meet performance requirements (thruput or "system"); and, of course, with desired objectives or goals and larger social policy (explicit or implicit as well as overt or covert reasons for building a dam in a given locality). In such a broad conceptualization, the affected human community perceived as an integrated system can help us, through a process of decomposition to provide pertinent variables and their more specific dimensions. To do so, however, it is also imperative to introduce the realm of public projects and their interrelationship and effects on the surrounding human communities.

PUBLIC PROJECTS AND HUMAN COMMUNITIES INTERFACE

Public Projects and the Technological Imperative

The central preoccupation of the present volume is the role, presence, and effects of public projects. Broadly stated, public projects refer to large-scale physical facilities, such as dams, highways, airports, buildings, power plants, etc. It is difficult indeed to provide both an adequate definition of what is meant by "public" and "project" and an appropriate classification scheme for orderly presentation. The introduction of this treatise and the Chapter on "Project Evolvement" have raised a host of issues connected with the nature and understanding of public projects. The reader is also advised to refer to Chapter 3 in order to see the range of public projects in the general areas of transportation, water, and power. Essentially, throughout this treatise continuous reference is made to a macro-scale infrastructure serving the utilitarian needs of society, an infrastructure that provides the means for the use of resources, for mobility and transportation, and for water and power generation and conveyance. Yet, even under such broad categorizations some areas of public projects have been more or less excluded, especially urban planning and its related activities. There is perhaps an underlying bias in both analysis and examples towards an open-space preoccupation and a non-metropolitan orientation. This is not difficult to understand when we observe that in the past planning was heavily committed to urban millieu, and that current environmental concern has drawn more attention to megastructures and physical facilities whose presence is assumed to have adverse environmental effects or major economic, ecological, or social consequences.*

Given the magnitude, intensity, and varying degrees of presence of various public projects it has always been difficult to describe, categorize, or otherwise classify and in a meaningful way evaluate the perturbations from such large-scale facilities. The principal concern in the past was with what one may describe as first order consequences, or immediate (primarily economic) benefits to be gained by developing and widely applying innovative technologies and large-scale facilities. As a matter of fact in pre-technological societies there was little, if any, concern with higher order and unanticipated consequences since most of the changes in the environment were incremental and gradual with few products introduced at a time. Today, however, in a highly technological society all conditions of the past have been demolished because of the unprecedented rate of change, visible negative effects, and formerly unrecognized harmful consequences of a seemingly good technology.

---

*After all, as of June 1, 1972 environmental impact statements filed in accordance to the provisions of the NEPA of 1970 came from two agencies, the Department of Transportation and the Army Corps of Engineers. Using the classification provided by the Environmental Council 49 percent of the reports deal with roads, 14 percent with watershed protection and flood control, 8 percent with airports, 6 percent with navigation, and 5 percent with power generation of all types.

If we use as a definition of technology Ellul's characterization of "the ensemble of practices by which one uses available resources in order to achieve certain valued ends," we may
ask for a minute why indeed have we become so preoccupied or
even "up-tight" over technology. The National Environmental
Policy Act may be used as an example and important document
exemplifying the shift and concern over an appropriate and
careful consideration of all environmental aspects of proposed
technological actions.   Such an act was made possible by the
facts that the negative effects of natural perturbations became
more visible, the concern over the high momentum and the fast
pace of technological changes has become widespread, and a pervasive feeling that each member of the public has been only a
secondary part to every decision on the exploitation of technology. At the same time, the present market forces did not seem
to be satisfactory in allocating secondary costs, and the legal
system seems to be relatively ineffective in coping with technological damages. The 70's, following the environmental fervor
of the late 60's, are shaping major environmental legislation
(as discussed in Chapter 8) and even more important, natural
resources policies which will help the quest towards comprehensive planning. Legal doctrines, administrative regulations, and
emerging policies are all part of a necessary reconsideration
of the prevailing notion of quite a few engineering designs
which were based upon the assumption of closed systems fully
described by the technological inner constraints and marked by
clear boundaries decoupling the systems from other parts of the
environment.

Public projects, therefore, are types of environmental intervention and change whose effects may be classified either as
impacts (i.e., short-range, primary) and consequences (i.e.,
long-range, higher order, intangible).   The overall scheme of
classifying in some elementary form the effects or outcomes
from the presence of a public project is summarized in Figure
6-1.   Knowing that any public project is an intervention or a
perturbatory event of ambient conditions, our challenge is of
describing:

1. How to balance in an equitable manner costs and benefits in
    the alteration of the environment.

2. How to make the appropriate change and transition to new
    states without unacceptable disruption to all systems.

3. How to measure in a valid, reliable, refined, comparable,
    and relevant manner project effects and, therefore, provide
    guidance for the previous two points.

If public projects involve the commitment of significant
amounts of resources, in addition to ecological perturbations
we need to see closer the positive or negative consequences on

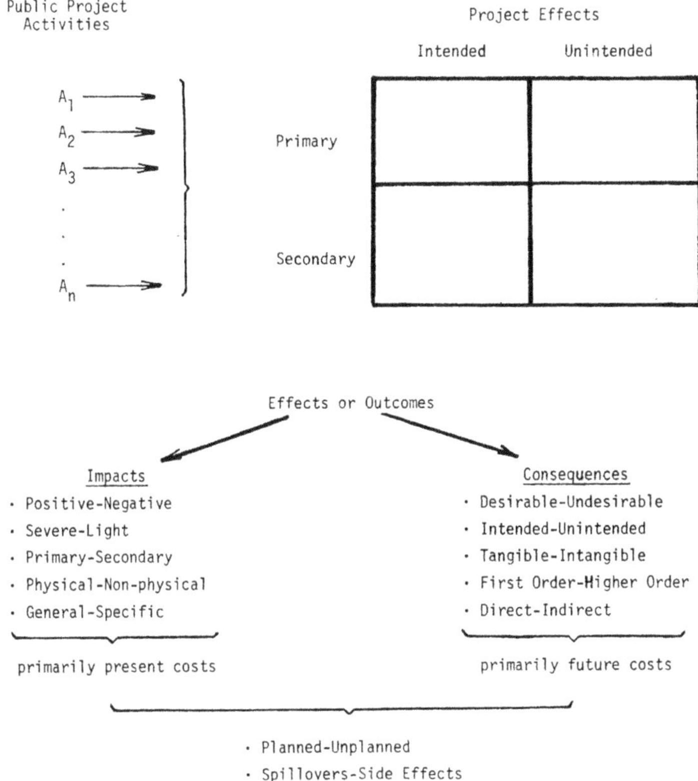

Figure 6-1.   Tentative classification of project impacts and consequences.

the social structure of given communities.   In this respect, then, we may raise some crucial questions: is the discussion of public projects directed towards an argumentation and evaluation of impact statements? Or, does such an activity require introduction of general principles of planning, especially of the mechanisms and epistemology of social planning? Or finally, does the discussion center around public projects as instruments and expressions of larger social policies, which would then guide our thinking towards a determination of social needs involving construction or non-construction options?

251

In trying to develop the interface between public projects and human community, we must turn attention to an environmental design process which incorporates both physical and non-physical dimensions. Given the variety and diversified scope of public projects, how does one systematically explore and account for the potential effect of such physical structures on the social environment? There is a sequence of questions that must be raised before we determine the basic variables of the human community that must be taken into account when a public project is contemplated, designed, or implemented. Four major phases characterize this sequence of questions:

    a. Conceptualization
    b. Design
    c. Evaluation
    d. Implementation

This particular sequence of reasoning is no different from all efforts of comprehensive planning and it will be later presented in detail in the final synthesis of the present chapter. What is emphasized here is the observation that a project has to be seen not as an isolated task but in terms of its relationship to various subsystems (human/social, physical) and as a continuous geographical and temporal inducement transcending immediate project boundaries. In addition, although a project may be very small in scale, it may be ultimately great through cumulative effects. In summary, the sequence of questions and the typical requirements of public project and human community interface unfold as follows:

1. What are the affected *environments?*
    which contribute to

2. Elaboration of *facilitators or constraints*
    incorporated in attempts of

3. *Comprehensive planning*
    which involves description and

4. Evaluation of potential or probable *effects*
    leading to questions of articulation and implementation of

5. General or specific *policies*
    aiming at the meeting of

6. Regional or national objectives or *goals.*

So far, almost exclusive attention has been paid to questions of public projects impacts rather than explicate the orderly and reasoned approach to the complex phenomena associated

with total environmental design. To cite a general example, the
typical requirements of impact statements referring to social
considerations for urban highways cluster around responses to:

1. *Description of the proposed project* and of ambient conditions, such as residential and neighborhood character, number of trips per day for the design year, etc.

2. *Probable impact of the proposed project*, such as impact on
   employment, impact on property values, division of existing
   uses, displacement of families and businesses, increased
   urban congestion, implications for population distribution,
   disruption of orderly planned development, impact on aesthetics or effect on natural and historic landmarks, etc.

3. *Probable adverse environmental effects* which cannot be
   avoided should the proposed project be implemented, such as
   noise generated, and the efforts to reduce effects.

4. *Range of alternatives*, with an estimation of probable beneficial and/or adverse effects of each alternative, estimated social costs, consideration of "do-nothing" alternative, or mass transit.

5. *Relationship between local short-term uses and long-term
   enhancement* and productivity, such as impact on changes in
   traffic patterns, compatibility of land uses and development configurations, etc.

6. Estimation of *any irreversible and irretrievable commitments of resources* involved, in case the proposed project
   will be implemented, such as blocking of future residential
   development in the area, or any other non-rescindable effect of a highway.

7. *Measures taken to minimize impact or mitigate adverse effects*, such as proper rehousing of people, operation of facilities during construction phase, generation of new
   transportation facilities, etc.

The examples offered above are only illustrative of the
types of concern appearing in typical environmental impact
statements. They are used here as a catalyst for our thinking
and as a beginning point for the eventual delineation of specific conditions when public projects are introduced in a given
social system.

More concretely, then, since we recognize the general need
for a more comprehensive sociological presence at all phases of
a public project, we must now generate specific lists of pertinent "social" variables. Of course, such variables will vary
from situation to situation and from time to time, given also

the type, intensity, and extent of the public project. What we
can propose, however, is the introduction of a *Generalized Protocol* (GP) containing elementary units and basic variables involved in most environmental designs with particular reference
to human community considerations.   The categories that follow
can be divided into three main clusters of concern: structural
parameters, value and attitudinal parameters, and processual
mechanisms.   These categories are only indicative of the proposed strategy and not all-inclusive of the sociological considerations in any public project:

Structural Parameters

These are the basic variables within a spatial-temporal
location and indicate the essential arena within which social
interaction takes place.   They are minimum variables and their
dimensions that we need to consider as important and as basic
for calculating effects and with differing degrees of intensity
on micro-, meso-, and macro-environments. These basic variables
can be clustered in the following categories:

1. Population characteristics and composition. The demographic
    characteristics are the obvious factors that must be taken
    into account, since they provide the immediate elements in
    the operation of any human community. Thus, the total number of people involved, density, the component characteristics (biological, ethnic, and cultural), vital rates and
    patterns of mobility, are among the first units to be taken
    into account in considering potential effects of any public
    project.   This implies that we need to know not only the
    number and composition of people affected, but infer from
    the given demographic profile the ability of any human community to be able to absorb or accommodate shocks induced
    by a public project inducement.

2. Spatial distribution.   This second major variable reflects
    the concern that not all population is evenly distributed
    over the geographic area where a public project takes place.
    Therefore, in public projects the spatial arrangement and
    the geographical orientation of the population is a vital
    variable in a successful combination of minimization of
    costs and maximization of social benefits.   For example,
    roads can be constructed in such a way as to provide minimal disturbance between and within existing communities,
    particularly by not bisecting existing neighborhoods, or
    other identifiable and closely-knit human aggregations.
    Severe disruptions of the physical framework may create
    hardship for either an entire community or diversified
    groups within the community.

3. Socio-economic characteristics. This basic structural par-
    ameter exemplifies the importance of the background charac-
    teristics of any impacted population. Communities as a
    whole, as well as parts of the community tend to be charac-
    terized by given populations which have certain social,
    ethnic, educational, occupational, religious, or economic
    backgrounds. Therefore, a public project such as let us say
    the construction of a dam may have differential effects if
    a community is mainly composed of retired people, as con-
    trasted to one primarily characterized by active farmers.
    The impact between these such categories of people, as for
    example among elderly retired persons and a rather youthful
    university town differs not only in the severity of econom-
    ic dislocation, but also by the ability of the affected
    population to absorb the shock of social adjustment from a
    public project that may alter or change the flow of commu-
    nity life.

4. Community organization and institutional networks. This
    general category of variables includes the whole network of
    specific social organizational arrangements in a given com-
    munity and reflects our concern with an understanding of
    the types of specific social relationships and institutions
    within a given social structure. The degree of integration
    or conflict, social power distribution, housing conditions,
    types of recreational facilities, public health and medical
    care factors, existing types and patterns of voluntary and
    non-voluntary associations are all important dimensions in
    the examination of a community social structure and its re-
    lationship to a public project construction. Some communi-
    ties have better organized facilities and institutional
    arrangements that seem to be able to handle more effective-
    ly the presence of a given public project; others, lack both
    the organizational background as well as the institutional
    structure (be it financial, educational or political) to be
    able to absorb the presence and consequences of a proposed
    public project. The construction of a power plant, for ex-
    ample, could serve to indicate the significance of communi-
    ty organization. Many times in various communities there
    do exist certain economic interest groups which seem to re-
    quire the presence of a power plant for the fulfillment of
    industrial purposes. At the same time, the welfare of the
    community may require the location of the plant at a place
    rather remote from the center of the community. To the ex-
    tent that dialogue is possible, that organizations exist,
    that intergroup relations have not been polarized, and that
    the overall community system is organized in such a way as
    to be able to carry out a cogent discussion on the impact
    of a proposed power plant, we may be able to generate a
    wider range of alternatives and, thus minimize potentially
    negative consequences of the proposed public project.

5. Social aggregates. In any human community not only do we
    have organizations and groups but there also exist, among
    others, such social aggregates as social class and ethnic
    or religious groups. We need to know in any social envi-
    ronment the presence and extent of such social aggregates
    which may be differentially affected by the introduction of
    a public project. For example, in the construction of a dam
    in a given locality we may have old historical landmarks
    reflecting the past of an Indian tribe and therefore it
    would be of great damage to violate sacred grounds of such
    a group. Social class, on the other hand, should be con-
    sidered in any public project, as for example, in the con-
    struction of highway when a by-pass may be constructed in
    such a way as to pass through the middle of a definable low
    income community. Often, in such cases the preoccupation
    with minimum economic cost provides little consideration of
    more severe and invisible costs of social dislocation and
    the destriction of a viable neighborhood.

All in all, communities as a whole, or parts of a community, are primarily characterized by populations of different social, educational, cultural, occupational, or institutional
composition. Therefore, a public project has also differential
potential for success and variegated impacts given the different socio-economic and cultural background of the surrounding
community.

Value and Attitudinal Parameters

These variables may be also labeled "valuative" variables
and reflect the attitudes and values held by people towards environmental conditions. In other words, while previously the
structural parameters reflect our concern with positions and
roles in the social structure and with the recurring patterns
of social interaction, the present category involves behavioral
aspects of perceptual characteristics shared by the members of
a social system. Such parameters include the intangible aspects
of any social system and they may be classified under three
major sub-categories of variables.

1. Social and cultural resources. These include the normative
    system, or the attitudes, values and ideological orienta-
    tions and goals as well as historical antecedents and cul-
    tural practices of a local population in a given community.
    The general community culture, the degree or extent of com-
    munal concensus, and the patterns of normative behavior are
    among the conditions to be considered when estimating the
    potential effects of a public project. In this category we
    need to develop all the types of communal and larger socie-
    tal values which may be effected by the introduction of a
    public project. A most poignant example would be the con-
    siderations that need to be taken into account if a public

project is to be introduced, let us say, in an Amish community or consider the construction of public projects in places where historical events are associated with certain sights, objects or structures. Such values may also include not only items related to religions and cultures but also certain cultural events and persons, architectural styles, and even such general considerations as "historical era" associated with certain parts of the country and the local community. One may use as an example here the construction of an airport that threatens the preservation of local traditions, the solidarity of the surrounding populations, and increased factionalization by pitting economic interests against a valued area envisaged as a major park.

2. Aesthetic considerations. This is a general but difficult category of variables which includes all factors involved in the creation of a total aesthetic effect in a given locality, the atmosphere of balance, and a sense of integration with nature. Here, more than anywhere else, personal opinion seems to be a determining factor for the evaluation of aesthetic effects of a given project. These are the categories of variables that evolve mostly around a measurement of individual and group responses towards an elusive quality of a pleasing environment in a given geographic area. Although individuals vary significantly to the external characteristics of the surrounding world (what is pleasing to one may be utter boredom to another) the sensitive recognition of a balanced, harmonious and aesthetically pleasing design, the attractivity of a site or area together with a recognition of other parameters of the social system, are becoming of fundamental importance for a well-integrated community and society.

3. Attitudes towards "environments." The last category of what we may now also declare as subjective or valuative parameters reflects the concern and reaction of individuals and collectivities towards the surrounding various environments. Such attitudes or predispositions spring from a framework of judgmental evaluation of the world around us. In considering the impact of a public project, such as for example the construction of a new airport, we need to solicit the opinions of the population involved in the immediate vicinity concerning their reaction to a new environment, to proposed alterations in the existing social environment, and their understanding of possible effects for both themselves and their community. Since each judgmental reaction will differ from group to group and from a social class to another, they will also act as additional sensitizing indicators of the potential effect of a public project for segments of a local population.

A generalized protocol of minimum social parameters should also involve a consideration of the mechanisms, procedures and processes for introducing and maintaining a public project in a given community. This category of minimal considerations of social aspects in the construction of a public project is of a different order from the previous two clusters. The previous structural and attitudinal parameters can be considered as delineating conditions, or as essential facilitators. The variables under these two categories may be used as "indicators" or crude evaluative signposts of the potential effects of a new public project. Crosscutting these two general parameters are the procedural requirements involved in the conception, maintenance and completion of a public project. Potential effects and consequences of a new public project can be alleviated or exacerbated by strong or weak procedural mechanisms and organizational arrangements, by the different networks of communication, and, finally, by established patterns of intergroup relations. Broadly, these processual mechanisms involve the following major categories:

1. Mechanisms of support and coordination, including the political networks and existing administrative apparatus. Any public project becomes part of larger competing thrusts which tend to emphasize diversified goals, such as preservation, development, and reclamation. Such general elements influence significantly the policy making and administrative agencies in a given community or society. They can be better understood under the concrete dimensions of administrative structure and organizational functioning. This simply means that if, for example, we are to introduce a new reclamation project close to an existing community, we may consider the differential effects of the project if the water agency is public or private, the extent and coordination of its legal and administrative forms, and the organizational preparedness as expressed in the form of available personnel, facilities and procedures in meeting the demands of the public project. Thus, communities which have developed mechanisms of support and coordination are better able to perceive and effectively organize themselves in meeting both the challenge and expected consequences of a public project. Here also, part of our generalized protocol is the examination of legal ocnstraints. Any water project in the West, for example, has to take into account the limitations of water rights and the significance of the prior appropriation doctrine. The construction of a canal system in the arid West requires determination of the basic water doctrine, the concept of ownership in the area, the appurtenancy requirements, permissibility of water transfers, etc. In addition, legal constraints refer also to procedural and

administrative details, such as enforcing rights or obtaining redress for their invasion. In our example it may very well be a procedural requirement that consciously or unconsciously creates efficiency impediments to the proposed project (concretely, lengthy and costly required litigation in order to effectuate changes may deter from the implementation of the project). In this respect we may point out how Environmental Impact Statements have provided a significant handle for citizen litigation. Citizen organizations are increasingly seeking court orders restraining government agencies or private corporations from a particular course of action.

2. Informal procedures. Here we may include a consideration of the informal communication network which may produce action in response to a public project in a given community. While in the first category one may use existing chains of command and control within organizations and the integration of the bureaucratic structure into the local community, informal linkages and mechanisms of intergroup relations reflect such items as the degree of public participation and individual input, emerging awareness, and consensus on the utility of the new work and overall community cooperation in the analysis and evaluation of a given public project.

Both in the Chapter on "Project Evolvement" and in Chapter 12, extensive discussion is made of the role of "public participation" in planning and in administrative processes. Citizen participation has become a major concern not only because in recent years environmental groups have certainly made their presence felt, but also because involvement of public, especially at the planning stage may act as powerful predictor of public response to future public works.

Yet, it should be realized that meaningful public participation should not be equated with the broad spectrum of public opinion. Who supports a project, may not necessarily coincide with those who benefit or pay the costs. Weighing the presence of special groups, recognizing power structures and interest coalitions, and being sensitive as to short versus long-range effects are important dimensions affecting the presence and nature of public participation.

What such a generalized protocol intends to convey is a first check list of major categories of sociological variables associated in any discussion of the relationship of a public project in the context of an affected human community. Most important, however, will be the attempt of developing categories and sub-categories with specific elements and perhaps quantifiable "indices," which may permit the public project practitioner to more easily identify major areas of concern and of possible negative manifestations in the social system. Ideally, various types of indicators and indices may provide a more concrete basis for "measuring" impact, for planning alternatives, and for evaluating the performance of the proposed new system. Table 6-1 summarizes the proposed categories of the Generalized Protocol and introduces the associated columns of indices of measurement, possible consequences, and level of impact. Each one of the cells created can be filled with the appropriate factor in order to provide an overview of a general social effects calculation per various environmental levels. It becomes imperative, therefore, at this point to turn our attention to the methodological issues involved in social effects' estimation and to the development of more specific measurement procedures for evaluating the impact of public projects on a human community.

EVALUATING PUBLIC PROJECTS EFFECTS: STRATEGIES AND TACTICS

The general argument about the human community and the preoccupation with the systems analysis outlined previously have been part of an effort for integrating physical and non-physical dimensions in environmental systems. Consideration was also given to the central theme of interdependence among various subsystems and to the need toward a holistic approach. Such an integrated approach implies that any change in either physical or social dimensions would have impact on each other. Change and its consequences are the ever-present characteristics of our dynamic environment. Thus, in the integrated system describing a public project, we need to be cognizant of the fact that each major component part is continuously affected by three types of changes:

1. Changes in the external environment which lead to changes in *inputs*, such as among other changes in the total number of people, changes in the economy, changes in technology, etc.

2. Changes in organizational structures and procedures, or changes in *thruput*, because of changes in size, capacities, different roles or organizational power, etc.

3. Changes in *output*, essentially described as "goal alterations," which result from goal displacement, new targets of society, or new policies of federal, state, or local authorities.

Since change, in any of the component parts and of the system as a whole, is constant any effort of evaluation will center in the determination of its sources and of the magnitude, severity, and importance of its consequences. The public project is a major form of change and an expression of expected

Table 6-1. Generalized Protocol for Social Effects Calculating.

CATEGORIES OF INDICATORS	INDICES OR UNITS OF MEASUREMENT			POSSIBLE NEGATIVE/ POSITIVE) CONSEQUENCES			LEVEL OF IMPACT (PRIMARY, SECONDARY, TERTIARY, ETC.)		
	Micro	Meso	Macro	Micro	Meso	Macro	Micro	Meso	Macro
A. Structural Parameters									

1. Demographic
   a. size of population
   b. density
   c. component characteristics
     (age, sex, etc.)
   d. vital rates
   e. mobility									

2. Spatial distribution
   a. rural-urban
   b. core-suburban
   c. planning space
   d. "neighborhoods"									

3. Socio-economic
   a. income distribution
   b. education
   c. occupation, employment
   d. industrial development
   e. households									

4. Community organization
   a. power groups, "elites"
   b. housing
   c. recreation facilities
   d. public health and medical care
   e. community services
   f. associations
   g. intergroup relations									

5. Social aggregates
   a. ethnic composition
   b. religion
   c. social class
   d. community diversity									

B. Attitudinal Parameters									

1. Normative resources
   a. expressed value system
   b. cultural practices
   c. degree of "social cohesiveness"
     or factionalization									

2. Aesthetic and legacy values
   a. quality and variety of natural areas
   b. historical antecedents, landmarks
   c. aesthetic integration
   d. "attractivity" and user satisfaction									

3. Attitudes towards "environments"
   a. towards physical environment
   b. towards social environment									

C. Processual Mechanisms									

1. Support and coordination
   a. political network and administrative
     structure
   b. legal constraints
   c. extent and nature of local autonomy									

2. Informal mechanisms
   a. public participation
   b. patterns of communication
   c. community cooperation									

261

alterations in all subsystems and their component parts. Similarly, we are interested in evaluating not only the effects
from the presence of a public project, but also the consequences
from non-intervention, i.e., the consequences for the surrounding environment from the non-construction of a public project
(such as e.g. the economic or social deterioration from the
non-exploitation of natural resources, the dislocation from the
absence of flood prevention structures, the decline of communities from the absence of adequate transportation network, and
similar cases of "do-nothing" policies). The crucial task will
be to compare present states and future states with the presence or absence of a public project and evaluate potential effects in some balance of costs and benefits for all environments involved. Further discussion of this important issue will
be made later within the context of a specific evaluative methodology. We are more interested here in introducing briefly the
general considerations of what project evaluation and assessment entail as broad issues for an appropriate social science
methodology.

General Remarks Towards an Assessment Methodology

Throughout the treatise, as well as in the general literature, criteria have been proposed and in many instances refined
indicators have been developed to measure the effects of public
projects on various systems (ecological, economic, etc.). For
example, in the construction of a dam repeated observations and
an accounting of important consequences have been made concerning the effects of the new water source on the aquatic life,
the displacement of other animal populations, the degree of ecological perturbations and the economic consequences for surrounding communities. To a lesser degree, however, there exist
indicators which attempt to account for the effects of a public
project on the social, cultural, and even aesthetic dimensions
of a human community. Once again the difficulty arises not only
out of a disagreement as to what are exactly the component
parts of a social system, but also because of the elusive and
subjective character of many of the scales used in measuring
social phenomena. There is always a temptation to equate cardinal with ordinal or metric scales without in many instances
clear understanding of the assumptions upon which a particular
scale of social effects has been constructed. At the same time,
the literature of social sciences, abounds in serious reserva-
tions and methodological cautions as to the very validity of
various survey instruments, the transference of physical science models to social phenomena, and on cautions concerning the
counterintuitive character of social systems.

Before we enter into a more specific discussion of strategies and tactics used by social scientists in order to assess
effects of changes upon the social structure, we need to indicate the concern for a multi-dimensional assessment model. We

are increasingly asked to describe an assessment methodology
which will enable us to identify the multi-faceted dimensions
of public projects, the temporality, and the wide array of units
involved in a total environmental system. Stated otherwise,
four dimensions characterize an integrated assessment methodology:

1. *Space* affected.   It may be convenient to conceive of pos-
    sible project impact in terms of:

    -   the area of primary impact (formal project boundaries)
    -   areas near or adjacent to the project (secondary zone)
    -   areas at a greater distance from the project (diffuse
        effects' zone).

    Another way of perceiving space affected by public project
    to visualize the differential effects from the local, to
    the regional, international, even global level (in terms of
    diminishing "ripple effects"). The focus of data gathering
    and analysis will depend upon the environmental level
    (micro to macro) and the type of project. An example may
    clarify the above. A small-scale irrigation canal system
    would have relatively localized impact (on both physical
    and social considerations) while an airport, a major power
    plant, or a major industrial facility can have far-reaching
    and complex effects.

2. *Units* or "substantive" component systems. As introduced
    before these may include, the physical, social, biological,
    and linkage systems which together provide the multi-
    component approach characterizing total environmental de-
    sign.   The present treatise's thrust is a prime example of
    this preoccupation.

3. *Time* perspective, with emphasis on impact chronology and
    expanded temporal horizons. It should be recalled, that a
    distinction was made between present impacts and long-range
    consequences. Diachronic considerations and the distinction
    between present and future are an important dimension to be
    incorporated in any cogent matrix of public project assess-
    ment and evaluation.

4. *Order* of effects. As contrasted to first order effects and
    the anticipated or expected results, we need to incorporate
    in our evaluation a system of accounting for second or
    higher order consequences, for the side-effects and by-
    products of a public project presence. A highway will not
    only provide fast transportation, but in doing so it will
    increase mobility of residents, higher isolation from imme-
    diate neighborhoods, and create a chain of side-effects
    that will disrupt present community practices. (Our sce-
    narios of chain effects do not need to be always negative

or disruptive. In our example higher mobility through a new transmountain highway, may increase recreational opportunities and bring more closely the family and friends together in joint leisure activities).

What is implied in all these dimensions is a much more complex strategy of assessing effects involving time, units and order and at the same time a differentiation between various impacted environments. Figure 6-2 summarizes in a graphic form the previous discussion. In this figure, we have extracted as an example a part of a four-dimensional matrix of environmental assessment model, reflecting a consideration of the impact of a public project on a given social system, in its macro-scale (e.g., the whole community), through a long-range perspective (e.g., estimation of consequences for the next ten years) with consideration of higher order side-effects (such as, for example, an attempt to examine effects of a highway when higher mobility of residents alters patterns of neighborhood interaction).

Despite the recognition of a demanding, complex overall system of environmental assessment, the specific scope of the methodological inquiry in the social sciences is not much different from the general quest of other scientists. Essentially, three major phases characterize the model of sociological inquiry:

1. Conceptual phase, or the definition of the problem and the development of not only concepts and descriptive hypotheses, but also of a general "model" incorporating the essential questions of the particular inquiry.

2. Empirical phase or field phase which incorporates all the efforts for finding the proper population, the strategies and tactics for the collection of data, and the specific techniques for acquiring valid and reliable data.

3. Interpretative phase which in addition to a general research analysis and classification of findings provides clues as to inferences that can be made about the data collected and an evaluation of the findings in the context of the problem formulated.

Figure 6-3 summarizes in abbreviated form the major steps involved in the research process as the social scientist formulates his problem, collects pertinent data, provides interpretation, and reports the findings.

Underlying such simplified diagrams are major problems on data requirements which seem to haunt social sciences and are especially important in any discussion or attempt to evaluate the effects of public projects on the social system. There are,

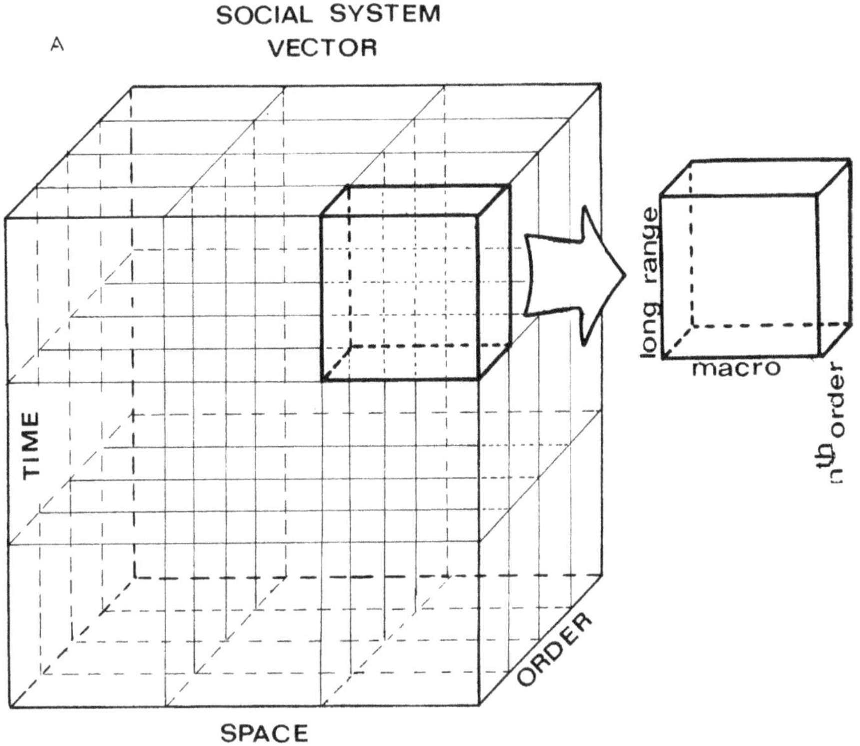

Figure 6-2.   A four-dimensional matrix   of   environmental assessment.

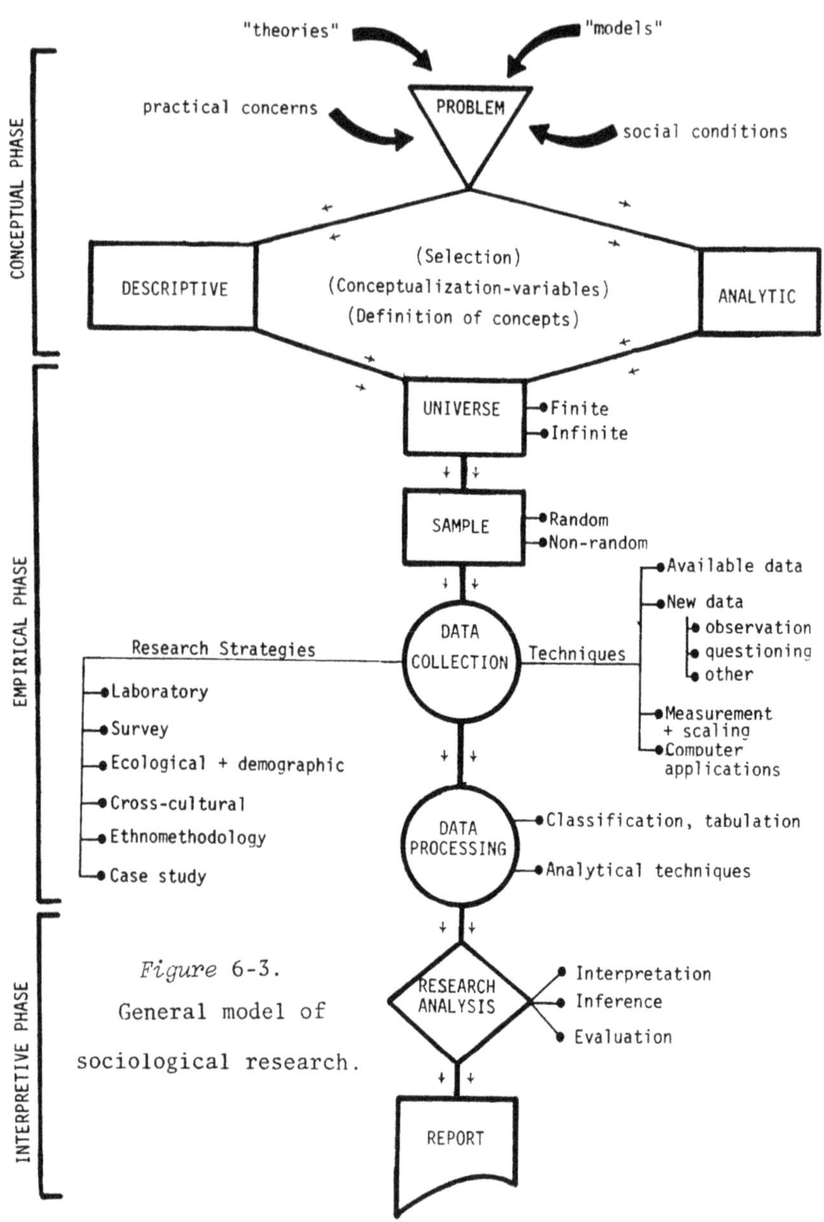

Figure 6-3. General model of sociological research.

indeed, a large number of methodological criteria and issues that are of great concern to any investigator in the social sciences. Although the discussion on such topics remains usually on an abstract level, the topic of public projects is among the concrete instances which prompt a closer scrutiny of the problems and inherent difficulties involved in attempts toward exactness or quantification in "soft sciences."

Let us, however, hasten to underline that the complexity
of total environmental analysis and project evaluation involves
both quantitative and qualitative dimensions. Our judgment
about effects will be based on data that ranges from purely *objective* to highly *subjective*. Objective evaluations are based
mostly on commonly accepted standards. (There are standards on
"water quality," for example, in physical terms but even "hard"
scientists do not seem to agree completely as to the exact
levels of such a measurement.) Subjective evaluation, on the
other hand, seems to be the order (and in many instances the
only desirable method) for such largely unquantifiable phenomena as community values,· aesthetic considerations, or other
dimensions of the surrounding environment that are not as yet
completely understood. To specify through an example: the
noise generated by through traffic or the discomfort from airplane take-offs can be measured objectively. But the total importance of noise for the human community needs both an objective (e.g. range of decibel over various areas and health hazards) and a subjective evaluation (such as e.g. the emotional
costs and the complex interrelationship between economic benefits and social discomfort).

Once again in a later section we will pick up the same
issue of the difficulty of estimating varied costs and benefits.
However, since this section is a brief introduction to the
methodological capabilities of sociology (without providing a
comprehensive presentation) we need to indicate three areas of
concern and considerations influencing our efforts to develop
systematic procedures for evaluating the role of public projects in the human community. These are: the sources of data
to be utilized in social research; questions of measurement,
i.e., criteria for valid and reliable indicators; and, finally,
some of the more widely used tactics or specific techniques for
generating appropriate data.

Sources of data. What are the types of sources needed to
be tapped in order to provide data concerning the presence
and/or effects of a public project? The sociologist has two
major types of sources that he continuously uses. First,
available data or existing information from various governmental, federal, state, and private reports or studies. This type
of data are particularly important for the structural parameters of the generalized protocol outlined above. They provide
the major categories for collective analysis, such as the results of censuses, vital statistics, economic indicators, and
other aggregate data.

The other major source of information may be characterized
as primary data. The origination of new data is especially important in the effort of locating information on the value and
attitudinal parameters of our generalized protocol. Such primary data are usually exemplified in the various surveys or

questionnaires which sociologists use in order to elicit information about attitudes towards the various environments.   It
should be noted, however, that primary data can be also collected on the structural dimensions of a given human community
especially through social reconnaissance studies.

Problems of measurement.   As in all fields, sociology is
continuously striving towards the development of techniques for
collecting data plus sets of rules for using these data.   We
raised previously the issue about the general difficulties of
quantification and subjective evaluation underlying the methodology of social sciences. The efforts for measuring or assessing effects will depend on the development of criteria that
will be characterized by:

    a. validity = criteria free of systematic error, or criteria that measure what is supposed to be measured.
    b. reliability = successive measures produce the same results (random error control)
    c. "refinability" = the criterion used is a sensitive one
        to be able to make fine distinctions among categories
        (precision)
    d. comparability = the definition of the criterion remains the same over time and space
    e. permissibility = range of error to be tolerated.

In addition two general conditions should also characterize criteria to be used in the evaluation of a project

    • relevance
    • timeliness.

The most important methodological problem seems to be the
presence of constants or errors referred to as validity.   Here
one finds all those factors which systematically effect the
characteristic being measured or the process of measurement itself.   Despite tremendous improvements in the methodology of
social science, there are still no definite answers as to what
constitutes a valid criterion or measure.

Field tactics and specific techniques.   The categories of
various procedures and techniques are many but it is important
to immediately distinguish between research strategies and
specific tactics. The research strategies involved a series or
research models for acquiring data, ranging from the well-
controlled laboratory conditions to the more loose "unobtrusive
methods" (i.e., any method of observation that directly removes
the observer from the events being studied). On the other hand,
specific tactics and techniques (as indicated in the previous
figure) have to do with four major specific ways of obtaining
concrete data (assuming a given research strategy): use of
available data, obtaining new data (either through observation

or through questioning), new data through measurement and scaling, and computer applications or simulation models. Some of the major types of activities involved in data collection and analysis can be summarized as follows:

<table>
<tr><td>Reconnaissance</td><td>Interpretation</td><td>Field Work</td></tr>
<tr><td>

• Maps
• Aerial photographs
• Available material
• Records

</td><td>

• Statistical analysis
• Evaluation
• Simulation

</td><td>

• Observation
• Key informants
• Life histories
  and diary
• Survey

</td></tr>
</table>

Each specific technique has its own requirements and relevance for the problem at hand. Most important, is to acquire, through living and working in the community, an appreciation of the area and people, of the range of problems and issues present in the social system. From such a sensitivity one is able to develop more quantifiable data and a sample survey reflecting experience with the people and their social environment. Despite the fashion to immediately constrict a questionnaire (as a fast and inexpensive way of generating relevant data) one should never depreciate the usefulness and, in many instances, necessity of qualitative research.

Another major concern of the social scientist in the area of field procedures and tactics is the question of sampling. Since most of our data is usually partial rather than complete and since we want to use this partial data to characterize the entire set, it is imperative to find out the types of representative population that would provide the requirements of good sampling. Given the two major distinctions between random and non-random sampling the basic requirements of good sampling are two: first, representativeness, i.e., representative of the entire set or of the universe; and second, adequacy, i.e., of sufficient size to allow confidence in the stability of its characteristics. Once again the social scientist is following here many of the principles established in other sciences in order to provide the samples that would permit confidence in generalizing to the population and to the problem at hand.

Evaluating Impacts and Consequences

The methodology outlined above corresponds to the classical tradition of conducting research. It refers to the general principles and methodological concerns of the social scientist in developing the design of any study. However, in evaluating the effects of public projects we need to adopt a much broader plan of attack and a systematic methodology utilizing a multidisciplinary model of evaluation and alternatives' assessment. In addition, recognition must be made of the systematic linkages of component parts within the system as well as between systems.

Evaluation as a concept is a derivative of "valuation," a term connoting not only a number of more or less objective considerations but also commitment to some normative purposes and definite value orientation. In a simpler language, we are asking the following question: "To what extent is the project succeeding in reaching its goals?" If we ignore for a minute a host of methodological questions as to the validity and reliability of measurement and other theoretical assumptions as to the specificity of goals, the evaluation process seems rather simple, namely:

1. Find out a project's goals.

2. Translate the goals into measurable indicators of goal achievement.

3. Collect pertinent data and/or set up experimental conditions.

4. Analyze and compare data on outcomes, participants, etc., with the goal criteria.

In practice, however, such a simple operation stumbles against a host of overt and covert considerations in the objectives and operation of a project. To mention a few of them:

1. Project goals are often ambiguous, elusive, or too broad to pin down. Even more, despite their obvious nature, there are doubts as to the content of stated goals. Are we, for example, interested in maximizing physical output in an efficient way and/or provide a more equitable distribution among project beneficiaries? Do we want to provide an expedient means of transportation, or are we aiming through the physical infrastructure in changing attitudes and behavior of people about their surrounding environment?

2. Projects do not accomplish only official goals. They also accomplish other things, sometimes in addition and sometimes instead. Unanticipated consequences and serendipitous effects tend to confound the evaluative effort. Timing is also an important factor since some projects may show immediate effects, while others create a type of a "sleeper" effect which shows up at a later time.

3. At which groups in the population is the project primarily aimed? Who are the potential "clients?" Are we seeking to influence or affect individuals, groups, or whole communities? And if so, how do we measure a series of changes and their differential effects among segments of the population?

4. Many times, projects aim at multiple effects rather than a unitary objective. How do we measure and evaluate effects differing, among others, in terms of type, order, magnitude, importance, timing, extent, and reversibility?

We need to recognize all the above issues and problems in an evaluation process for two reasons: first, in order to emphasize the key role of goal specification in any evaluation effort; and, second, in order to indicate that even under the best circumstances and despite continuous advancements in the field, good evaluation studies of projects are relatively rare. Furthermore, even if we come close to learning how well a project is reaching its goals, participants or members of a project can have widely differing expectations of the kinds of answers that will be produced. With due recognition, therefore, of its "valuative" character an assessment methodology and a strategy of evaluating environmental perturbations or consequences of a public project are characterized by the following preoccupations:

1. *What* are we going to evaluate. Any public project in order to be assessed for its consequences and effects on the social structure needs to have a delineation of the goals or the purposes to be achieved, the overt and even covert, conditions for which the public project has been proposed or is constructed.

2. *How* are we going to evaluate. This implies some indicators and specific techniques by which perhaps a quantification, a judgment, or some weighting procedure will attempt to put more definite parameters in the assessment procedure. The nature of evaluation depends largely on the type of measure and on indicators available for the determination of the attainment of our objectives.

3. *When* are we going to evaluate. The evaluation may come at any point of an on-going project. It is important to know if the assessment of consequences and impacts will be taking place at a pre-planned period, during the discussion of the public project, in the last formulative stages of the public project, during the actual construction of the public project, or finally after the public project has been completed. The earlier the evaluation takes place the less the potential conflict and the better the feedback signals for correcting aspects of the proposed project. Indeed, a number of strong criticisms have been made against evaluations that usually take place after the project is completed. Public participation has been proposed as a mechanism for meeting this criticism.

In the context of our overall assessment strategy and with the repeated recognition of a systems approach, we may be able to outline the basic rationale of an evaluative design. The conditions and basic dimensions of our model may be graphically depicted in Figure 6-4. This figure repeats a general argument, namely that no event or output has a single cause and that each event has multiple effects or consequences. All events and outputs are interrelated in a complex causal system, with no single factor being a necessary and sufficient cause of any other factor.

Three different, but interrelated conditions of project evaluation are explicated in Figure 6-4:

    a. Traditionally, the most widely used term in evaluating both performance and effects of a project has been that of *efficiency*. Efficiency attempts to relate in simple, economic benefit-cost analysis the relationship between resources (input) and proposed goals or attempted targets (output). Efficiency is concerned with the evaluation of alternative paths or methods in terms of costs. It represents a ratio between input and output or the capacity of a project to produce results in proportion to the effort expended. Efficiency means minimization of costs and maximization of benefits, always expressed in terms of dollar values. Example: money spent for an airport related to volume of traffic.

    b. The term *effectiveness* can be used mostly in terms of the organizational performance or the meeting of purely organizational goals, i.e., the relationship between a given organizational structure (thruput) and perceived goals (output). Such a term has also been used in the context of an overall measurement of achievement for a system, derived from its subsystem's performance, or related to its interactions with other systems. Example: private, semi-private, or public agencies contrasted in their meeting of a stated project goal.

    c. Finally, a new term is coming more and more into use, namely, *efficacy* which attempts to incorporate the meaning of social goals and a much more comprehensive relationship between input, thruput, and output. Efficacy attempts to move beyond purely economic considerations or criteria of organizational effectiveness by relating how a particular system can efficiently, effectively, and guided by principles of social awareness, meet the goals of a given social system. As a matter of fact, the term efficacy brings forward the increasing consideration of all those intangible benefits to be accrued from a given public project and which cannot be directly measured by existing economic, technical, or other quantitative criteria. This, however, implies qualitative criteria and the consideration of social goals which transcend purely utilitarian

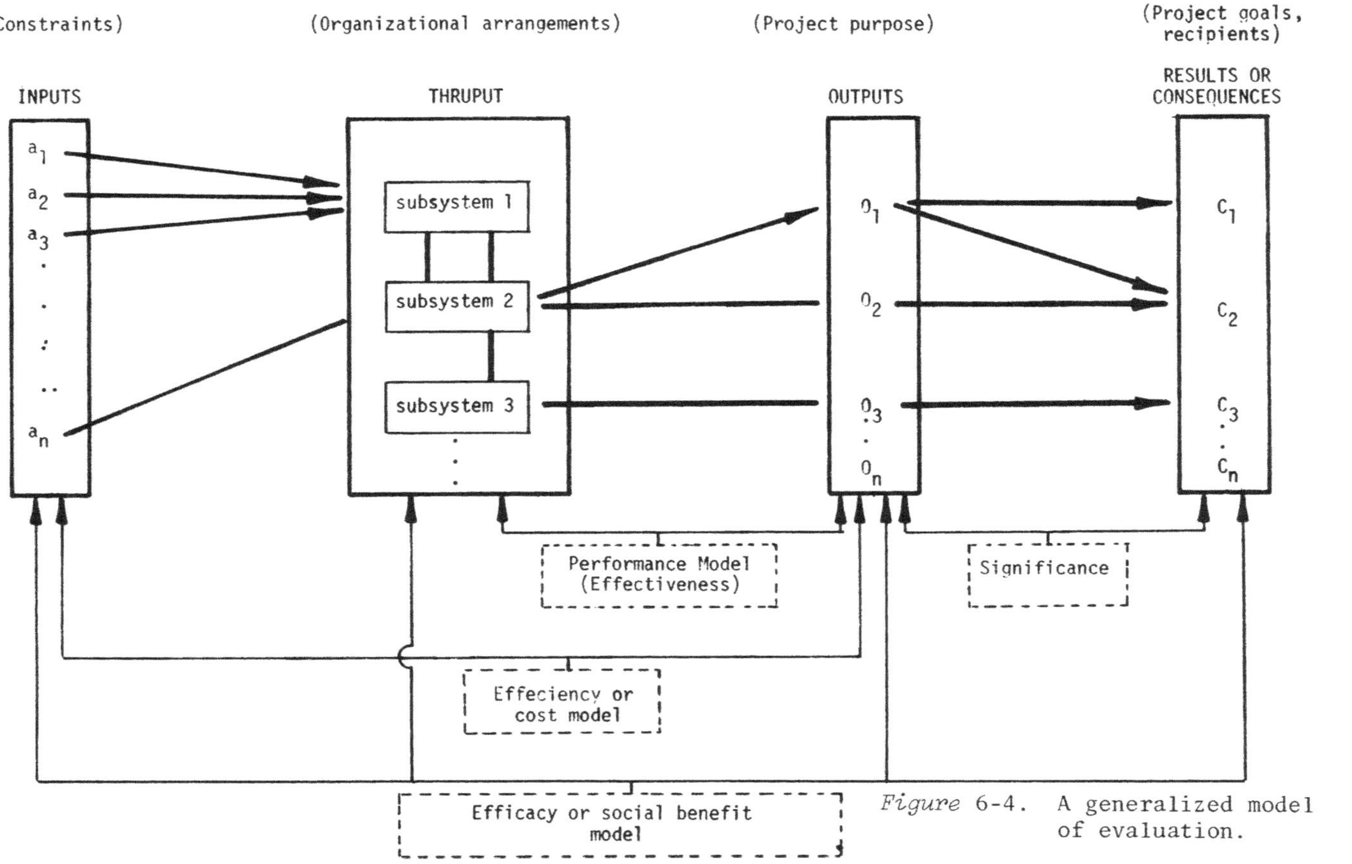

Figure 6-4. A generalized model of evaluation.

calculations of immediate cost/benefit analysis, Efficacy, therefore, has to do with the power to produce intended results and refers especially to long-range consequences and benefits for the larger population, including considerations of equity. This type of evaluation intends to move beyond simple accounting and organizational procedures by analyzing the how and why of the program behind the project, especially by locating causes of failure even when efficiency and performance seem to be adequate. This last category requires a specification of additional dimensions in evaluation and such qualitative considerations as:

- the affected recipients of the project (target populations);
- the conditions of project operation (time, locale, etc.);
- differential effects (multiple effects, side-effects, duration of effects, etc.).

Thus, this "efficacy" model becomes more pertinent, if we add the dimension of *significance* or the extent to which achievement of objectives contributes to broader goals.

We are faced, therefore, with the increasingly difficult task of trying to strike a balance of fulfilling technological and economic goals in expedient feasible ways, but at the same time answer questions of social policies and of total environmental accounting. This simply implies that we need to introduce not only effective criteria that would permit us to measure effects and consequences (mostly of lower order) for our social system, but also long-range policies for a social use of natural resources.

From such general notions and with a distinction between efficiency, effectiveness and efficacy we may then proceed to evaluate in a systematic way the presence of a public project. Using as a backdrop the systems approach we can contrast input constraints, limitations of the organizational structure, and stated objectives or explicated goals. We need to develop some meaningful matrices which may not only point out problematic situations and areas of concern, but would also help us develop specific indicators or clues as to the performance, impacts, and consequences of a public project. Figure 6-5 illustrates in a summary form an attempt for a systematic mapping of persistent problems, desirable criteria and feasible alternatives in the component parts of a social system. While such a general figure illustrates the general approach, we may use as a specific example the construction of a highway to illustrate pertinent variables and criteria. Table 6-2 contains an indicative number of criteria used in the context of an analysis of forces within the human community, criteria which in some instances

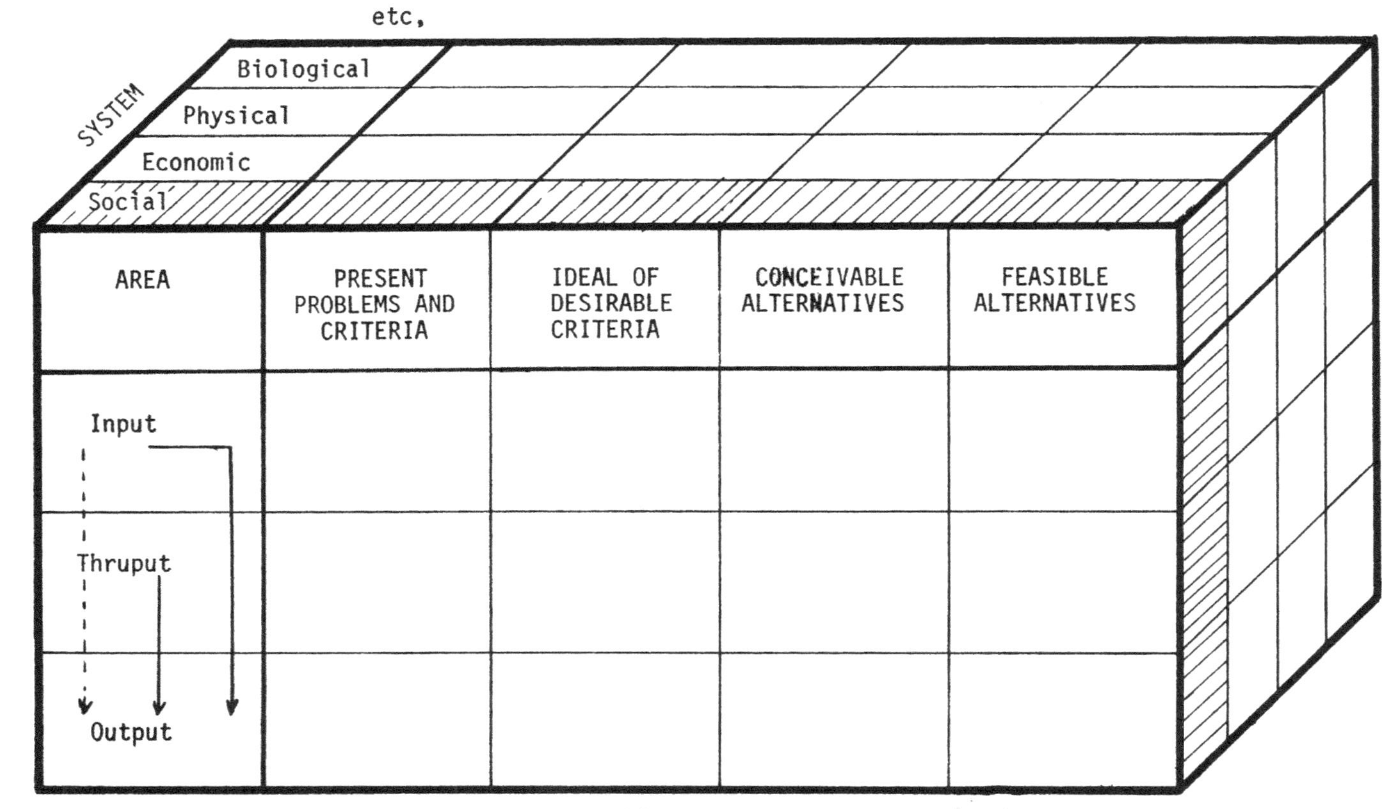

Figure 6-5.   Evaluation matrix for a systematic analysis of a public project.

	Present Problems and Criteria	Ideal or Desirable Criteria	Conceivable Alternatives	Feasible Alternatives
INPUT CONSIDERATIONS				
1. Physical Environment and natural resources	- site factors - accessibility - spatial organization			
2. Demographic characteristics	- size of population - composition - rates of growth			
3. Normative resources and socio-cultural limitations	- social values not clearly defined - problems of equity - aesthetic considerations not defined			
4. Nature of legal system and legal constraints	- problems of content - lag between environmental law and public projects operation			
5. Economics	- availability of financing - availability of materials and products			
6. Political network	- political interference - lack of organized pressure groups - political expediency			
7. State of technology	- availability of equipment and techniques - availability of skills - lag between general state of technology and environmental considerations			
THRUPUT CONSIDERATIONS				
1. Personnel	- labor constraints - quality of personnel - discipline and morale - interdisciplinary focus			
2. Facilities and infrastructure	- means of resource allocation - need for technological innovations - existing facilities and need for integration			
3. Procedures and rules	- project organization - project management - relations with the community			
OUTPUT CONSIDERATIONS				
1. Enhancement of living environment	- diversified goals of the community - lack of definition - problems of measurement			
2. Well-being of people	- abstractness of category - lack of definition - lack of moral consensus			
3. Long-range land development	- no comprehensive policy - legal loopholes - haphazard growth			
4. Preservation	- no clear definition - private rights - excessive protection			
5. Transportation and mobility	- lack of transit system - private cars - polluting practices - integration with the surrounding community -aesthetic factors			

overlap with variables of other related systems or "environments." Ultimately such a mapping, used in conjunction with the
Generalized Protocol, can provide a strategy for assessing the
presence and effects of a public project.

Before we continue with some specific suggestions as to
indicators for calculating social effects of a public project
we need to remind ourselves of the limitations and distortions
in constructing indices of environmental quality. Various
authors have attempted recently to develop specific criteria
and matrices of environmental quality which incorporate also
social, cultural, aesthetic and similar "human" considerations.
Throughout such attempts, however, there are widespread methodological difficulties concerning an adequate or agreed upon
system of indicators and indices such as:

1. *Invalidity*, which reflects the perennial methodological
   problem involved in the correspondence between phenomeno-
   logical meaning and judgment about a particular indicator.

2. *Perception*. The perception of the environment varies not
   only from person to person, but it is also subject to
   change by the person himself in accordance with changing
   situations. Even more, a person's perception and attitude
   towards the various environments is influenced by both
   socio-cultural processes as well as by specific situations.

3. *Lack of data* and inaccuracy. Not only we may have errors of
   measurement, sampling, and enumeration, but contradictory
   assertions may also exist whenever we have no usable statistical data to support critical statements about the
   state of a given community or society.

4. *Reversibility*. Another factor which effects the construction of indices of environmental quality refers to those
   circumstances where evaluation is based on the possibility
   of action, i.e., certain situations which have the same effect may be evaluated differently according to different
   degrees of reversibility or retrievability. The above implies that the higher the degree of reversibility, such as
   for example the invention of new, cheaper, and/or more efficient techniques, the higher the degree of acceptance of
   a legitimate risk in the introduction of a given public
   project.

5. *Time preferences*, or those environmental situations in
   which improvements of present comfort can be obtained only
   at the price of giving up future comfort and vice versa.
   Various societies and communities have different "discount-
   factors" for evaluating future situations which depend on
   their time horizon, the acceptability of risk, and their
   commitment to long-range planning. For example, the

exploitation of a given natural resource such as a forest
may be much more important for the survival of a human community, as contrasted to long-range benefits to be accrued
from the existence of an aesthetically and ecologically
pleasing environment for future generations.

6. *Value consensus*.   Here the difficulty exists in that in
order to construct an environmental quality index we have
to ask ourselves whose values are we taking as a basis.
Indeed, there is limited agreement on preferences, standards, taste, and other abstractly conceived community
values.   Conditions which are perceived as "good" or "bad"
depend very much on the consensus of given groups in society and are many times a reflection of not fully compatible
values in the social system.   An example here could be the
distinction between efficiency and equity where the construction of a recreational center, let us say, in an efficient way may clash with conditions of equity.   In other
words, growth or material development (efficiency) may conflict with fair access of resources and consumption of different elements in the population (equity).   Who gets what
(equity) interacts with how much there is (efficiency),
thus imposing the requirement that in any public project we
need to consider the trade-offs between efficiency and
equity having in mind at the same time ecological constraints which will provide the framework for rational
trade-offs.   Simply, if we are to build a park, a lake,
etc., and efficient operation of this particular work may
require limited access, thus, raising the question "who
will be the persons benefiting from such a project and how
do we guarantee equitable access?"

7. *Incompatible conceptual systems*.   Indices of environmental
quality are always calculated in relationship to concepts
and models about man and society and are explicitly or implicitly related to some theoretical relationships as to
the structure and functioning of social ysstems.   It has
been posited above that various strategies in the interpretation of man, human community, or society provide us with
an array of concepts and indicators as to what are the component parts and essential parameters of a social system.
Diversified theoretical approaches have contributed to a
proliferation of concepts used in the context of any discussion about human community and social structure. Despite
valiant efforts for introducing fundamental dimensions and
variables in the examination of social systems there still
exist incompatible conceptual models and contending theories
of society and a bewildering number of key concepts regarded by various social scientists as essential in the discussion of society and social systems.

In conclusion, we may indicate first of all that environmental quality indicators now being used are mostly expressions of physical conditions and not as directly meaningful for social effect considerations. Second, no good calculation exists dealing with net social benefits (benefits minus costs or losses in some sense). Third, decisions relating to environmental quality are based currently on processes where social benefits cannot be adequately measured in engineering or economic terms (see also discussion on comparability later on). Fourth, not only absolute goals for environmental quality, but also directions and rates of change need additional indicators in assessing environmental quality. Finally, in recognizing social welfare indicators we need also to integrate them in the available quantities of resources and services for the good life in society (trade-offs between efficiency and equity).

The state-of-the-art in environmental quality indices and pertinent measurement technology in social sciences is characterized by a highly demanding task of "instrumentation." This task is complicated by the overall problematic situation of trying to bring together three non-comparable items.

1. Cost
2. Desirability
3. Impact.

Our challenge is to develop appropriate techniques and a way of measuring the combined effect that reflect some average value between these three items. In particular, the problem of calculating "social costs" is taxing the ingenuity of authors in developing scales for transferring essentially qualitative aspects of a project. It is not only that social costs are difficult to assess directly; often, they reflect to larger questions of social policy transcending the immediate project and affecting wider segments of population.

Similarly, desirability and goal fulfillment is based on criteria that are difficult to be quantified and even more difficult to be compared on parallel scales with various costs and diversified impacts. Take, for example, the case of calculating aesthetic integration of a reservoir and the types of indicators needed in scenic analysis and community satisfaction. Two models exist for non-monetary evaluation: *attractivity* models and *aesthetic* measure models. Attractivity models are based on the demand for an environmental experience as measured in user participation or satisfaction; aesthetic measures, on the other hand, are based on more elusive criteria of measuring the psychology of stimulation derived from the physical environment, (such as subjective point-scale ratings, or quantifications dependent on degree of uniqueness).*

More important for our discussion are the *social indicators* which are increasingly referred to in the literature, especially in an estimation of possible adverse effects of a public project in a community. Pure economic and technical indicators have concentrated on the quantity of goods and services. Social indicators, on the other hand, are increasingly concentrating on questions about quality of life, on expanding the basis of decision making, and on anticipating (through a social accounting structure) the secondary effects of programs and projects. Social indicators have been increasingly recognized as important forms of evidence for assessing social goals, for evaluating specific programs and projects and for determining their effects and consequences. Such important socio-economic indicators require selection of the unit of analysis, specification of the areas of concern for which indicators are to be developed, and collection and aggregation of basic statistics for the construction of sensitive indices.

A number of studies have developed rather intricate systems for assigning values on various social indicators. If we assume that we can quantify some of the dimensions involved, we may, then, be able to provide a value analysis which would incorporate both a benefit curve and a detriment curve and an optimum way of calculating the "total efficacy score" to be gained in a juxtaposition of various parameters in total environmental design. (The Generalized Protocol of Table 1 can be used towards such a purpose.)

In summary, the argument of the strategy and tactics of methodology and the quest for obtaining valid and reliable data for evaluating effects of public projects on the human community tends to follow two major directions:

1. Towards an attempt to construct social indicators which would reflect mostly the preoccupation with the structural parameters of the system. Such attempts of social accounting are increasingly utilized, especially in macro-studies of national social welfare. This is a slow methodological process, since it involves a tremendous number of difficulties both in conceptualizing the types of social indicators needed and in obtaining the types of data that would fit the parameters of the new methodology.

---

*See in this connection the cogent discussion in Chapter III of *County of San Diego Regional Issues; Vol. 1: Environmental Quality Index. A Feasibility Study*. San Diego: Environmental Development Agency, 1972.

2. Towards an understanding of the subjective parts of the
    system, in unearthing the normative dimensions of a commu-
    nity, and the attitudes of individual community members (or
    various groups) towards their surrounding environment. Such
    an effort is mostly incorporated in a large variety of
    scales of measurement which attempt to elicit responses and
    feeling and other unobtrusive methods in order to make in-
    ferences about social phenomena.

All in all, the social scientist employs a combination of
both strategies and tactics and available and/or primary data
to increase his sensitivity in providing answers as to questions on the measurement of the surrounding social world. In
doing so, however, one should always be aware of the limitations and of the potentialities involved in conducting social
research. More than anything else, however, there is urgent
need in developing not only indices, which will vary from cultural set-up to cultural set-up and from time to time, but also
procedures of science which will incorporate physical and non-
physical dimensions involved in the study of a human community.
Above all, there must be a correspondence between quantifying
and qualifying techniques and a correspondence between our
models and the real world. Thus, in addition to the pressing
demand for providing a methodology for assessing the immediate
consequences of public projects, we are increasingly asked to
provide imaginative and future-oriented thinking with expanded
time-horizons and with multi-dimensional considerations of alternative models of living, acting and behaving. It is imperative, therefore, that we conclude this chapter by providing a
more general commentary on holistic planning, a synthesis for
the required methodology, and a concluding statement as to the
type of thinking that should guide our efforts for understanding the interrelationship of public projects and human communities.

PLANNING AND THE HUMAN COMMUNITY: REMARKS TOWARDS SYNTHESIS

It is apparent by now that in discussing the impact and
consequences of a public project on a human community, we must
also introduce some larger principles and dimensions concerning
the social components of total environmental design. The attempts toward a measurement of the quality of life, the extent
and severity of various effects, as well as the question of who
benefits from the construction of a public project lead us to a
much broader framework of systematic description and analysis.
These general premises of social concern may be more important
than the preoccupation with quantification and the elusive pursuit of indices of environmental quality. In this respect, many
sociologists would point out, again, that given the subjective
character of social values and the complexities of social systems, it might be impossible, even futile, to pursue the

question of being able to measure or quantify the effects and long-range social consequences of a given public project.

Thus, it may be more useful at this point to introduce some general principles of social planning if we are to understand the larger arena and forces impinging upon public projects.   Let us start by providing a short definition of planning: a means of directing social change and social relationships towards the ultimate objective of orderly and harmonious community process.   As part of this general process of change a public project becomes a concrete means for exemplifying and realizing goals and for improving the welfare of individuals and collectivities within a given community, region, state, or the nation.   By using a public project as a focal activity we can trace in a simplifying figure the major concerns which seem to run through a total environmental design effort (Figure 6-6).

In this generalized diagram the underlining proposition is that a public project, in affecting primarily the economic, social, and physical life of the community, is directed towards a conscious changing or integration of two major environments, namely the physical and "the institutional." A proposed public project provides and facilitates the type of activities, densities, and intensities which have impact not only on the physical arena within which social life takes place, but also on people and organizational structures (social, political, cultural, etc. life of the community). Such activities are related to, coincide, or may even conflict with values, goals, and objectives of the local as well as of the larger society.   In practical terms, these goals and objectives may be translated to norms and standards of a more technical level.   Norms and standards may be incorporated in specific plans or programs, providing, thus, the basis for the execution, alteration, or even abandonment of a proposed public project. To continue the explication of Figure 6-6, the public project is also affected by such external considerations as state and federal policies, the availability of resources and the policies for their allocation and use, the state of technology (the ability of meeting stated objectives within existing technical capabilities), and, finally, the considerations and constraints of the poliΛical culture (or, the recognition of political power and expediency).

All the above remarks imply, then, that in considering the effects and consequences of public projects on human communities we are guided at the same time by a sensitivity and awareness as to what public projects do for the general welfare to society.   The last brings forward the topic of social planning and policy intervention.   But the epistemology of social planning requires a consideration of definite steps in meeting the demands of larger social policy.   These considerations follow approximately this order:

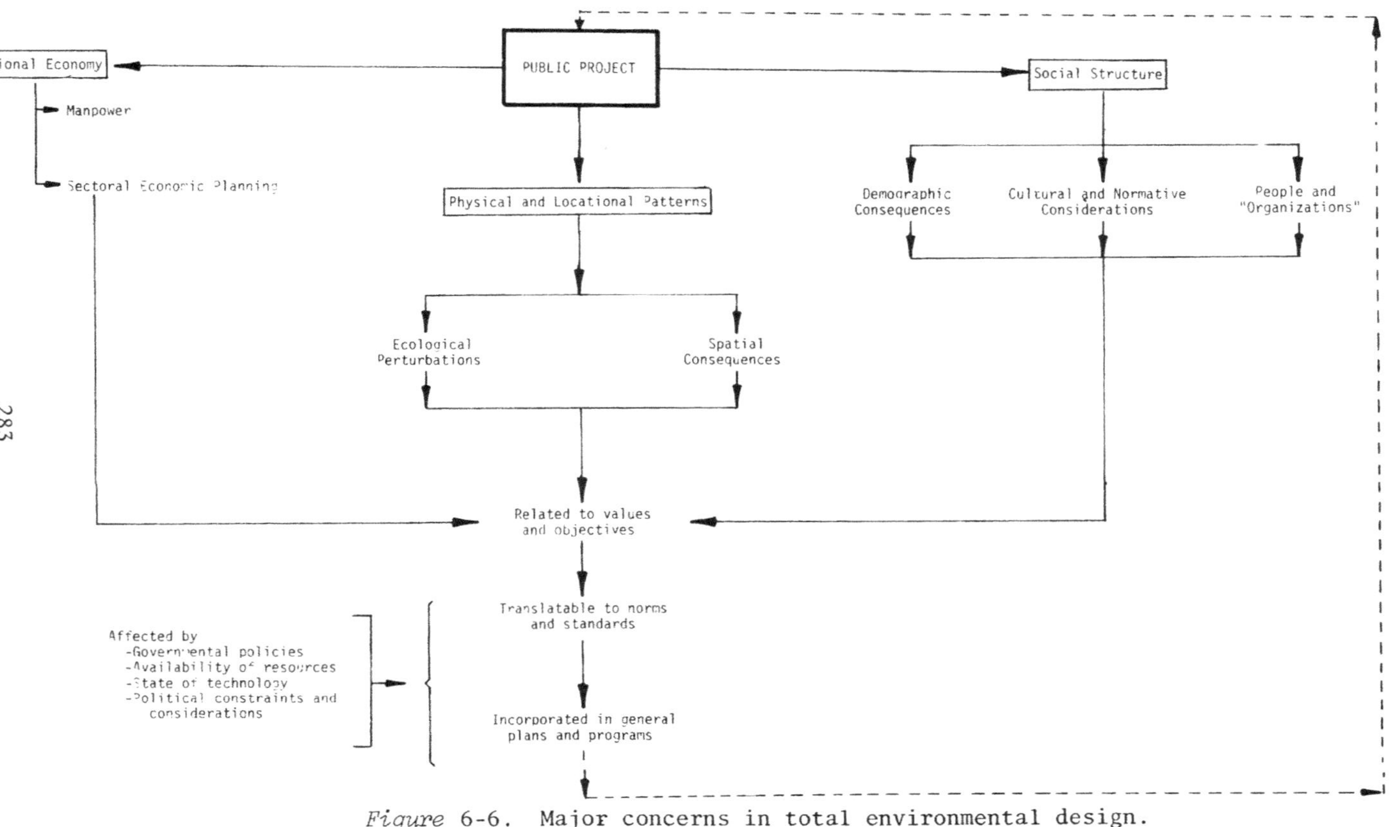

Figure 6-6. Major concerns in total environmental design.

1. Definition of the problem.
2. Determination of the community system.
3. Exploration and screening of choices and alternatives.
4. Formulation of the intervention plan.
5. Synthesis and project implementation.
6. Monitoring, evaluation, and continuous feedback.

It has been repeatedly emphasized that any public work produces disruptions in the land structures and in the collective life of the people. In considering, then, both an attempt to evaluate the consequences of such a disruption as well as in building in a more systematic way the potentialities on the welfare of the people incorporated in a public project we need to be cognizant of four phases in planning and the typical planning activities pertaining to each of these (Figure 6-7).

1. *The systemic mapping phase*, which in addition to describing needs, concerns and resource constraints, incorporates, also questions and problems of priorities and a determination of the community system. This descriptive phase concentrates on problem formulation through an analysis of what *is*.

2. *Valuative or normative phase* which includes all the questions reflecting the quest for social goals, desirable targets, definition of objectives, and the generation of alternatives through a screening process of values and policy priorities. The possible approaches to attaining objectives can be achieved through a value, political, resource and reality screening in the context of a cogent intervention policy for improving the welfare of the community (summarized into what *should* be done, or if one wants to attach a moral commitment, what ought to be done).

3. *The decision making phase* or the selection of specific targets and the evaluation of various alternatives. This phase incorporates the considerations about the feasibility of the project, evaluation of ecological and socio-economic consequences, the mechanisms and criteria for the selection of specific targets, and the decision-making process or trade-off studies. In emphasizing feasibility in the selection of specific targets we incorporate essentially the term *could* and its correspondence between what actually exists and what is desirable.

4. *The implementation phase* with the specific plan of action, design specifications, the management of the project, the specific techniques and technological breakthroughs for the actual construction and operation of the public project and, finally, the measurement of impacts and consequences for the given public project.

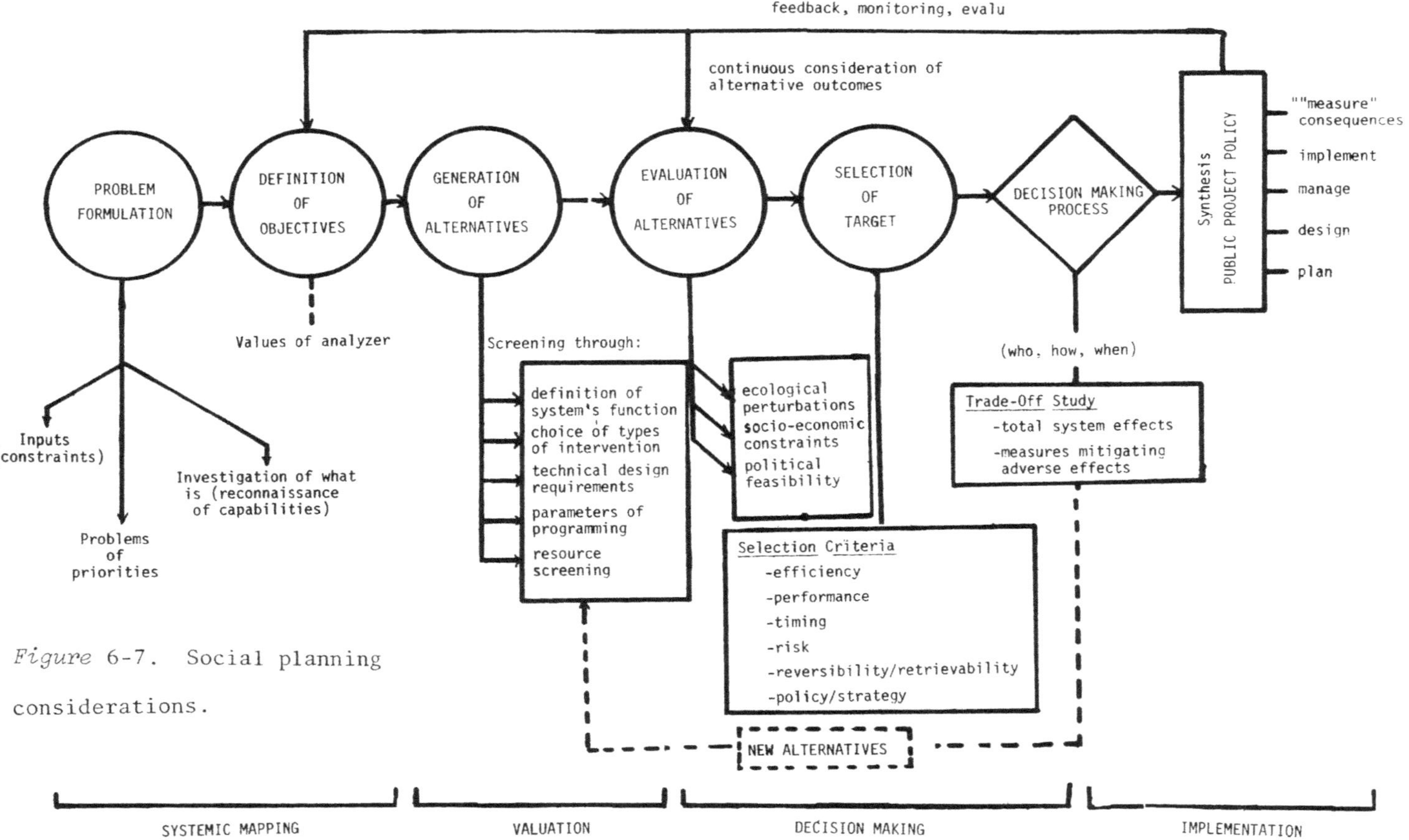

Figure 6-7. Social planning considerations.

In describing and mapping out dimensions of the planning process and in providing evaluation through a process of decision-making one must be aware of not only who is benefiting and who will be paying the cost for the improvement of the general social environment, but also of other trends and disruptions occurring in the larger society (such as uneven population distribution, differential patterns in consumption and affluence, and general forces of political and economic change) which may accentuate the effects of a number of public projects.

What can we do then to provide a specific methodology for judging alternative decisions and policies as well as to be able to see the social consequences of a public project in the context of the measurement of other environmental impacts such as ecological, economic, aesthetic, etc.?

There are two ways of proceeding in our effort for assessing the technological imperative of a public project. First, provide the framework for developing strategies of planning and evaluation. Second, delineate a proposed methodology that can provide some dimensions of "measurement" and evaluative criteria for alternative consequences. Parts of the general premises of these two approaches were discussed throughout the previous pages. To add, however, some specificity to the proposed analysis, we are utilizing a case study concerning the construction of a highway by-pass at the edge of a medium size town in the Western United States.

A few details about our hypothetical community will help orientation towards the problem. Ramburg is a rapidly growing community with a changing economic character from a primarily agricultural to a more economically diversified city. In order to accommodate rapid population growth and facilitate easy access to the attractive hinterland for large numbers of tourists (a lake by the mountains attracts thousands of visitors annually) a beltway transportation system has been proposed, essentially circling the city, passing through open land and reconnecting with the existing highway close to Lake Lafayette (Figure 6-8). The proposed by-pass corridor has been challenged by environmentalists in the city. A team of consultants have been hired by the State Highway Commission to assess potential consequences of the proposed by-pass, with special emphasis on ecological and social costs. The sociologist in the team is faced with the following expressed arguments:

Pro	*Con*
1. Traffic alleviation and accident reduction	1. A viable Chicano community is bisected
2. Pupil safety of a nearby school	2. A major proposed greenbelt area will be destroyed

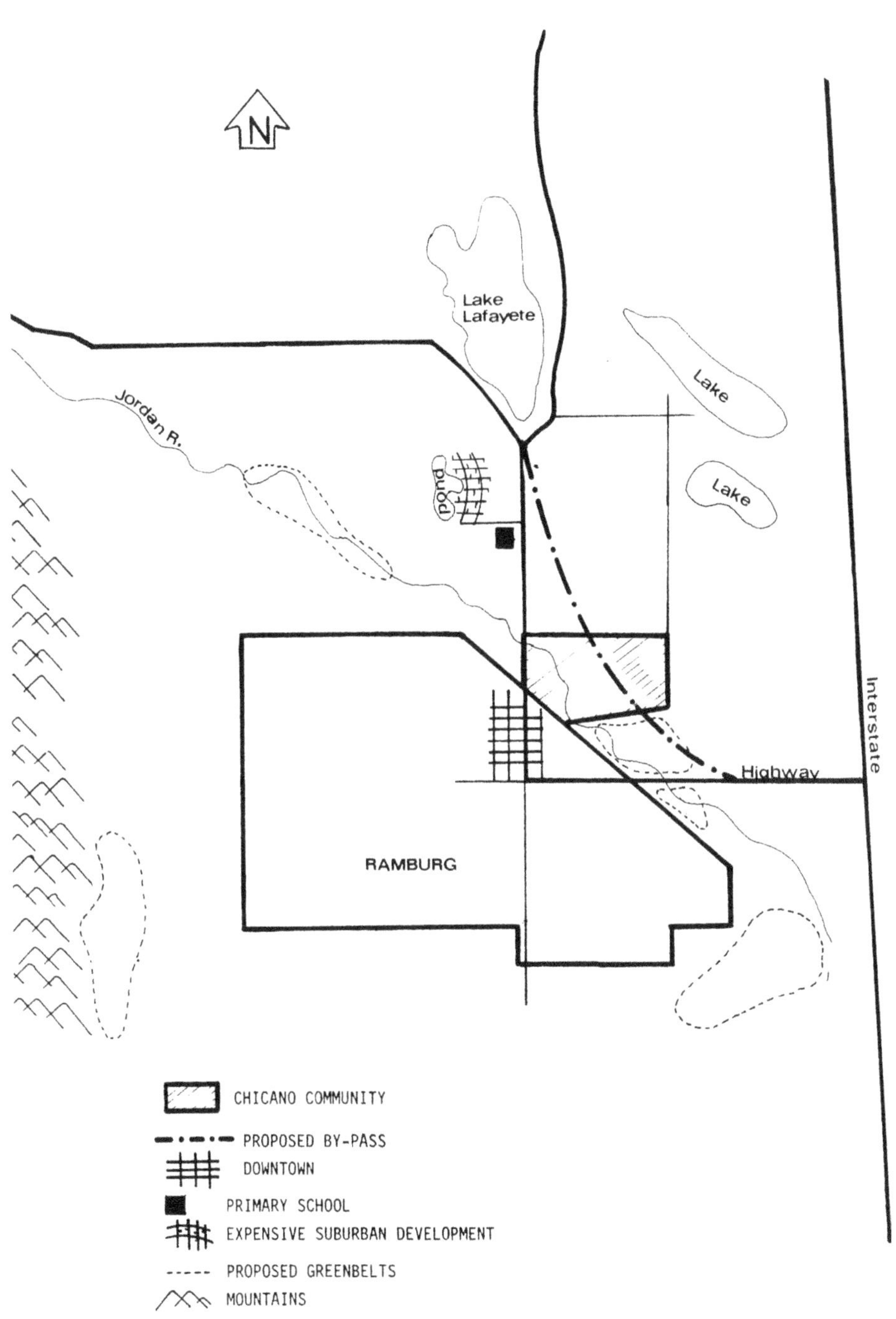

Figure 6-8. Hypothetical case study of a highway by-pass.

3. Encouragement of city development to the north

3. Encroachment of the river
streambed accentuates pollution

4. Retardation of overdevelopment to the south

4. The river is impaired as a
recreational resource

5. Stimulation of downtown
business (arrest present
downtown decay due to sub-
urbanization).

Needless to say, there are also a number of hidden factors
in the case, including the significance of a 90% Federal participation in the project, strong local land interests, state
plans for a highway system, and some considerations of political expediency. However, since it is rather difficult to estimate covert considerations, our analysis will concentrate on
the more or less overt dimensions of the problem at hand.

At this point we may begin by considering the various approaches suggested above such as the Generalized Protocol
(Table 6-1) or the Input-Thruput-Output problematization suggested in Figure 6-5 (and Table 6-2), and lay out the research
strategy suggested in the assessment methodology. Although all
the tables and figures are part of our eventual methodological
armory, we may start here by delineating a general strategy for
problem identification (Figure 6-9).

In the first column of the proposed problem identification strategy, we may consider the goals, needs or actions of
the community and at the same time larger regional and national
trends which may be producing the problematic situation of, let
us say, traffic bottlenecks and transportation inefficiencies.
This category of our map coincides with the earlier described
part of planning procedures, i.e., the systemic or descriptive
phase.

In a second column we can elaborate the causes or factors
for the appearance of the particular problem. The causes of the
problem may be as stated, or different from what appear in the
official pronouncements. In the case of the proposed by-pass we
may include such items as the high level traffic, the degree of
accidents, the need for improving a particular area of the town,
the generation of income, etc.

A third column incorporates the responses to the problem
as conditioned by the goals, values, and objectives that we try
to accomplish in introducing the public project. Essentially,
here we are talking about two major types of responses: letting
things be as they are (do-nothing option) or intervening. In

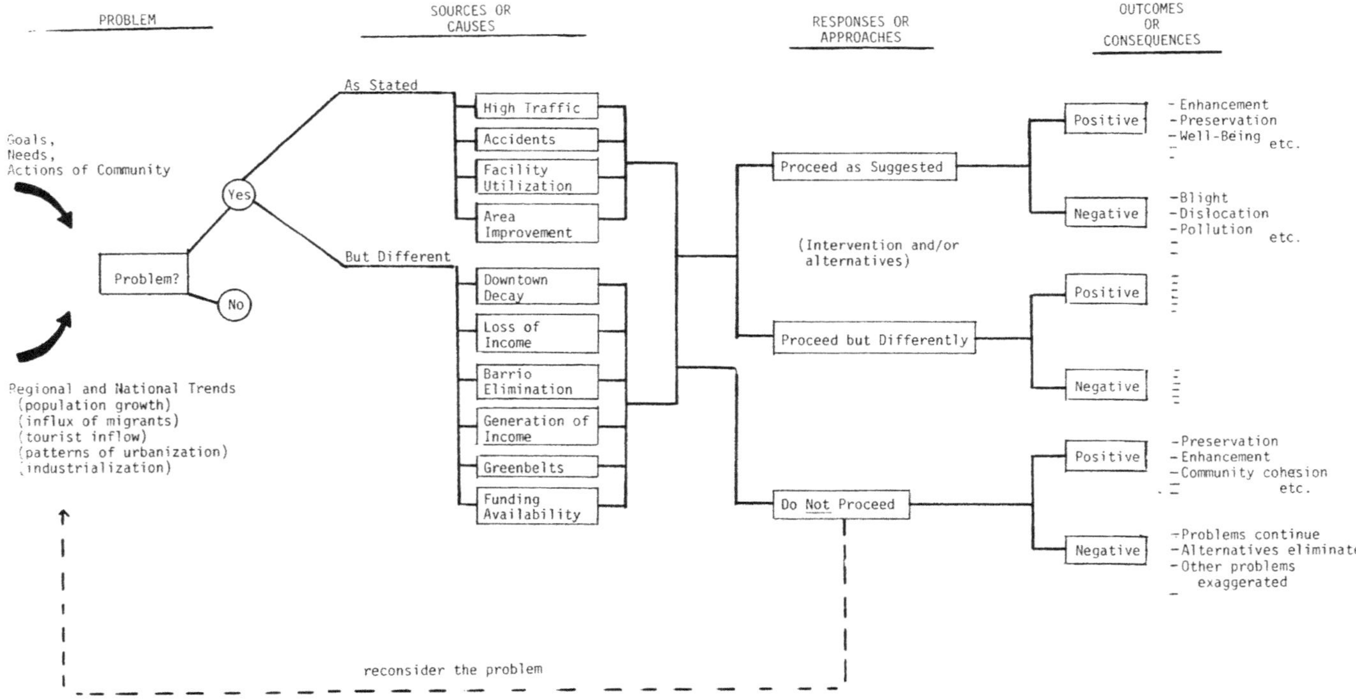

Figure 6-9. General strategy of problem identification.

the second case, i.e., the decision to proceed with the construction of the by-pass, we may want to consider responses as presently suggested, or other alternatives of construction, each to be determined by an evaluation of potential effects (parallel to the evaluation of alternatives operation in the social planning process of Figure 6-6). At the same time, in the case of non-construction we may be referred back for reconsideration of the problem (consequences for the community from the absence of the project).

The fourth column of our diagram refers to the consequences or outcomes if the particular public project will or will not be introduced in the community under examination. Given the problem, the causes, and the type of response we have certain consequences. Our thinking here is guided not only by the specific envisaged effects of the by-pass in our community, but also by the results of a large number of other highway impact studies. While we recognize in our diagram some broad outcomes of highway construction (or non-construction) past literature can give us a range of impacts expected such as:*

- new highways often hasten economic changes that were previously underway;
- the joint development of highways and residential, recreational, or commercial facilities is one method of insuring compatibility of the total environment;
- highways sometimes affect local businesses adversely;
- bypass highways have had some adverse effects but the net effect on the bypassed area has tended to be neutral or beneficial;
- residential properties have apparently been more often benefited than damaged by nearby highways;
- dislocated residents of urban areas who are poor, old, or non-white have serious economic and social problems in relocating their houses;
- residents relocated from right-of-way areas have typically improved their living accommodations, though at increased costs to them, etc., etc.

It can be certainly observed that in most of the above random items of past research the emphasis is on economic costs, short-range effects, with little attention to the more diffuse "social costs" of disrupted neighborhood networks, unsightly structures, and interference with old established patterns of social interaction. We can extend our thinking in the example of Ramburg by following through the various methods outlined

---

*This list of impact was taken from a long list of studies provided in the publication of the Federal Highway Administration, U. S. Dept. of Transportation, *Economic and Social Effects of Highways* (Wash., D.C., 1972).

above and in the context of the systematic problem identification scheme in Figure 6-9 alternative outcomes from the construction of a bypass that intersects the identifiable Chicano
neighborhood.   Thus, the consequences would not only be the
ecological perturbations of having a highway by the streambed
with assorted pollution levels, and not only economic in that
it may be relatively efficient to construct a route through a
rather inexpensive land use area. The destruction of community
life and the displacement of people who can ill afford expensive housing in other parts of the city may have a much more
severe cost.   On the other hand, the non-construction of this
bypass may have significant consequences for the rest of the
community in that it may increase downtown decay, or through
increased traffic congestion contribute to more accidents.   Far
reaching consequences may also result from the continuous sprawl
of the city to other undesirable directions, strip-city growth,
and elimination of future opportunities to plan towards better-
ment of an increasingly isolated Chicano community.

Given the above general cognitive map of problem identification and with the background of all previous attempts to explicate the dimensions of human community affected by a public
project, we must consider some specific methodology that would
do a number of things for us:

1. "Measure" responses and consequences, or develop indicators
    sensitive enough to provide us with various degrees of the
    effects of the public projects.

2. Provide criteria of assessment following the "measurement"
    of various alternatives and a procedure for deciding among
    various policy interventions.

3. Define categories of recommendations for certain action
    options.

We have seen already that the both general methodology and
specific indicators are in their infancy stage vis-a-vis public
projects and human community. The Generalized Protocol proposed
is at best a sensitizing instrument as to minimum sociological
categories and variables of concern and a first attempt of generating some pertinent "indices."   Together with the impact-
thruput-output scheme of problematization (Figure 6-5) they are
essentially checklists against which we can provide some initial estimates of social costs, or effects of the project on
the social system.

But let us speculate for a minute as to what an ideal social cost estimation might involve.   To be able to develop a
methodological scheme or a matrix that would reflect our concern for evaluating impacts and assessing the technological imperative on human community, we need to arrange in some meaningful way the following items:

1. Desired goals.
2. Predicted effects (impacts and consequences).
3. Evaluative criteria for estimating effects.
4. Ranking priorities of the evaluative criteria used.
5. Probability of occurrence of predicted effects.

Perhaps the best way of showing the outline of such a future methodology is the following scheme relating criteria and effects (Figure 6-10).* The procedure can be briefly described as follows: If we are to pursue a given goal (let us say goal A, such as the construction of a bypass to alleviate traffic congestion) the horizontal axis shows a number of predicted effects if this particular public project is to be pursued. On the vertical side we may want to develop the evaluative criteria or specific indicators against which "measurement" can be applied to the calculation of predicted effects. What we need next in our matrix, is a set of numbers that make is possible to measure the impact of any evaluative criterion on the predicted effects, i.e., a value that could be placed in any given cell of our particular matrix. Two additional operations would make our attempt more sensitive in the judging of effects in the construction of a public project. We need a numerical ranking of the evaluative criteria, or an ordering which will indicate the relative worth of each criterion used in our total environmental design. Finally, it would be important to estimate the probability with which the predicted effects may occur. Thus, the above five conditions (enumeration of predicted effects, list of evaluative criteria, cross-impacts of effects and criteria, ranking of evaluative criteria, and probability of predicted effects) may provide us with different measures or sensitive indices for deciding the effects of a proposed project. We may think, for example, of additional indices in the interaction between Effects x Criterion, or Priorities x Effects, or Probability of Effects x Priority, etc.

What we might envisage at this point is the development of a Grand Index, or perhaps a "total worth index" which is indicated in our figure by the grand sum ($\Sigma\Sigma$). We may also think of partial indices in the comparison of various single criteria per a whole range of effects (Σ). To expand the argument even further, rather than a single goal analysis, we may even think of a methodology that could incorporate different matrices and appropriate comparisons in the pursuit of various goals, as suggested in the lower part of Figure 6-10.

---

*The impetus for such a model was provided from notes taken at a presentation of B. Crawford at the 1972 Rocky Mountain Social Sciences Assoc. Meeting in his paper "Social Values and Environmental Policy: An Evaluative Model."

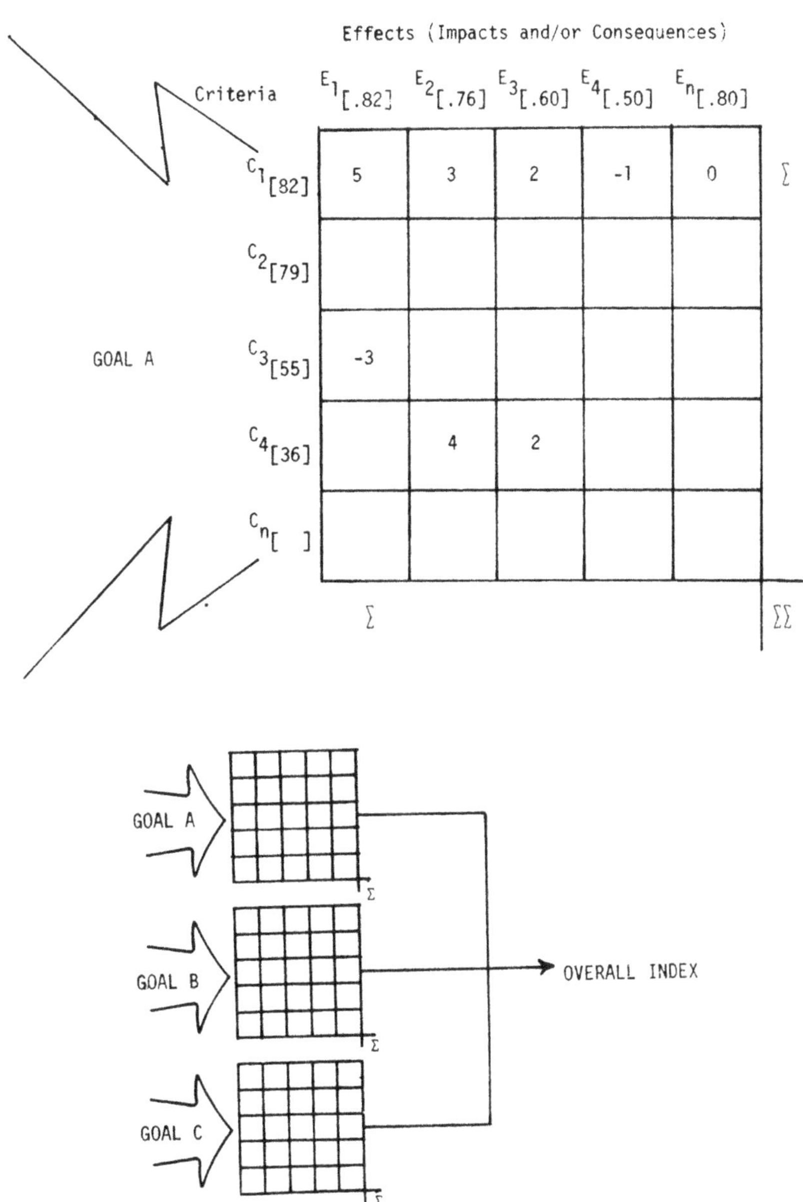

Figure 6-10.   Idealized scheme of social effects calculation.

Such devices and indices, although desirable and pointing to the direction of comprehensive strategies in total environmental design suffer, of course, from the general problem introduced earlier in our discussion of social sciences methodology and data requirements. Not only the types of data available for more sensitive evaluative criteria are limited, but even data at hand are often characterized by serious problems of validity and reliability. Also, the existing field procedures and techniques for obtaining appropriate data are crude and in great need of improvement at least in the field of measuring environmental impacts on the human community. We need not only the classical methodology of obtaining structural data and the attitudinal surveys which map the reactions and feelings of people vis-a-vis, various environments but also new types of forecasting techniques. Such promising techniques involve a host of "consensus studies," the most prominent of which are the Delphi studies. The Delphi technique is directed toward the systematic solicitation of expert opinion where rather than achieving a consensus by open discussion, it uses a carefully designed program of sequential individual interrogations interspersed with information and opinion feedback derived from consensuses which are computed from earlier parts of the technique. Other non-traditional approaches of use for a far-sighted methodology of technology assessment and social forecasting include scenarios, simulation models, dynamic modeling, cross-impact matrix methods, decision-trees and other techniques which may provide sensitivity to questions of subjective values and to normative forecasting.

By now, we have come a long way from the earlier discussion of trying to delineate human community and the effects of a public project on a given social structure. Throughout the previous pages it has become apparent that there is a dramatic quest toward conceptual redefinition, the creation of cognitive maps, and more than anything else development of methodologies that would permit us to assess more precisely and with higher sensitivity the extent to which public projects affect individual and communal life. In such a discussion it has become obvious what already has been expressed in quite a number of studies and what has become almost an intuitive wish of most practitioners in the field: namely, the desire to develop much more concrete indices for being able to understand and assess the social costs involved in any discussion of public projects. There is a tremendous challenge in raising much more demanding questions toward people who profess an interest and commitment to the understanding of social systems. We need to move from a generalized social criticism and from a diffused affective thinking about the environment, to a sober discussion of questions of social policy, delineation of alternatives, and thoughtful development of appropriate indices (quantitative and qualitative) measuring social effects of public projects.

Before closing this section a concluding digression becomes necessary. There is an overriding need for both a holistic and future-oriented thinking. One may venture the term metaplanning as a model that can move us beyond routine responses that tend to characterize limited impact analysis. It would be, then, appropriate to hope that, as various authors throughout the treatise have also hinted, we can develop a thinking and a planning commitment characterized by the following dimensions:

1. *Teleological* (Normative)

We need to know the aims, goals and targets that we try to achieve with a given public project and the normative commitment towards present and future goals. A "goaless" system as well as a "goaless" society are bound to be wasteful and impossible to evaluate.

2. *Hierarchical*

Our thinking should also involve considerations of social priorities and a system of ranking in the establishment of probable, possible, and plausible, goals and alternatives in our society. Such thinking implies not only questions of social policy but also an understanding of political constraints and the exigencies of the political system that affect the implementation of a public project.

3. *Systematic* (Integrative)

There should be full awareness of the open character of the total environmental system, where everything interacts with everything else in an interlocking series of sub-systems. Such thinking not only tries to establish compatibility of objectives between and within various component systems and sub-systems but also a clear understanding of the multidimensionality of our model of thinking.

4. *Diachronic* (Anticipatory)

Our thinking should also be guided by an understanding of continuous, on-going change and a recognition of the fact that many consequences of public projects have long-time lags before they become visible and significant. We need to avoid entrapments into expedient present projects which may foreclose our options for the future and, thus, be cognizant of fundamental conflicts between short-term gains and long-term consequences.

5. *Ecumenical* (Holistic)

Because of the growing number of problems and the recognition of the interrelationship between micro-, meso-, and macro-environments we need to adopt a cosmopolitan attitude about the

world around us.   The essence of such thinking is based on the
lack of parochialism, a global outlook, and on a very clear understanding that impacted or affected environments have spill-
over effects on larger social systems.

The above rather diffuse principles and the increased concern with the rational management of our world also provide the
social scientist with heavy but exciting responsibilities.   We
have the tremendous opportunity for being not only in the fore-
front of technological innovations, but also in transforming
social awareness and a moral imperative into an operational
philosophy of total environmental design.   A design which may
successfully incorporate physical, ecological, and socio-
cultural demands in a cogent model and provide guidelines for
planning social policy aiming to a better future for all mankind.

GLOSSARY

Attitude	A predisposition toward action or parts of the social system; probability that given behaviors will be manifested in defined situations.
Community	A combination of social units and systems which perform the major social functions having locality relevance; a combination of people, territorial base, identity, and common culture.
Consensus	Condition of agreement or shared under-standing within a social unit.
Criterion	An operational means that can be applied to specific objectives or goals.
Culture	Standardized ways of feeling, thinking, and acting that man acquires as a member of society; a group's total man-made en-vironment or social heritage.
Delphi	A technique for the systematic solicita-tion of expert opinion and the deriva-tion of group consensus.
Effectiveness	Organizational performance, or the rela-tionship between organizational structure and perceived goals.
Efficacy	The overall relationship between input, thruput, and output in a social system

296

	and the calculation of social costs thereof.
Efficiency	A ratio between input and output; primarily, benefit cost analysis.
Environment	The system of spatial, temporal, and social regularities which influence the biological and behavioral processes of a given population; the sum total of the external factors that affect an organism.
Equilibrium	The tendency of a system towards balance among parts and forces operating within and upon it; a tendency of a social system to remain the same.
Equity	Fair access of resources and consumption by different elements of the population.
Evaluation	The process of assessing the extent to which a given project or program succeeds in meeting stated goals.
Folkways	The most spontaneous and unconscious adaptations, or usages and patterns of social relations; a type of norm.
Forecasting	The probabilistic assessment of future alternatives; the comparative evaluation of alternative technological options.
Function	Observed consequences that make for the adjustment of a given system; the effect or contribution that one unit of society has on another unit or on the society as a whole.
Goal	Change which a person or group intends to effect in any given situation; things worth striving for.
Index	A ratio or composite reflecting on-going changes in the environment; a numerical measure for ranking alternatives.
Indicator	A signal of a current condition in the environment.
Institution	Formal and enduring system of norms, growing out of fundamental individual interests, implemented by organizations,

aiming at the attainment of important societal goals; crystallized ways of doing things.

Interaction — Behavior based on taking into account reciprocal actions and reactions of both individuals and groups; reciprocal influence of phenomena on one another.

Mores — Social norms that provide the moral standards of a society and which are strictly enforced.

Norm — Pattern of social behavior, of what is socially required, expected, or usually done; rules that specify appropriate and inappropriate behavior; behavioral expectations.

Normative Order — The system of norms and values governing behavior in a given society, and/or orderliness resulting from behavior so governed.

Organization, Formal — A structural group having explicit objectives, rules and procedures, and specifically defined roles, each with designated rights and duties.

Power Structure — Pattern of arrangement and exercise of authority within a social system.

Reification — Confusing a concept with the phenomenon the concept's supposed to symbolize; (fallacy of): When a concept is treated as if it were real-life situation.

Reliability — Random error control in measurement procedures; tendency to produce same results over time

Role (or Social Role) — Expected pattern of behavior of an individual or a group, attached to a given position incorporated within a social system.

Sanction — A reward or punishment intended to encourage or discourage certain behavior.

Social Environment — The surroundings affecting behavior, including physical environment, social interaction and institutional networks, and common ties and normative system.

Social Order	The stable, systematic pattern of social life, society, and the social system; the total of social relationships and of culture.
Social Planning	A means of directing change and social relationships toward the ultimate objective of orderly and harmonious community processes.
Social System	A network of social interaction, or any unit of social organization composed of a set of component parts and a complex pattern of operation; a collection of people, devices, and procedures intended to perform some function.
Society	The most inclusive, complex, and self-sufficient type of social grouping; a complex web of social relationships.
Status	Place of an individual within a structure of positions; the resulting prestige and esteem of a given position.
Structure	A high degree of continuity and a recurring pattern of interaction over a considerable period of time; patterned arrangements among parts of a system.
Subculture	The culture of a certain segment of society, such as a social group; the normative system of a group smaller than the total society.
System	A complex of elements or a configuration of parts that are in a relationship of interdependence through a period of time.
Technology	Knowledge directed toward practical applications in the physical and social world; the collection of practices by which one uses available resources in order to achieve certain valued ends; the body of knowledge and the tools available for the production and distribution of goods and services.
Technology Assessment	The evaluation and assessment of either new technological innovations or the applications of alternative technologies and proposed approaches of existing problem areas.

Validity	Criteria of measurement free of systematic error; the degree to which a scale measures what it purports to measure.
Value	A generalized conception of what is good, beneficial, desirable, or worthwhile; a desired goal.
Value System	A set of priorities in various dimensions of a social system that forms the basis for the establishment of criteria.
Variable	Any changeable element of the social system, capable of serving as either cause or effect; any phenomenon that can change.

BIBLIOGRAPHICAL COMMENTARY

The published literature that bears, directly or indirectly, on the subject matter of the topic "Human Community," as well as on social aspects of the environment, is not only vast, but grows daily by leaps and bounds. What is intended in the following text is a brief overview of some key words (primarily books) that they have guided the thinking and the writing of the previous pages. Essentially, the bibliographical argument will follow the sequence of presentation in Chapter 6--noting key words that the reader may consult for further commentary.

There are many books, professional and otherwise, that touch upon the study of human communities and social systems. If one is to start, any basic textbook in sociology (and there are plenty) may provide clues as to the typical argumentation concerning society and social phenomena. Perhaps one of the better books is Everett K. Willson's *Rules, Roles and Relationships* (Homewood, Illinois: The Dorsey Press, 1971). Other favorite introductory textbooks are also Robert Bierstedt's *The Social Order* (3rd edition. N.Y.: McGraw Hill, 1970) and Robert A. Niesbet's, *The Social Bond* (N.Y.: Alfred A. Knopf, 1970). Beyond such general books, a wide variety of specialized volumes discuss aspects of society, social problems, and questions of social organization.

In terms of human community, the field has early been represented in various works, especially in studies of small towns and communities all over the United States. The reader may want to start with Roland L. Warner's *Studying Your Community*, N.Y.: The Free Press, 1965 (a paperback bargain, indeed), which sets out useful guidelines as to the kinds of questions to be asked in the understanding of a given community. For a more general discussion of the field of human ecology, the classical work remains that of Amos H. Hawley, *Human Ecology* (N.Y.: Ronald Press, 1950), although it is antiquated and quite replete with sometimes sweeping biological metaphors or analogies. The basic book on community is that of Irwin T. Sanders, *The Community* (2nd edition. N.Y.: Ronald Press, 1966). More recently, the work of Dennis E. Poplin, *Communities; A Survey of Theories and Methods of Research* (N.Y.: Macmillan, 1972) provides a most useful inventory of current research and recent findings on human settlements. (Equally useful is also the discussion of the author on conceptual differentiations in the study of community.

Social systems analysis has descended on the field with unusual intensity. An essential book here is the descriptive but insightful little volume by C. West Churchman, *The Systems Approach* (N.Y.: Delta Books, 1968), which although popular and addressing the general public, states succinctly the major premises of the systems approach. Useful are also such books as Walter Buckley's *Sociology and Modern Systems Theory* (Englewood Cliffs: Prentice Hall, 1967) and his edited volume, *Modern Systems Research for the Behavioral Scientist* (Chicago: Aldine Press, 1968). (See also the succinct article by Robert R. Mayer, "Social System Models for Planners" *American Institute of Planners, Journal,* 38 (1972) pp. 130-139). Other works on the systems approach include also George H. Monane, *A Sociology of Human Systems* (New York: Appleton Century Crofts, 1967), Ervin Laszlo's overview of "Systems Thinking" in *The Systems View of the World* (New York: Basic Books, 1972), Robert Bognslaw's, *The New Utopians; A Study of System Design and Social Change* (Englewood Cliffs: Prentice-Hall, 1965), and the recent commentary and succinct insights of Ida R. Hoos, in her *Systems Analysis in Public Policy: A Critique* (Berkeley: University of California Press, 1972) which puts into a proper perspective both the strengths and weaknesses of systems analysis in the social sciences.

At this point one may also want to turn attention to the tremendous spate of books that appeared with the "environmental" movement. The bibliography here is more than abundant although descriptive and characterized by a "doomsday" approach. One may want to start with a bibliographical compendium by Robert Durrenberger, *Environment and Man* (Palo Alto: National Press Book, 1970). Among other volumes on the relationship between social and natural aspects of the surrounding environment that have been useful for the formulation of the argument in the chapter are Sterling Brubaker, *To Live on Earth; Man and his Environment in Perspective* (Baltimore: The John Hopkins Press, 1972); John McHale, *The Ecological Context* (New York: George Braziller, 1970); W. W. Murdoch (editor), *Environment: Resources, Pollution and Society* (Stamford: Sinanez Associates, 1971); Richard H. Wagner, *Environmental and Man* (New York: W.W. Norton, 1971; William R. Burch, Jr. *et al.* (eds), *Social*

Behavior, Natural Resources and the Environment (New York: Harper and Row, 1972); and, similar books, articles, and collections which increasingly bring attention to the interconnection between man and his surrounding physical world.

The interface between public projects and human community has not been adequately explored. More discussion has taken place around the role of technology in society, rather than specified aspects of public project inducements. Perhaps, more important here are the arguments developed in the context of social planning and the interrelationship between environmental design and social systems. For the technological imperative, one may find most useful the large volume edited by M. T. Favar and J. P. Milton, *The Careless Technology* (Garden City: The Natural History Press, 1972). A whole number of books are also discussing problems of technology and man's future such as Jack D. Douglas (ed.), *The Technological Threat* (Englewood Cliffs: Prentice Hall, 1971), the small volume by Emmanuel G. Mesthene, *Technological Change; Its Impact on Man and Society* (Cambridge: Harvard Univ. Press, 1970), as well as the reader by Carl Mitchum and Robert McKay, *Philosophy and Technology* (N.Y.: The Free Press, 1972).

Much more important today is the increasing preoccupation with the problems of technology assessment and social forecasting. Extremely valuable here would be such works as Olaf Helmer's *Social Technology* (New York: Basic Books, 1966), the primer by H. W. Lanford *Technological Forecasting Methodologies: A Synthesis* (American Management Association, 1972) and the six volumes by Mitre Corporation on technology assessment.

How does one go about developing guidelines for comprehensive environmental planning from the sociological point of view? The literature here is almost non-existent. However, general books on introductory sociology, social systems analysis, etc. have all lists of important dimensions, factors, variables, etc., involved in any discussion of environmental design and social systems. Most important here are not so much existing books in social sciences, but specialized studies, as well as environmental impact statements, in which authors working out in the field have been trying various checklists, models, or matrices that reflect concern with sociological dimensions of environmental design.

It will be more appropriate to talk about the general epistemology of social planning rather than specific studies. On this topic the works of Alfred J. Kahn, *Theory and Practice of Social Planning* (New York: Russell Sage Foundation, 1966), Harvey Perloff (ed.), *The Quality of the Urban Environment* (Baltimore: The Johns Hopkins Press, 1969), and, Mellvile C. Branch, *Planning: Aspects and Applications* (N.Y.: John Wiley and Sons, 1966), are books that can be immediately used for the general guidelines of social planning. Very useful for a general sensitivity can be also quite a few urban planning books. A favorite here is J. Brian McLoughlin's *Urban and Regional Planning; A Systems Approach* (New York: Praeger, 1969) and the vast collection of articles edited by Bernard J.Friedan and Robert Morris, *Urban Planning and Social Policy* (New York: Basic Books, 1968).

Turning to more specific efforts concerning an assessment methodology, social scientists, as indicated above, were suddenly hit by the requirements of NEPA. There is an abundance of books on the methodology of social sciences, but virtually nothing with the specificity required by EIS. The reader will find very useful for an understanding of methodological strategies and tactics such general works in sociology as Bernard Phillips' *Social Research, Strategy and Tactics* (2nd edition. New York: Macmillan, 1971); William J. Goode and Paul K. Hatt, *Methods in Social Research* (New York: McGraw-Hill, 1952); and, Norman K. Dezin, *The Research Act* (Chicago: Aldine Press, 1970). For many of the issues discussed in Chapter 6, such as sources of data, problems of measurement, and specific techniques in sociology, one may want to consult such specialized volumes as Earl R. Babbie, *Survey Research Methods* (Belmont: Wadsworth, 1973), Morris Rosenberg, *The Logic of Survey Analysis* (New York: Basic Books, 1968), Peter Abell, *Model Building in Sociology* (New York: Schocken Books, 1971), Delbert C. Miller, *Handbook of Research Design and Social Measurement* (2nd edition. New York: David McKay, 1970), and a whole host of similar books on specialized aspects of field work in social sciences.

In terms of evaluating the impact and consequences of any large scale project, there are a number of interesting books discussing concepts and general premise of evaluation methodology. The classical book in this area is Edward A. Suchman's *Evaluative Research* (New York: Russell Sage Foundation, 1967) and a new volume by Arnold C. Harberger, *Project Evaluation* (Chicago: Markham Publishing Co., 1973). Two important readers in the area may be also of interest: the work edited by Peter Rossi and Walter Williams, *Evaluating Social Programs* (N.Y. Seminar Press, 1972) and Francis G. Caro (ed.), *Readings in Evaluation Research* (N.Y.: Russell Sage, 1971).

A very important issue in the methodology of evaluation and assessment is the increasing use of social indicators. The classical study here is the Report by the United States Department of Health, Education and Welfare, *Towards a Social Report* (Ann Arbor: University of Michigan Press, 1968). The book edited by Raymond A. Buaer, *Social Indicators* (Cambridge: MIT Press, 1966), set out the stage on the field and was then succeeded by a number of recent works on the topic, such as Eleanor B. Sheldon and Wilber E. Moore (eds.), *Indicators of Social Change* (New York: Russell Sage Foundation, 1968), and, Angus

Cambell and Phillip E. Converse (eds.), *The Human Meaning of Social Change* (New York: Russell Sage Foundation, 1972). One of the better collections is the vast bibliography provided in the recent volume by Leslie D. Wilcox *et al.*, *Social Indicators and Societal Monitoring* (San Francisco: Jossey-Bass, Inc., 1972).

As the field of environmental sociology is expanding it is expected that the literature will rise to the occasion. More than anything else, the reader will also benefit from the books that they come under the general label of futurism where they discuss the question of technology assessment and long-range consequences of present public projects. Such books would include the work edited by Kurt Baier and Nicholas Rescher, *Values and the Future* (N.Y.: The Free Press, 1969), as well as Arthur Bronwell (ed.), *Science and Technology in the World of the Future* (N.Y.: John Wiley, 1971), Stewart Chase, *The Most Probable World* (N.Y.: Harper and Row, 1968), William R. Ewald, (ed.) *Environment and Change; The Next Fifty Years* (Bloomington: Indiana University Press, 1968 -- together with the two other edited volumes in this series *Environment for Man and Environment and Policy*), Donald N. Michael, *The Unprepared Society; Planning for a Precarious Future* (N.Y.: Harper-Colophon Booles, 1968) and an increasing number of similar books.

We could conclude by indicating that quite a number of propositions concerning environmental assessment and planning are borrowed from parallel literature in other fields or in the areas of urban planning and social policy. The reader would greatly benefit from the perusal of literature in the general area of social policy and here a very useful small book is the paperback by Howard E. Freeman and Clarence C. Sherwood, *Social Research and Social Policy* (Englewood Cliffs: Prentice Hall, 1970). This small, but admirably well-written volume tackles many of the methodological and larger questions concerning public policy, the connection of research and planning, and the dynamics of the social research task. On the other hand extremely useful material is to be found daily in the periodical literature, governmental reports, and in the continuously improving environmental impact statements.

AUTHOR NOTES

Evan Vlachos has become increasingly involved over the past several years in the applications of sociology relating to large scale engineering projects and resource development. This involvement is reflected in his research and teaching in the areas of futurism, technology assessment, and studies of irrigation institutions, which compliment his work in the broader areas of demography, urbanization and methodology. Consulting assignments have ranged from urban planning for the City of Athens to diversified environmental-social assessments for many

state and federal agencies.   His numerous books and   articles
reflect this wide range of interests.

    In addition to his doctorate in Sociology, he also holds a
law degree from the University of Athens and a Certificate of
Russian Studies from Indiana University. He is Professor of
Sociology at Colorado State University.

VISUAL HARMONY

BY DWAYNE C. NUZUM

Public projects are a physical means by which a culture can express its attitudes and aspirations. Do the public projects being built today express our concerns for our land, our sense of design, or our commitment to total environmental design?

The objective of this chapter is to provide an understanding of the concept and significance of visual harmony relating to landscape design and physical (structures) design--emphasis will be on the appropriateness or esthetics of a design project and the means or methodology by which this can be attained. Within each culture the esthetics of that culture may vary from the past culture. Therefore, an analysis of past cultures is necessary to understand changing values as a concept. If the relationship between form and culture can be established, then appropriate landscape design and structure design principles can also be established.

Little written work exists on the visual aspects of public projects since many designers think design is intuitive instead of a decision making process. However, recent research is being done on the visual aspects of urban design projects and architecture projects.

This chapter will also attempt to define the basic aesthetic (with variations) that would seem to be appropriate to public works projects in their visual design. That is, they can be regarded as a popular art *form* in the best sense of the word that are perceived continuously by a multitude of people. These artifacts are exposed to the public, but they do not necessarily cry for deliberate aesthetic appreciation. In many cases, inoffensiveness is an adequate goal, but the tremendous potential to achieve mass sensitization to good design has to be recognized and exploited. The upgrading of general public taste

is a responsibility that should not be avoided.   Beauty in the
classical sense may be a wrong criterion since much more is involved.   But a contemporary public works aesthetic has to be
developed for the public project.

HISTORICAL VISUAL ASPECTS

The Greeks (if any culture) are best known for their sense
of visual harmony.   The design of their buildings, site plans
and cities is based upon the sense of the finite and the sense
of simplicity.   Each landscape design or building design was
based upon its relationship to the surrounding landscape including buildings. Although little evidence exists that planned
growth systems were developed, each additional physical unit
was designed, built and rebuilt until it met the functional requirements and the visual requirements--harmony between man and
nature--as experienced by the user.

ACROPOLIS

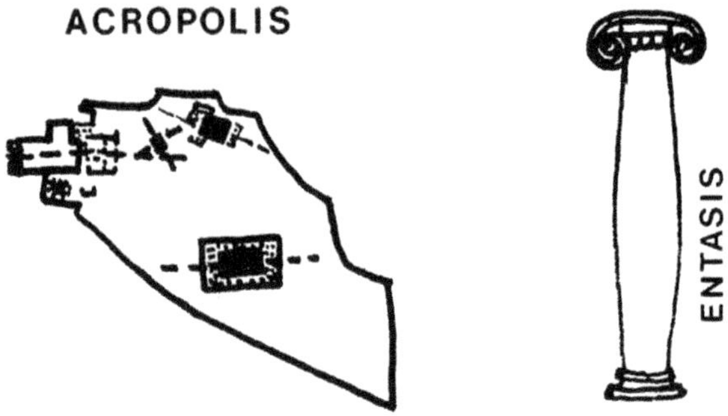

Another Greek visual design requirement was the utilization of optimum size of physical objects as defined by comprehension, function and scale with man as the measure. This was a
culture in which beauty was important and criteria were established by which to measure it.

The *Roman Empire* was motivated by political power, organization and technology.   Although we marvel at the engineering
feats of the Romans, few praise the visual aspects of their
aqueducts, roads, arches and public facilities. These physical
structures make marvelous ruins, but today, we question their
appropriateness to the scale of man and the adjoining landscape.
The Roman engineering took precedence over esthetics and emphasized the visual power of the emperors.

As the Roman Empire declined, over a long period of time,
western civilization began to develop regional towns.   These
medieval town designs were a direct response to the life of the

community. The major shopping-market space was dominated by the church (or City Hall) at the edge.

The form of the town was dictated by logical growth systems and public works projects consisting of defensive systems, minimal sewage and drainage facilities and circulation patterns. Today the forms of the medieval cities are picturesque, however, even though they were an appropriate concept of life for that

era, today this is not totally true. The star shapes and circular forms of the medieval city were a direct response to the armament and protection devices of that time.   Stone walls were constructed as a means of protection utilizing available materials or in areas where stone was not available, heavy thick brick walls could be utilized for protection. The points of the star in the star shaped cities were located such that all of the area between the points could be reached with bows and arrows and other artillery devices of that time. Although sewage systems are known from the Agean era, the medieval city started to incorporate sanitary sewer and drainage systems within the street layout.

Usually, each city was a day's ride from the next and the size was set by the amount of food grown that could support the population within a half day's ride. Medieval building was characterized by buildings that were usually human in scale and built of similar material.   Saarinen, the Finnish architect, believed that there

309

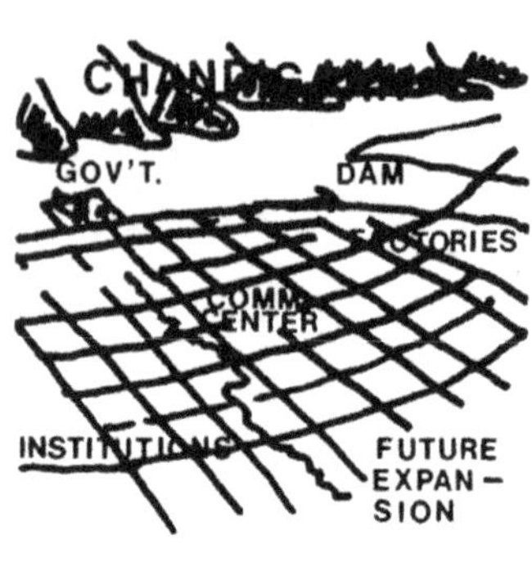

is a direct visual relationship between the medieval city and present day planning objectives. Especially the concern of growth systems in which a community is a "total" entity during the growth process. Many present day projects are unfit to live in until totally completed. Examples include Brazilia and Chandigarh.

The Renaissance designer was interested in the new tool of space simulation perspective and the cultural interest in past Greek and Roman eras. These interests coupled with the power of church and Pope brought about the rebuilding of Rome in the new style. Problems of traffic, defense, sanitary facilities (public works) forced the Popes to undertake civil improvement projects. These projects also became visual nodes to strengthen the overall city circulation concept. This is a good example of public works and civic works projects being coordinated. The use of perspective as a design tool allowed the designer to explore visual deception. In Michelangelo's capital project the entry stairs and buildings are not parallel to change the perceived scale. St. Peter's square uses the opposite technique for opposite reasons. Both are appropriate for the stated objectives.

MICHELANGELO'S CAPITAL   ST. PETER'S

The Renaissance designer also became involved in the design of rural areas for the nobility of the era. In designing hunting grounds, it was found that by cutting long straight

clearings in the forests which intersected at high ground
points, animals could be spotted as they moved from one forest
to another.   This design concept became the model for the town
and park of Richelieu.   Later in time, this same design concept
was utilized by Pierre L'Enfant for the design of a new capital,
Washington, D.C.

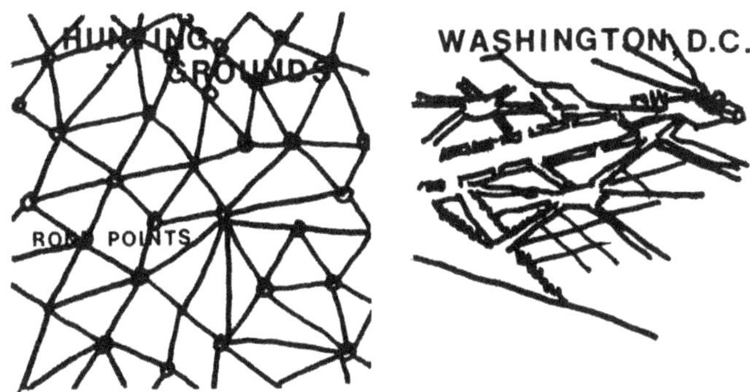

The different design ideas of the French and English
should also be compared. The French thought that their natural
landscape was inappropriate to man.   Therefore, they tried to
remake nature into primarily geometric patterns.   Shrubs were
trimmed into cubes, trees into conical shapes, and flowers were
planted into mosaic patterns.   The English after experimenting
with the French forms, developed a "picturesque" landscape system.   In essence this was the taming or altering of the natural
landscape into a man-made design. Examples would include removing bushes from under trees to give accent to the trees or
the planting of bushes along a prominent topography to accent
it.   As Ian McHarg points out in his lectures, the history of
western man shows few instances, if any, in which man has hon-
ored the natural environment.   Man has always tried to change
the environment to better his way of life.   In the future, public works projects will have to respond to the question, "How
much of the natural environment is being changed to man-made

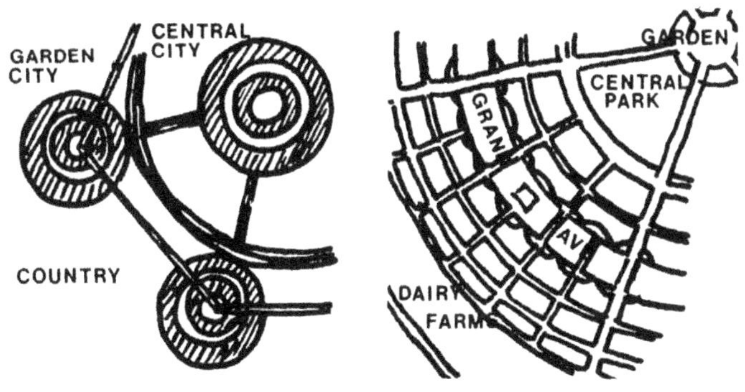

environment and what are the positive trade-offs in this change?" The modern designer has been primarily influenced by rapidly changing technology. In planning, the Garden City concept of Ebenezer Howard was a direct response to technology. Howard's concept tried to incorporate the best of both environments, the country and the town, he tried to design a community that could produce goods and services without the pollution, poor working conditions and a poor health environment of the factory towns. Technology was to provide mass transit facilities to link optimum sized communities, to provide pollution free industries, and to provide moderate cost housing with individual gardens. The Garden City concept is the standard objective of town design today. Although the forms appear different in Le Corbusier's and Wright's projected Garden City concepts, the forms represented the new technology--simplicity, repetition, use of new materials and mass transit. Wright realized that people still required contact with nature therefore his forms tried to exploit nature by building a house over a waterfall, or around a tree or on the side of a hill instead of the top. Today's designers are primarily interested in "expressed function," the design of the project should clearly state the use to the observer, articulate the mechanical and structural systems utilized and express the use of contemporary materials in a natural way.

CONTEMPORARY VISUAL IMPLICATIONS

A recent example of visual concern was shown in the People's Park episode in Berkeley, California. The "street" people objected to the University parking facility, therefore, they tried to create their own garden-park. The design was mundane, but the involvement in the design process was critical to all of those involved. The idea of process and involvement has been important to most contemporary artists. Painters, such as Robert Motherwell or Jackson Pollack, have expressed content through spontaneity of process. They are capturing on canvas their interior feeling for that time. This kind of philosophy has invaded our culture. The short term "visable city," the hippie light shows, the paintings upon construction fences are all part of today's visual world. The idea of permanence in today's man-made landscape design and structures design is suspect.

POP ART

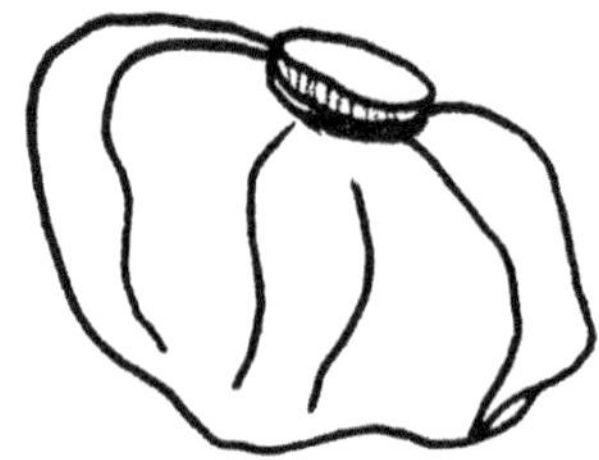

Pop-art is another art form which relates contemporary society to its perceived physical objects. Is a large ice-bag or a coke bottle unattractive when put in the proper context? Is not the functional beautiful? The pop-art movement requires the viewer to re-evaluate his preconceptions. Are all billboards and

signs bad?   Do all wires   have to be underground?   Should all
highway projects be stopped? Such questions force the viewer to
distinguish between acceptable and unacceptable design questions.

A critical problem for the designer is "The Media is the
Message" syndrome. Television has changed the concepts of time
and has condensed visual experiences so that past visual processes are no longer valid. T.V. provides information and complex solutions within a one-half to one hour time frame. Therefore, the planting of small trees is no longer acceptable since
immediate solutions are expected by the user.   A similar problem is the accelerated visual simulation provided by T.V. and
movies.   Man-made projects which do not have a similar visual
simulation tend to bore the user.   T.V. is also the "great
equalizer."   A successful public facility built in South Caro-
lina will be copied in Northern Montana Within a year's time
frame.   Therefore, the indigenous criteria of regional design
is being lost for the sake of national image building via the
communication systems.

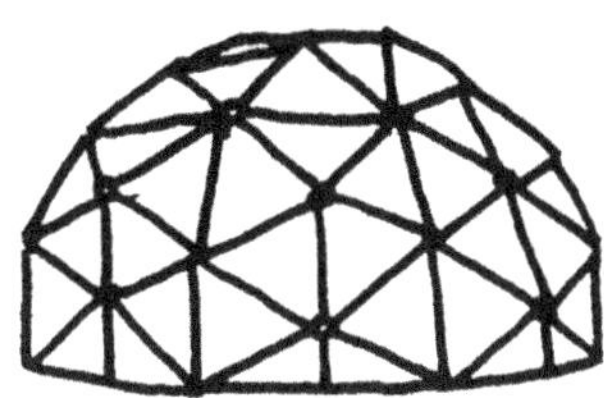

Another design philosophy developing today is the "Whole Earth Catalog." Young people are wanting to
acquire simplicity and determination
of their own life patterns.   The do-
it-yourself thing upon your own land
has become an alternative way of life.
The visual results, at times startling, are unsophisticated, simple
and lacking in craftsmanship. This
attitude of simplicity, and involvement questions most public projects
since they are designed and developed
by professionals.

Within the context of the Whole Earth Catalog is the concept of reuse of existing materials.   This concept has been
utilized in the development of Drop City in southern Colorado.

DROP CITY

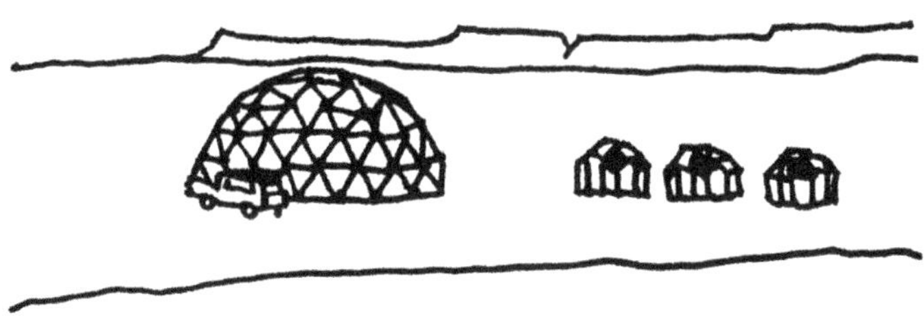

Mashed car hoods are utilized as the structural portions in the
development of Buckminster Fuller domes. The total effect is
one of a brightly colored junk yard which has become a functional entity for a different way of life. However, this concept is not new since Mr. Fuller has been talking about the in-
availability of enough materials to house the world effectively
for some time. The resultant projected physical effect may not
be what Bucky Fuller anticipated but his concept will become
more valid as the population of the world increases, and its
resources decline.

Another growing concern among designers is known as the
"Future Shock" syndrome. Technology is advancing at such a
rapid pace that many people feel a need for an environment they
can relate with. People's psychic is demanding a visual experience related to their past experiences. In many instances, the
visual form is assuming historical aspects. For example, houses
are designed in colonial, early Spanish, Georgian, and Western
Ranch. All of these styles are trying to relate the user to the
past, to something he can relate to.

VISUAL DESIGN DISCIPLINE INPUTS THEORETICAL BASE

Kevin Lynch's book, *The Image of the City* has set general
guidelines for establishing the "imageability" of an area. The
idea of imageability has come about by recognizing that people
image their environment into basic elements. If an area has a
high degree of imageability, the area usually has a high level
of visual quality and vice versa.

Included in Lynch's book is a section on form qualities.
This section deals with appropriate design criteria that could
be utilized in the public works projects. These design criteria
might be summarized as follows:

1. *Singularity*. The quality of a
   public project that identifies it
   as a remarkable, noticeable or
   vivid project is singularity. For
   example, a water tower serving a
   small community may in itself make
   the community unique although the
   tower is similar to others. Another example would be a well designed highway around a city which
   not only provides circulation for
   the community but also develops an
   edge or boundary for the community.
   A dam project will provide not
   only contrasting surface, but it
   can change the form of the land

and increase the recreational use, a singularity of image
to the observer and/or user is therefore created by its
uniqueness.

2. *Form Simplicity*. Most people organize physical form into a
   geometric order. They also tend to limit the parts. Psychology has found that geometric forms are basic to image
   development in western culture. School children begin their
   education with geometric blocks and living environments
   utilize rectangular furniture and cabinets. Therefore,
   people tend to construct aesthetic systems into geometric
   patterns which establish a form simplicity. An example of
   this is the user of a mass transportation system who requires some type of form simplicity to assist the rider in
   getting on and off the right vehicle. In Boston the subway
   traveler travels from square to square since the pattern is
   of irregular form. However, in New York City the traveler
   travels by streets and avenues which are laid out in a re-
   cognizable grid pattern.

   Since the natural earth forms and ecosystems are not
   directly geometric physical forms, simplified physical form
   as a contrast is appropriate.

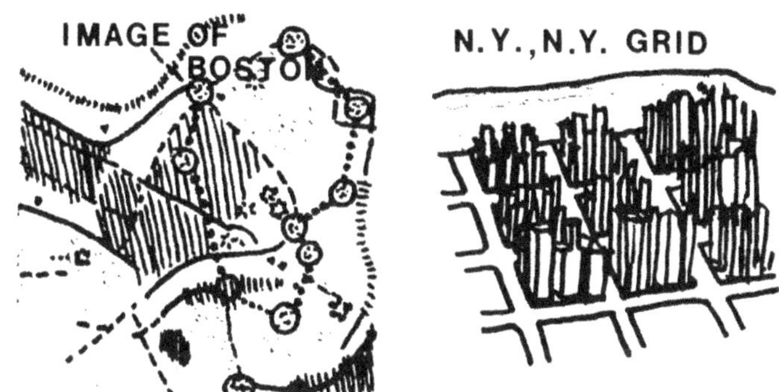

3. *Continuity*. A public project should *not* provide a discon-
   tinuity of physical edge or surface. The designer should be
   aware of the existing rhythms of the visual environment in
   which the public project will be placed, and utilize these
   rhythms in his design.

4. *Dominance*. Design projects require a dominance or hierarchy
   of parts. This is achieved by varying the size, intensity
   or interest of the parts. A hierarchy of parts allows the
   observer to have a basic understanding of the public project as they experience it. For example, a power plant when
   first seen should look like a power plant with radiating
   power lines. As the observer nears the plant the generators, power source and process are recognized. Upon entry

into the plant each physical section of the plant continues
the hierarchic system experience.

5. *Clarity of Joint*.   When two physical or visual systems interact, the point of interaction should be highly percep-
   tible to add to the clarity of the visual image.

6. *Directional Differentiation*.   A visual system should try
   to express or differentiate one end from another.   For example, a road along a dam should either go parallel along
   the lake edge created by the dam or else radiate at a right
   angle away from the dam.   Or as in ski resort areas the
   buildings should be on one side of the slope for the southern sun while the ski slopes should be on the northern side
   to keep the snow from melting.   In large public works projects the use of directional differentiation can be used to
   structure the image of the project.

7. *Visual Scope*.   The utilization of techniques which can increase the range of penetration of vision either actual or
   symbolic are important to the visual systems. In the public
   project the use of vistas and panoramas can assist in the
   articulation of the total concept. For example, the design
   of scenic roads should include the engineering aspects and
   the visual aspects of the user. This might include locating
   the road at strategic scenic vistas, even though the road
   becomes longer.

8. *Motion Awareness*. In public projects which involve movement
   of people the visual system should be designed to allow the
   observer to experience the visual and kinetic aspects of
   movement through space. This can be accomplished by articulation of road curves to give a sense of movement, and
   utilization of perspective to show speed and distance.
   Previous examples of form simplicity, dominance, clarity of
   joint and directional differentiation should be utilized in
   establishing motion awareness.

9. *Time Series*.   Design sequences which are sensed over time.
   The concept of time as a dimension of design is not always
   recognized in public project design. Why not design a path
   as a melodic in nature.   For example, a series of visual
   landmarks would increase in intensity of form until a cli-
   max point was reached--the end.   These are sophisticated
   concepts.   However, the continuing concern for the visual
   environment may provide the catalyst for such design research.

10. *Names and Meanings*. Non-visual or physical characteristics
    may enhance the imageability of a public project also.
    Names related to physical form or direction will increase
    the visual experience many times.

As in all synthesis processes, the above-mentioned qualities must work in an appropriate balance with themselves and
the other concerns related to the public project.

Many designers have used Lynch's theories in the development of time--physical space sequence systems. The objectives
of these systems are to design highways, pedestrian ways,
buildings, etc. by optimizing the visual experiences instead of
the functional experiences. For example, today the normal
highway is designed to traverse the distance between two points
in the shortest most economical way.

The alternative could be to design a highway that would
optimize the potential visual experience between two points. An
analogy would be music which is orchestrated for the ear versus
"motation systems" which are orchestrated for the eye. Phillip
Theil, Lawrence Halprin and Donald Appleyard have been instrumental in the development of these concepts.

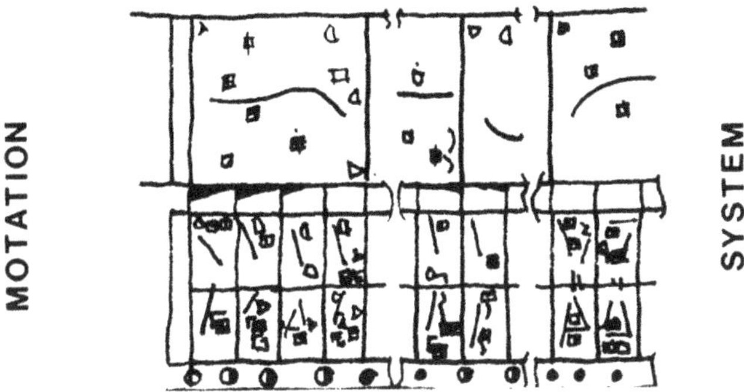

The English town planners developed the concept of "townscape" in which the physical elements are evaluated by their
relationship to the land.

METHODOLOGY OF PHYSICAL-VISUAL ASPECTS

As in other forms of problem solving. a generalized methodology or approach must be utilized. However, for each specific problem, a specific methodology develops as part of the problem solving process.

A generalized *methodology diagram* for the physical-visual aspects could be shown as follows:

DESIGN PROCESS

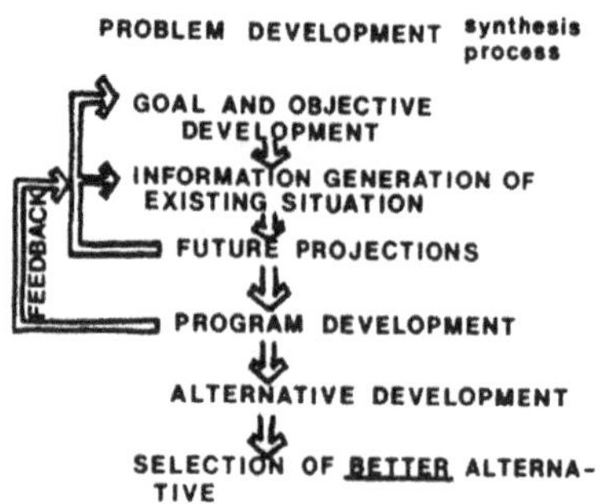

318

PROBLEM DEVELOPMENT

The problem development or statement is one of the most critical steps. In many cases it is not what you do but what you choose to do that is important. For example, a community may be using all of its existing water sources. It will assume that it should acquire additional water rights and build additional facilities and the problem would be how to best develop these additional water resources. An alternative not mentioned would be to reduce the present usage. Each of these alternatives has many physical-visual implications; however, the selection of one of these problem statements will reduce the alternatives by half before the study has been started--in fact, today many designers believe that problem solving may not be as useful as problem creating. What is the transportation problem? Not how can we achieve a balanced transportation solution.

What is the Power Problem? Not how can we build a *better* power generation facility. In the same context, the development of goals and objectives in an honest way are very important. Many projects become unacceptable solutions because they were based upon assumed goals or objectives, not real goals or objectives. For example, most communities would state that a housing objective or goal would be to provide adequate housing for all of the citizens employed within the community. However, when the housing authority starts to select sites for moderate and low income resident housing, the community's goals change. The real goal was that the community felt a need for such housing as long as it did not affect the community directly. The problem of "platitude" goal development can become a trap for the designer.

VISUAL AND PHYSICAL DESIGN GOALS

The visual quality or appropriateness of a project is primarily the relationship or harmony between nature and the landscape design. This quality is achieved by the utilization of proper scale in relationship to the surrounding environment, by utilizing colors and materials which are in harmony with the surrounding environment, by developing forms which seem proper for the function of the public project, and designing physical projects which do the least amount of damage to the "environment."

The first visual goal is to *minimize the visual change* resulting from the public project as compared with the existing natural setting. Obvious as this goal seems, if this were the only criteria a designer used, a majority of the visual problems related to public projects would not exist today.

The second visual goal is to relate the public project with the *indigenous visual aspects* of the area. Most regions of the United States have developed physical solutions utilizing local materials, physical form responses to local climate conditions and electric forms related to the heritage of the residents. An understanding of these indigenous forms and their derivation is needed by the designer of a public project.

The third visual goal of a public project is *appropriate scale*. A physical entity must be in scale with its surroundings. A road that is *too* big for its canyon, or a power plant without bold forms to relate to its desert rock forms are examples of scale. The forms relation to function may also be a concern of scale. A dam should look strong enough to hold the water, a bridge should give the illusion that it will serve its function.

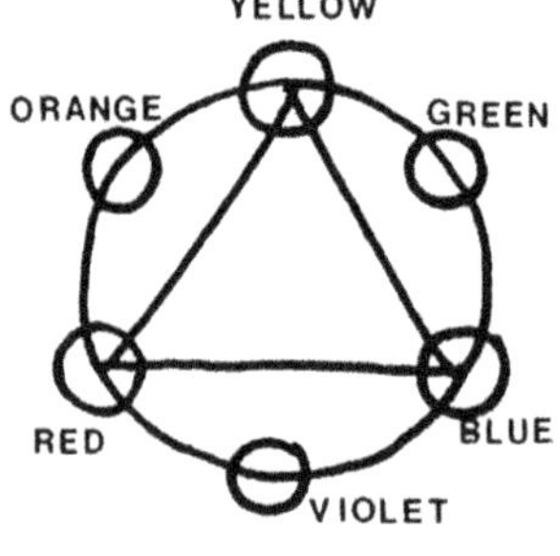

The fourth goal is the *utilization of proper color*. Color is a form giver as is physical space. Color can change the scale of a building. Light colors come forward and dark colors recede. Earth colors tend to belnd in with the environment while bright colors tend to isolate the object.

The fifth goal is the understanding of *time-speed relationship* to public projects. This is especially true in transportation projects. The design of a highway should include a visual experience at 60 m.p.h., as well as a pedestrian visual experience at 4 m.p.h. Bridges, signs, landscaping and safety equipment should all be designed within a time-speed framework. This goal implies that a physical-visual project may have two or more time speed-scales.

The sixth goal is the *understanding of growth (expansion) and change*. One of the design problems of most public projects is expansion. Most designs do not adequately meet the criteria of providing a complete project at the end of each phase if expansion is a criteria. This problem is part of the lack of projections within the concept of change. For example, what happens if the first phase of the project is the only part completed? Is it visually complete? Frank Lloyd Wright called his

DESIGN

DESIGN

DESIGN

architecture organic since he considered it to be capable of addition and being complete at every phase.

The last goal is *good design*. A public project that has been designed within a synthesis process will usually possess visual harmony.

INFORMATION GENERATION OF THE EXISTING VISUAL AND PHYSICAL DESIGN SITUATION

In developing physical-visual data for a public project the following visual information is usually needed:

Land Features (Natural). Land features include the topography of the area, plant material, rock outcroppings and water areas. When recording land feature information, the unique characteristics should be noted. For example, water ways, unique plants, special views or vistas, high noise levels, wildlife areas and migration paths, special geological formations and other unique features should be recorded. Much of this information has been developed in previous chapters.

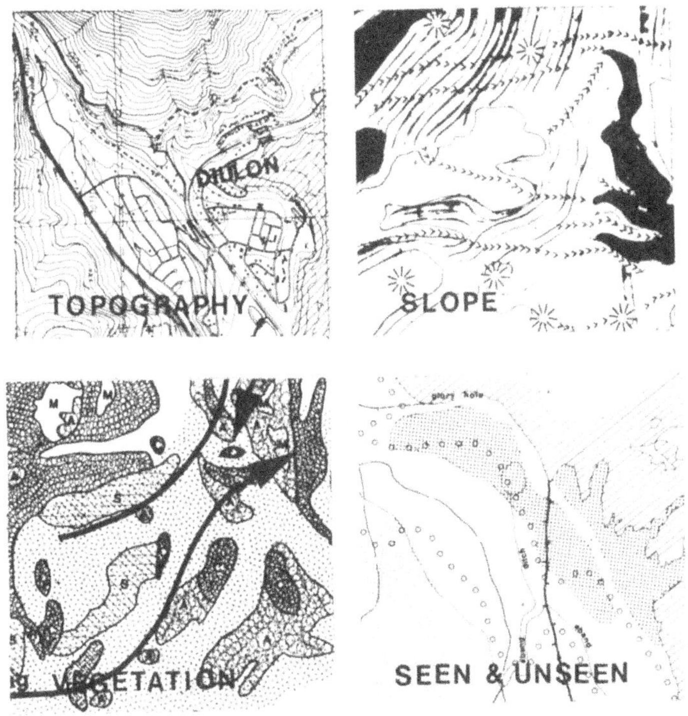

"Imageability" Studies. In larger public projects, an
"imageability" study is recommended.   Base information can be
gathered by the use of interviews and site inspection. The interview method would consist of half-hour interviews in which
the local residents are asked to draw maps of the project area.
With a series of maps, a general image map of the area can be
developed.   It will articulate the important visual-image aspects of the project site in relation to the local perceptions.
The interview will also help assure the area residents that
their concerns will be taken into account.   A minimum of 10-12
interviews is necessary to develop a general image map.   The
people interviewed should represent a wide spectrum of residents including life style and income. Examples of an interview
map and the resultant image map are below.

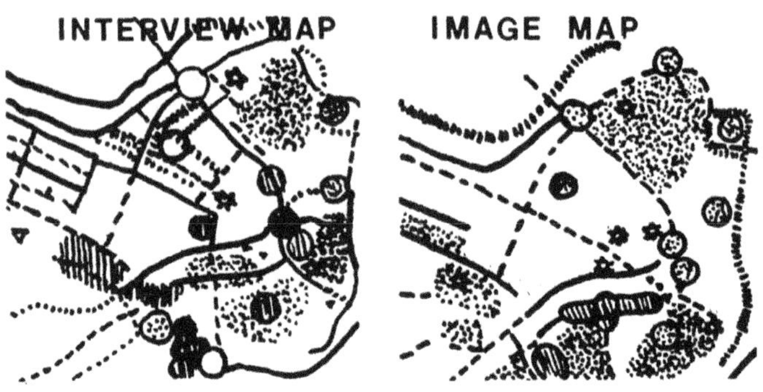

     The site inspection map is developed by utilizing existing
maps and either walking the land or driving over it. It should
include the following visual elements as developed by Professor
Lynch:

> *Paths* - All major and minor circulation systems which are
>   used by people to move about.
> *Nodes* - Those activity centers in which the area users
>   participate.
> *Landmarks* - The prominent visual features of the area are
>   its landmarks.
> *Edges* - Physical, functional, or psychological edges which
>   form like things into districts.
> *Districts* - Areas of like things or potential like things
>   which can be used as a planning unit.

By combining the site inspection map and the interview maps, a
visual image map can be developed to assist the designer as an
information base.

     Land Use Maps.   Maps should be developed showing existing
uses including transportation and utilities.   Land ownership
maps should be developed since potential development can be a

function of existing land ownership patterns. An existing
zoning map (if any) should also be developed and analyzed as to
potential changes (positive or negative) that the public project might bring about.

View Analysis Maps. View analysis maps should include the
totally visible area, the partially seen area and the unseen
area. The view analysis might include the vegetation enclosure
system and foreground and middle ground visual areas.

Historical Designation Map. What is a historical unit?
The designation and possible preservation of historical units
is a complicated decision. In designating the preserving of a
historical unit, the following concerns must be considered:

- Is the unit an integral part of the area's historical development?

- Will it be an "appropriate" element in the proposed public
  project? For example, an old railroad station without a
  railroad may not be an appropriate physical structure for
  a sewage plant office.

- Should the historical unit be retained because of its
  quality or an anti-change attitude?

- Is the historical unit of "real" historical value?

Especially important is not the physical unit itself, but the
historical development process and how it can become an integral part of the public project. It is not enough to record
history; consider how it can become a part of the educational
process as a public works development.

PROJECTED VISUAL AND PHYSICAL DESIGN IMPLICATIONS

A difficult portion of the design process is the development of projected implications of alternative public projects.
The question--what will happen within the area if the project
is implemented?--should always be asked and an answer supplied.

The establishment of maximums and minimums within the area
of visual harmony is unknown. Today, the designer responds to
a design program by proposing the better alternative. The designer rarely asks the question about the optimum visual aspects. For example, what is the optimum number, type and design
of camp sites for a specific project based upon design alone.
Usually the number and type is set by the economics, land saturation factors or maybe even the social factors but rarely the
optimum number which appropriately can be handled by the visual
implications. For example, the location of camp sites might be

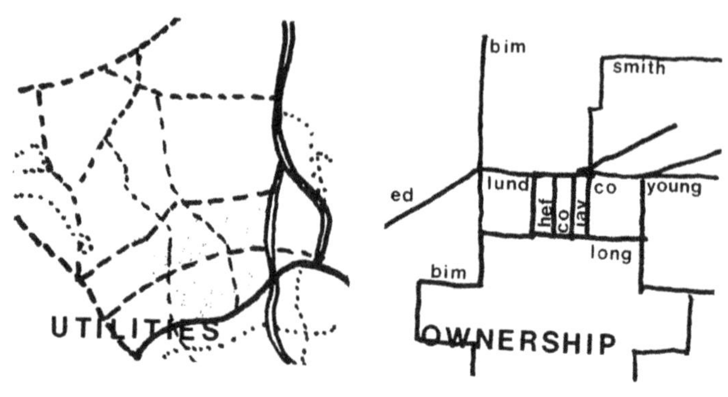

UTILITIES
bim
smith
ed
lund
hef
co
ray
co
young
long
bim
OWNERSHIP

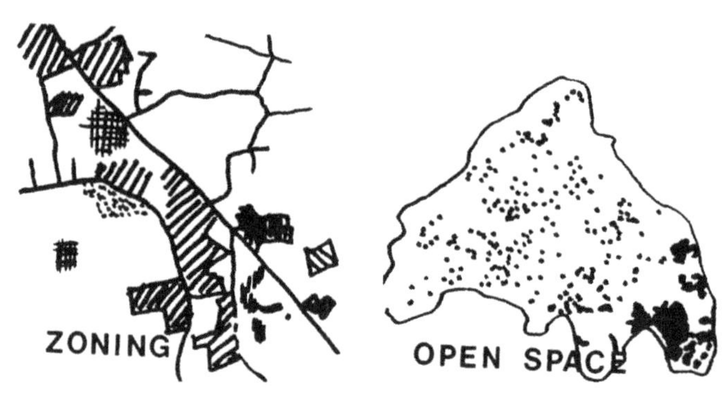

ZONING
OPEN SPACE

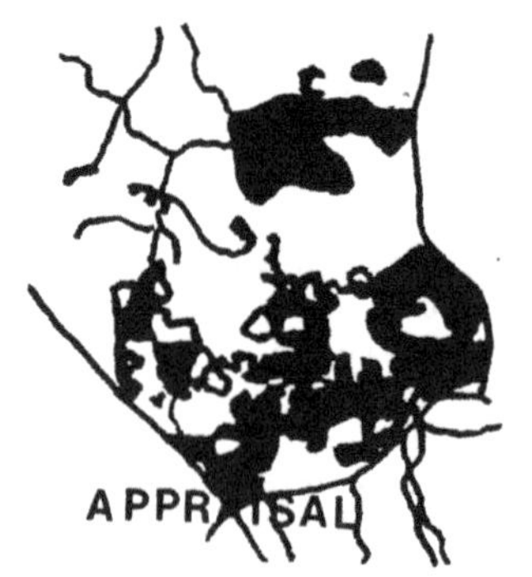

APPRAISAL

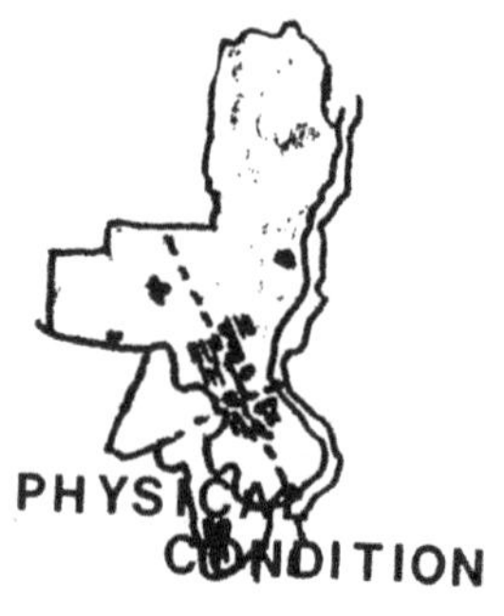

PHYSICAL
CONDITION

dictated by the screening from view by natural features. Another example might be the height of a physical project in relationship to the surrounding area. This height could dictate the overall size and scope of the proposed project, or the dictated design and quantity of a parking facility may dictate the total project scale including the number of users, and the projects location.

In evaluating the visual alternatives, various visual simulation techniques are available to the public project designer.

1. View analysis maps with projected alternatives overlaid will usually show the visual impact of a project.

2. Photographic slide sequence studies showing the affected area before the development of the public project and the visual implications after development.

This simulation technique is done by projecting the existing sequence of slides on paper pad and drawing the public project over the existing situation. Photographic slides are made of the drawings and a simulation of visual aspects of the project can be obtained.

3. Scale models of the projected project are one of the best means of simulating the proposal. Large scale models with people and scale objects photographed onto slides can simulate the real world. Color is important to this technique.

4. In highway design, computer programs are available in which the visual experience of the trip can be simulated on a T.V. screen. This physical simulation usually includes the built aspects of the highway only and does not include the adjacent natural features.

Notation systems which utilize symbols for visual experiences have been developed by Halprin and others. These systems are presently used by designers but are not universal enough to be used as simulation techniques.

PROGRAM DEVELOPMENT

The total program is a resultant of the visual and physical design goals and objectives, the generation of existing information, and future projections of the visual and physical design implications. During this process these become entangled.

The "optimum" program of the client may not be the same as the community's goal, therefore, a composite program must develop. The statement "the design if finished when the program

	slope	crown	curve	drainage	shoulder					

MATRIX PROGRAM

is finished" seems appropriate. However, an inadequate program will usually lead to an inadequate solution. Since the design program will include much more than the visual aspects, it is suggested that the visual aspects be also correlated with related program elements. A useful tool is to develop interrelation chart similar to that developed in the book *Community and Privacy*. Computer programs are now available which can group like things to assist the designer in decision making.

ALTERNATIVE DESIGN DEVELOPMENT AND SELECTION

The synthesis process takes place in the development of visual and physical alternatives to meet the program. The synthesis process is a result of the designer's input in conjunction with the program. Unless the program is so restrictive that no alternatives are possible, a series of alternatives can be developed. This set of alternatives is influenced by the designer's past experiences and additional research in the specific area of the public project. If possible, three or more alternatives with combinations are recommended. Each alternative should be evaluated to establish an order of priority. If implementation feasibility has not been a goal or objective, then the best solution which can be implemented becomes the selected "better" alternative. The selection process should involve both affected citizen participation and a weighted evaluation matrix (the Battele Institute Matrix is a good total example).

The following visual and physical guidelines outline criteria for power lines, transportation systems and waterways.

SPECIFIC GUIDELINES FOR THE PHYSICAL VISUAL
ASPECTS OF POWER TRANSMISSION LINES

(Note the following information is taken from the Federal Power Commission's Brochure "Electric Power Transmission and the Environment.")

326

1. Rights-of-way should be selected with the purpose of minimizing conflict between the rights-of-way and present and prospective uses of the land on which they are to be located. To this end, existing rights-of-way should be given priority as the locations for additions to existing transmission facilities, and the joint use of existing rights-of-way by different kinds of utility services should be considered.

2. Rights-of-way should avoid the natural historic places, natural landmarks, parks, scenic, wildlife and recreational lands, officially designated by duly constituted public authorities. If rights-of-way must be routed through such historic places, parks, wildlife or scenic areas, they should be located in areas or placed in a manner so as to be least *visible* from areas of public view and so far as possible in a manner designed to preserve the character of the area.

3. Rights-of-way should avoid scenic or prime timbered areas, steep slopes and proximity to main highways where practical. In some situations scenic values would emphasize locating rights-of-way remote from highways while in others where scenic values are less important rights-of-way along highways in timbered areas would achieve desirable conservation of existing forest lands.

4. In scenic and residential areas clearing of natural vegetation should be limited to that material which poses a hazard to the transmission line. Determination of a hazard in critical areas such as park and forest lands should be a joint endeavor of the utility company and the land manager in keeping with the National Electric Safety Code, state or other electric safety and reliability requirements.

5. Long tunnel views of transmission lines crossing highways in wooded areas, down canyons and valleys or up ridges and hills should be avoided. This can be accomplished by having the lines change alignment in making the crossing, or in other situations by concealment of terrain or by judicious use of screen planting.

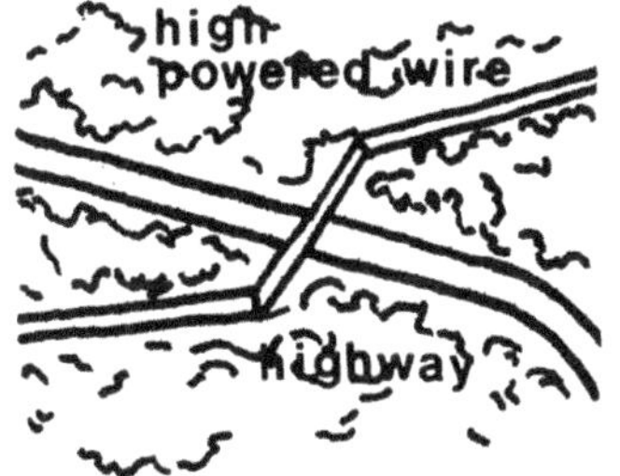

6. Rights-of-way clearing should be kept to the minimum width necessary to prevent interference of trees and other vegetation with the proposed transmission facilities. In scenic or urban areas trees which would interfere with the proposed transmission facilities and those

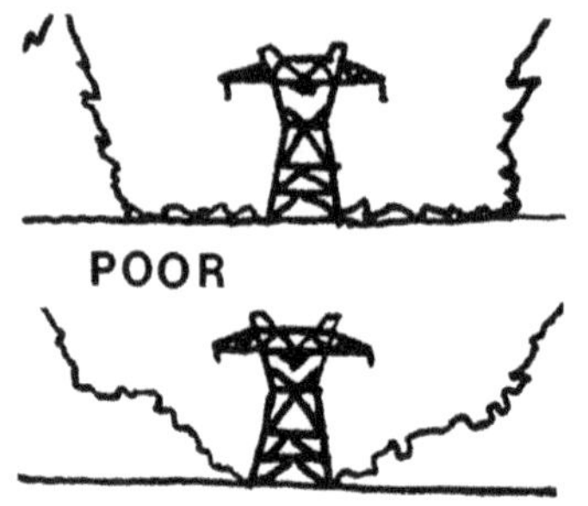

which could cause damage if fallen should be selectively cut and removed.

7. The time and method of clearing rights-of-way should take into account matters of soil stability, the protection of natural vegetation and the protection of adjacent resources.

8. The use of helicopters for the construction and maintenance on rights-of-way should be considered in mountainous and scenic areas. This would permit rights-of-way to be located in more remote areas and would reduce disturbance of the ground and the number of access roads.

9. Unsightly tree stumps which are adjacent to roads and other areas of public view should be cut close to the ground or removed.

10. Scars on the surface of the ground should be repaired with top soil and replanted with appropriate vegetation or otherwise conformed to local, natural conditions. Grading generally should not be done on slopes where the scars cannot be repaired without creating an erosion problem.

11. In scenic areas visible to the public, rights-of-way strips through forest and timber areas should be deflected occasionally and should follow irregular patterns or be suitably screened to prevent the rights-of-way from appearing as tunnels cut through the timber.

12. At road crossings or other special locations of high visibility rights-of-way strips through forest and timber areas should be cleared with varying alignment to comport with the topography of the terrain. In such locations also where rights-of-way enter dense timber from a meadow or other clearing, trees should be feathered in the entrance of the timber for a distance of 150-200 yards. Small trees and plants should be used for transition from natural ground cover to larger areas.

LOCATION AND DESIGN FACTORS FOR TRANSMISSION
TOWERS AND OVERHEAD LINES

1. If an overhead line must be routed across uniquely scenic, recreational or historic areas or rivers, the feasibility of placing the lower voltage line underground should be considered. If the line must be placed overhead, it should be located on a right-of-way least visible from areas of public view.

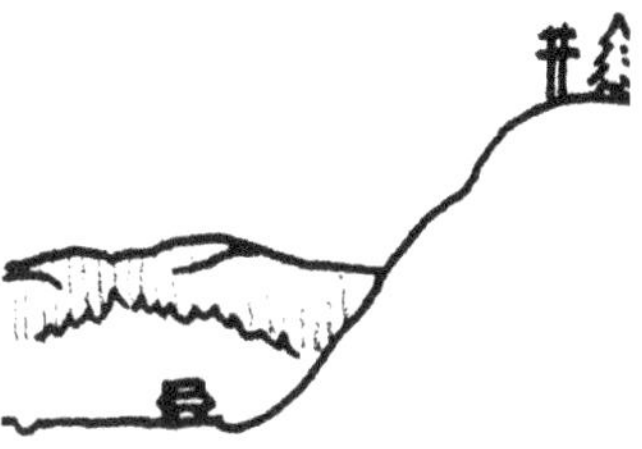

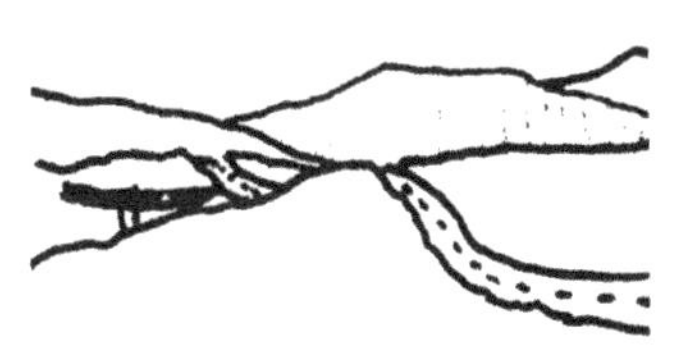

2. Transmission facilities should be located with a background of topography and natural cover where possible. Vegetation and terrain should be used to screen these facilities from highways and other areas of public view.

3. Where transmission facilities must be placed on slopes which parallel highways or other areas of public view, they should be located approximately two-thirds the distance up the slopes where feasible. With the slopes as background, the presence of the facilities would be less noticeable.

4. Transmission line rights-of-way should not cross hills and other high points at the crests and when possible should avoid placing a transmission tower at the crest of a ridge or hill. Towers should be spaced below the crest to carry the line over the ridge or hill, and the profile of the facilities should present a minimum silhouette against the sky.

5. Transmission lines should not cross highways at the crest of a road.

6. Long views of transmission lines parallel to highways should be avoided where possible. This may be accomplished by overhead lines being placed beyond ridges or timber areas.

7. Transmission lines should cross canyons upslope from roads which traverse the canyon basins if the terrain permits.

8. When crossing canyons in a forest, high-longspan towers should be used to keep the power lines above the trees and to eliminate the need to clear all vegetation from below the lines. Only as much vegetation as is necessary to string the line should be cut.

9. Where ridges or timber areas are adjacent to highways or
    other areas of public view, overhead lines should be placed
    beyond the ridges or timber areas.

10. In forest or timber areas, high longspan towers should be
    used to cross highways in order to retain much of the natu-
    ral growth along the highways.

11. Native shrubs and trees should be left in place or planted
    at random, with the necessary allowance for safety, near
    the edges of rights-of-way adjacent to roads.

12. Simple, but functional, designs of towers and poles should
    be used. Illustrations of these kinds of structures can be
    found in the book *Electric Transmission Structures*, spon-
    sored by the Electric Research Council.

13. The use of poles designed without cross-arms for electric
    transmission lines of 138 kv and below and communications
    cables should be considered.

14. The materials used to construct transmission towers and the
    colors of the components of the towers should comport with
    the natural surroundings.

15. In addition to steel and aluminum transmission towers, the
    use of towers constructed of fiberglass, reinforced plastic,
    laminated wood, concrete, and other materials should be
    considered.

16. The use of treated single or double wood poles should be
    considered in forest or timber areas.

17. The use of weathered galvanized steel structures should be
    considered when transmission towers are to be silhouetted
    against the sky.

18. Where two or more circuits are required at high crossings,
    the use of multiple circuit towers should be considered
    where it is consistent with adequate reliability.

When considering the visual aspects of power lines, the
maintenance of such lines should be a consideration in the design. Access road and service road design should meet the same
visual criteria as the power lines.

POWER LINE RIGHTS-OF-WAY

One of the potential benefits of power line routes is that
clearings at safe distances adjacent to transmission facilities
may be used for secondary purposes. Consistent with general

safety factors the following should be considered as possible
secondary uses of rights-of-way to the extent permitted by the
property interest involved:

- Equestrian paths, bicycle paths, ski touring, snow mo-
  biling,

- Cultivation of Christmas trees, elderberry and huckleberry
  bushes, and other nursery stock,

- Parks

- Golf courses

- Game refuses

- Hiking trail routes

- General agriculture

- Orchards

POWER LINE DESIGN CRITERIA

The location of related above ground facilities should include the following concerns:

1. Unobtrusive sites should be selected whenever possible.

2. The bulk of the facility should be kept to a minimum, if
   possible.

3. The designs of the exteriors of the facilities should be
   in harmony with its surroundings. This should include concern for color, like materials, scale and appropriate landscaping.

4. Whenever possible the facilities should be placed underground.

5. Storage tanks and other large structures should be painted
   a neutral harmonizing color or painted a color which will
   reflect the changing natural color. (Sometimes a silver
   reflective surface is the best alternative.)

SPECIFIC GUIDELINES FOR THE PHYSICAL-VISUAL
ASPECTS OF TRANSPORTATION SYSTEMS

All transportation systems utilize corridors for movement
and require physical responses for intersections of conflict.

Therefore, no matter what the mode of transportation the guidelines are similar.

SELECTION OF RIGHT-OF-WAY ROUTES

1. Transportation rights-of-way should be selected to conserve
    and protect natural and scenic values.

2. Rights-of-way selected should not conflict with the exist-
    ing terrain or existing landscape if possible.

3. Specific highway location should be a function of the scenic
    quality and visual experience to the user.

4. Where possible, long straight stretches of road should be
    avoided since they are monotonous and tiring to the eye.
    The curvilinear alignment is preferred.

5. Routes should be along lines of demarkation or of boundaries
    of different land uses, e.g., between forest and farmland,
    orchards and hill pasture, or urban developments and green-
    belts.

6. Rights-of-way can provide land for other recreation uses,
    the combination of use can be visually beneficial for the
    mixed uses. An example is the use of a bikeway system sep-
    arated by an earth mound from the highway system within the
    same R.O.W.

LOCATION AND DESIGN FACTORS FOR TRANSPORTATION SYSTEMS

1. Selective cutting or thinning of woody growth along road-
    sides or in medians should establish or re-establish a
    scenic value.

2. Scars resulting from construction should be repaired to
    conform with appropriate natural conditions.

3. Cuttings and embankments should be rounded at top and bot-
    tom and be softened in form by planting local plants.

4. Boundary walls or fences should *not* be parallel to the
    transportation route for long stretches.

5. Walls and fences should be made from local materials or
    utilize colors that blend with the natural environment,
    (example is the Vail Pass Environmental Study in Colorado).

6. The architectural appearance of bridges and other major
    structures should be simple, clean and functional design.

332

7. The roadway section design can reduce negative sight impact. This
can be accomplished using depressed
roadway sections or banked earth
mounds.

8. The scale of the corridor, the relationship of its width and height
to its immediate surroundings is
of primary importance. An example
would be separating opposing lanes
to provide a smaller scale, visual
interest and variety.

9. Most traffic barriers are hazards in themselves and reducing
the number of such installations is justified both visually
and as a safety need.

10. Adequate screening between highway right-of-way and camp
ground, rest or recreational areas should be provided. This
screening will provide possible quiet and inspiration with
the natural environment.

11. The physical form should express the speed and excitement
of movement.

SPECIFIC GUIDELINES FOR THE PHYSICAL-VISUAL
ASPECTS OF WATER RESOURCE DEVELOPMENT

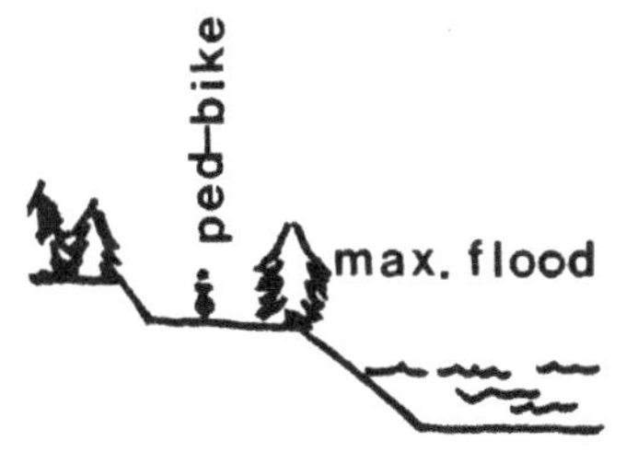

1. River basins and river systems
should be developed to conserve and
protect natural and scenic values.

2. When possible, flood channels
should be developed for secondary
uses such as bikeways and pedestrian paths in an indigenous manner.

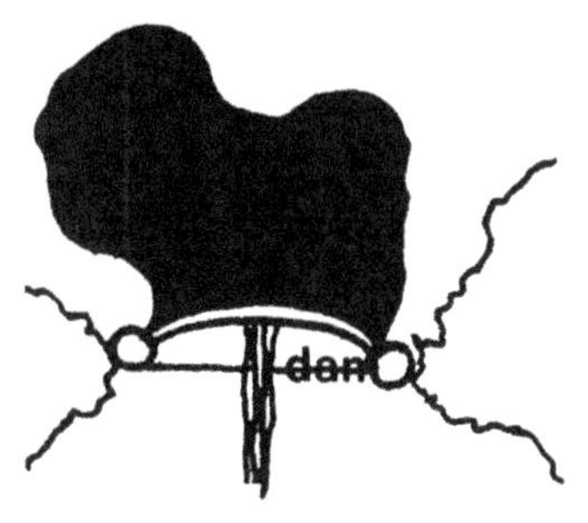
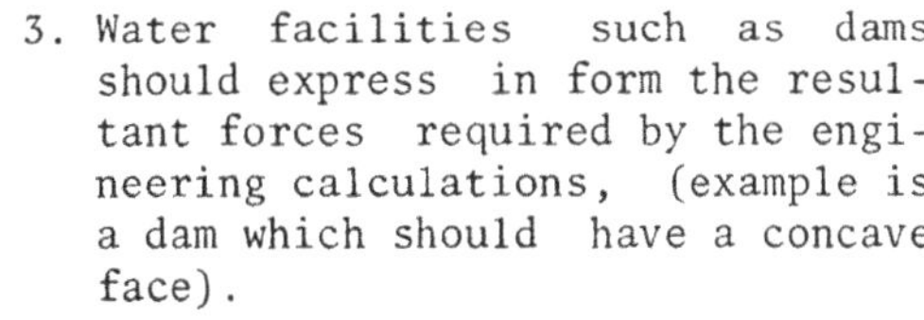

3. Water facilities such as dams
should express in form the resultant forces required by the engineering calculations, (example is
a dam which should have a concave
face).

4. Water facilities should exploit
the excitement of splashing water.

5. Water culverts or pipelines should
avoid scenic or prime timbered

333

areas, steep slopes and proximity to major circulation routes.

6. Long tunnel views of open culverts should be avoided by slight curves or other realignments.

In conclusion, designing the physical-visual environment is minimizing discord. While it will be difficult, if not impossible to satisfy the divergent public, an awareness of the problem will help solve it.

REFERENCES

1. Spreiregen, Paul D., Urban design: *The Architecture of Towns and Cities*. McGraw Hill Book Company, 1965, New York.

2. Bacon, Edmund N., *Design of Cities*, The Viking Press, 1967, New York.

3. Rubenstein, Harvey M., *A Guide to Site and Environmental Planning*, John Wiley and Sons, Inc., 1969, New York.

4. Blake, Peter, *God's Own Junkyard*, Holt, Rinehart and Winston, 1964, New York.

5. Lynch, Kevin, *The Image of the City*, M.I.T. Press, 1960, Cambridge, Massachusetts.

6. Shahn, Ben, *The Shape of Content*, Vintage Books, 1960, New York, N. Y.

7. Cullen, Gordon, *Townscape*, The Architecture Press, 1963, London.

8. Thiel, Phillip. "A Sequence Experience Notation for Architectural and Urban Spaces," Town Planning Review 32 (1): 33-52, April 1961.

9. Appleyard, Lynch and Myer, *The View from the Road*, The M.I.T. Press, 1964, Cambridge, Massachusetts.

10. Hornbeck, Peter L. et al. *Highway Esthetics* Harvard Graduate School of Design, 1968, Cambridge, Massachusetts.

11. Reekie, R. Fraser, *Design in the Built Environment*, Crane, Russak & Company, Inc., 1972, New York.

12. Isaacson, Larry, *Park and Recreational Facilities*, U. S. Printing Office, 1971.

13. U. S. Dept. of Agriculture, Forest Service, Northern Region, *Forest Landscape Management*, Revised 1972.

14. President's Council on Recreation and Natural Beauty, *A Proposed Program for Roads and Parkways*, U. S. Department of Commerce, 1966.

15. Taliesin Associated Architects et al., *Vail Pass Environmental Study*, 1972.

AUTHOR NOTES

Dwayne Nuzum is the Dean of the College of Environmental Design at the University of Colorado. He teaches courses in urban design and urban planning in the architecture program at the University. Before joining the faculty at the University of Colorado, he taught at the University of Virginia and the Rhode Island School of Design. His doctoral work was done at the Delft Technical University in the Netherlands under the direction of C. Van Eesteren, the planner for Amsterdam and former deStijl member. He also has a consulting practice which has assisted in the planning of the various development projects in Texas and Colorado. He also collaborated on a study for the State of Virginia on highway beautification and screening techniques. Dr. Nuzum also serves as Councilman and Acting Mayor for the City of Boulder, Colorado.

The chapter was developed with consulting assistance from Harvey Atchison, a landscape architect and Chief of the Environmental Division of the Colorado State Highway Department.

PART III

SYNTHESIS AND EVALUATION

This part provides the synthesis of linkage mechanisms connecting the substantive areas and large scale projects through an evaluation of economic trade-offs and legal constraints. Specific mechanisms maximizing project evaluation and assessment include the methodological help of systems analysis, various planning techniques, the coordination of activities through appropriate management methods, and last, but not least public involvement as a crucial element for linking projects to the desires of the populations to be served.

8

ENVIRONMENTAL LAW

BY NORMAN I. WENGERT

No one involved with planning and implementing public projects, no matter at what hierarchical level, whether he is working in the public or the private sector, can avoid being involved with the law and with legal relationships. This chapter seeks to provide a kind of overview of the subject, emphasizing particularly the relatively new field called "environmental law."

Public projects are in the first instance authorized by law. Statutes enacted by the U.S. Congress and by state legislatures, or ordinances passed by county boards or city councils provide the authority and the appropriations for most public projects. Others receive authorization from school boards or other special districts, or in a few cases from popular referenda. In some cases the enactments are very specific; in others they establish broad programs within which the public project is authorized by executive or administrative action. Although a prerequisite for planning and implementing public project proposals, this kind of authorizing law is not the focus of this chapter. Instead, attention will be directed to a narrower and more specialized meaning of law, namely, the policies and procedures which govern relationships among individuals, corporations, and government agencies, and which are ultimately the concern of the courts and the judicial system.

It is not that every issue or every conflict ends up in court, but rather that much activity goes on under the umbrella of the judicial system. For example, thousands more contracts are complied with than are ever involved in law suits, but behind all contracts and behind the legal relationships they involve is the possibility of ultimate resort to litigation, and the applicability of concepts and principles developed by the courts.

It is these aspects of law that are the primary focus of
this chapter especially as related to environmental law and its
effect upon public projects. It should be noted that the literature on environmental law is already large and growing rapidly! Some of it is directed primarily to the lawyer; some to
the interested citizen. But in any case, the review in this
chapter can only touch on some of the highlights and emerging
problems.

Most traditional fields (or subjects) of law may be involved in public projects. Even the smallest project requires
a variety of contractual relationships. There are employment
contracts, design contracts, materials supply contracts, to
mention but a few. Construction activities, moreover, run the
risks of tort actions flowing from negligence, personal injuries, auto accidents, and professional misconduct. Many public projects will involve both personal and real property law
since real property may have to be condemned, access rights acquired, and personal property utilized. The private practitioner may be concerned about the law of business organizations,
partnerships and corporations. Some public projects, moreover,
may in one way or another, involve financial laws, including
such problems as floating bond issues and tax limitations or
loopholes.

There are a host of statutes which apply to public projects in a variety of ways. Of a general character are laws requiring bid procedures, workmen's compensation laws, related
safe places statutes, labor law, and a variety of anti-
discrimination statutes. Special statutes have been enacted,
some of which apply to public projects generally and others
which apply specifically to highway projects, to flood control
and other water projects, to airport construction projects, and
so forth.

A substantial body of administrative rules and regulations,
having the full effect of law, govern relationships arising out
of both planning and implementing public projects. And in some
cases, initial responsibility may lie with the administrative
agency rather than with the regular courts. A substantial number of "environmental protection" policies may be contained in
administrative regulations, project specifications, and other
day-to-day administrative actions.

The growing complexity of the legal situation, as it relates to public projects, to a large extent simply mirrors the
complexity of society, the size of public operations, the sophisticated technology, and consequent patterns of relationships. In the face of this complex structure of laws and legal
constraints the question arises "How can the practitioner not
trained in the law cope with the resulting problems?" The obvious answer is to have access to good legal advice.

To be most useful, such advice should help the practitioner avoid legal complications. Increasingly, where the firm or agency perceives the problems clearly (and can afford the cost) employing a staff lawyer can be of great utility. The staff attorney provides at least 3 types of legal advice:

1. That which might be called *preventive* advice in which he seeks to guide his client or employer to avoid legal complications;

2. That which involves guiding the client or employer in *complying* with the host of applicable statutes and administrative rules and regulations;

3. That which arises out of *litigation*.

If litigation is involved, it is often desirable to secure assistance of lawyers skilled in trial work, rather than simply relying upon the contributions of staff attorneys. In any case, if a staff attorney has been given the opportunity to perform effectively with respect to prevention of legal problems and compliance with legal rules and regulations, the risk of litigation may be minimized.

THE AMERICAN LEGAL SYSTEM[2]

The American legal system evolved from the English Common Law. A primary characteristic of the Common Law is that it is judge-made law rather than legislated or statutory law. Although statutory enactments and administrative rules and regulations have become increasingly important in the last 50 years, American lawyers still spend much of their time on case law, which is judge-made law. Even where statutes are involved, judicial interpretations are often of major importance. In few legal systems, in fact, are the interpretations of judges more important, and in a sense a major function of lawyers is the prediction of how the courts (i.e., the judges) are likely to interpret the rules if and when a particular issue gets into litigation.

It is still important to distinguish between Common Law and Equity Law actions. Both are parts of the larger Common Law System, but they differ with respect to procedure and with respect to the remedies available to the litigants. Following the pattern in England, many American states originally had established separate courts of Equity and of Common Law. The federal Constitution recognized these two kinds of law, but assigned jurisdiction to a single federal court system. Today, all but one or two states have similarly consolidated the courts, but still recognize the procedural and substantive differences in the two bodies of law--in both of which the judge and his interpretations are of major significance.

Litigation may involve issues of law and issues of fact (and sometimes combined issues of law and fact). A distinguishing feature of an action at Common Law is that issues of fact are determined by a jury (unless a jury is waived), while in Equity, fact issues as well as law issues, are determined by the judge. There is no jury in an Equity case. Common Law and Equity actions also differ with respect to the relief or remedies which the litigants may seek. Actions in Common Law are primarily concerned with money damages or restitution of property. Actions in Equity seek injunctions or specific performance, i.e., orders from the court restraining certain types of action or requiring other actions.

It is a curious consequence of this dual system, that although we have abolished imprisonment for debt, a litigant who fails to heed the order of an Equity court to pay a certain sum may find himself imprisoned, not for failing to pay, but for failing to obey the court order.

In most situations, these distinctions between Common Law actions and Equity actions are not of primary concern to the layman since his attorney will act for him as the situation requires, but this brief reference to the historical evolution of American law is meant *first*, to emphasize the importance of case law and the judicial function, and *second* to stress the power and creative initiative which American judges continue to exercise. Case law or "judge made law," whether Common Law or Equity, rests on the concept of precedent (*stare decisis*). But past precedents are not an absolute constraint on any judge. Through various devices, such as distinguishing the case at hand from the precedents, or even by deliberately reversing past holdings, American judges have considerable freedom to avoid rigid traditions of the past, and to adjust the law to new situations, often by building on the doctrines of the past.

CONSTITUTIONS AND STATUTES

In the broadest sense, the American legal system gets its authority from state and federal constitutions and from statutes enacted by state legislatures and the congress, as well as from English Common Law traditions.

The Constitutions are important, *first*, because they establish or authorize the court system, both state and federal; and *second*, because of the limitations which they contain. Among the limitations of most importance, perhaps, are the bills of rights and other specific provisions protecting individual liberties. Statutes are important because of the procedures they establish and because of the substantive goals and policies they express which apply to many of the suits coming before the courts.

The twentieth century has seen a substantial growth in
statutory law, so that today statutes are increasingly more important than case law as compared to the situation in 1900.
Many statutes have systematized and codified the Common Law
case rules and precedents. And as the problems dealt with and
the attention span of government have become broader and more
and more complex, there has been an increasing reliance on what
is called administrative law, which consists of the rules and
regulations issued by administrative agencies in accordance
with statutes, and also their rules, orders, adjudications,
licenses and permits. And as administrative law has grown in
importance, so has the issue of bureaucratic accountability. Of
major importance, in this connection, is the Federal Administrative Procedures Act[3] which regularized administrative procedure of federal agencies.

As previously noted, important to public projects are
those statutes which have established programs and outlined
policies and procedures for implementing such programs. Formal
administrative law may not be involved, but the issues of accountability remain important for agencies are left to adminis-
ter policies and programs based sometimes on the most general
legislative directives. Thus the practitioners dealing with
public projects must know both the statutory basis as well as
administrative policies and procedures.

Something needs also to be said about the hierarchical
structure and intergovernmental relations within the American
legal system. To a large extent, the federal and the state
systems are separate. The federal government operates under the
Federal Constitution, and as such is still considered a
"limited" government having those powers specifically granted
to it or necessary and proper to implement specific powers. The
limitations on the federal government have, however, been obscured by the fact that the "spending power" of the Federal
Constitution has become the basis for many federal programs,
and most constitutional lawyers regard the spending power as
unlimited except by specific provisions in the Federal Constitution itself, such as the Bill of Rights. Reliance on the
spending power has, in turn, resulted in many Federal programs
being formulated in terms of using grants and contracts to accomplish purposes not specifically delegated to the national
government.

But while the federal government can only do those things
spelled out in the Constitution or necessary to the realization
of constitutional purposes, state governments are technically
unlimited. The people of any one state may put into their constitution whatever they choose, so long as it does not violate
provisions of the Federal Constitution. Thus, for example, it
would theoretically be possible for an American state to establish a socialist governmental and economic system within its

borders by "nationalizing" the means of production so long as
just compensation were paid in the expropriation process.   In
actual fact, state constitutions generally contain many more
limitations on legislative activity than does the Federal Constitution.

It should be emphasized again that much of the expansion
of federal activities, including those involving public projects, has been based on the spending power.   When relying on
this power the federal government, in effect, buys compliance
from the states, or from private corporations and individuals,
saying that grants of such-and-such magnitude are available if
the terms and conditions of the program are met.

The highway program is one example of the grant approach,
and it is also found in federal programs dealing with the control of water and air pollution.   Water resource development by
the Corps of Engineers and the Bureau of Reclamation, in contrast, is a direct federal activity, the intergovernmental relations characteristic of these programs being matters of administrative convenience or policy.

This dual structure of federal and state laws also shows
up with respect to the way in which the federal and state
courts operate.   In general terms, the federal courts consider
two classes of cases: *first*, those which can be designated
federal questions, i.e., cases arising under the Constitution,
Statutes and Treaties of the United States, and, *second*, cases
involving diversity of citizenship, which simply means that the
parties live in different states.   In the latter type of case,
the law which the federal courts apply is generally state law
(case law or statutes).   The diversity of citizenship authority
of the federal courts reflects the fears of the founding fa-
thers that litigants from other states would not receive fair
consideration, and so the Constitution specifically provided
for the use of the federal courts, when the parties to an action had residence in different states.   In contrast, the state
courts have virtually unlimited jurisdiction, including authority to apply federal law to the cases before them.   There are
certain rights of appeal from state to federal courts, but
these tend to represent special situations. To summarize, then,
the existence of our federal system has resulted in a duality
of responsibility with respect to litigation and statutory interpretation, as well as with respect to court jurisdiction.

PUBLIC PROJECTS AND PUBLIC LAW

Although the judicial precedents of the Common Law are not
unimportant for public projects, particularly in providing useful concepts and rationales in certain kinds of cases, it is
clear that today statutory enactments are generally of greater

importance. At the same time, Common Law, Statutes, and Administrative Law are a kind of seamless web.   For example, to understand the effects of statutes it is necessary also to consider the interpretative role and function of the courts with
respect to statutes, for there are few statutes which are completely self-explanatory and self-executing and the process of
interpretation often rests on Common Law doctrines and principles.   To put it another way, most statutory language at one
time or another requires interpretation. In the first instance,
the executive agency interprets the statutes in its area of
responsibility. But if challenged, the courts are involved. In
fact statutory interpretation has become a major function of
courts, especially in connection with the application of general policies and guidelines to particular situations including
environmental protection. This is a major aspect of what is
called environmental law, where the courts are engaged not only
in applying long established principles of the Common Law to
environmental situations, but also in interpreting and applying
provisions of newer statutes. Even where there is an extensive
body of statutory law, the courts thus still may have a major
role to play in interpreting and applying statutory provisions
to particular situations and in protecting the rights established by statutes.

     Statutes, as interpreted by the courts, spell out the re-
lationships of individuals and corporations to government. With
respect to public projects this is a dominant pattern for the
simple reason that in the United States most public projects
are carried out by private sector firms under contract to the
government.   Conversely, the government builds very few public
projects by "force account."   The outstanding exception in recent decades has been the Tennessee Valley Authority which constructed many of its projects with its own employees.

ENVIRONMENTAL LAW

     It is not possible in this treatise to review in detail
all aspects of the relationship between law and public projects.
Perhaps of greatest interest in that developing field of law
designated environmental law is the use of suits (litigation)
to protect environmental values, and it is to this subject
which we now turn.

     If this treatise had been written ten years ago, a chapter
on "environmental law" would not have seemed important.   But
today environmental law is of major interest to the practitioner concerned with public projects because much of what he
does may be affected by developing doctrines in this field, and
it is not impossible that the projects in which he is interested
may be involved in litigation. Many law schools are teaching
courses in this field; text and case books are being published

345

on the subject; and a number of lawyers are specializing in the
field.   The American Bar Association has a section on Natural
Resources Law, which today is predominantly concerned with environmental questions.[4]

What then has happened to identify the "environment" as a
field of legal interest and concern?   What has brought this
subject to public attention?   Why is it desirable in 1973 to
devote a chapter to this aspect of law?   And just what is the
significance of the developing principles of environmental law
to public projects and to the professionals concerned with the
formulation, design, and construction of public projects, particularly in the fields of transportation, water, and power,
which are of principal concern in this treatise?

As commonly used, the term environmental law refers primarily to that body of law which seeks to regulate individual,
corporate, and governmental behavior in order to prevent adverse environmental consequences. It includes both substantive
program content (The Wilderness Act, the Clean Air Act) and administrative procedure.   It is based on legislative enactments
as well as judge made law resulting from specific litigation.
It is, thus, a form of social control having as its objective
the preservation of environmental values such as ecological
balance, natural beauty, and resource productivity, in order to
maintain a stable and satisfactory level of living, and quality
of life for present and future generations. It recognizes certain individual and community rights as against other individuals, firms and corporations, and government agencies engaged
in activities which may be construed as adversely affecting the
environment and thus diminishing these rights.

As previously indicated, it is not possible to review the
substantive statutes which establish a variety of environmental
policies; there are just too many of these and they are being
changed and added to almost daily. The concern here is with
litigation, with suits seeking to protect or establish basic
environmental rights and duties. This means, in many cases,
that the practitioner involved with public projects may be the
defendent in an environmental suit. The reason for this is
simply that at this stage in the development of environmental
law the emphasis is on stopping or redirecting public action.
No action is not a threat to the environment, whereas almost
any public project may be alleged to be a threat to the environment.

Thus the primary object of environmental law is the individual, firm, or government agency whose actions may be alleged
or construed to pose a threat of damage to the environment. At
the same time, the simple fact of an alleged threat to the environment is not sufficient as a basis for legal action. There
must be a real injury as well as an accepted basis for suit,

i.e., either at Common Law or established by statute. In addition, principles of environmental law, however derived, are subject to applicable Constitutional constraints and are limited by societal values which are important components of American legal doctrines. Illustrative of these constraints and limitations are concepts of "due process" and "equal protection of the law," as well as deep-seated values with respect to individual freedom and property rights.

Environmental law is not so much "new" law, as it is a new combination of old law and old legal principles with new concerns, policies, and interests expressed by judges handling ordinary Common Law cases, as well as by congress and state legislatures in statutes, and by administrative agencies in rules and regulations affecting or governing individual, corporate and agency behavior.

At the same time, although the roots of environmental law lie in well-established case law, in statutory law, and in administrative regulations, it is important to recognize that the amalgam of case, statutory, and administrative law now designated environmental law owes much to the creative inputs of those lawyers and legal scholars who recognized the need for developing legal theories and concepts to control and prevent what have been called "insults" to the environment, and who sought to express these ideas in litigation, in treatises, in law review articles and in other scholarly works (which influence judges, and practicing members of the legal profession, as well as other scholars).

Creative acts can also be credited to judges who responded to the arguments of environmentally concerned lawyers and who have begun to incorporate some of these ideas in their opinions to explain and justify judicial intervention into both public and private spheres of action in order to protect the environment.

It is a truism that judges, even when blessed with life tenure as are federal judges, "follow the election returns," i.e., respond to currents of opinion expressed in society generally. Judges, too, are a part of American life and society, and in this respect are aware of the concerns over water and air pollution, and of other real or imagined threats to the environment. Although judicial responses to societal concerns may, at times, be slower than those of legislators, elected officials, and bureaucrats (if only because the emphasis on precedents provides a built-in kind of gradualism), nevertheless changes and new emphases do develop. And in the case of environmental law the Courts have evidenced an amazing willingness to move in new directions and to formulate new principles.

There are probably many explanations, and certainly no
single factor can account for the present interest in the subject. Undoubtedly a widespread disillusionment with government,
with the bureaucracy, and with "the system" is partly involved.
Undoubtedly, too, the activism of Ralph Nader and other young
lawyers has contributed to the development of environmental law.
Finally, it is clear that the concern for what seems to be hap-
pening to the environment has encouraged a re-examination of
many devices and techniques and out of this combination of interests has emerged the field of environmental law.

COMMON LAW AND CONSTITUTIONAL CONSTRAINTS

It is often alleged that when Americans are confronted
with a public problem, the solution proposed is, simply, "pass
a law." Certainly, today, most changes in social policy take
the form of legislative enactments.   And while legislating new
policies and principles seems a natural and logical way to
solve public problems, it is important, especially in the field
of environmental law, to recognize some of the constraints on
legislative action which flow from well-ingrained Common Law
and constitutional principles.

As suggested in the previous definition, environmental law
is to a large extent concerned with social control of individuals, groups (e.g. corporate entities), and government agencies.
It deals with how people behave, with how they use (or abuse)
natural resources. Hence, it is most generally concerned with
restricting human action and activity, and for this reason the
practitioner responsible for public projects more often than
not finds himself in the role of the defendant--forced to justify his actions and to explain the public interests in what he
is doing so that courts may engage in balancing the positive
against the adverse effects as well as consider compliance with
statutes and judicial principles.   Environmental law tends to
focus on the "Public Interest," rather than on individual interests or rights.   As such, environmental law, in most of its
aspects, seeks to control and constrain individual behavior[5]
The result may be conflict with established constitutional and
Common Law principles, which in turn may require re-thinking
and interpretation in the context of present-day problems and
needs.

One such conflict is that with concepts of private property in land and the rights of ownership which have evolved from
feudal doctrines.   Rights in landed property in the United
States have always been highly regarded.   Although in the Declaration of Independence Jefferson substituted the phrase "pursuit of happiness" for the more common "property," both the
Virginia Bill of Rights and the Massachusetts Bill of Rights

included strong references to *property* stressing that "enjoyment of life and liberty, with the means of aspiring and possessing property" was a basic right of citizens. The connection between liberty and ownership of private property was taken for granted, since it was felt that without property, one could not resist the encroachments of an autocratic monarch[6]

Sir William Blackstone, whose *Commentaries on the Laws of England*[7] were basic to the training of 19th century American lawyers, articulated ideas of private property which became part of the warp and woof of American practical thinking and acting. Blackstone's phrase suggesting that the owner of land in fee simple had "sole and despotic dominion" over his land, continues to limit approaches to environmental problems concerned with land. The absence of liability for what economists call "externalities," contributed significantly to obscuring the interests of the general public in how land and other resources were used and managed. Such dominant ideas with respect to property rights in land were reinforced by provisions of the U.S. Constitution with respect to due process in and just compensation for taking private property[8]

Undoubtedly the simplistic concepts with respect to land rights espoused historically by many lawyers and judges in the United States were also strongly reinforced by frontier values, and by the fact that land was plentiful. Without responsibility to a superior lord in the medieval pattern, owners of land in the New World believed that they could, in fact, do with their land what they saw fit, so long as no *direct* damage to others resulted. Thus very little of the concept of social obligation, which accompanied the use of land in a feudal society, carried over in this country.

One scholar has stated that: "in the case of feudalism, it is regrettable that there could not have been preserved the idea that all property was held subject to the performance of duties--not a few of them public."[9]

The absolute control over land-based resources was reinforced by the failure of Congress to deal with some of the problems posed in new territory (particularly that ceded by Mexico after the Mexican War) as settlement moved westward. Thus the concept of *occupation* as the means by which title was to be secured was accepted, particularly in two important areas of environmental law: western mineral and water law[10] (The emphasis on *occupation* was also an important element of the Homestead Act[11]) And as a result, Western mineral and water law evolved from primitive legal arrangements formulated by community agreement often in advance of governmental enactments. In the case of mining, miners formed extra legal associations which established procedures and policies under which those first on the scene of discovery gained title. These associational rules specified the size of claim to which each miner was entitled, determined the amount of work which must be done each year on the claim, and provided for record books in which the regulations of the association and a description of each claim was recorded. A similar pattern of development occurred with respect to water rights in the West, where the doctrine of "Prior Appropriation" recognized that the first man who took water from a stream to use it, either for irrigation or in a mining operation (beneficial use) established his right to the water so long as he continued to use it. In both cases, ownership was established by occupancy and use, and legislative recognition of these practices (legitimizing them) came after the fact. Ultimately, both practice and doctrine supported ideas of absolute ownership of resources and such *laissez faire* ideas continue to constrain approaches to environmental problems.

Property rights are shaped and controlled by law. Law in turn is shaped and controlled by societal values. Pragmatic necessity may result in institutionalizing practices and philosophies which serve later as constraints on changes which society may wish to attempt. Hence to change traditions deeply imbedded in constitutional provisions, statutory enactments, and judicial interpretations, is no simple matter.

The 19th century concept of "sole and despotic dominion" was theoretically limited by "police power" doctrines permitting governmental intervention for the health, welfare, and safety of citizens. But the courts required a specific justification of uses of the "police power," which in any case was held not to be a federal power. Hence the need for and difficulties in formulating concepts of public duty and of public interest in the delimitation of principles of environmental law.

Even though the courts are beginning to struggle with externalities, spill-over effects, and common property values, and some scholars are urging a doctrine of "public trust"[12] with respect to resource use and environmental values, the strong respect for private property rights has not been abrogated. Thus governmental action to protect the environment in the public interest may necessitate restriction of private rights in land, and could thus come under constitutional requirements of just compensation for taking of private property, or could conflict with constitutional principles of equal protection of the laws. These constraints are most evident in attempts to plan and control uses of private lands, and while requirements for procedural due process may readily be met, problems of compensation for taking property and of equal protection are far more difficult--and expensive. For these reasons, among others, the National Park Service in developing the Mississippi River Parkway which is to run from Minnesota to the Gulf of Mexico is relying on scenic easements (paying landowners for specified restrictions on the free use of their lands) rather than on simple land use controls![13]

In addition to limitations on legislation discussed in the previous section, it should be emphasized that the simple act of passing a law may often not solve the problems to which such enactments are directed. Among the more obvious constraints on legislative effectiveness are those which might be designated "managerial." Primarily these pose the question of whether the means authorized in fact achieve the ends sought. Limitations may be technological or sociological.

If the state of technology is such that legislated goals cannot be achieved, there is little that can be done except to seek to improve technology. The goal may be sound, but intermediate goals (e.g. research and development) are required.

Sociological limitations include such questions as public acceptance and understanding. Reliance on statutory enactments may often place too much emphasis upon command, and not enough upon understanding of causes and of the need to motivate individuals. Law, to be effective, must reflect the values, hopes, and aspirations of substantial segments of society (legitimation). Even a dictator cannot simply command; his wishes will not be implemented unless his orders to some degree coincide with what his public expects and understands. Changes in human behavior rest on changes in values and expectations, which are accomplished through socialization processes.

Regarding the law as an expression of societal values, rather than simply the decrees of the sovereign, has been a primary characteristic of the Common Law system. Thus the great, historic, statutory enactments of Parliament (e.g. the Bill of Rights) reflected the expectations of significant segments of the English people, almost in the sense of Rousseau's General Will. And it is just for this reason that the evolvement of environmental law must confront long-established doctrines of the Common Law, for the establishment of new policies and the determination of new directions in law is not simply a matter of new legislative enactment. It involves dealing with old traditions, grafting on them new ideas, and developing approaches which are consistent with the desires and understanding of the public generally.

But a word of caution is in order. Environmental law is loaded with exciting policy issues; the procedural niceties are fascinating; the stakes for both plaintiffs and defendants are often large. At the same time, the temptation for amateurs to play at practicing law is great. The consequences can only be tragic, for if the environment is truly threatened, amateurs at environmental law can do no better than amateurs in construction of public projects. Environmental law is sophisticated law; it deserves the best legal minds available.

COMMON LAW ACTIONS PROTECTING THE ENVIRONMENT

Concern for the environment was not a dominant feature of American Common Law as it developed through judicial interpretation in the 19th and early 20th centuries. Yet there was some litigation which had the effect (if not the purpose) of protecting environmental values.[14] Most of this litigation was based on doctrines of nuisance and trespass and involved suits between private parties in which plaintiffs sought money damages or judicial restraints. Essentially, the private nuisance doctrine stated that if one carried on activities which damaged one's neighbor, the neighbor could seek redress. Trespass litigation rested on the concept that if one's activities invaded another's property or person, a suit for damages might result. In some situations, where money damages were inadequate or the damage was irreparable, equitable relief (an injunction against the nuisance or trespass) might have been available.

In these situations, the concern was to protect individual rights, rather than to establish a public interest in the quality of the environment. In some cases, the latter was kind of a by-product. And while both types of action were used to deal with air and water pollution, the problems of proof of damage were often difficult, especially where spill-over effects (externalities) were involved. Thus a plaintiff living immediately next to a cement plant polluting the air with harmful dust might be able to show damage, a plaintiff living a mile away where some dispersion would already have occurred would have more difficulty.

Of a somewhat different character is the *public* nuisance suit where action is initiated by governmental authorities and only rarely by private citizens. A public nuisance is generally defined as conduct causing "an unreasonable interference with a right common to the general public."[15]

While nuisance and trespass doctrines have been available for centuries, they are to some extent becoming tools for environmental protection particularly because the courts have relaxed their definitions of who can sue, and also have simplified the proof requirements for such suits.

In the 19th century the concern of the courts for natural resources and the environment was largely incidental to private litigation, or to theories of private damage. Only in very recent times have problems of the environment been phrased in public interest terms resulting in increasing legislative, judicial, executive and administrative interest and activity. But the influences of judicial precedents and traditional practices continue to have an impact on the way in which we define our problems today. And, of course, Common Law concepts have been supplemented by a growing body of federal and state statutory law.[16]

PROCEDURAL DEVELOPMENTS

A dramatic development of the last ten years has been the growth of environmental litigation initiated by a range of citizen organizations which have been able to use the courts to challenge and often to force government agencies as well as private companies to change their courses of action.

But the extensive litigation to control the environment in the last few years would not have been possible had not some fundamental procedural changes occurred![17] One of the most significant of these changes concerns the question of *standing to sue*. The Constitution gives to the federal courts jurisdiction over "cases and controversies." The traditional viewpoint with respect to this language had been that to constitute a "case or controversy" the parties to the dispute had to have a clear and distinct interest in the subject matter, and this was usually interpreted to mean an economic interest. Thus, for example, it had not been possible for a taxpayer to sue the federal government simply because he was a taxpayer. Paying taxes was not regarded as giving a sufficient interest to qualify as a litigant. But in the past decade, courts have been reconsidering the concept of standing to sue. Now, rather than simply a mechanical test, the courts have begun to examine the characteristics of the agrieved person, what the nature of the legal interests entitled to protection may be, emphasizing public policy as expressed in legislation. While the courts have not departed from the basic concept that the judicial function is the decision of concrete cases and the granting of specific remedies to those claiming they have been specifically wronged, nevertheless they have tended to use broadened criteria to determine the nature of the interests at stake and thus to search for some basis for establishing standing to sue.

Federal courts have been construing many statutes establishing public programs, giving particular beneficiaries an adequate basis for asserting claims of one sort or another. In a recent case, the U.S. Supreme Court supported this view, stating that the party seeking relief must "have such a personal stake in the outcome...as to assure the concrete adverseness which sharpens the presentation of issues upon which the court so largely depends for illumination."[18]

Professor Louis L. Jaffee has commented that the issue of a proper plaintiff can be viewed in two ways: one is that the plaintiff must be seeking protection for a right granted to him by statute, the Constitution, or the Common Law; the other and less rigid approach is simply to require that the plaintiff have a sufficient interest to assure proper presentation of the case![19]

In a recent environmental case[20] the court permitted the Scenic Hudson Preservation Conference, which was an unincorporated association, to sue to set aside a permit granted by the Federal Power Commission on the grounds that the Commission had not satisfied its statutory duty to consider the impact of the proposed power plant on the scenery and historic significance of Storm King Mountain. In dealing with the problem of standing to sue the Federal Circuit Court indicated that the Scenic Hudson Preservation Conference was in a class of "agrieved parties" who under the statute are entitled to seek review of agency action by a court.

It might be noted, incidentally, that some states are more liberal in permitting persons to sue whose interests are somewhat less clearly defined than the traditional situation in the federal courts.

In interpreting the present trend Professor Jaffee has said:

> "There is no reason, however, to predict that courts or agencies will impose strict or unsympathetic standards. They will probably demand no more than to be assured that there is a legitimate controversy of general public interest, and that the plaintiffs are responsible parties, able effectively to present the case."[21]

It should be noted that the results in Scenic Hudson and other environmental cases which have recently been considered by the courts, have been largely procedural rather than substantive. Thus the responsibility for making policy decisions remains with the agency or with the legislature. The concern of the courts has simply been whether the agency has given careful enough attention to all the factors which statutes require should properly be considered.

In the recent "Mineral King" case[22] the U.S. Supreme Court redefined the basis for standing to sue somewhat more narrowly, but most commentators are not concerned that the Mineral King limitations will be a serious impediment to environmental suits. In that case the Court rejected a suit by the Sierra Club on the grounds that the club had not asserted a sufficient stake in the preservation of the area concerned, observing that "the impact of the proposed changes in the environment of Mineral King will not fall indiscriminately upon every citizen,"[22a] but only on those who use the area. Hence it seems that only such users and organizations representing such users have a sufficient interest to challenge the developmental decisions of the Federal agencies responsible for the area. Although the Court rejected the broadest possible kind of interest, it clearly stated that "Aesthetic and environmental well-being, like economic well-being, are important ingredients of the quality of

life in our society" and deserving of legal protection. The
Sierra Club, following the suggestion of the Court, has now
amended its complaint to claim a more direct injury and is pur-
suing the issue in the lower federal courts.

A significant statutory development was put into the new
water pollution control law enacted by Congress in the closing
days of the 1972 session overriding the veto of President
Nixon.[23] Essentially Congress wrote into that law a citizen
standing to sue provision quite similar to that established in
the Mineral King case, broadening the right to sue to permit
actions against private parties as well as government agencies.
The full impact and meaning of this new venture into citizen
suits remains to be seen.

Another procedural development, important to facilitating
environmental litigation, concerns so-called "discovery rules."
Partly to speed up judicial action, and partly as a matter of
basic justice, courts have for some time been trying to limit
"surprise elements" in civil suits. The greater use of pre-
trial concerences and of depositions are a part of this trend.
In addition, discovery procedures permit a plaintiff to have
access to certain information and data which is in possession
of the defendant and vice versa. In environmental litigation
this can be of critical importance, since often only the defendants will have the operating data (e.g., in an air pollution
suit) which will indicate the extent to which pollution is in
fact occurring. The hard, expert, data required in many environmental suits can only be secured in this way.

Although not enacted as environmental law, the *Freedom of
Information Act*[24] (1966) provides for citizen access to much
federal government data, exceptions being proprietary information and internal papers dealing with matters of policy.

The trend is clear: if citizens and citizen groups can
establish their standing to sue, they will have access to much
better data and information relevant to the suit than had been
the case some years ago.

Related to discovery procedures is the problem of expert
testimony. This can often be one of the most difficult aspects
of environmental litigation for the individual or group seeking
to protect the environment, since all of the "expertness" may
be in the hands of government or industry. The practitioner
should also note that the judicial requirements for establishing "expertise" are more rigorous than those required of legislative witnesses, where testimony may go unchallenged. Clearly,
many specialists who have testified on various environmental
problems before Congress or state legislatures might quickly be
disqualified as experts in a judicial proceeding. While the
subject to which an expert may testify and the scope of his

testimony is rigorously constrained in a judicial proceeding, before a legislative committee the witness is free to say almost anything he may wish. The practitioner should also note that although he may be employed by the defendant. he may be required to testify as an expert witness for the plaintiff.

Another procedural problem is that of "burden of proof." Professor James E. Krier has stated that the rule with respect to burden of proof may be "a serious obstacle to successful environmental litigation...," pointing out that "it is one thing to provide that an enterprise causing environmental damage is liable only for fault; it may be quite another to place the burden of proof of the issue of fault on the complainant rather than the enterprise."[25] The difficulty is particularly serious in cases where litigation seeks to prevent threatened damage to the environment, such as *might* result from construction (of a highway, an airport, a powerplant, and so forth).

Although this is a rather technical aspect of trial procedure, the burden of proof responsibilities on the plaintiff seeking to raise environmental questions have been eased considerably by statute and by judicial interpretation. For example, amendments to the Clean Air Act in 1970[25a] authorized the administrator of the Environmental Protection Agency to establish clean air standards. This has had the effect of shifting the burden of proof from plaintiffs seeking to restrain air pollution to the defendants, for now the latter, who object to the standards, must prove that their emissions are *not* hazardous. A similar emphasis is contained in a number of pending bills dealing with various other aspects of pollution.

The willingness of some courts to ease the burden of proof requirements is suggested by a recent New Jersey case[26] in which the court placed considerable emphasis on the responsibility of the defendant (namely the gas transmission company) to show that its proposed actions were reasonable and not capricious. In summing up this case, Professor Krier has stated:

> Once a reasonable showing is made that a proposed course of action poses a probable threat of significant environmental damage, the body desiring to initiate that action should be required to come forward with evidence on the likelihood of such damage, the unavailability or unfeasibility of alternatives, and the justification for its activities[27]

It is still too early to predict the extent to which other courts will follow this line of reasoning, but it is noteworthy that a number of state statutes have been enacted or are being considered which would guarantee this approach where a challenge is made to actions which threaten the environment.

As stated earlier, with a few important exceptions, the
courts have not been anxious to move into substantive questions
directly. At the same time, articles and books are appearing
which are beginning to suggest theories of law (substantive
legal doctrines) which courts might apply in dealing with environmental matters. It is not unreasonable to expect, therefore,
that many courts will be willing to consider substantive questions, *at least to the extent of determining the reasonableness
of agency or private action which threatens the environment*. As
will be indicated below, it is this emphasis in the National
Environmental Policy Act of 1969[28] (NEPA) which may well be its
most significant contribution to the field of environmental
protection and enhancement.   But before considering this landmark legislation, it is useful to examine a number of other
legislative enactments which have set policies and provided a
basis for environmental litigation.

ENVIRONMENTAL LEGISLATION AND PUBLIC PROJECTS

Both state and federal governments have recently enacted
statutes dealing with environmental protection which affect the
way in which public projects are planned, developed and constructed.

Statutes may do three things of importance to public projects (in addition to simply authorizing programs): *first*, they
establish policies and programs with respect to environmental
protection; *second* they identify specific beneficiaries with
interests in environmental protection and enhancement,   and
third, they establish procedures (guidelines) which in turn
serve as the basis for administrative rules and regulations.
Statutes in all three of these categories represent important
inputs to environmental law and may be used to constrain and
limit how public projects are developed.   [It should be noted
that under the U. S. Constitution International treaties (and
the statutes enacted to implement them) are "the law of the
land," and may have important consequences for preventing environmental degradation.]

As already noted, statutes like the *Administrative Procedures Act*[29] and the *Freedom of Information Act of 1967*[30] have
substantial procedural significance for environmental action.

At the next level of specificity are basic program and
agency statutes which often include policy provisions expressing congressional intent with respect to environmental matters.
A few of the classic statutes in this category are:

The Federal Highway Act, as amended, Title 23, USC;
The Federal Power Act of 1920, as amended, Title 16 USC
    791 ff.;

The Atomic Energy Act of 1954, as amended, Title 42 USC
   2011 ff.;
The Water Resources Research Act of 1964, 42 USC 1961;
The Water Resources Planning Act of 1965, 42 USC 1962;
Federal Water Pollution Control Act, as amended, Title 33
   USC;
The Clean Air Act of 1963, as amended, Title 42 USC;
The Solid Waste Disposal Act of 1965, Title 42 USC.

Congress has, understandably, enacted many statutes governing the administration of federal lands (the "Public Domain",
as well as lands dedicated to particular purposes). Many of
these statutes are codified in Title 16 of the U. S. Code, including:

The Organic Act of the Park Service; the Historic Sites
Act; the Endangered Species Act; the Wildlife Restoration
Act; the Migratory Bird Treaty Act; the Fish and Wildlife·
Coordination Act of 1965; the Wilderness Act; the Multiple
Use and Sustained Yield Act; the Wet Lands Act; the Estu-
aries Act; and many others.

Beginning in 1926, Congress provided for periodic codifi-
cation of the general and permanent laws in a set of volumes
known as the United States Code (USC). The first edition of the
United States Code was issued on June 30, 1926. The second
edition was released in 1934 and the seventh in 1964. The Code
is divided into 50 titles, and the titles, in turn, are divided
into sections. Each title contains most of the laws relating to
a particular subject.

Several words of caution: *first*, if the Code varies from
the Statutes at Large in a significant way, the wording of the
Statutes govern; *second*, although the codifiers have done a
good job of consolidating related legislation into the appropriate titles, it is obvious that some enactments might reasonably have been placed in more than one Title, so the index
should be consulted with care; *third*, many statutes have been
interpreted by the Courts, so that judicial opinions become a
part of the statute's legal meaning; and *fourth*, many statutes
must be implemented and interpreted by those responsible for
their administration.

Formal administrative interpretations and implementation
actions are published in the *Federal Register*, which appears
five days each week. The continuing rules and regulations published in the *Federal Register*, serially as issued, are then
subsequently codified in the *Code of Federal Regulations* (CFR).
It should be emphasized that administrative rules and regulations are generally just as much "law" as are the statutes on
which they are based.

It should be clear from the preceding discussion that federal statutes contain many provisions concerned with protecting the environment, and these provisions, in turn, have resulted in a body of administrative law interpreting and implementing the statutes. Where statutory provisions or administrative regulations and policies pursuant to statutory grants of authority and responsibility have been involved in litigation, judicial opinions are also part of the total body of law. To discover these requires special research techniques and of course access to a good law library. Interpretations of the provisions of the U.S. Code may be found in the U.S. Code Annotated (USCA), which is organized just like the U.S. Code (USC).

A similar situation exists at the state level--with fifty possible variations in policy, procedures, programs, and goals. A variety of constitutional and statutory provisions have been enacted by the states dealing with environmental problems. New York and Illinois have amended their constitutions, adding what is popularly called a "conservation bill of rights." Pennsylvania and Rhode Island have added similar declarations into their fundamental law. A number of states, following the lead of the federal government, have enacted environmental policy acts which require the preparation of impact statements similar to the federal requirement. Montana has enacted one of the most comprehensive provision of this kind[31] Michigan adopted an Environmental Protection Act which gives to individual citizens the right to sue state agencies adversely affecting the environment or failing to carry out their responsibilities to protect the environment. Somewhat similar laws have been adopted in Connecticut, Indiana and Minnesota[32]

THE NATIONAL ENVIRONMENTAL POLICY ACT

Unquestionably, the most important recent legislation dealing with the environment has been the National Environmental Policy Act of 1969 (NEPA) (reproduced in Appendix A) which created the Council on Environmental Quality, and established a number of important policies[33]

───────────────────
Note on Sources for Federal Statutes

The laws enacted by Congress are published in a set of volumes called the *U. S. Statutes at Large*, cited in various ways including "*Stat at L*," or just "*Stat.*" Currently one volume of the *U. S. Statutes at Large* is issued each year (sometimes in several parts) and contains the laws enacted during the calendar year. The entire set includes all laws enacted by Congress from 1789, printed serially in the time sequence in which they were enacted.

The National Environmental Policy Act is not a long statute. Its language is general, which gives flexibility, but also
establishes the need for judicial interpretation.

The purposes of the Act are stated:

To declare a national policy which will encourage productive and enjoyable harmony between man and his environment;
to promote efforts which will prevent or eliminate damage
to the environment and biosphere and stimulate the health
and welfare of man; to enrich the understanding of the
ecological systems and natural resources important to the
nation; and to establish a council on environmental quality.

In Section 101 (a), the Act indicates the premises for
congressional action as follows:

The Congress, recognizing the profound impact of man's activity on the interrelations of all the components of the
natural environment, particularly the profound influences
of population growth, high density urbanization, industrial expansion, resource exploitation, and new and expanding technological advances, and recognizing further
the critical importance of restoring the maintaining environmental quality to the overall welfare of man, declares
that it is the continuing policy of the Federal Government,
in cooperation with state and local governments, and other
concerned public and private organizations, to use all
practical means and measures, including financial and
technical assistance, in a manner calculated to foster and
promote the general welfare, to create and maintain conditions under which man and nature can exist in productive
harmony, and fulfill the social, economic, and other requirements of present and future generations of Americans.

The act also expresses concern for succeeding generations,
for esthetically and culturally pleasing surroundings, for preventing degradation of the human environment, for preserving
historic cultural and natural resources, for encouraging diversity, for seeking a balance between population and resources,
and for enhancing the quality of renewable resources.

The legislative history of this act indicates that only
casual attention had been given to how it would work, or what
its effects would be.   There is reason to believe that many a
congressman (and the act passed with substantial majorities in
both houses) felt that he was simply supporting a general policy
declaration which, like motherhood, would find few objectors.
To be sure the act required *all* federal agencies to be concerned
with the effects of their activities on the environment, and

while the Council on Environmental Quality was given some responsibility for monitoring agency activities, there was no
suggestion that this would be a large task. Reflecting this
view the Nixon Administration limited council staff to about 50
employees, which remains the size of the agency today.

The surprising development in the act (the "sleeper" among
its high-sounding provisions) was section 102. This section
emphasized the need for a systematic, interdisciplinary approach based upon the natural and social sciences, and the environmental design arts in planning and in decision making. By
itself this language seemed harmless enough, but in the context
of another subsection of section 102, it turned out to have
tremendous significance. That subsection provided for the
preparation by federal agencies of "Environmental Impact Statements" reviewing proposals for legislation and for major federal actions which were likely to significantly affect the
human environment. These "Impact Statements," the act indicated,
were to explore the environmental impact of the proposed action
indicating adverse consequences for the human environment, outlining alternatives, distinguishing between short-term and
long-term consequences, and honestly facing up to "irreversible
and irretrievable" commitments of resources. This section also
required consultation by each agency with all other agencies
which had jurisdiction by law or had special expertise with
respect to any environmental impact involved. But the section,
departing from federal agency preoccupation with "coordination,"
recognized the primary responsibility of particular agencies,
and these set the stage for a new level of accountability.
Guidelines issued by CEQ and published in the *Federal Register*
have developed these ideas in even greater detail. The major
agencies, in turn, have issued their policies and procedures
which are also published in the *Federal Register*.[34]

Section 102 is staggering in its implications for federal
administrative practice. And it has thus been interpreted.

Quantitatively the result has been completely unexpected.
Through May 31, 1972, 2,933 impact statements have been submitted to the Council on Environmental Quality (see Figures 8-1
and 8-2) and it very soon became apparent that the council with
its small staff, could not hope to review any but the politically sensitive statements. As a result, the statements are
filed, printed, listed in various sources,[35] and made available
on request to anyone who may be interested (at a relatively low
cost). In addition to the quantity of impact statements, their
length is also significant. Many run to hundreds of pages, and
some (e.g., the statement on the proposed Alaska pipeline) fill
volumes, running to thousands of pages.

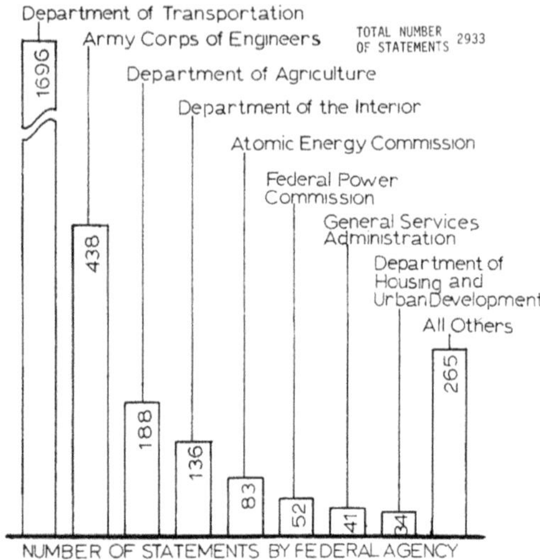

Figure 8-1. Environmental impact statements filed with the Council on Environmental Quality through May 1972 by agency.*

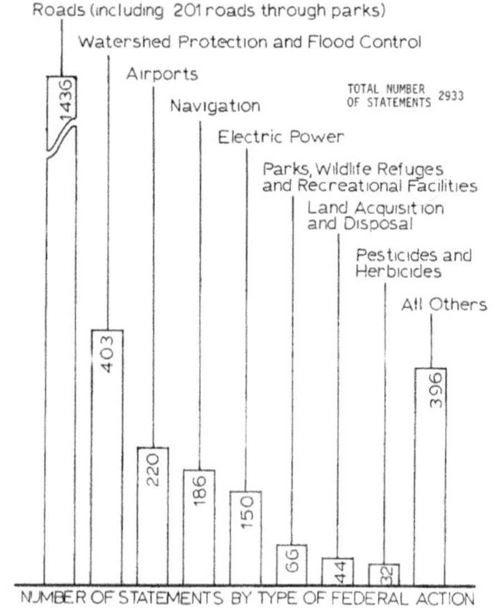

Figure 8-2. Environmental impact statements filed with the Council on Environmental Quality through May 1972 by type of federal action.*

---

*Includes all final statements and draft statements for actions on which a final statement has not yet been filed. (From *Environmental Quality*, Third Annual Report of the Council on Environmental Quality, Washington, D. C., August 1972, pp.248-249.)

But since the Council on Environmental Quality does not
systematically review the impact statements, the question naturally arises "what is their utility?" For the time being the
best answer to this question seems to be that the preparation
of impact statements has served as a focus for more effective
interagency coordinating. The statements themselves have provided a significant handle for citizen litigation. In these law
suits, citizens and citizen organizations are generally seeking
court orders restraining government agencies or private corporations from a particular course of action. As of December 31,
1971, 48 Federal District Court opinions on NEPA had been rendered, 15 Federal Courts of Appeals decisions had been issued,
and 3 discussions of the Act had occurred in Supreme Court dis-
sents. And the flood of litigation based upon NEPA is continuing. The Council on Environmental Quality, in its third annual
report stated that over 200 suits have been initiated under
NEPA.[36]

This development, also largely unanticipated, reflects the
present receptiveness of the federal courts to suits by a variety of individuals and groups seeking to protect the environment by challenging the impact of both federal and state actions. It must be emphasized, however, that although some of
the opinions have involved interpretations of NEPA, the cases,
almost without exception, are being settled on a procedural
basis, the courts referring the issues back to the responsible
agencies for such action as would reflect more effective compliance with the procedural requirements of Section 102, including more effective coordination, or a broader consideration
of alternatives, or, by implication at least, more emphasis on
interdisciplinary analysis.

One of the landmark cases, Calvert Cliffs,[36a] provided
Judge Skelly Wright the opportunity to use impressive rhetoric
with respect to the environment and its preservation. But the
result of this case was that the court referred the matter back
to the Atomic Energy Commission for further action consistent
with the views expressed in the opinion and the standards of
NEPA. In this connection, it resulted in major changes in the
procedures of the Atomic Energy Commission.[37] Since the Calvert
Cliffs plant was well over 50% completed at the time of the
Circuit Court of Appeals decision, it must realistically be
concluded that the plant will be completed and put in operation,
so that the effect of Judge Wright's decision has been to delay,
rather than halt, the construction of this atomic power plant
on the Chesapeake Bay.

But this emphasis on procedure is not unimportant because
in these decisions new criteria have been established which
agencies must follow, and they are being held accountable to
the procedures of NEPA itself. There is convincing evidence
that as a result of the way in which NEPA has been interpreted,

and the extent to which the courts have permitted citizen litigation based upon NEPA, a new multi-disciplinary influence has been imposed on program agencies, although at a substantial administrative cost.

The courts have stated that agencies cannot rely on impact statements prepared by the private beneficiaries but must prepare their own analyses. This has contributed to slowing up statement preparation.

Other agencies are finding the requirements of NEPA, as interpreted by the courts, a burden to their customary administrative procedures. Some observers have asserted that there seems to be developing what a writer in *Science News* calls "signs of backlash."[38]

In many cases, the immediate effect of NEPA has been to flood agencies with paper work, bringing many activities to a virtual halt. And a major battle is developing in Congress over attempts to amend NEPA so as, in one way or another, to exempt certain activities or agencies. One example of this effort, the results of which remain to be seen, was the enactment, over the president's veto, of the Federal Water Pollution Control Act Amendments of 1972 (October 18, 1972)[39] which repeated the Refuse Act of 1899[40] and changed the permit procedures explicitly removing the requirement for impact statements. This requirement had created a backlog of 20,000 permit requests in the U. S. Army Corps of Engineers under the 1899 law.

But it is clear that NEPA would not be having this far-reaching effect, if it were not that the courts have been willing to entertain private suits by various environmental groups. Thus the procedural changes discussed earlier in this chapter (standing, burden of proof, and so forth) have been of critical importance to the development of NEPA and the use of the impact statements in litigation questioning government action.

The scope of NEPA litigation is suggested in a thoughtful review by Harold P. Green entitled *The National Environmental Policy Act in the Courts*[41] in which he reviews jurisdictional questions of standing and sovereign immunity, procedural questions concerning necessary allegations in initiating a suit, discovery procedures, indispensable parties, the requirement for a security deposit, and restraint of administrative proceedings or review of them. He explores the questions of when a "102 Statement" is required, who must prepare the statements, what is meant by compliance, the extent to which NEPA is a full disclosure law, and how a NEPA statement is to be used. He also reviews the relief available where there is a NEPA violation and the role of public participation. Obviously few statutes only two years old (when Green wrote his article) have created so many problems and raised so many issues.

Reference should also be made to the excellent reviews of
the developing situation in the second and third annual reports
of the Council on Environmental Quality (each entitled simply
Environmental Quality)[42] Chapter five of the second report and
chapter seven of the third are of particular usefulness in
giving background information on environmental litigation and
on the development of NEPA and of CEQ under its provisions.
Other chapters deal with state and local developments, as well
as with the international situation.   These reports are of
course "in house" documents and as such are not expected to be
critical.   To date the critical literature is limited, but one
thoughtful paper raising many important questions was delivered
in February 1972 by Roger Hansen, a noted environmental attorney, and Executive Director of the Rocky Mountain Center on the
Environment[43] While it is not possible to outline all of Hansen's concerns, it is important to note his doubts that private
litigation will prove an adequate instrument to protect the en-
vironment--the cost in time and money is just too high. Thus he
urges an agency which would have the task of reviewing all "102
statements," with the power of vetoing proposed actions which
in its judgment endangered the environment.

Thus it is worth emphasizing that while NEPA does represent a significant advance, and while agency administrative
procedures have been altered in response to NEPA, there is reason to believe that in the long run substantive program issues
will need to be dealt with. And in its present form, it is not
at all clear that NEPA is adequate for this purpose.

In this context attention needs to be directed briefly to
the range of other techniques available to federal and state
governments. As indicated earlier in this chapter, the constitutional doctrines with respect to federal interest in environmental problems are very limited. The spending power continues
to be the major approach of the Federal government for protecting environmental values. Subsidies, and tax credits are among
important techniques being considered. Grants to states, in
addition, provide incentives for them to act in ways desired by
the federal government.   Thus, state penalties and charges may
be encouraged by federal action requiring pollutors, for example, to internalize what are now external costs.   In addition,
grant programs will undoubtedly continue to urge the states
into more effective state and local planning, including land
use control.   Independently, as suggested earlier, many states
are also responding to the various conservation and environmental pressure groups, and are enacting new laws which may have a
profound influence on environmental protection.

Perhaps the 1970's may be designated as the "Era of Environmental Protection."   But it is not clear how far this trend
will continue. An important question concerns the distribution

of costs connected with environmental protection. Most observers agree that the environmental protection movement to date has been primarily an upper class, intellectual, movement, and as such it may in due course encounter serious citizen backlash. The forecast for the future must, therefore, be a forecast for considerable political tension and conflict, with the ultimate resolution of the issues not at all clear. There is reason to fear that many environmental costs may fall most heavily on lower income groups. In some cases, the immediate cost of environmental protection may involve increases in unemployment. In other cases, the effect may be more subtle, but the burden will nevertheless be identifiable in increased taxes and perhaps lower standards of living. In many cases, the issue may boil down to *who* pays the costs and *who* receives the benefits. It is neither reasonable nor just to have certain groups pay for benefits received by others, particularly when such allocations are the result of regulatory action. If such effects are clearly evident and unequal, claims of violating the "equal protection" clause of the Constitution are almost certain to develop. Only time and much litigation will resolve issues of this sort.

NEW DOCTRINES

Some authorities recommend that an environmental bill of rights must be enacted or put into constitutional form in order to guarantee the public interest (as against individual rights and interests) in a quality environment. On a more modest basis, other scholars have begun to examine older doctrines to see whether they might be revived and reformulated to protect the public interest in the environment. One ancient concept is that of "common property," developed in Medieval times in England and applied to certain land known as "The Commons," which were to be available to the public generally and particularly to those who did not have feudal land rights. The enclosure movements in England (16th and 19th centuries) converted most of The Commons into private property, driving the landless into the cities, and eroding the legal principles which applied to The Commons. The doctrine of common property survived in vestigal form with respect to land covered by water (fishing grounds)[44] and in a special sense with respect to wild animals, over which invidual property rights cannot be established.

A broader approach, being developed by a number of legal scholars including Professor Joseph Sax of the University of Michigan Law School, has been designated *The Public Trust Doctrine*.[45]

It must be noted that The Public Trust Doctrine is still largely unsupported by explicit judicial opinion, although it may ultimately be applied to environmental litigation when some court chooses to do so. It would seem to be most appropriate as a theory for restraining government action where public property is involved (parks and forest reserves) or mandating government action to protect public interests in particular resources. It may, also, be argued that the *Public Trust Doctrine* could be applied to property owned by private individuals and about to be used in a manner inconsistent with the public interest. The Public Trust Doctrine might, for example, be used to prevent changes in existing patterns of land use, but the precedents which prohibit "taking of private land without just compensation" (reinforced by constitutional provisions) raise serious question as to its applicability without further statutory or constitutional clarification. The most difficult situation would be that in which private property is involved and where there has been no apparent dedication of its use to the public interest. Attempts to control such used would quickly raise the issue of "taking" and thus would require compensation.

But it is just at this point where the deeply rooted regard for individual private property ownership clashes with emerging public interests in environmental protection. In the opinion of many legal scholars, for example, there is a very limited basis for controlling agricultural and other open land uses. Only where *specific* adverse impacts on others (involving nuisance or trespass) can be shown, are there clear precedents for public control. In general, courts have been reluctant to recognize scenic values and maintenance of an attractive environment as grounds for public control, even though these values are beginning to be recognized as a basis for restraining proposed public agency actions. Some argue that these values can only be achieved through purchase of the land in fee or of an easement.[46]

It would not seem politically feasible at this time to amend the Federal Constitution so as to modify the provision against the taking of property without just compensation in order to achieve better land use controls. Perhaps it is just for this reason that recent legislative enactments have dealt to a major extent with procedure rather than with substance. Or at least it might be said that the significance of recent enactments tends to be more procedural than substantive.

END NOTES

1. The literature on environmental law (books, periodicals, monographs, etc.) is growing at a rapid pace. A few titles and sources are listed here to be suggestive of the variety of publications; they are not meant to be a selected bibliography!

Two so-called "reporters" are of special note; both are designed for use by lawyers; both are loose-leaf:

a. *Environmental Law Reporter* (ELR), now in its third year,
   is published by the Environmental Law Institute (Dupont
   Circle Building, 1346 Connecticut Avenue, N.W., Washington, D. C. 20036) in twelve monthly issues. Its content
   is divided into the following sections: Summary & Comments; Litigation; Administrative Proceedings; Statutory
   & Administrative Materials; Articles & Notes; Bibliography & Facsimile Service; Indexes.

b. *Environmental Reporter* (ER) published by the Bureau of
   National Affairs, Inc. (1231 - 25 Street N.W., Washing-
   ton, D.C. 20037) in six volumes, entitled respectively:
   Current Developments; Federal Laws; State Air Laws;
   State Water Laws; Decisions; and Monographs.

The following cautionary statement adapted from a "Declar-
ation of Principles" jointly approved by a Committee of the
American Bar Association and a Committee of Publishers and
Associations applies to both of these services:

> These publications are designed to provide accurate and
> authoritative information...They are sold with the under-
> standing that the publishers are not engaged in rendering
> legal, accounting, or other professional services.   If
> legal advice is required, the services of a competent pro-
> fessional person should be sought.

A number of periodicals have been initiated dealing with
general aspects of environmental law and litigation, or with
special aspects of the subject.   In its 13th year, the *Natural
Resources Journal*, published by the University of New Mexico
School of Law, emphasizes policy issues as well as the legal
aspects of environmental problems.

Environment Law Review is an annual volume of essays published by Sage Hill Publishers, Inc. (Albany, New York) and
Clark Boardman Co., Ltd. (New York City).   Other periodicals
include: *Natural Resources Lawyer*, published quarterly by the
American Bar Association Section of Natural Resources Law; the
Land and Water Law Review; the *Atomic Energy Law Journal*;   and
of course the more general law reviews have been devoting considerable space to the subject of environmental law.   Searches
of periodical literature should begin with the *Index to Legal
Periodicals* available in most larger libraries.

A substantial number of text and case books are also appearing, e.g., Norman J. Torskan and Paul D. Rheingold, *The Environmental Law Handbook*, Friends of the Earth/Ballatine Books
(1971); Malcolm Baldwin and James K. Page, Jr. (eds) *Law and*

the Environment, Walker and Co. (1970); Joseph L. Sax, *Defending the Environment*, Knoff (1970); James B. Machondt and John E. Conway, *Environmental Litigation*, University of Wisconsin Extension (1972); Colorado Bar Association, Emission Law Handbook (1971).

2. There are many sources in the American legal system, including articles in good encyclopedias. A brief reference is James L. Houghtling, Jr. *The Dynamics of Law*, Harcourt, Brace & World (1968).

3. Federal Administration Purchase Act, 5 USC 551.

4. See note 1.

5. See Charles A. Reich, "The New Property," 73 *Yale Law Journal* 733 (April 1964).

6. Vernon Carstenson, "Our Fundamental Cultural Attitudes Toward Land," in Governor's Conference on Rural Land Use in an Urban Environment, Pub. #416, Congressional Extension Service, College of Ag., University of Mass. (Aug. 1964).

7. See Sir William Blackstone, *Commentaries on the Laws of England*, 4 volumes (1765-1769). A recent standard edition is by W. D. Lewis, 2 volumes, Philadelphia 1922.

8. John E. Cribbit,"Changing Concepts in the Law of Land Use," 5 volumes *Iowa Law Review* 245 (1965).

9. Francis S. Philbrick, "Changing Conceptions of Property in Law," from 86 U. Pr. L. Rev. 691 (1938) as cited by Cribbit, *op cit* n. 8.

10. See Carsension, *op cit* n 6.

11. The Homestead Act, _______ *Stat. at L.* _______.

12. Joseph L. Sax, *Defending the Environment*, op cit n. 1; also Joseph L. Sax, "The Public Trust Doctrine in Natural Resource Law: Effective Judicial Intervention," 68 Michigan Law Rev. 471 (1970).

13. William H. Whyte, *The Last Landscape*, Doubleday and W. (1968); Kamnowski versus the State, 31 Wis. 2nd, 265, 142 NW 2nd, 793 (1966).

14. While the early twentieth century saw the articulation of the first conservation movement (Pinchot/Roosevelt), the emphasis tended to be on development and misuse, rather than on regulation and control.

15. American Law Institute, *Resettlement 2nd - The Law of Torts*, sec. 821b-1 (tentative draft No. 17, April 26, 1971).

16. For a thorough review of the basis for legal action see Macdonald and Conway *Environmental Litigation*, op cit n. 1.

17. For a discussion of the professional developments see *Ibid*. Also see *Baldwin and Page*, op cit no. 1, pp. 105-133. See also Council on Environmental Quality (CEQ) *Second Annual Report: Environmental Quality* (1971) esp. Ch. 5 and *Third Annual Report: Environmental Quality* (1972), esp. ch. 7. See further Council on Environmental Quality, *102 Monitor* various issues; Harold P. Green, *The National Environmental Policy Act in the Courts*, The Conservation Foundations, Washington, D.C. (1971).

18. Flast V. Cohen, 392 US 83 (1968).

19. Louis L. Jaffe, "Standing to Sue in Conservation Suits," in *Balkin and Page op cit* n. 1, pp. 123-133.

20. Scenic Hudson Preservation Conference versus Federal Power Commission, 354 F 2nd 608 (2nd Circuit, 1965).

21. Jaffe, *op cit* n. 19.

22. Sierra Club versus Morton 2 ELR 20192 (U.S. April 19, 1972).

22a.*Ibid.*, at 20194.

23. P.L. 92-500 (Oct. 18, 1972) Sec. 505 deals generally with the right to sue and subsection (g) states simply that the citizen plaintiff must have an "interest which is or may be adversely affected."

24. 5 USC 552 (1967).

25. James E. Krier, "Environmental Litigation and the Burden of Proof" in Baldwin and Page, *op cit* n. 1, p. 108.

25a.42 USC 1857.

26. Texas East Transmission Corp. versus Wildlife Preserves Inc., 48 N.J. 261, 225 A.2d 130 (1966) and 49 N.J. 403, 703 A.2d 505 (1967).

27. Krier, *op cit* n. 25, p. 115.

28. 42 USC 4371; see appendix to this chapter.

29. 5 USC 551.

30. 5 USC 552.

31. "Montana Environmental Policy Act" (En. Sec. 1, Ch. 238, *Lewis 1971*) Revised Code of Montana Sec. 69-6501-69-6517.

32. Michigan Environmental Protection Act of 1970 (Act 127 P.A. 1970, Mich. Comp. Laws Ann. Sec. 691-1201-1207). See also CEQ, *Second Annual Rept. op cit* n. 17, Ch. 5.

33. 42 USC 4321 (PL 91-190, 83 Stat 852).

34. 36 *Federal Register* (No. 239) Part II, Dec. 11, 1971. For Agency procedures see also 1 ELR 46049-46139.

35. The principle source is entitled "The 102 Monitor" published periodically by CEQ. From time-to-time Congressmen have introduced lists of impact statements into the Congressional Record. In addition the monthly listing of environmental impact statements in the *102 Monitor* CEQ provides a similar weekly listing in the *Federal Register* each Saturday.

36. CEQ, *Third Annual Report op cit* n. 17. See also 102 Monitor, e.g., Vol. 1, No. 12, January 1972.

37. See "Calvert Cliffs Revisited" in *102 Monitor* Vol. 2, No. 8 September 1972 where a speech by AEC Commission William O. Donb outlines the changes in AEC procedure.

38. Richard H. Gillerly, "Man and Environment: Fighting the Backlash," *Science Views*, Vol. 100, Dec. 18, 1971 p. 410. Also "Environmental Action Organizations are Suffering from Money Shortages, Slump in Public Commitment," *Science*, Vol. 175, p. 394, 28 Jan. 1972.

39. PL 92-500.

40. Text of Refuse Act of 1899: Sec. 13, River and Harbor Act of 1899; March 3, 1899, c 425, 30 *Stat* 1152, 33 USC 407. See House Report No. 92-1333, 92 Cong. 2d Sess."Enforcement of the Reform Act of 1899," Aug. 14, 1972.

41. Green, *op cit* no. 17.

42. CEQ, *Second* and *Third Annual Reports op cit* n. 17.

43. Roger P. Hansen, "NEPA: Problems and Outer Limits," Rocky Mountain Mineral Law Foundation, National Resources Environmental Law Institute, University of Denver Law Center, Februrary 26, 1972.

44. Illinois Central Railroad Company versus Illinois, 146 US
    387 (1892).

45. Sax *op cit* n. 12.

46. Whyte, *op cit* n 13.

AUTHOR NOTES

Dr. Wengert is also the author of Chapter 2, The Evolvement Process and author notes will be found at the end of that
chapter.

APPENDIX A: NATIONAL ENVIRONMENTAL

POLICY ACT OF 1969

42 U.S.C. § 4321 et seq. (originally enacted as Act
of Jan. 1, 1970, Pub. L. No. 91-190, 83 Stat. 852)

*Be it enacted by the Senate and House of Representatives
of the United States of America in Congress assembled,* That
this Act may be cited as the "National Environmental Policy Act
of 1969."

Purpose

Sec. 2.   The purposes of this Act are: To declare a national policy which will encourage productive and enjoyable
harmony between man and his environment; to promote efforts
which will prevent or eliminate damage to the environment and
biosphere and stimulate the health and welfare of man; to enrich the understanding of the ecological systems and natural
resources important to the Nation; and to establish a Council
on Environmental Quality.

TITLE I

Declaration of National Environmental Policy

Sec. 101.   (a) The Congress, recognizing the profound impact of man's activity on the interrelations of all components
of the natural environment, particularly the profound influences of population growth, high-density urbanization, industrial expansion, resource exploitation, and new and expanding

technological advances and recognizing further the critical importance of restoring and maintaining environmental quality to the overall welfare and development of man, declares that it is the continuing policy of the Federal Government in cooperation with State and local governments, and other concerned public and private organizations, to use all practicable means and measures, including financial and technical assistance, in a manner calculated to foster and promote the general welfare, to create and maintain conditions under which man and nature can exist in productive harmony, and fulfill the social, economic, and other requirements of present and future generations of Americans.

(b) In order to carry out the policy set forth in this Act, it is the continuing responsibility of the Federal Government to use all practicable means, consistent with other essential considerations of national policy, to improve and coordinate Federal plans, functions, programs, and resources to the end that the Nation may--

(1) fulfill the responsibilities of each generation as trustee of the environment for succeeding generations;

(2) assure for all Americans safe, healthful, productive, and esthetically and culturally pleasing surroundings;

(3) attain the widest range of beneficial uses of the environment without degradation, risk to health or safety, or other undesirable and unintended consequences;

(4) preserve important historic, cultural, and natural aspects of our national heritage, and maintain, wherever possible, an environment which supports diversity, and variety of individual choice;

(5) achieve a balance between population and resource use which will permit high standards of living and a wide sharing of life's amenities; and

(6) enhance the quality of renewable resources and approach the maximum attainable recycling of depletable resources.

(c) The Congress recognizes that each person should enjoy a healthful environment and that each person has a responsibility to contribute to the preservation and enhancement of the environment.

Sec. 102. The Congress authorizes and directs that, to the fullest extent possible: (1) the policies, regulations, and public laws of the United States shall be interpreted and administered in accordance with the policies set forth in this Act, and (2) all agencies of the Federal Government shall--

(A) utilize a systematic, interdisciplinary approach which
will insure the integrated use of the natural and social sciences and the environmental design arts in
planning and in decisionmaking which may have an impact on man's environment;
(B) identify and develop methods and procedures, in consultation with the Council on Environmental Quality
established by title II of this Act, which will insure
that presently unquantified environmental amenities
and values may be given appropriate consideration in
decisionmaking along with economic and technical considerations;
(C) include in every recommendation or report on proposals
for legislation and other major Federal actions significantly affecting the quality of the human environment, a detailed statement by the responsible official
on--

   (i) the environmental impact of the proposed action,
  (ii) any adverse environmental effects which cannot
be avoided should the proposal be implemented,
 (iii) alternatives to the proposed action,
  (iv) the relationship between local short-term uses
of man's environment and the maintenance and
enhancement of long-term productivity, and
   (v) any irreversible and irretrievable commitments
of resources which would be involved in the proposed action should it be implemented.

Prior to making any detailed statement, the responsible Federal
official shall consult with and obtain the comments of any Federal agency which has jurisdiction by law or special expertise
with respect to any environmental impact involved. Copies of
such statement and the comments and views of the appropriate
Federal, State, and local agencies,. which are authorized to develop and enforce environmental standards, shall be made available to the President, the Council on Environmental Quality and
to the public as provided by section 552 of title 5, United
States Code, and shall accompany the proposal through the existing agency review processes;

(D) study, develop, and describe appropriate alternatives
to recommended courses of action in any proposal which
involves unresolved conflicts concerning alternative
uses of available resources;
(E) recognize the worldwide and long-range character of
environmental problems and, where consistent with the
foreign policy of the United States, lend appropriate
support to initiatives, resolutions, and programs designed to maximize international cooperation in anticipating and preventing a decline in the quality of
mankind's world environment;

(F) make available to States, countries, municipalities, institutions, and individuals, advice and information useful in restoring, maintaining, and enhancing the quality of the environment;

(G) initiate and utilize ecological information in the planning and development of resource-oriented projects; and

(H) assist the Council on Environmental Quality established by title II of this Act.

Sec. 103. All agencies of the Federal Government shall review their present statutory authority, administrative regulations, and current policies and procedures for the purpose of determining whether there are any deficiencies or inconsistencies therein which prohibit full compliance with the purposes and provisions of this Act and shall propose to the President not later than July 1, 1971, such measures as may be necessary to bring their authority and policies into conformity with the intent, purposes, and procedures set forth in this Act.

Sec. 104. Nothing in section 102 or 103 shall in any way affect the specific statutory obligations of any Federal agency (1) to comply with criteria or standards of environmental quality, (2) to coordinate or consult with any other Federal or State agency, or (3) to act, or refrain from acting contingent upon the recommendations or certification of any other Federal or State agency.

Sec. 105. The policies and goals set forth in this Act are supplementary to those set forth in existing authorizations of Federal agencies.

TITLE II

Council on Environmental Quality

Sec. 201. The President shall transmit to the Congress annually beginning July 1, 1970, an Environmental Quality Report (hereinafter referred to as the "report") which shall set forth (1) the status and condition of the major natural, man-made, or altered environmental classes of the Nation, including, but not limited to, the air, the aquatic, including marine, estuarine, and fresh water, and the terrestrial environment, including, but not limited to, the forest, dryland, wetland, range, urban, suburban, and rural environment; (2) current and foreseeable trends in the quality, management and utilization of such environments and the effects of those trends on the social, economic, and other requirements of the Nation; (3) the adequacy of available natural resources for fulfilling human and economic requirements of the Nation in the light of expected population pressures; (4) a review of the programs and activities (including regulatory activities) of the Federal

375

Government, the State and local governments, and nongovernmental entities or individuals, with particular reference to their effect on the environment and on the conservation, development and utilization of natural resources; and (5) a program for remedying the deficiencies of existing programs and activities, together with recommendations for legislation.

Sec. 202. There is created in the Executive Office of the President a Council on Environmental Quality (hereinafter referred to as the "Council").   The Council shall be composed of three members who shall be appointed by the President to serve at his pleasure, by and with the advice and consent of the Senate.   The President shall designate one of the members of the Council to serve as Chairman.   Each member shall be a person who, as a result of his training, experience, and attainments, is exceptionally well qualified to analyze and interpret environmental trends and information of all kinds; to appraise programs and activities of the Federal Government in the light of the policy set forth in title I of this Act; to be conscious of and responsive to the scientific, economic, social, esthetic, and cultural needs and interests of the Nation; and to formulate and recommend national policies to promote the improvement of the quality of the environment.

Sec. 203.   The Council may employ such officers and employees as may be necessary to carry out its functions under this Act.   In addition, the Council may employ and fix the compensation of such experts and consultants as may be necessary for the carrying out of its functions under this Act, in accordance with section 3109 of title 5, United States Code (but without regard to the last sentence thereof).

Sec. 204.   It shall be the duty and function of the Council--

(1) to assist and advise the President in the preparation of the Environmental Quality Report required by section 201;

(2) to gather timely and authoritative information concerning the conditions and trends in the quality of the environment both current and prospective, to analyze and interpret such information for the purpose of determining whether such conditions and trends are interfering, or are likely to interfere, with the achievement of the policy set forth in title I of this Act, and to compile and submit to the President studies relating to such conditions and trends;

(3) to review and appraise the various programs and activities of the Federal Government in the light of the policy set forth in title I of this Act for the purpose of determining the extent to which such programs and activities are contributing to the achievement of such policy, and to make recommendations to the President with respect thereto.

(4) to develop and recommend to the President national policies
    to foster and promote the improvement of environmental
    quality to meet the conservation, social, economic, health,
    and other requirements and goals of the Nation;

(5) to conduct investigations, studies, surveys, research, and
    analyses relating to ecological systems and environmental
    quality;

(6) to document and define changes in the natural environment,
    including the plant and animal systems, and to accumulate
    necessary data and other information for a continuing anal-
    ysis of these changes or trends and an interpretation of
    their underlying causes;

(7) to report at least once each year to the President on the
    state and condition of the environment; and

(8) to make and furnish such studies, reports thereon, and re-
    commendations with respect to matters of policy and legis-
    lation as the President may request.

Sec. 205. In exercising its powers, functions, and duties
under this Act, the Council shall--

(1) consult with the Citizens' Advisory Committee on Environ-
    mental Quality established by Executive Order numbered
    11472, dated May 29, 1969, and with such representatives of
    science, industry, agriculture, labor, conservation organi-
    zations, State and local governments and other groups, as
    it deems advisable; and

(2) utilize, to the fullest extent possible, the services, fa-
    cilities, and information (including statistical informa-
    tion) of public and private agencies and organizations, and
    individuals, in order that duplication of effort and ex-
    pense may be avoided, thus assuring that the Council's ac-
    tivities will not unnecessarily overlap or conflict with
    similar activities authorized by law and performed by es-
    tablished agencies.

Sec. 206. Members of the Council shall serve full time
and the Chairman of the Council shall be compensated at the
rate provided for Level II of the Executive Schedule Pay Rates
(5 U.S.C. 5313). The other members of the Council shall be
compensated at the rate provided for Level IV of the Executive
Schedule Pay Rates (5 U.S.C. 5315).

Sec. 207. There are authorized to be appropriated to
carry out the provisions of this Act not to exceed $300,000 for
fiscal year 1970, $700,000 for fiscal year 1971, and $1,000,000
for each fiscal year thereafter. Approved January 1, 1970.

ENVIRONMENTAL QUALITY IMPROVEMENT ACT OF 1970

42 U.S.C. §§ 4371-4374 (originally enacted as Act of
April 3, 1970, Pub. L. 84 Stat. 91)

Title II - Environmental Quality
(of the Water Quality Improvement Act of 1970)

Short Title

Sec. 201.   This title may be cited as the "Environmental
Quality Improvement Act of 1970."

Findings, Declarations, and Purposes

Sec. 202. (a) The Congress finds--

(1) that man has caused changes in the environment;

(2) that many of these changes may affect the relationship between man and his environment; and

(3) that population increases and urban concentration contribute directly to pollution and the degradation of our environment.

(b) (1) The Congress declares that there is a national policy
for the environment which provides for the enhancement of
environmental quality. This policy is evidenced by statutes
heretofore enacted relating to the prevention, abatement,
and control of environmental pollution, water and land resources, transportation, and economic and regional development.

(2) The primary responsibility for implementing this policy
rests with State and local governments.

(3) The Federal Government encourages and supports implementation of this policy through appropriate regional org niza-
tions established under existing law.

(c) The purposes of this title are--

(1) to assure that each Federal department and agency conducting or supporting public works activities which affect the
environment shall implement the policies established under
existing law; and

(2) to authorize an Office of Environmental Quality, which,
notwithstanding any other provision of law, shall provide
the professional and administrative staff for the Council
on Environmental Quality established by Public Law 91-190.

Office of Environmental Quality

Sec. 203. (a) There is established in the Executive Office
of the President an office to be known as the Office of Environmental Quality (hereafter in this title referred to as the
"Office"). The Chairman of the Council on Environmental Quality
established by Public Law 91-190 shall be the Director of the
Office.   There shall be in the Office a Deputy Director who
shall be appointed by the President, by and with the advice and
consent of the Senate.

(b) The compensation of the Deputy Director shall be fixed
by the President at a rate not in excess of the annual rate of
compensation payable to the Deputy Director of the Bureau of
the Budget.

(c) The Director is authorized to employ such officers and
employees (including experts and consultants) as may be necessary to enable the Office to carry out its functions under this
title and Public Law 91-190, except that he may employ no more
than ten specialists and other experts without regard to the
provisions of title 5, United States Code, governing appoint-
ments in the competitive service, and pay such specialists and
experts without regard to the provisions of chapter 51 and sub-
chapter 111 of chapter 53 of such title relating to classification and General Schedule pay rates, but no such specialist or
expert shall be paid at a rate in excess of the maximum rate
for GS-18 of the General Schedule under section 5330 of title 5.

(d) In carrying out his functions the director shall assist and advise the President on policies and programs of the
Federal Government affecting environmental quality by--

(1) Providing the professional and administrative staff and
    support for the Council on Environmental Quality estab-
    lished by Public Law 91-190;

(2) assisting the Federal agencies and departments in apprais-
    ing the effectiveness of existing and proposed facilities,
    programs, policies, and activities of the Federal Govern-
    ment, and those specific major projects designated by the
    President which do not require individual project authori-
    zation by Congress, which affect environmental quality;

(3) reviewing the adequacy of existing systems for monitoring
    and predicting environmental changes in order to achieve
    effective coverage and efficient use of research facilities
    and other resources;

(4) promoting the advancement of scientific knowledge of the
    effects of actions and technology on the environment and
    encourage the development of the means to prevent or reduce

adverse effects that endanger the health and well-being of
man;

(5) assisting in coordinating among the Federal departments and
    agencies those programs and activities which affect, pro-
    tect, and improve environmental quality;

(6) assisting the Federal departments and agencies in the de-
    velopment and interrelationship of environmental quality
    criteria and standards established through the Federal
    Government;

(7) collecting, collating, analyzing, and interpreting data and
    information on environmental quality, ecological research,
    and evaluation.

(e) The Director is authorized to contract with public or
private agencies, institutions, and organizations and with individuals without regard to sections 3618 and 3709 of the Revised Statutes (31 U.S.C. 529; 41 U.S.C. 5) in carrying out his
functions.

Report

Sec. 204.   Each Environmental Quality Report required by
Public Law 91-190 shall, upon transmittal to Congress, be referred to each standing committee having jurisdiction over any
part of the subject matter of the Report.

Authorization

Sec. 205.   There are hereby authorized to be appropriated
not to exceed $500,000 for the fiscal year ending June 30,
1970, not to exceed $750,000 for the fiscal year ending June 30,
1971, not to exceed $1,250,000 for the fiscal year ending
June 30, 1972, and not to exceed $1,500,000 for the fiscal year
ending June 30, 1973.   These authorizations are in addition to
those contained in Public Law 91-190. Approved April 3, 1970.

APPENDIX B

EXCERPTS FROM CALVERT CLIFFS COORDINATING COMMITTEE VS. ATOMIC
ENERGY COMMISSION, 449 F 2d 1102 (D.C. Circuit, 1971); 2 ER
1799; 1 ELR 20346.

[Parts of the opinion are omitted; and only those footnotes
relevant to the reproduced material are included with the original numbering.]

380

Before Wright, Tamm and Robinson, *Circuit Judges*.

Wright, *Circuit Judge:* These cases are only the beginning of what promises to become a flood of new litigation -- litigation seeking judicial assistance in protecting our natural environment. Several recently enacted statutes attest to the commitment of the government to control, at long last, the destructive engine of material "progress."[1] But it remains to be seen whether the promise of this legislation will become a reality. Therein lies the judicial role. In these cases, we must for the first time interpret the broadest and perhaps most important of the recent statutes: the National Environmental Policy Act of 1969 (NEPA)[2]. We must assess claims that one of the agencies charged with its administration has failed to live up to the congressional mandate. Our duty, in short, is to see that important legislative purposes, heralded in the halls of Congress, are not lost or misdirected in the vast hallways of the federal bureaucracy.

NEPA, like so much other reform legislation of the last 40 years, is cast in terms of a general mandate and broad delegation of authority to new and old administrative agenices. It takes the major step of requiring all federal agencies to consider values of environmental preservation in their spheres of activity, and it prescribes certain procedural measures to ensure that those values are in fact fully respected. Petitioners argue that rules recently adopted by the Atomic Energy Commission to govern consideration of environmental matters fail to satisfy the rigor demanded by NEPA. The Commission, on the other hand, contends that the vagueness of the NEPA mandate and delegation leaves much room for discretion and that the rules challenged by petitioners fall well within the broad scope of the Act. We find the policies embodied in NEPA to be a good deal clearer and more demanding than does the Commission. We conclude that the Commission's procedural rules do not comply with the congressional policy. Hence we remand these cases for further rule making.

---

[1]See, e.g., Environmental Education Act, 20 U.S.C.A. § 1531 (1971 Pocket Part), Air Quality Act of 1967, 42 U.S.C. § 1857 (Supp. V 1965-1969); Environmental Quality Improvement Act of 1970, 42 U.S.C.A. §§ 4372-4374 (1971 Pocket Part); Water and Environmental Quality Improvement Act of 1970, Pub. L. 91-224, 91st Cong., 2d Sess. (1970).

[2]42 U.S.C.A. § 4321 *et seq.* (1971 Pocket Part).

I

We begin our analysis with an examination of NEPA's structure and approach and of the Atomic Energy Commission rules which are said to conflict with the requirements of the Act. The relevant portion of NEPA is Title I, consisting of five sections.[3] Section 101 sets forth the Act's basic substantive policy: that the federal government "use all practicable means and measures" to protect environmental values. Congress did not establish environmental protection as an exclusive goal; rather, it desired a reordering of priorities, so that environmental costs and benefits will assume their proper place along with other considerations. In Section 101(b), imposing an explicit duty on federal officials, the Act provides that "it is the continuing responsibility of the Federal Government to use all practicable means, consistent with other essential considerations of national policy," to avoid environmental degradation, preserve "historic, cultural, and natural" resources, and promote "the widest range of beneficial uses of the environment without *** undesirable and unintended consequences."

Thus the general substantive policy of the Act is a flexible one. It leaves room for a responsible exercise of discretion and may not require particular substantive results in particular problematic instances. However, the Act also contains very important "procedural" provisions--provisions which are designed to see that all federal agencies do in fact exercise the substantive discretion given them. These provisions are not highly flexible. Indeed, they establish a strict standard of compliance.

NEPA, first of all, makes environmental protection a part of the mandate of every federal agency and department. The Atomic Energy Commission, for example, had continually asserted, prior to NEPA, that it had no statutory authority to concern itself with the adverse environmental effects of its actions.[4] Now, however, its hands are no longer tied. It is not only permitted, but compelled, to take environmental values into account. Perhaps the greatest importance of NEPA is to require

---

[3]The full text of Title I is printed as an appendix to this opinion. (See 102 Monitor, Vol. 1, No. 1, pp. 3-5, for reprint of NEPA's Title I).

[4]Before the enactment of NEPA, the Commission did recognize its separate statutory mandate to consider the specific radiological hazards caused by its actions; but it argued that it could not consider broader environmental impacts. Its position was upheld in *State of New Hampshire* v. *Atomic Energy Commission*, 1 Cir., 406 F. 2d 170, *cert. denied*, 395 U.S. 962 (1969).

382

the Atomic Energy Commission and other agencies to *consider* environmental issues just as they consider other matters within
their mandates. This compulsion is most plainly stated in Section 102[5] ... . Senator Jackson, NEPA's principal sponsor,
stated that "[n]o agency will [now] be able to maintain that it
has no mandate or no requirement to consider the environmental
consequences of its actions."[6] He characterized the requirements of Section 102 as "action-forcing" and stated that

---

[5]Only once -- in § 120(2)(B) -- does the Act state, in terms,
that federal agencies must give full "consideration" to environmental impact as part of their decision making processes.
However, a requirement of consideration is clearly implicit in
the substantive mandate of § 101, in the requirement of §
120(1) that all laws and regulations be "interpreted and administered" in accord with that mandate, and in the other specific procedural measures compelled by § 102(2).   The only
circuit to interpret NEPA to date has said that "[t]his Act
essentially states that every federal agency shall consider
ecological factors when dealing with activities which may have
an impact on man's environment." *Zabel* v. *Tabb*, 5 Cir., 430
F.2d 199, 211 (1970). Thus a purely mechanical compliance with
the particular measures required in § 102(2)(C) & (D) will not
satisfy the Act if they do not amount to full good faith *consideration* of the environment. *See* text at pages 14-18 *infra.*
The requirements of § 102(2) must not be read so narrowly as
to erase the general import of §§ 101, 102(1)(A) & (B).

On April 23, 1971, the Council on Environmental Quality--
established by NEPA -- issued Guidelines for federal agencies
on compliance with the Act.   36 FED. REG. 7723 (April 23,
1971).   The Council stated that "[t]he objective of section
102(2)(C) of the Act and of these guidelines is to build into
the agency decision making process an appropriate and careful
consideration of the environmental aspects of proposed action
***." *Id.* at 7724.

[6]*Hearings on S.1075, S. 237 and S. 1752 Before Senate Committee
on Interior and Insular Affairs,* 91st Cong., 1st Sess. 206
(1969). Just before the Senate finally approved NEPA, Senator
Jackson said on the floor that the Act "directs all agencies
to assure consideration of the environmental impact of their
actions in decisionmaking." 115 CONG. REC. (Part 30) 40416
(1969).

"[o]therwise, these lofty declarations [in Section 101] are
nothing more than that."[7]

The sort of consideration of environmental values which
NEPA compels is clarified in Section 102(2)(A) and (B). In
general, all agencies must use a "systematic, interdisciplinary
approach" to environmental planning and evaluation "in decisionmaking which may have an impact on man's environment." In
order to include all possible environmental factors in the decisional equation, agencies must "identify and develop methods
and procedures *** which will insure that presently unquantified environmental amenities and values may be given appropriate consideration in decisionmaking along with economic and
technical considerations."[8] "Environmental amenities" will
often be in conflict with "economic and technical considerations." To "consider" the former "along with" the latter must
involve a balancing process. In some instances environmental
costs may outweigh economic and technical benefits and in other

---

[7]*Hearings on S. 1075, supra Note 6*, at 116. Again, the Senator
re-emphasized his point on the floor of the Senate, saying:
"To insure that the policies and goals defined in this act are
infused into the on-going programs and actions of the Federal
Government, the act also established some important 'action-
forcing' procedures." 115 CONG. REC. (part 30) at 40416. The
Senate Committee on Interior and Insular Affairs Committee Report on NEPA also stressed the importance of the "action-
forcing" provisions which require full and rigorous consideration of environmental values as an integral part of agency decision making. S. Rep. No. 91-296, 91st Cong., 1st Sess.
(1969).

[8]The word "appropriate" in § 102(2)(B) cannot be interpreted to
blunt the thrust of the whole Act or to give agencies broad
discretion to downplay environmental factors in their decision
making processes. The Act requires consideration "appropriate"
to the problem of protecting our threatened environment, not
consideration "appropriate" to the whims, habits or other particular concerns of federal agencies. *See* Note 5 *supra*.

instances they may not. But NEPA mandates a rather finely tuned and "systematic" balancing analysis in each instance.[9]

To ensure that the balancing analysis is carried out and given full effect, Section 102(2)(C) requires that responsible officials of all agencies prepare a "detailed statement" covering the impact of particular actions on the environment, the environmental costs which might be avoided, and alternative measures which might alter the cost-benefit equation. The apparent purpose of the "detailed statement" is to aid in the agencies' own decision making process and to advise other interested agencies and the public of the environmental consequences of planned federal action. Beyond the "detailed statement," Section 102(2)(D) requires all agencies specifically to "study, develop, and describe appropriate alternatives to recommended courses of action in any proposal which involves unresolved conflicts concerning alternative uses of available resources." This requirement, like the "detailed statement" requirement, seeks to ensure that each agency decision maker has before him and takes into proper account all possible approaches to a particular project (including total abandonment of the project) which would alter the environmental impact and the cost-benefit balance. Only in that fashion is it likely that the most intelligent, optimally beneficial decision will ultimately be made. Moreover, by compelling a formal "detailed statement" and a description of alternatives, NEPA provides evidence that the mandated decision making process has in fact taken place and, most importantly, allows those removed from the initial process to evaluate and balance the factors on their own.

Of course, all of these Section 102 duties are qualified by the phrase "to the fullest extent possible." We must stress as forcefully as possible that this language does not provide an escape hatch for footdragging agencies; it does not make NEPA's procedural requirements somehow "discretionary." Congress did not intend the Act to be such a paper tiger. Indeed,

---

[9]Senator Jackson specifically recognized the requirement of a balancing judgment. He said on the floor of the Senate: "Subsection 102(b) requires the development of procedures designed to insure that all relevant environmental values and amenities are considered in the calculus of project development and decisionmaking. Subsection 102(c) establishes a procedure designed to insure that in instances where a proposed major Federal action would have a significant impact on the environment that the impact has in fact been considered, that any adverse affects which cannot be avoided are justified by some other stated consideration of national policy, that short-term uses are consistent with long-term productivity, and that any irreversible and irretrievable commitments of resources are warranted." 115 CONG. REC. (part 21) 29055 (1969).

the requirement of environmental consideration "to the fullest
extent possible" sets a high standard for the agencies, a
standard which must be rigorously enforced by the reviewing
courts.

Unlike the substantive duties of Section 101(B), which require agencies to "use all practicable means consistent with
other essential considerations," the procedural duties of Section 102 must be fulfilled to the "fullest extent possible."[10]

Thus the Section 102 duties are not inherently flexible.
They must be complied with to the fullest extent, unless there
is a clear conflict of *statutory* authority.[11] Considerations of

---

[10]The Commission, arguing before this court has mistakenly confused the two standards, using the § 101(B) language to suggest that it has broad discretion in performance of § 102
procedural duties. We stress the necessity to separate the
two, substantive and procedural, standards. *See* text at page
37 *infra*.

[11]§ Section 104 of NEPA provides that the Act does not eliminate any duties already imposed by other "specific statutory
obligations." Only when such specific obligations conflict
with NEPA do agencies have a right under § 104 and the "fullest extent possible" language to dilute their compliance
with the full letter and spirit of the Act. *See* text at pages
28-35 *infra*. Sections 103 and 105 also support the general
interpretation that the "fullest extent possible" language
exempts agencies from full compliance only when there is a
conflict of statutory obligations. Section 103 provides for
agency review of existing obligations in order to discover
and, if possible, correct any conflicts. *See* text at pages
21-22 *infra*. And § 105 provides that "[t]he policies and
goals set forth in this Act are supplementary to those set
forth in existing authorizations of Federal agencies." The
report of the House conferees states that § 105 "does not ***
obviate the requirement that the Federal agencies conduct
their activities in accordance with the provisions of this
bill unless to do so would clearly violate their existing
statutory obligations." 115 CONG. REC. (Part 29) at 39703.
The section-by-section analysis by the Senate conferees makes
exactly the same point in slightly different language. 115
CONG. REC. (Part 30) at 40418. The guidelines published by
the Council on Environmental Quality state that "[t]he phrase
'to the fullest extent possible' *** is meant to make clear
that each agency of the Federal Government shall comply with
the requirement unless existing law applicable to the agency's
operations expressly prohibits or makes compliance impossible." 36 FED. REG. at 7724.

administrative difficulty, delay or economic cost will not suffice to strip the section of its fundamental importance.

We conclude, then, that Section 102 of NEPA mandates a
particular sort of careful and informed decisionmaking process
and creates judicially enforceable duties. The reviewing courts
probably cannot reverse a substantive decision on its merits,
under Section 101, unless it be shown that the actual balance
of costs and benefits that was struck was arbitrary or clearly
gave insufficient weight to environmental values. But if the
decision was reached procedurally without individualized consideration and balancing of environmental factors--conducted
fully and in good faith--it is the responsibility of the courts
to reverse. ... .

... . NEPA went into effect on January 1, 1970. On April
2, 1970--three months later--the Commission issued its first,
short policy statement on implementation of the Act's procedural
provisions.[12] After another span of two months, the Commission
published a notice of proposed rule making in the Federal Register.[13] Petitioners submitted substantial comments critical of
the proposed rules. Finally, on December 3, 1970, the Commission terminated its long rule making proceeding by issuing a
formal amendment, labeled Appendix D, to its governing regulations.[14] Appendix D is a somewhat revised version of the earlier
proposal and, at last, commits the Commission to consider environmental impact in its decision making process.

The procedure for environmental study and consideration
set up by the Appendix D rules is as follows: Each applicant
for an initial construction permit must submit to the Commission his own "environmental report," presenting his assessment
of the environmental impact of the planned facility and possible alternatives which would alter the impact. When construction is completed and the applicant applies for a license
to operate the new facility, he must again submit an "environmental report" noting any factors which have changed since the
original report. At each stage, the Commission's regulatory
staff must take the applicant's report and prepare its own "detailed statement" of environmental costs, benefits and alternatives. The statement will then be circulated to other interested and responsible agencies and made available to the public.
After comments are received from those sources, the staff must

---

[12] 35 FED. REG. 5463 (April 2, 1970).

[13] 35 FED. REG. 8594 (June 3, 1970).

[14] 35 FED. REG. 18469 (December 4, 1970). The version of the
rules finally adopted is not printed in 10 C.F.R. § 50, App.
D. pp. 246-250 (1971).

prepare a final "detailed statement" and make a final recommendation on the application for a construction permit or operating license. ... .

II

NEPA makes only one specific reference to consideration of environmental values in agency review processes. Section 102(2)(C) provides that copies of the staff's "detailed statement" and comments thereon "shall accompany the proposal through the existing agency review processes." The Atomic Energy Commission's rules may seem in technical compliance with the letter of that provision. ... . The question here is whether the Commission is correct in thinking that its NEPA responsibilities may "be carried out in toto outside the hearing process" -- whether it is enough that environmental data and evaluations merely "accompany" an application through the review process, but receive no consideration whatever from the hearing board.

We believe that the Commission's crabbed interpretation of NEPA makes a mockery of the Act. What possible purpose could there be in the Section 102(2)(C) requirement (that the "detailed statement" accompany proposals through agency review processes) if "accompany" means no more than physical proximity --mandating no more than the physical act of passing certain folders and papers, unopened, to reviewing officials along with other folders and papers? What possible purpose could there be in requiring the "detailed statement" to be before hearing boards, if the boards are free to ignore entirely the contents of the statement? NEPA was meant to do more than regulate the flow of papers in the federal bureaucracy. The word "accompany" in Section 102(2)(C) must not be read so narrowly as to make the Act ludicrous. It must, rather, be read to indicate a congressional intent that environmental factors, as compiled in the "detailed statement," be *considered* through agency review processes.[15]

---

[15]The guidelines issued by the Council on Environmental Quality emphasize the importance of consideration of alternatives to staff recommendations during the agency review process: "A rigorous exploration and objective evaluation of alternative actions that might avoid some or all of the adverse environmental effects is essential. Sufficient analysis of such alternatives and their costs and impact on the environment should accompany the proposed action through the agency review process in order not to foreclose prematurely options which might have less detrimental effects." 36 FED. REG. at 7725. The Council also states that an objective of its guidelines is "to assist agencies in implementing not only the letter, but the spirit, of the Act." *Id.* at 7724.

388

Beyond Section 102(2)(C), NEPA requires that agencies consider the environmental impact of their actions "to the fullest extent possible." The Act is addressed to agencies as a whole, not only to their professional staffs. Compliance to the "*fullest*" possible extent would seem to demand that environmental issues be considered at every important stage in the decision making process concerning a particular action--at every stage where an overall balancing of environmental and nonenvironmental factors is appropriate and where alterations might be made in the proposed action to minimize environmental costs. ... .

The Commission's regulations provide that in an uncontested proceeding the hearing board shall on its own "determine whether the application and the record of the proceeding contain sufficient information, and the review of the application by the Commission's regulatory staff has been adequate, to support affirmative findings on" various nonenvironmental factors. NEPA requires at least as much automatic consideration of environmental factors. ... .

... . NEPA establishes environmental protection as an integral part of the Atomic Energy Commission's basic mandate. The primary responsibility for fulfilling that mandate lies with the Commission. Its responsibility is not simply to sit back, like an umpire, and resolve adversary contentions at the hearing stage. Rather, it must itself take the initiative of considering environmental values at every distinctive and comprehensive stage of the process beyond the staff's evaluation and recommendation. In fact, in recent years, the courts have become increasingly strict in requiring that federal agencies live up to their mandates to consider the public interest. They have become increasingly impatient with agencies which attempt to avoid or dilute their statutorily imposed role as protectors of public interest values beyond the narrow concerns of industries being regulated.

... . The Act, it is true, lacks an "inflexible timetable" for its emplementation. But it does have a clear effective date, consistently enforced by reviewing courts up to now. Every federal court having faced the issues has held that the procedural requirements of NEPA must be met in order to uphold federal action taken after January 1, 1970.[16] The absence of a

---

[16]In some cases, the courts have had a difficult time determining whether particular federal actions were "taken" before or after January 1, 1970. But they have all started from the basic rule that any action taken after that date must comply with NEPA's procedural requirements. *See* Note, *Retroactive Application of the National Environmental Policy Act of 1969*, 69 MICH. L. REV. 732 (1971), and cases cited therein. Clearly, any hearing held between January 1, 1970 and March 4, 1971 which culminates in the grant of a permit or license is a federal action taken after the Act's effective date.

"timetable" for compliance has never been held sufficient, in itself, to put off the date on which a congressional mandate takes effect. The absence of a "timetable," rather, indicates that compliance is required forthwith. ... .

No doubt the process of formulating procedural rules to implement NEPA takes some time. Congress cannot have expected that federal agencies would immediately begin considering environmental issues on January 1, 1970. But the effective date of the Act does set a time for agencies to begin adopting rules and it demands that they strive, "to the fullest extent possible," to be prompt in the process. ... .

Even if the long delay had been necessary, however, the Commission would not be relieved of all NEPA responsibility to hold public hearings on the environmental consequences of actions taken between January 1, 1970 and final adoption of the rules. Although the Act's effective date may not require instant compliance, it must at least require that NEPA procedures, once established, be applied to consider prompt alterations in the plans or operations of facilities approved without compliance. Yet the Commission's rules contain no such provision. Indeed, they do not even apply to the hearings still being conducted at the time of their adoption on December 3, 1970--. ... The delayed compliance date of March 4, 1971, then, cannot be justified by the Commission's long drawn out rule making process. ... .

In the end, the Commission's long delay seems based upon what it believes to be a pressing national power crisis. Inclusion of environmental issues in pre-March 4, 1971 hearings might have held up the licensing of some power plants for a time. But the very purpose of NEPA was to tell federal agencies that environmental protection is as much a part of their responsibility as is protection and promotion of the industries they regulate. Whether or not the spectre of a national power crisis is as real as the Commission apparently believes, it must not be used to create a blackout of environmental consideration in the agency review process. NEPA compels a case-by-case examination and balancing of discrete factors. Perhaps there may be cases in which the need for rapid licensing of a particular facility would justify a strict time limit on a hearing board's review of environmental issues; but a blanket banning of such issues until March 4, 1971 is impermissible under NEPA.

IV

The sweep of NEPA is extraordinarily broad, compelling consideration of any and all types of environmental impact of federal action. However, the Atomic Energy Commission's rules specifically exclude from full consideration a wide variety of

environmental issues. ... . The upshot is that the NEPA procedures, viewed by the Commission as superfluous, will wither
away in disuse, applied only to those environmental issues
wholly unregulated by any other federal, state or regional body.

... . NEPA mandates a case-by-case balancing judgment on
the part of federal agencies. In each individual case, the
particular economic and technical benefits of planned action
must be assessed and then weighed against the environmental
costs; alternatives must be considered which would affect the
balance of values. *See* text at pages 7-9 *Supra*. The magnitude
of possible benefits and possible costs may lie anywhere on a
broad spectrum. Much will depend on the particular magnitudes
involved in particular cases. In some cases, the benefits will
be great enough to justify a certain quantum of environmental
costs; in other cases, they will not be so great and the proposed action may have to be abandoned or significantly altered
so as to bring the benefits and costs into a proper balance.
The point of the individualized balancing analysis is to ensure
that, with possible alterations, the optimally beneficial action is finally taken.

Certification by another agency that its own environmental
standards are satisfied involves an entirely different kind of
judgment. Such agencies, without overall responsibility for the
particular federal action in question, attend only to one aspect of the problem: the magnitude of certain environmental
costs. They simply determine whether those costs exceed an allowable amount. Their certification does not mean that they
found no environmental damage whatever. In fact, there may be
significant environemtnal damage (e.g., water pollution), but
not quite enough to violate applicable (e.g., water quality)
standards. Certifying agencies do not attempt to weigh that
damage against the opposing benefits. Thus the balancing analysis remains to be done. It may be that the environmental
costs, though passing prescribed standards, are nonetheless
great enough to outweigh the particular economic and technical
benefits involved in the planned action. The only agency in a
position to make such a judgment is the agency with overall responsibility for the proposed federal action--the agency to
which NEPA is specifically directed.

The Atomic Energy Commission, abdicating entirely to other
agencies' certifications, neglects the mandated balancing analysis. Concerned members of the public are thereby precluded
from raising a wide range of environmental issues in order to
affect particular Commission decisions. And the special purpose
of NEPA is subverted. ... .

Of course, federal agencies such as the Atomic Energy Commission may have specific duties, under acts other than NEPA,
to obey particular environmental standards. Section 104 of NEPA
makes clear that such duties are not to be ignored:

> "Nothing in Section 102 or 103 shall in any way affect the
> specific statutory obligations of any Federal agency (1)
> to comply with criteria or standards of environmental
> quality, (2) to coordinate or consult with any other Fed-
> eral or State agency, or (3) to act, or refrain from act-
> ing contingent upon the recommendations or certification
> of any other Federal or State agency."

On its face, Section 104 seems quite unextraordinary, intended
only to see that the general procedural reforms achieved in
NEPA do not wipe out the more specific environmental controls
imposed by other statutes. Ironically, however, the Commission
argues that Section 104 in fact allows other statutes to wipe
out NEPA.

... . Thus Section 104 applies in some fashion to consideration of water quality matters. But it definitely cannot
support--indeed, it is not even relevant to -- the Commission's
wholesale abdication to the standards and certifications of any
and all federal, state and local agencies dealing with matters
other than water quality.

As to water quality, Section 104 and WQIA clearly require
obedience to standards set by other agenices. But obedience
does not imply total abdication. Certainly, the language of
Section 104 does not authorize an abdication. It does not suggest that other "specific statutory obligations" will entirely
replace NEPA. Rather, it ensures that three sorts of "obligations" will not be undermined by NEPA: (1) the obligation to
"comply" with certain standards, (2) the obligation to "coordinate" or "consult" with certain agencies, and (3) the obligation to "act, or refrain from acting contingent upon" a certification from certain agencies. WQIA [Water Quality Improvement
Act] imposes the third sort of obligation. It makes the granting of a license by the Commission "contingent upon" a water
quality certification. But it does not *require* the Commission
to grant a license once a certification has been issued. It
does not preclude the Commission from demanding water pollution
controls from its licensees which are *more strict* than those

demanded by the applicable water quality standards of the certifying agency.[17] It is very important to understand these facts
about WQIA.   For all that Section 104 of NEPA does is to reaf-
firm other "specific statutory obligations."   Unless those obligations are plainly mutually exclusive with the requirements
of NEPA, the specific mandate of NEPA must remain in force.   In
other words, Section 104 can operate to relieve an agency of
its NEPA duties only if other "specific statutory obligations"
clearly preclude performance of those duties.

Obedience to water quality certifications under WQIA is
not mutually exclusive with the NEPA procedures.   It does not
preclude performance of the NEPA duties.   Water quality certifications essentially establish a *minimum condition* for the
granting of a license.   But they need not end the matter.   The
Commission can then go on to perform the very different operation of balancing the overall benefits and costs of a particular proposed project, and consider alterations (above and beyond the applicable water quality standards) which would further reduce environmental damage.   Because the Commission *can*
still conduct the NEPA balancing analysis, consistent with WQIA,
Section 104 does not exempt it from doing so.   And it, therefore, *must* conduct the obligatory analysis under the prescribed
procedures. ... .

This rather meager legislative history, in our view, cannot radically transform the purport of the plain words of Section 104.   Had the Senate sponsors fully intended to allow a
total abdication of NEPA responsibilities in water quality
matters--rather than a supplementing of them by strict obedience to the specific standards of WQIA--the language of Section
104 could easily have been changed.   As the Supreme Court often
has said, the legislative history of a statute (particularly

---

[17]The relevant language in WQIA seems carefully to avoid any
such restrictive implication.   It provides that "each Federal agency *** shall *** insure compliance with applicable
water quality standards ***."   33 U.S.C.A. § 1171 (a).   It
also provides that "[n]o license or permit shall be granted
until the certification required by this section has been obtained or has been waived *** .   No license or permit shall
be granted if certification has been denied ***." 33 U.S.C.A.
§ 1171 (b)(1).   Nowhere does it indicate that certification
must be the final and only protection against unjustified
water pollution--a fully sufficient as well as a necessary
condition for issuance of a federal license or permit.
    We also take note of §21(c) of WQIA, which states:"Nothing
in this section shall be construed to limit the authority of
any department or agency pursuant to any other provision of
law to require compliance with applicable water quality
standards. ***" 33 U.S.C.A. § 1171(c).

such relatively meager and vague history as we have here) cannot radically affect its interpretation if the language of the statute is clear. *See, e.g., Packard Motor Car Co. v. NLRB*, 330 U. S. 485 (1947); *Kuchner v. Irving Trust Co.*, 299 U. S. 445 (1937); *Fairport Painesville & Eastern R. Co. v. Meredith*, 292 U. S. 589 (1934); *Wilbur v. United States ex rel. Vindicator Consolidated Gold Mining Co.*, 284 U.S. 231 (1931). In a recent case interpreting a veterans' act, the Court set down the principle which must govern our approach to the case before us:

> "Having concluded that the provision of § 1 are clear and unequivocal on their face, we find no need to resort to the legislative history of the Act. Since the State has placed such heavy reliance upon that history, however, we do deem it appropriate to point out that this history is at best inconclusive. It is true, as the State points out, the Representative Rankin, as Chairman of the Committee handling the bill on the floor of the House, expressed his view during the course of discussion of the bill on the floor that the 1941 Act would not apply to [the sort of case in question] ***. But such statements, even when they stand alone, have never been regarded as sufficiently compelling to justify deviation from the plain language of a statute. ***"

United States v. Oregon, 366 U.S. 634 (1961). (Footnotes omitted.) It is, after all, the plain language of the statute which *all* the members of both houses of Congress must approve or disapprove. The courts should not allow that language to be significantly undercut. In cases such as this one, the most we should do to interpret clear statutory wording is to see that the *overriding purpose* behind the wording supports its plain meaning. We have done that here. And we conclude that Section 104 of NEPA does not permit the sort of total abdication of responsibility practiced by the Atomic Energy Commission.

V

Petitioners' final attack is on the Commission's rules governing a particular set of nuclear facilities: those for which construction permits were granted without consideration of environmental issues, but for which operating licenses have yet to be issued. These facilities, still in varying stages of construction, include the one of most immediate concern to one of the petitioners: The Calvert Cliffs nuclear power plant on Chesapeake Bay in Maryland.

The Commission's rules recognize that the granting of a construction permit before NEPA's effective date does not justify bland inattention to environmental consequences until the operating license proceedings, perhaps far in the future. The

394

rules require that measures be taken *now* for environmental protection. Specifically, the Commission has provided for three such measures during the pre-operating license stage.   First, it has required that a condition be added to all construction permits, "whenever issued," which would oblige the holders of the permits to observe all applicable environmental standards imposed by federal or state law. Second, it has required permit holders to submit their own environmental report on the facility under construction.   And third, it has initiated procedures for the drafting of its staff's "detailed environmental statement" in advance of operating license proceedings.[18]

The one thing the Commission has refused to do is take any independent action based upon the material in the environmental reports and "detailed statements." Whatever environmental damage the report and statements may reveal, the Commission will allow construction to proceed on the original plans.   It will not even consider requiring alterations in those plans (beyond compliance with external standards which would be binding in any event), though the "detailed statements" must contain an analysis of possible alternatives and may suggest relatively inexpensive but high highly beneficial changes.   Moreover, the Commission has, as a blanket policy, refused to consider the possibility of temporarily halting construction in particular cases pending a full study of a facility's enviornmental impact. It has also refused to weigh the pros and cons of "backfitting" for particular facilities (alteration of already constructed portions of the facilities in order to incorporate new technological developments designed to protect the environment). Thus reports and statements will be produced, but nothing will be done with them.   Once again, the Commission seems to believe that the mere drafting and filing of papers is enough to satisfy NEPA.

The Commission appears to recognize the severe limitation which its rules impose on environmental protection. Yet it argues that full NEPA consideration of alternatives and independent action would cause too much delay at the pre-operating license stage.   It justifies its rules as the most that is "practicable, in the light of environmental needs and 'other essential considerations of national policy'."[19]   It cites, in particular, the "national power crisis" as a consideration of national policy militating against delay in construction of nuclear power facilities.

---

[18]10 C.F.R. § 50, App. D, ¶¶ 1, 14.

[19]Brief for respondents in No. 24,871 at 59.

The Commission relies upon the flexible NEPA mandate to "use all practicable means consistent with other essential considerations of national policy." As we have previously pointed out, however, that mandate applies only to the substantive guidelines set forth in Section 10. of the Act. *See* pages 9-10 *supra*. The procedural duties, the duties to give full *consideration* to environmental protection, are subject to a much more strict standard of compliance. By now, the applicable principle should be absolutely clear. NEPA requires that an agency must-- to the *fullest* extent possible under its other statutory obligations--consider alternatives to its actions which would reduce environmental damage. That principle establishes that consideration of environmental matters must be more than a *pro forma* ritual. Clearly, it is pointless to "consider" environmental costs without also seriously considering action to avoid them. Such a full exercise of substantive discretion is required at every important, appropriate and nonduplicative stage of an agency's proceedings. ... .

A full NEPA consideration of alterations in the original plans of a facility, then, is both important and appropriate well before the operating license proceedings. It is not duplicative if environmental issues were not considered in granting the construction permit. And it need not be duplicated, absent new information or new developments, at the operating license stage. In order that the pre-operating license review be as effective as possible, the Commission should consider very seriously the requirement of a temporary halt in construction pending its review and the "backfitting" of technological innovations. For no action which might minimize environmental damage may be dismissed out of hand. Of course, final operation of the facility may be delayed thereby. But some delay is inherent whenever the NEPA consideration is conducted--whether before or at the license proceedings. It is far more consistent with the purposes of the Act to delay operation at a stage where real environmental protection may come about than at a stage where corrective action may be so costly as to be impossible.

Thus we conclude that the Commission must go farther than it has in its present rules. It must consider action, as well as file reports and papers, at the pre-operating license stage. As the Commission candidly admits, such consideration does not amount to a retroactive application of NEPA. Although the projects in question may have been commenced and initially approved before January 1, 1970, the Act clearly applies to them

since they must still pass muster before going into full operation[20]   All we demand is that the environmental review be as
full and fruitful as possible.

We hold that, in the four respects detailed above, the
Commission must revise its rules governing consideration of environmental issues.   We do not impose a harsh burden on the
Commission.   For we require only an exercise of substantive
discretion which will protect the environment "to the fullest
extent possible."   No less is required if the grand congressional purposes underlying NEPA are to become a reality. Re-
manded for proceedings consistent with this opinion.

The AEC announced that it would not seek reconsideration
of the *Calvert Cliffs* decision by the Court of Appeals and that
it would recommend to the Justice Department that the Government not seek Supreme Court review.   In response to the *Calvert
Cliffs* decision the AEC issued revised NEPA procedures.

---

[20]The courts which have held NEPA to be nonretroactive have not
faced situations like the one before us here -- situations
where there are two, distinct stages of federal approval, one
occurring before the Act's effective date and one after that
date. *See* Note, *supra* Note 25.
The guidelines issued by the Council on Environmental
Quality urge agencies to employ NEPA procedures to minimize
environmental damage, even when approval of particular projects was given before January 1, 1970:   "To the maximum extent practicable the section 102(2)(C) procedure should be
applied to further major Federal actions having a significant
effect on the environment even though they arise from projects or programs initiated prior to enactment of [NEPA] on
January 1, 1970. Where it is not practicable to reassess the
basic course of action, it is still important that further
incremental major actions be shaped so as to minimize adverse
environmental consequences.   It is also important in further
action that account be taken of environmental consequences
not fully evaluated at the outset of the project or program."
36 FED. REG. at 7727.

9

THE ECONOMIC SYNTAX

BY R. BURNELL HELD

Public works projects do not emerge immediately as mature entities but develop as a consequence of decisions which involve economic analysis of various types at several levels. Although this book is chiefly concerned with the problems which face the design and planning team of a public project, the stage on which that team performs, particularly for projects in the field broadly defined as water resources, has been set for them by earlier decisions and policies of which they should have some awareness.

The economic principles employed are basically simple. The difficulties encountered are often those of gaining general acceptance of the procedures that will be used and the standards that will be set. Because applied economics depends to such a degree on the comparison of various values, there is often a real problem in obtaining measures of all the values which are involved, and expressed in units that can be compared.

THE NATURE AND SCOPE OF ECONOMICS

Economists are not in agreement as to the scope of economics. Kenneth Boulding, a past president of the American Economic Association, deplores what he terms "the imperialistic tendencies" in the profession, that is, the attempts by some economists to encompass *all* areas of decision-making. Boulding chooses to define economics by limiting it to "that part of the total social system which is organized through exchange and which deals with exchangeables."[2] His concept of economics, however, is not narrow as any who are acquainted with his many books and journal articles can attest.

V. C. Walsh and others see economics in a different light. The core of the pure science in economics, he asserts, is

choice theory.   The concern in economics is one of making the
best possible choices[3]   Walsh's definition has the merit of
setting economics apart from narrow pecuniary considerations.

Economic Efficiency

The resources available to a society are seldom sufficient
to satisfy all the needs that the members of the society can
identify.   Thus, the primary concern of economics is that of
satisfying human wants, which tend to know no end, with a stock
of resources which is finite.   Such an assignment calls for
careful calculation and comparisons of values derived from particular combinations of goods and services.   The allocation of
resources necessary to produce these goods and services is also
important.   Economic efficiency becomes the key in converting
resources into the quantity and combination of goods and services which will produce the greatest possible consumer satisfaction.

The method of resolving the question of how those resources will be used is to allocate them to those uses and activities which generate the greatest net benefit. This solution
says nothing about how benefits are to be determined, nor anything about the relative distribution of income and wealth in
the community.   A society in which there are great disparities
in income and wealth is likely to be one in which luxury goods
for a few receive priority over the production of a greater
volume of subsistence goods for the many.   A society which recognizes only pecuniary values will likely be one in which the
greatest net benefit is defined in terms of the traditional
goods and services obtained through the market. Those goods and
services of a public nature, which are seldom provided by private enterprise, tend to be left out of the calculus of what
constitutes the greatest good.

While economics may seem to be more closely involved with
the private sector than with the public sector of the economy,
this is only because of the greater activity within the private
sector.   Similarly, economics is traditionally associated with
market transactions and money values again because of the volume of activity that occurs within the market, but an important
function of economic analysis is the evaluation of public policies to determine their probable impact on the well-being or
welfare of people. While the welfare of people is determined in
part by the availability in the market of the services and
material goods they desire, the market does not provide them
with everything necessary for their health and safety, the open
spaces needed to ventilate congested urban areas, and a long
list of other things that cannot be purchased individually, but
can be provided only through joint effort by government on behalf of its people.

Only in recent years has there been any concerted effort to apply economics to value decisions which lie beyond the realm of the market.   This has come about because society is increasingly willing to support efforts to provide resources for so-called "non-productive" purposes--recreation, beautification, environmental values, preservation of rare and endangered species or unusual biotic communities.   Such activities receive relatively broad support even when there are conflicts with others who wish to use the same resources for the production of the goods and services of commerce.   Interestingly, "non-productive" activities often appear to be a kind which involve the consumer directly and which render him immediate satisfaction and pleasure without the necessity of creating anything tangible. The same activity becomes productive as soon as someone is paid to do it for the enjoyment of another. Contrast, for example, a game of football for fun with a college or the professional team game where spectators are entertained.   The growing public acceptance of this type of resource use suggests that values have changed.   The values which have been so difficult to quantify in the past that economists have found ways of ignoring them can no longer be ignored.   Economists are beginning to address these problems using various techniques to derive values that can be compared with the more familiar values of the commercial sector.

Economists and Their Critics

It is not surprising, however, to find views such as Ian McHarg's which suggest a lack of understanding of economics and its role:

> The economists, with a few exceptions, are the merchants'
> minions and together they ask with the most bare-faced ef-
> frontery that we accommodate our value system to theirs.
> Neither love nor compassion, health nor beauty, dignity
> nor freedom, grace nor delight are important unless they
> can be priced.   If they are non-price benefits or costs
> they are relegated to inconsequence...the components which
> the (economic) model excludes are the most important human
> ambitions and accomplishments and the requirements for
> survival[4]

Because it is possible for statements like the above to be written and accepted without question, it is important to understand that economics can be put to a variety of uses. It can be as readily employed when human welfare is concerned as with the concern of the market place.   In this regard it is interesting to note that Adam Smith was professor of moral philosophy when he wrote *The Wealth of Nations*.   Thomas Malthus was a minister whose stern and pessimistic views concerning population

were nevertheless primarily a concern with the human condition.
The terms, "pamphleteers," "reformers," and "special pleaders,"
have been used to describe other economists whose concerns in
developing theories of economic behavior seem to have been motivated by an intense desire to bring about a change in existing social conditions.

Some of the harshest criticism of the narrow concerns and
views which some economists take relative to *money, value,* and
utility comes from within the profession itself.   One critic,
Joan Robinson, an eminent English economist, notes that while
the concept of *utility* was evolved to help overcome the bias of
money values, *"utility* cannot be measured, while money values
can, and economists have a bias in favor of the measurable..."[5]
An additional bias to monetary values comes about because uses
of national income statistics, such as gross national product,
(GNP), which tend to be looked upon as the real measure of our
collective welfare. This type of accounting fails to reflect the
total utility which consumers derive from the available goods
and services. Because it is a measure of gross rather than net
production, it includes expenditures made for reconstruction
after a disaster has destroyed property in a community.   The
disutility, rather than the utility of the community, is measured by the economic activity which is generated to replace
these buildings and reflected in the GNP account as a gain although the community is poorer because of its loss.

There are also the pressures in modern economic life for
the production of

> goods with a sales value against those which are free. The
> fight that has to be put up, for instance, to keep wild
> country from being exploited for money profit is made more
> difficult because its defenders can be represented as
> standing up for 'non-economic' values (which is considered
> soft-headed, foolish and unpatriotic) though the economists
> should have been the first to point out that *utility* not
> money, is economic value and that the *utility* of goods is
> not measured by their prices[6]

Miss Robinson also asserts that the argument "that public
investment, however beneficial, must be less eligible from a
national view than any private investment, merely because it is
public, has no logical basis; it is just a hang-over from
laisser-faire ideology.   She then follows the line of thinking
developed by Kenneth Galbraith and asks, "Would a greater contribution of human welfare be made by an investment in capacity
to produce knick-knacks that have to be advertised in order to
be sold or an investment in improving the health service."[7] It
is a question, she concludes, that the *laisser-faire* ideology
answers by avoiding it in the first instance.

Galbraith has challenged traditional thinking, or conventional wisdom, that all producers of goods and services in the economy respond in every instance to the desires of consumers.[8] The truth, he asserts, is that consumer sovereignty is a fiction in those areas of the private sector where advertising campaigns are mounted to create consumer needs which the consumer was previously unaware he had. This continuing pressure on the consumer to spend his income for the knick-knacks together with the idea that the public sector is incapable of producing anything comparable to the wealth generated by the private sector leads, Galbraith says, to an imbalance or distortion in the economy. There is a complementary relationship between certain goods provided by the private sector and other goods and services provided by the public sector. Automobiles, for example, cannot be used in large numbers without providing the necessary streets and highways, the traffic police and highway patrolmen, and more recently, air pollution control measures. These are services which must be provided from the public sector. While the ratio that must be maintained between these two sets of goods and services may not be fixed, the limits within which they may be combined is likely to be narrow.

The substance of an economist's concerns today cannot be limited to that which he observes in the market economy if, as Mason Gaffney has asserted, "direct damage to human health and happiness is more directly 'economic', ... than damage to property which is simply an intermediate means to health and happiness."[9] The economist who subscribes to this point of view can hardly be described as someone unable to see anything in front of him except dollar signs. He may still be concerned with economic efficiency, but only as the means to the achievement of higher human ends.

There *is* a bias among some economists toward things that can be quantified. While there are opportunities for professional employment where it may not be necessary to deal with non-monetary values, the economist who is a member of the staff of a public works project decision-maker must be prepared to include non-monetary values in his analyses.

Unlimited Desires but Limited Resources

A word concerning *economic efficiency* is in order. It has, and continues to be, a rightful concern of economists who see the inability of a society to meet all of its needs to the fullest with limited resources, but who know that many more needs can be met if the available resources are carefully husbanded. Mentioning economic efficiency as a decision criterion immediately raises a red flag for some. This comes about because of the common misunderstanding that in applying such a criterion, non-monetary values automatically go to the bottom of the list. The criterion of economic efficiency, however, cannot be applied to anything unless a value of some kind can be determined.

In a society which is capable of meeting the basic material needs of its people with more to spare, where surplus agricultural production is an embarrassment and where service workers now outnumber persons engaged in the production and processing of material goods, it is possible to forget that this has not always been the case. A society that "has everything" of a material nature is more likely to be ready to consider values of a non-material nature than one which is struggling to meet its basic needs. Economists have traditionally been concerned with the ways and means to convert scarce resources into forms which would yield a maximum of satisfaction to members of the society. That role has not changed but society's idea of the kind and combination of goods and services which are necessary for its well-being has changed. McHarg's apparent failure to see this surely accounts for his intemperate statement.

Non-Market Values and Economic Man

Anthropologists and historians have ample evidence that economics has been and can be organized and operated effectively within social systems different from that of the United States.[10] Karl Polanyi described the market economy with its self-regulating markets as something entirely unprecedented prior to the nineteenth century.

> ...gain and profit never before played an important part in human economy . . . The outstanding discovery of recent historical and anthropological research is that man's economy, as a rule, is submerged in his social relationships. He does not act so as to safeguard his individual interest in the possession of material goods; he acts so as to safeguard his social standing, his social claims, his social assets. He values material goods only insofar as they serve this end. Neither the process of production nor that of distribution is linked to specific economic interests attached to the possession of goods; but every single step in that process is geared to a number of social interests which eventually insure that the required step be taken. These interests will be very different in a small hunting or fishing community from those in a vast despotic society but in either case the economic system will run on non-economic motives.[11]

Every culture has some set of beliefs and common values or it is not, by definition, a culture, Boulding has argued.[12] The classical economists invented the concept of *economic man* and those cultures under western European influence accepted and came to believe it. *Economic man*, although a recognized fiction, was a useful link in the argument that there was a basic harmony between the welfare of society as a whole and the cold calculations of the individual, concerned with nothing else but his purely selfish ends. That harmony could exist as long as none of the individual actors were strong enough to have an effect on the behavior of others and as long as external economies or diseconomies (spill-over effects, externalities, or third-person effects) could be ignored. Giving free reign to individual economic self-interest will no longer bring a society with a developed industrial economy to the highest state of well-being. No economist should be under the illusion that he is necessarily serving the ends of society if his primary concern is for the well-being of a powerful producer or group of producers. I believe the same can be said for the economist involved in public works project planning.

The narrow orientation that some give to economics disappears when the choice-theory approach is emphasized, and when the realm of decision-making, though confined to exchangeables as Boulding has suggested, is not confined exclusively to the transactions taking place through the market or only to those values expressed in monetary units. It would be more convenient if all values could be expressed in monetary units but the inability to do this in no way lessens the importance of dealing with the extra-market values as well as the market values. The tendency of some economists to recognize no values as legitimate unless they are or can be quantified is offset by non-economists who contend that the objects they value are priceless and that no one should even think of trying to quantify the value. McHarg, regardless of what he may say, recognizes the necessity of finding common ground on which disparate values can be measured and compared. His mapping technique, described briefly elsewhere in this book, establishes values and compares them but does so in other than monetary measures.

Admittedly, the non-economic values are difficult to determine, but the problem is now being attacked with fervor, particularly with respect to valuation of experience users derive from recreation opportunities, the value of wildlife habitat, the value of water in certain uses, the value of open space in urban areas, etc. This was an area virtually neglected by economists until the late 1950's, but it has received much attention since then.[13] The approach that is most often used is that of determining, by some means, an estimate of what users would be willing to pay for access to these resources or the services from them, rather than to forego their use. Just as the proof of the pudding is in the eating, one measure of the value we place on an object or an experience is what we are willing to give up to obtain it, or to obtain various quantities (days of use) of it.

If a particular landscape yields satisfaction to those who view it, it has a value to them. If they are also willing to give up something else of value, if need be, to assure that the landscape is protected, the value is just as legitimate as the value assigned to the trees growing upon the site which may essentially become pulp for paper. It is *not* an impossible task to determine a dollar value for a landscape.

Economic efficiency simply raises the question of whether available resources have been used where they produce the greatest utility for consumers. Finding it impossible to measure utility itself, economists have settled for marginal utility, or price. The manufacturing process itself may be the epitome of physical efficiency and yet not be an example of economic efficiency. Assume for instance that the plant has produced a large quantity of hoola hoops and that these no longer have a market and must be given away. Suppose that there is an unsatisfied demand for bicycles. The same human effort and material that was devoted to the manufacture of hoola hoops might have been put into the manufacture of a much smaller number of bicycles. If this had happened, the reallocation of resources from the manufacture of hoola hoops to the manufacture of bicycles would have represented an increase in economic efficiency. As an alternative, the money value of those resources might also have been transferred from hoola hoop manufacturing and used instead to create a neighborhood park in a ghetto area. If greater satisfaction was then gained from the services rendered by the park than from the use of a million hoola hoops, there was also a gain in economic efficiency.

While the interest of consumers in most goods and services does not fluctuate as rapidly as for fad items, changes do take place. If the economy fails to increase the production of the goods and services for which consumer desire has increased and continues to offer more goods and services of another type to which consumers now attach less value, a condition of economic inefficiency exists. An optimal situation exists when the overall collection of goods and services being produced yields a higher value, when measured in prevailing prices, than any other collection of goods and services that could be produced from the same set of resources. If this has been achieved, then it can be said that maximum economic efficiency has been obtained in the use of these resources.

Economic efficiency, even though it may not be the only criterion to be met in a public project, is nevertheless an important criterion. It is particularly difficult to determine the degree of economic efficiency that exists where public works projects are involved because they produce goods and services which for the most part are not valued on the market. In times of changing public values, public works planners may get sufficiently out of touch with what the public wants and what

it objects to that projects may be undertaken which result in
the inefficient use of resources, given society's current desires and preferences.

    An example of a situation in which the values of project
planners have been at odds with those of the community is to be
found in the controversy between San Francisco's Board of Su-
pervisors and California's State Division of Highways.   The
abrupt termination of the Embarcadaro Freeway on the San Francisco waterfront makes the constructed portion unusable for
traffic.   It stands as a monument to the determination of the
city that other values, which would have been reduced or lost,
had the freeway been completed, were of greater value to San
Francisco that the moneys invested in the portion already constructed and of the benefits to those who would have driven the
freeway. Figure 9-1 is a photograph showing a portion of the
freeway. The project was one of six freeways which were stopped
by the city in the 1960's.   Another effort to extend the Embarcadaro Freeway was stopped in 1966. The people of San Francisco believed that the social costs of extending the freeway,
not all of them monetary, would exceed the benefits the freeway
would produce.

Figure 9-1. San Francisco's Embarcadaro Freeway was terminated abruptly at the wharf when citizens of the city became
aroused.

Economic analysis used as a means for choice determination is a must given the decisions which must be made in the economy where the possibility exists to produce for a far greater quantity of goods and services of all kinds than can possibly be achieved given a particular state of technology and a given stock of resources. How to find that happy combination of all goods and services which will maximize the public welfare is certainly within the scope of economics. This statement is made recognizing the fact that in addition to the traditional goods and services, the extra-market values of landscapes and recreation, and other environmental values such as air and water quality, food stuffs free of residual pesticides, freedom from excessive noise, and ecosystems where a rich diversity of life forms have been preserved and protected are included.

THE ROLES OF ECONOMISTS

There are basically two roles for economists in the formulation of public works projects. One examines the consequences of the options open to the decision maker in terms of the welfare of society. The other analyzes the options open in the planning and design phase and the operating phase to assure the economic efficiency of the project.

In a sense the distinctions here may be artificial. The difference between the two is largely one of relative emphasis. It is also likely that the analyses of the economist serving in the first role will have been largely completed before the project is assigned to the team which will develop the proposal in detail. The results of his work will be embodied in the decisions provided as guidance to the project planners. Much of the work done at this level, moreover, may have been done without reference to specific projects.

The second role may be compared to the role that an economist might play as a consultant or a staff economist with a private enterprise, for his primary concern is with the resource allocation opportunities on the project. It is also in the second role that the economist may, in a joint effort with others, use the methods of operations research and systems analysis which are discussed in detail elsewhere in this book.

Members of the project planning and design team who work on water and natural resource public works projects will be acquainted in general with the work that has been underway for at least thirty years in developing the framework within which public investment in projects is evaluated. The highway field, however, does not have as well developed an analytical base for appraising investment.

Project evaluation is an activity of considerable importance when one considers the magnitude of the resources involved. Public works projects in the United States account for a substantial portion of total government expenditures at the state and local levels. While public elementary and secondary education and public institutions of higher education take nearly 36 percent of the aggregated budgets of the states and local governments, highway construction is the next largest category of public spending.[14] Nearly twelve percent of state and local budgets are spent for these purposes. The federal government recognizes both of these activities to be primarily the responsibility of the states and, with some exceptions, does not become directly involved in these efforts except through grants. The federal contribution appears to be nearly thirty percent of the total outlays for highways in the United States, with grant monies for that category totaling more than 4.6 billion dollars in recent years and total expenditures exceeding 15.4 billion dollars.[15]

In the area of natural resources and water projects, federal involvement is generally direct, although grants are involved to a small extent. Federal expenditures for public works of this kind are estimated to have totaled 2.6 billion dollars in fiscal year 1971.[16]

The truly noteworthy fact about these statistics is the extent to which investments are made in the American economy through means other than the much vaunted, and sometimes reverenced, free market. Some ten billion dollars in federal outlays were made in fiscal year 1971 and the outlays for all public works for the five years beginning with 1967 totaled in excess of 42 billion dollars.[17] Investments of such magnitude are not without consequences as far as the national welfare is concerned. The decisions made as to how these monies were to be allocated were, in the last analysis, often political. And while a decision may ultimately be made on political grounds, wherein considerations beyond those which the economist can take account of are given weight, the point of departure in the decision may rest on the analysis of the economist, quantifying and bringing together as best he can the values implicit in the claims and desires of different interest groups. His role as arbiter of these differences has its limits. The ultimate compromises are assigned to those who practice the art of politics.

PUBLIC WORKS PROJECTS IN A MIXED ECONOMY

The free market has been one of the distinguishing characteristics of classic capitalism together with a government

which maintained the law and order necessary for private property to exist, which enforced contracts, controlled the coin-
age of money and after that had little or no role to play in
internal economic affairs. The United States of America in the
1970's does not fit that model and the correspondence to the
model was not perfect even at its birth. The present "mixed
economy" with its public schools, colleges and universities,
public lands and forests, public steam and hydroelectric plants
does not preclude private activity in these areas although that
is the case in the economy of the Soviet Union and other nations which have adopted economic systems similar to theirs.
However, even in the Soviet Union, collective agriculture does
not prevail completely. Small plots of private land have been
permitted that provide a tiny vestige of capitalism.

The economy of the United States and that of every major
nation in the world today combines private and public economic
enterprise to some extent. Ideological differences between societies account for the different emphases that private and
public enterprise are given, but there are other reasons for
the differences too. A major reason for a government to assume
an economic function is that the activity, though of value to
society, would be performed by no one if not performed by the
government. An English economist has pointed out that in this
respect, the agenda of the state as proposed by Adam Smith in
the *Wealth of Nations* and that proposed by Keynes, while not
identical in content, do have a formal similarity which "indicates the essential continuity of thought in the tradition of
economic liberalism concerning the positive nature of the cooperation between the state and the individual."[19]

Departures from the Competitive Model

Economists have been aware for some time that there are
serious differences between conditions which prevail in the
real world and the assumptions about the real world which are
required by the competitive model. Even Adam Smith and the
other classical economists must have been cognizant of these
problems or they would not have allowed any role for government.
In their time, however, the latitude in which the market system could function effectively in a self-regulating fashion
without governmental interference was substantial because of
the prevalence of small firms and a much less sophisticated
economy than exists today.

"Market failure" is the expression used to refer to those
situations in which public intervention in economic activity is
necessary. It is one of the reasons for justifying government
involvement in public works projects. One of the causes of
market failure is the existence of a category of goods and services termed "collective goods." The definitions of a collective good vary, but generally embody the following ideas:[19] A

collective good or service is one which, if provided to one
person is also available to others without the need for them to
pay a price to enjoy it. It is impossible to provide such services to anyone without it becoming available to others as well.
For example, a homeowner may beautify the public space around
his home. The results of his effort are then open to all to
view and thus to enjoy unless he erects a wall to prevent this.
Or, a television signal, unless electronically scrambled, is
available in intelligible form to any receiver without payment
to the sender. Furthermore, as is true of many such goods, the
act of viewing the lawn by one person or the use of the signal
by one household does not reduce the availability of either
service to others.

Still another view of the goods produced by the public
sector is that:

> any publicly induced or provided collective good is a pub-
> lic good...Collective goods arise whenever some segment of
> the public collectively wants and is prepared to pay for a
> bundle of goods and services other than what the unham-
> pered market will produce.[20]

The private sector may have the ability to produce a similar
good or service but be unwilling or unable to produce the quantity or the quality of the good or service desired by consumers.
Highways, public housing and public supported education are
often cited as examples of such goods that are not supplied to
the extent desired when market prices are relied upon alone. In
some instances this condition prevails because while a need may
well exist, there is no *effective demand* for it. That is, those
who would willingly pay the price to obtain the good or service
lack the funds to pay for it.

Externalities

While most consumers know what it is to have wants greater
than their means and learn to adjust to the situation, the members of our society share the belief that the well being of all
members of society is enhanced if a minimum level of certain
goods and services is made available to all regardless of the
ability to pay. The public schools are an example of such a
service which is provided outside the market. Interestingly, it
is an example where coercion is applied if necessary to assure
that use is made of the service. Whether the parents or guard-
ians of a child wish him to be in school or not, attendance is
compulsory until a given age.

Long before the term externality was used by economists,
the concept was recognized by society. The preceding reference
to education is a good example of a positive externality. The
accepted rationale for providing public support to education is

that society itself benefits from having a citizenry that has
been educated at least to some minimum level.   The same philosophy has been applied to the provision of certain health services, recreation facilities and the like.

Exchange may well be the common thread running through
economic activity, but the existence of externalities indicates
that an exchange in some instances involves more than the traditional two parties or two groups. The market institution, as
it is now structured, is incapable of dealing with externalities, or the equally descriptive term, third-party effects, unless they can be internalized, that is, reflected as a cost or
benefit to the appropriate party to the exchange.

Public works projects often involve externalities, which
is one of the reasons they are undertaken by the public rather
than the private sector. Benefits often accrue to parties other
than the primary beneficiaries but they may be too widespread
or too small in individual instances to identify so that all
costs are reimbursed, an equivalent proportion of the costs of
the project might be assumed as the public share of the project
cost.   Revenues from the sale of power, irrigation water or
other services are usually expected to be generated and to re-
imburse the government in part or in full for project costs.[21]

Declining Costs

Under a competitive situation, prices reflect the marginal
cost of the level of output. Figure 9-2 may be taken as illustrative of this. A firm produces that quantity of goods which
can be sold at such a price that the cost of producing the marginal unit of output is equal to the marginal dollar of revenue
received for it.   Thus, in Figure 9-2, marginal cost and marginal revenue are equal with an output of OM which can be sold
for the price OL.   At this price, the production costs are
covered.

Figure 9-3 represents another type of activity.   It is
similar to that found in certain large public projects, inland
waterways for example. The cost of operating the system does
not vary significantly as the volume or traffic increases. The
requirement for a Pareto optimum (which is discussed later in
this chapter) is that the price of the good or the service equal
the cost of producing the marginal unit of output and the marginal output is determined by the point where marginal cost and
marginal revenue are equal.   In the situation represented in
Figure 9-3, marginal cost, while low does not yield a price,
OL, sufficiently high to generate revenue sufficient to cover
average costs at that point, OK.   No private firm could undertake such a project.

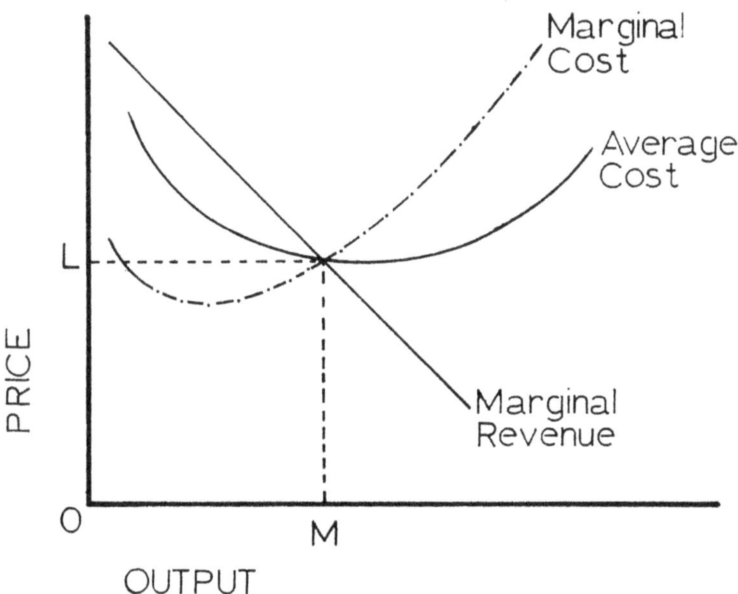

Figure 9-2.   Marginal cost pricing under conditions of constant and increasing average cost is adequate to cover average production costs.

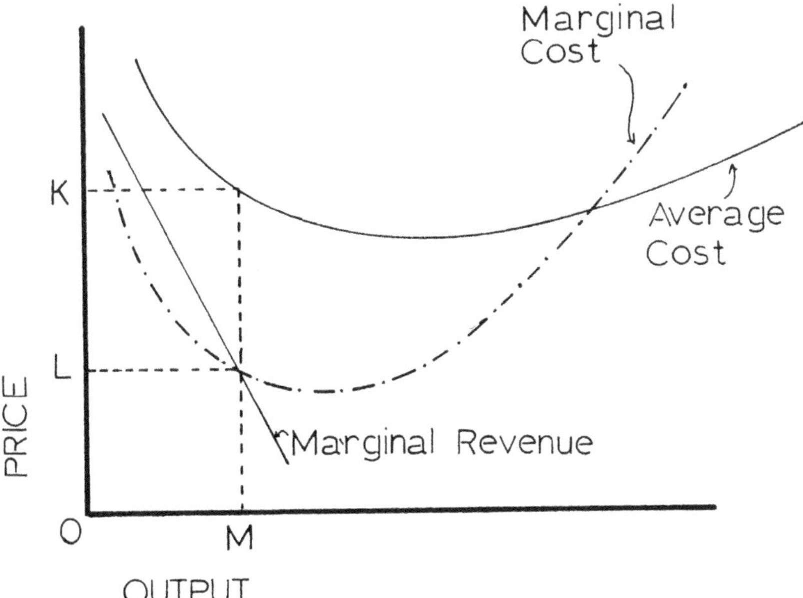

Figure 9-3.   Under conditions of decreasing average costs the prices established at the level of output where marginal cost equals marginal revenue is too low to cover average costs.

The necessary conditions if the competitive model is to
produce the optimum combination and quantity of various goods
and services are violated in a number of other respects too.
The relative immobility of labor when more attractive opportunities become available elsewhere, underemployment and unemployment, capital rationing, imperfect knowledge, interest
rates which are influenced by fiscal and monetary policy of the
government and do not reflect the preference of the present
over future consumption--these are all factors which suggest
that even in the market sector of the economy, prices may fail
to reflect consumer preferences truly.   These various examples
of market failure suggest that the quantity and quality of certain types of goods and services, if produced only by the private sector, may be inadequate in terms of what would be required if an optimum level of well being were achieved for society.

PUBLIC WORKS PROJECTS AND EFFICIENT USE OF RESOURCES

Because of market failures and the inability of the private sector to meet society's needs, the public sector is usually called upon to make up the deficiency with public services
and public works. The same guides of resource efficiency are as
applicable to the public sector as they are to private enterprise. Absolute efficiency in resource use is little more than
a general goal, however, both in the private and public sector.
While considerable emphasis is placed on economic efficiency in
economics, it is naive to suppose that anything closely approx-
imating theoretical efficiency will be achieved in either sector.   Some of the reasons for this have been alluded to in the
previous section and there are other reasons for expecting this
as well.

Multiple Goals

To acknowledge that the construct, *economic man*, is not
valid is to recognize, as Polanyi did, that those who produce
economic goods and services seek objectives in addition to monetary returns.   In fact, it is more accurate to think of the
typical decision maker as responding to multiple objectives,
where he seeks to find an optimum solution, given the constraints of attempting to achieve certain goals in which maximization of one precludes the maximization of the others, but a
point short of maximization of any one of the objectives may
yield the highest possible satisfaction.

The emphasis on cost-benefit analyses in certain public
projects may obscure the fact that certain projects are undertaken to achieve goals other than supplying those goods or services which the private sector of the economy is unable to satisfy.   For example, Richard Musgrave has pointed out that most

governmental expenditures can be classified within one of three broad categories.[22] The first category is the one which has just been discussed--the goods and services which the private sector is not able to provide in part or in whole. The second category covers those activities which consist of income transfers and are designed to change the income distribution to persons defined by their level of income, their age, their health, their occupation, their geographic location, etc. The third category of budgetary expenditures are those used to raise or lower the level of aggregate demand in the economy in an effort to maintain full employment and avoid inflation.

Individual public works projects may fit under any or all of the above categories. Whether the mingling of basic purposes is a matter of accident or design, the results cannot be assessed in terms of a single criterion, which some critics want to do.[23] Examples of the deliberate mixing of objectives can be found in the rationale for the reclamation projects in Western United States, where the objective was the economic development of the arid regions of the West. The reclamation program, like many others, involves income transfers. Revenues from the projects repay some but not all of the costs involved. The assumption is that in time the welfare of the entire economy will be greater than it would have been had there not been such a transfer. This may not always be true for the provisions which are made, or not made, to recover the public investment, and, the rate of interest are important. Reclamation projects have been responsible for depressing the price of certain agricultural commodities because of the overall increase in output which they have made possible. Similarly, navigation projects undertaken by the U. S. Army Corps of Engineers, in addition to providing navigation improvements which would not be provided by the private sector, often have the effect of bringing about a reduction in rail freight rates for those areas served both by rail and the improved waterway.

The primary purpose of the public works projects of many types that were undertaken during the New Deal period of the 1930's was to provide job opportunities for the unemployed. The program of accelerated public works program of the early 1960's had a similar purpose. Public works are used on a smaller scale in an effort to assist people living in economically depressed areas. They are often promoted by congressmen to benefit their districts even though the income transfers to the particular district may have little or no economic justification. On the other hand, projects have been restricted by orders saying "no new starts" when inflationary pressures have been great.

While Congress may insist that project benefits at least equal project costs, it is by no means certain that a project with a considerably larger margin between benefits and costs will be given priority. One economist has summed up congressional behavior as follows:

Of the various motives, strict economic efficiency appears
to be the least outstanding and most uncertain. A fair
appraisal would seem to take the following approach: Although projects with high benefit cost ratios tend to be
chosen before projects with low ratios, no consistent or
persistent pattern is manifest. In fact, substantial
amounts continue to be spent on projects with low or marginal ratios, when, in fact, projects with higher ratios
are available.

Second, a somewhat stronger and more consistent mo-
tive seems to be present in both the force of aid to low
income and depressed areas and the drive to exploit development potential in areas of substantial opportunities for
productive resource investment. As the evidence has shown
not only have Congressmen often spoken of such goals but
also their actions have to a large extent been consistent
with their utterances. Income has, in fact, been redis-
tributed. Certain geographic regions have been aided to
the relative neglect of others, and a significant favoring
of both low income and high potential regions and states
has been noted. Finally, evidence of a strong tendency
toward the equalization of state per capita incomes has
been presented.[24]

The decisions of Congress noted above seem to say that
while the economic efficiency criterion is important, it is not
ruling. Congress is willing to give equal or even greater
weight to equity considerations. Although the benefit-cost
analysis will undoubtedly remain a required part of project
proposals submitted to Congress, that analysis will be only one
of several measures which Congress will use in determining
whether it will approve a project.

THE NEED FOR VALID CHOICE INDICATORS

An implicit assumption of this book is that public works
projects will continue. The failure of the market under the
circumstances previously outlined, together with the need for
the goods and services not provided, suggests that such projects will of necessity continue. A difficulty is encountered
here, however. One of the advantages claimed for the untram-
meled market is that it provides the equivalent of a continuing
consumer poll of their preferences. The consumer "votes" by
making purchases with his income and producers are said to take
their cues for the type and quantity of their output from this
information.

Galbraith, as has been previously indicated, challenges
the contention that the consumer is really sovereign anymore in
these areas of the market which are dominated by a relatively
small number of producers. Research and development efforts may

lead to the development of new consumer products. These efforts
are supported by marketing research to test consumer acceptance
of new products, and by advertising campaigns to develop new
tastes and to convince consumers of their desire for the product. Whether this type of activity is desirable or not is not
the issue. The point is that consumption patterns and preferences are being determined so much *for* consumers rather than *by*
consumers that the basic logic on which welfare economics was
built is being swept away. If individual consumers preferences
are not independent of each other and if they can be influenced
not only the other consumers but also by producers, can anything be said that is both objective and meaningful about the
effects of a change in the mix of goods and services available
in the economy? Are more consumers better off after the change
than before? Are others worse off? More will be said about
this later.

Decisions by Interest Groups

While consumers, to an undetermined extent, do exercise
some influence on how the resources of the economy are allocated, that influence does not appear to be what it once was.
Public works projects are justified on the basis that they produce the goods and services which consumers desire but that are
not produced by the private sector. But even here, are consumers sovereign? To what extent were projects such as the
Interstate Highway System, or the ill-fated supersonic transport organized and promoted in Congress by consumer groups? To
what extent are inland navigation projects, reclamation projects, flood control projects, etc. organized and supported by
broad-based consumer groups? To be fair to all concerned, it
must be recognized that seldom is there a broad base of consumer support behind any major proposal involving public financing, be it for public housing, schools, parks, highways or
what have you. The system does not work that way but through
the efforts of various special interest groups. This is one of
the reasons that public involvement in project development,
discussed in an earlier chapter has become a matter of growing
interest, particularly when basic project values threaten to
conflict with other values.

This situation places a heavy responsibility on planners
and decision makers, at the project authorization level and at
the project planning level, to be certain that the project provides services that consumers value and to resolve the conflicts which may arise with other values. The problem is the
lack of choice indicators, prices or some reasonable substitute,
that will provide some measure of the intensity of desire of
users for the services provided. Given the opportunity to avail
themselves of the service without paying a fee, or more than a
nominal fee, the facilities will sooner or later be overwhelmed
by users. There is sometimes a reluctance to charge users a fee

for the services they enjoy at a public facility. It is reasoned that as taxpayers they have already paid for it and are
entitled to the services available without additional cost.
Sometimes tradition or just good public relations dictates such
a policy be followed. In other instances, the policy is formalized by statute.

Measuring Desires Through the Use of Fees

What happens then when a service is provided that is paid
for out of tax revenues rather than user fees which measure the
marginal cost of supplying the last unit of the good or service?
This situation can be found in certain public park and recreation areas, certain highways and certain inland waterways. The
words to describe what happens are congestion and crowding. The
facilities are so completely used that the. quality of the service they render is diminished. But does this mean that additional facilities should be provided? The expenditure of more
funds for similar projects may or may not be warranted, at
least from the standpoint of economic efficiency. That is,
there is the possibility that other uses of the same funds
might generate greater benefits for society.

Prices as a Rationing Device

This situation points up one of the valuable functions
performed by prices in the economy--they serve to ration items
that are scarce. The theory is that he who can make the most
effective use of the items, or receive the greatest satisfaction from them, should have them. Being unable to determine who
that person is, the decision is made through competitive bid-
ding. The procedure is acknowledged to be effective and fair
if there are no major differences in the wealth of the competing bidders. Society does not often intervene in these matters
but it may on occasion. In time of war or under other circumstances when certain items become especially scarce, societies
have acted to prevent the wealthiest persons from using their
wealth to acquire more than what would be a fair share of the
item. Thus rationing systems are established which override the
price system and determine how the goods and services will be
divided.

A rationing system has merit in those situations where the
income and wealth of users differs greatly and where it is in
the best interest of society that all have at least a minimum
opportunity. Under other circumstances in which income and
wealth differences are of less consequence, user fees may be
the most appropriate rationing device available. The fees,
however, should be designed to make fullest use of the services
offered without creating congestion and a loss of satisfaction
to all users. Furthermore, the fees will be most useful if they
are varied with the highest rates associated with periods of

peak use and the lowest, and perhaps no fee at all, charged
during the periods of low use.   The fees may be of less importance for the revenues they generate than for the control over
use that they provide.   When user fees are imposed it is up to
the user himself to decide whether the service rendered is of
sufficient value to him to warrant payment of the fee.   From a
planning point of view, if it becomes necessary to increase the
user fee in order to maintain a level of use within the limits
that the facility can accommodate, there is good reason to believe that expansion of facilities there or elsewhere would be
justifiable.

Balancing use with service capacity through pricing may
not be a simple matter in practice, but it deserves consideration. The planner, concerned with the question of whether additional capacity can be justified, is not really helped when
told that existing facilities are overtaxed because they are
available at little or no cost to present users. If pricing is
not to be used, methods must be developed to measure, if possible, the probable willingness of users to pay for services
even though no charges are exacted.   Here, the techniques mentioned previously that have been pioneered in the field of outdoor recreation may be of value in other situations.

THE GENERAL WELFARE

One of the purposes for forming a federal union, according
to the Preamble of the United States Constitution, was to "promote the general welfare."   The same phrase, *the general welfare*, is often found in the initial sentences of federal legislation authorizing new programs where legislative intent and
purpose is spelled out.   The exact meaning of the phrase is
elusive although it is generally understood to refer to the
well-being of the members of society and to refer to such general conditions as health, happiness and prosperity.

The general welfare requires more than the availability of
goods and services, but these would seem to be a necessary part
of any effort to promote it. Thus, there is an economic dimension to be considered, and a special field, welfare economics,
has developed as an analytical system for the evaluation of
public policies.   An effort has been made to give specific
meaning and content to the concept of welfare.   It starts with
the notion of the ultimate but subjective measure of value to
the consumer, utility. Utility is satisfaction derived from the
consumption of a good or use of a service.   The welfare of a
person is said to have increased when the level of his utility
is increased.   Conversely, a decline in utility represents a
loss of welfare.

Boulding defines the objectives of welfare economics in the following manner;[25]

> Welfare economics...tries to set up standards of judgment by which events and policies can be judged as 'economically' desirable, even though on other grounds (political, national, ethical) they might be judged undesirable. The search for such a standard of judgment leads to a further search for a definition of an economic 'optimum', this being the position of all economic variables at which riches are at a maximum, the test of economic desirability thus being an increase in riches. Of two events or policies then, the one which increases riches more would be judged economically the more desirable.

In addition, a great effort has been made to develop propositions which are "scientifically" free of ethical judgments, but which make it possible to delimit the area in which the final ethical judgment, with respect to alternatives being considered, can be made.

The value of the welfare economics approach lies in its usefulness in appraising the merits of a reorganization of resources. Public works projects and other public investments usually involve such reorganizations. They are justified in terms of the benefits they produce for certain members of society. Do these benefits represent net gains for society? The measure of the relative desirability of a public project is not profitability, the measure that a private firm would use in assessing a business investment, but the total welfare to those involved, taking account of all costs and benefits.

The Welfare Optimum

The key concept in welfare economics is the Pareto, or welfare optimum. An economy which is Pareto optimal is said to be efficient. As long as it is possible through bargaining and exchange or through a reallocation of resources from one use to another and through similar economic reorganizations to make anyone better off than he was before, such trades and reallocations are desirable. It is only when the possibility for gain comes at the expense of someone else that voluntary exchange will cease. The welfare optimum for society has been reached when an additional change, while benefiting one would impose a cost on another.

Welfare economics has been the subject of discussion and controversy within the economics profession for a number of reasons which will be mentioned briefly. Although it is not possible to make it operationally practical, it introduces some useful concepts that can be applied.

Individual welfare or well-being does not as yet lend itself to unbiased and accurate objective measurement.   There is
no way to measure utility in cardinal terms so that one might
say that before a particular change, A and B both had ten units
of utility or satisfaction. Nor can the gains and losses A and
B experience as the result of a change be measured in these
terms. For example, let the change be the construction of a
high speed, limited access highway running along properties
owned by A and B. After the highway has been built, A must use
a service road and drive two more miles to reach his property
from the new highway. B can reach his property immediately from
the interchange. A has lost utility; B has gained utility. We
cannot be sure exactly how much gain or loss has taken place,
however.

Decreasing Satisfaction from Increased Consumption

It is generally assumed that the more one has of a given
good, the better off and happier he will be. This may be true,
but with certain important qualifications. If money is excluded
from consideration, and only one good or service is considered
at a time, eventually a point will be reached at which there is
no desire for additional quantities of the item for consumption
during a particular time period. The satisfaction produced by
the unit consumed is likely to be the greatest. Additional
units consumed, while producing satisfaction, will yield less
satisfaction as the initial appetite or desire becomes satiated.
Thus, the thirsty traveler will find the greatest satisfaction
from his first cup of water and will make the greatest effort
to obtain it. Each cup he consumes thereafter will yield him
decreasing satisfaction. Finally he reaches a point at which he
has no interest whatsoever in having more water to drink. This
illustrates the principle of diminishing marginal utility.

A second concept of importance relates to the fact that
the fullest satisfaction of human desires requires a wide variety of goods and services. When given a choice between two sets
of goods, one much more limited in variety than the other, most
people are likely to select the set offering the greatest variety. Further, some are likely to be willing to accept a set
with smaller quantity of goods in order to obtain a variety of
goods rather than have little or no variety. This leads to the
concept of the preference function or the indifference curve,
an analytical tool developed by Vilfredo Pareto.

The Indifference Curve

The preference function or indifference curve is usually
shown diagrammatically as a situation in which a consumer is
confronted with alternative combinations of two goods.   This
oversimplification of reality is necessary because the diagram
is confined to two dimensions and a diagram for $\underline{n}$ commodities

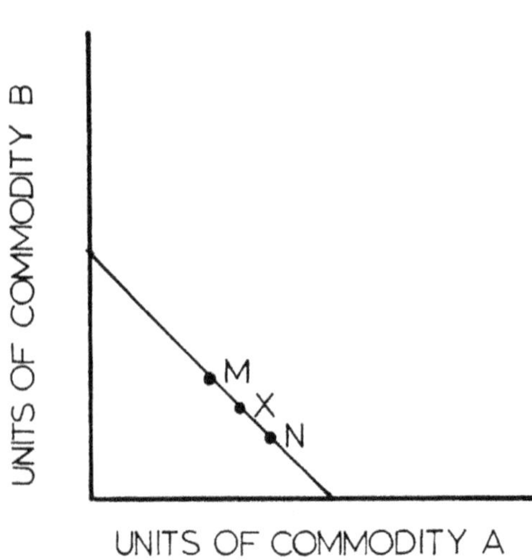

Figure 9-4. Constant rate of substitution between two goods.

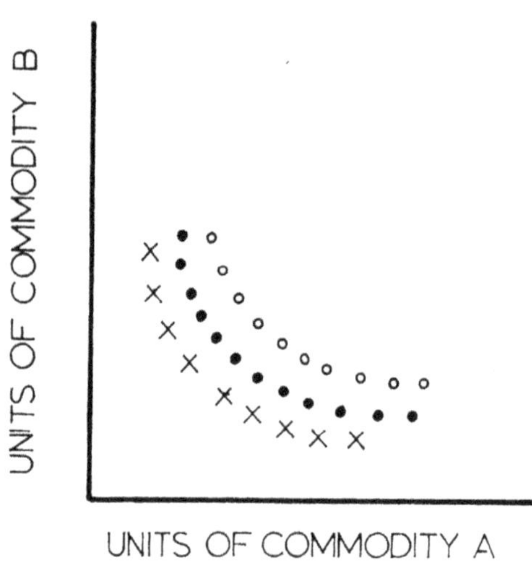

Figure 9-5. Derivation of a consumer preference function.

would require a dimension for each. The two-commodity example, however, is applicable to any number of commodities.

The preference function specifies the various combinations and quantities of goods which are capable of yielding equal satisfaction or utility to the user. The concept of decreasing marginal utility is important in understanding the logic of the function. A preference function can be derived experimentally in the following manner: The experimenter has a supply of two different objects of nearly equal unit value. These objects can also be divided fractionally, if necessary. The experimenter gives the subject an initial allotment of the objects, A and B. The subject is told, however, that to obtain additional units of B, he must give up units of A and vice versa.

Starting with equal numbers of goods A and B, or point X on Figure 9-4, the subject might be willing to give up one unit of A to obtain an additional unit of B, point M on the figure. Conversely, he might also be willing to give up a unit of B to obtain another unit of A. It is unlikely that he would want to continue trading for long on such a basis

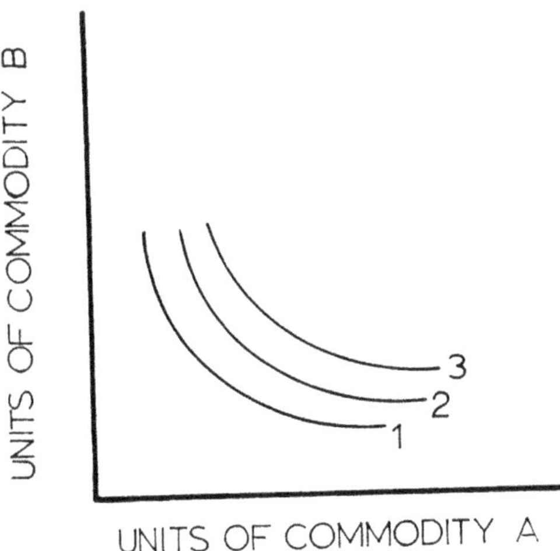

UNITS OF COMMODITY B
UNITS OF COMMODITY A

Figure 9-6. A map of consumer preference functions ordered in terms of increasing welfare.

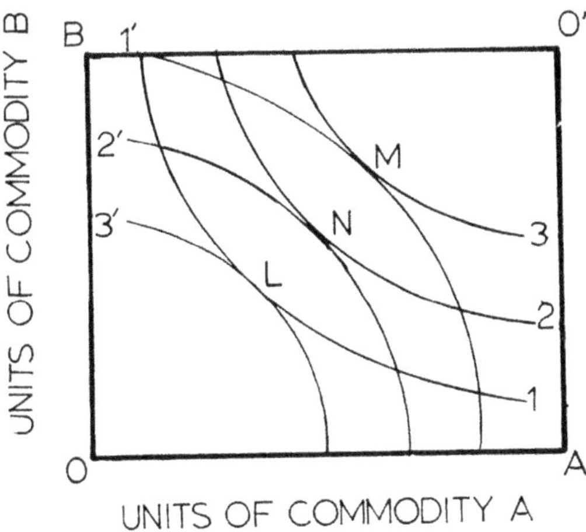

UNITS OF COMMODITY B
UNITS OF COMMODITY A

Figure 9-7. The Edgeworth Box with two goods and the consumer preference functions of two consumers.

for the value to him of an additional unit of one good would be less than the value of the unit he traded because of the diminishing utility of additional units. Therefore, if he wished to maintain himself at a comparable level of satisfaction, he would seek trades in which he might gain a whole unit but give up a partial unit.

Returning to the experiment, let us assume that the subject has no power to determine the terms of the trade but must trade on the terms set by the experimenter. The experimenter, however, for purposes of the experiment, varies the terms of trade from time to time. He also asks the subject to indicate after each swap whether he feels that he is better off than before, worse off, or no different. These responses are recorded in that order as +, -, and 0, in Figure 9-5. Figure 9-6 translates the responses shown in Figure 9-5 into three curves, each of which is defined by points of equal satisfaction for the subject, but curve 1 represents the results of a series of trades which leave the subject worse off than he was initially while curve 3 represents the results of a series of trades in which the subject is better off. By definition, the subject

423

is indifferent to any combination of the two goods that is found on the same curve. However, if he has the choice, he prefers a position on curve 3 to either curve 2 or curve 1 and prefers curve 2 positions to curve 1. Thus, although the experimenter has not measured utility in cardinal numbers, he has an objective measure in ordinal terms of the subject's preferences. He can say that any position on curve 3 is preferred to curves 2 or 1. The welfare of the subject is not changed as his position on any one of the indifference curves changes, but his welfare does increase if he is able to move from a lower to a higher indifference curve.

The concept of the indifference curve can be used to show in diagrammatic form the possibilities of economic reorganization of various types to achieve a welfare optimum. The technique is known as the Edgeworth Box and is shown in Figure 9-7. OA represents the entire stock of commodity A available to consumers total while OB is the total stock of commodity B. For simplicity of exposition, the number of consumers is limited to two. The lines 1, 2, and 3 of the figure, concave to the origin at point O represent only three of a larger set of indifference curves for the first consumer. Lines 1', 2', and 3' in the figure represent a portion of all the indifference curves for the second consumer. They have been inverted, however, and originate from point O'. The welfare for the person at point O increases as his indifference curves move from left to right, while the welfare of the person at point O' increases as his indifference curves move from right to left. In both situations curves 3 and 3' represent higher levels of satisfaction to the respective consumers than do curves 1 and 1'.

Let us assume that a welfare optimum has not been achieved and also that both parties are on their lowest indifference curve. Note that curves 1 and 1' intersect at both the top and bottom of the box. If by chance both parties are at this intersection, it is an indication that they are both consuming quantities of one commodity but none of the other. If the consumption patterns were changed, it would be possible to immediately increase the welfare of one or both the parties. Consider first the points L and M on the diagram in Figure 9-7. At L, the welfare of the first consumer remains the same but the welfare of the second consumer is greatly enhanced as he moves from curve 1' to 3'. This is possible because with a more even mix of commodities, the first consumer gets just as much satisfaction from a lesser quantity of goods as he did by concentrating his consumption at one or the other extreme. At Point M the situation is just the reverse with no decrease in welfare for the second consumer but a great increase for the first consumer. Point N represents a situation in which both parties experience a gain in welfare. It should be kept in mind that the total output of goods has not changed. A reorganization of the division of the goods available has taken place, however.

The situation used in the illustration involved the division of goods between consumers. Similar reorganizations in the
allocation of production resources between producers, etc.,
might also be cited.

Suppose now that a decision is to be made with respect to
the design of an engineering project. The project will produce
two types of output, X and Y.   During the design phase it becomes obvious that there are design alternatives, all of which
will produce X and Y but will produce them in different proportions.   Assuming that it were possible to obtain information
from each person in the community regarding his preferences of
different quantities of X and Y, the next step would be to sum
these preferences to obtain an indifference function for the
community similar to that for an individual. In Figure 9-8 this
type of function is indicated by curve C.   In addition the figure shows another curve, T.   This represents the different outputs of X and Y that the project can produce.   At one extreme
the project might be designed to produce all X and no Y. At the
other extreme, it might produce all Y and no X.

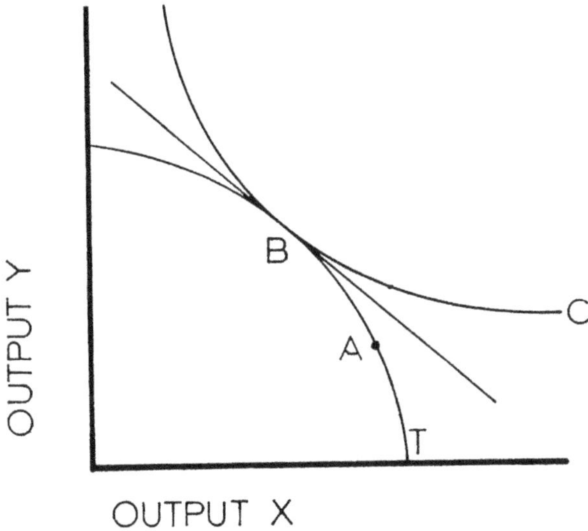

Figure 9-8. The relation between the production possibility function for two outputs, a social preference function for
the same two outputs and the price ratio of the two commodities.

A family of these curves can be imagined but each curve
that represents a greater output necessarily calls for a greater
budget.   (The budget line in this example is the straight line
between the two curves, C and T).   In Figure 9-8 point A is
identified on the production possibility curve T.   A project
designed to produce the mix of X and Y indicated by that

point would not satisfy the community. The community's preference function indicates, however, that the output of X and Y at
point B would be satisfactory and the production possibility
curve indicates that such a plant can be designed at the given
budgetary level.

Howe discusses the situation just illustrated and suggests
that X and Y may represent hydro-peaking power and environmental-
esthetic value.[26] Peaking power and esthetics must be traded
off as water is drawn down to generate power, the esthetic values at the reservoir decline, but to have a visually more
pleasing reservoir site means giving up some power production.
The resolution of the problem is found by selecting the combination of peaking power and esthetic values represented by the
point of tangency between the two curves.

Determining the output curve is a relatively simple task.
Determining the indifference curve is the real problem.   The
market provides information relative to the value that is placed
on power, but the determination of society's willingness to
pay for the esthetic values is a far more difficult problem to
solve.

Assumptions of Welfare Economics

In order to eliminate as much bias as possible, certain
assumptions were made and restrictions were imposed on the welfare economics analysis.   Not only was there no attempt to
measure utility, but interpersonal comparisons of utility were
ruled out. That is, analysts are unwilling to say that a person
with a low income would derive greater utility from an added
dollar of income than a person with a high income because there
is no objective evidence to this effect although logic would
suggest it.

Another assumption necessary to make the system work is
that the personal preferences and utility of each person must
be considered to be independent of those of other members of
the community. This simply means that A's behavior and attitude remain unchanged even after he knows that B has been given
a substantial salary increase while A has received none.   If C
devotes a larger proportion of his income to transportation and
a lesser proportion to savings and to other consumer goods and
buys an expensive sports car, we are to believe that his friend
D does not try to emulate this behavior.

Related to this assumption is another, that the consumer
is truly in the position to determine the mix of products produced in the economy.   If this is true, we must believe that A
is not in the least influenced in his buying habits and patterns, for *he* has made his decision independently of the massive advertising campaigns.   Another major assumption is that
free and perfect markets exist.

A further assumption is implicit, too. The complex model which has been developed calls for finely tuned adjustments in seven different marginal relationships. Marginal rates of substitution must equal the ratio of prices of the items in the transaction such as in the exchange of goods, in the reallocation of production between producers, the reallocation of the factors of production between producers, the substitution of factors of production between producers, the substitution of products between producers and consumers, the substitution of leisure for product, and unhindered lending and borrowing.

A diagrammatic example of this will be easier to understand.   Returning to Figure 9-8, we assume that the community desires some combination of the two goods X and Y, and that a project can be designed to produce varying quantities of these same two goods.   While the project design that is accepted may be one which produces the combination of outputs such as that indicated by point A, the welfare of the community would be increased if the project design made possible the output of X and Y shown at point B. Here, the rates of substitution between the X and Y in the production process and the diagonal line in the diagram, which represents the ratio of the price of X to the price of Y, is also tangent at this point.   The other marginal rates of substitution could be illustrated similarly.

When in each of these relationships, all parties in the transaction have reached their most favorable position, the conditions have been achieved for a welfare or Paretian optimum and no further improvement is theoretically possible without putting someone in a less favorable position than he previously enjoyed. But the optimum is attainable only if all the marginal conditions are fulfilled.   This state of bliss, however, would be next to impossible to achieve even if all the assumptions could be accepted, for the real world is a world in constant flux in which relatively long term decisions· must be made. Once made, there may be little opportunity to adjust to the changes which occur thereafter. When adjustment opportunities do arise, they are not likely to occur simultaneously in all the relationships which must be in adjustment. Thus it becomes next to impossible for the necessary fine tuning to ever occur.   While a "second best" solution may be the answer, it has been argued that these can be obtained only by departing from all the other optimum conditions[27]

The Income Distribution

The Pareto or welfare optimum takes the existing distribution of wealth and income as given. Because of the avoidance of interpersonal comparisons, it has nothing to say about the "goodness" or "badness" of that income distribution.   As noted earlier in this chapter, one of the categories of governmental activity is that of attempting to make certain changes in the

income distribution in society and some public works projects
have this effect. Where society has determined through political action that projects which have an income redistribution
effect are desirable, the economic analyst has nothing, and
should have nothing, to say. Once the desired income distribution has been attained, welfare economics offers the standards
by which to determine whether the economy is making the most
effective use it possibly can from its resources.

Far greater effort has gone into developing the theoretical structure of welfare economics than into making it operationally useful. The previously mentioned problems with some of
the basic assumptions on which the theory rests probably account for most of the difficulty, but an even more obvious difficulty is that of obtaining individual, much less, collective
utility functions.   Indifference functions have been obtained
experimentally, but economic analysis has large data requirements. Many of the data are generated automatically, as through
the market.   As yet there is no economically feasible way to
obtain the data required. If we can be content with a workable
but less complete and less elegant model, many of the ideas behind welfare economics can be of value.

BENEFIT-COST ANALYSIS

The benefit-cost analysis is the operational response to
the need for a welfare economics approach to public investments.
Congress called for it in the Flood Control Act of 1936, indicating that a project should be considered justified if "...the
benefits to whomever they accrue exceed the costs."   The inci-
dence of the costs and benefits *does* make a difference, and
this primitive approach to welfare has changed substantially
over time as the various federal agencies with water resource
development and management programs have worked to develop
evaluation procedures which were consistent with economic logic,
which worked, and which were uniform from one agency to the
next. The so-called Green Book, *Proposed Practices for Economic Analysis of River Basin Projects*, which was printed in 1950,
represented the fruits of the first efforts of the agencies.
The evolution of planning procedures can be traced through
Bureau of the Budget Circular A-47, December 31, 1952, the pub-
lication as Senate Document 97 of the Ad Hoc Water Resources
Council's *Policy, Standards and Procedures in the Formulation,
Evaluation, and Review of Plans for Use and Development of Water
and Related Land Resources*, March 1962, and Supplement Number 1,
Evaluation Standards for *Primary Outdoor Recreation Benefits,*
March 1964.   The most recent proposal of the Water Resources
Council is the *Proposed Principles and Standards for Planning
Water and Related Land Resources*.   The document, as revised,
approved, and published in September 1973 replaces Senate Docu-
ment 97 and its amendment.[28]

428

Benefit-cost analysis is but one part of the planning process. The preceding documents were written as guides in that process. The newest document defines *principles* as "the broad policy framework for planning activities" which "include the conceptual basis for planning." The principles reflect major public policy and investment theory and will change and evolve slowly. The *standards* "provide for uniformity and consistency in comparing, measuring, and judging the beneficial and adverse effects of alternative plans." They represent "the best available techniques for the application of Principles and will change as "progress in the development of planning and evaluation techniques takes place." The *procedures* "provide more detailed methods for carrying out the various levels of planning activities, including the selection of objectives, the measurement of beneficial and adverse effects, and the comparison of alternative plans for action." Procedures are subject to "frequent revision as experience, research, and planning conditions require such revision."

Multi-Objective Planning

Senate Document 97 set forth three major objectives of planning: (1) national economic growth and development and development of each region within the country, (2) preservation, protection and rehabilitation of natural resources with a special emphasis given to recreation purposes, and recognition of values of natural beauty and places of historical and scientific interest, and (3) the well-being of all people with care to be taken "to avoid resource use and development for the benefit of a few or the disadvantage of many."

The current document made some changes in the multiple objectives of planning. National economic efficiency and development remained, but regional development was emphasized by treating it as a separate objective. The conservation objective was rephrased to emphasize enhancement of environmental quality. The fourth objective, the well-being of all people, disappeared as a separate and distinct objective. However, it remains an explicit objective for in a section of the 1970 Rivers and Harbors Act (84 Stat 1818), signed by the president on December 31, 1970, the following statement of congressional intent is made:

It is the intent of Congress that the objectives of enhancing regional economic development, the quality of the total environment, including its protection and improvement, the well-being of the people of the United States, and the national economic development are the objectives to be included in federally financed water resource projects, and in the evaluation of benefits and costs attributable thereto.

Thus, general welfare considerations are to be given attention in public works projects relating to water resources. To some people the phrase "the well-being of all people" is redundant. This may be true if the other objectives go far enough in what they seek to accomplish. However, even in the area of economic concerns, which should not be looked upon as encompassing the whole of human welfare, income redistribution effects have not been squarely faced. It is not the role of economics to prescribe what changes in income distribution are desirable, but economic analysis can and should predict what changes in income distribution will likely occur as a consequence of a project.

Economics Changing Role

The role of economics in public works projects has changed over the years. Initially, in some instances, it appears that economics was largely ignored until the last step when some hurried calculations were made to justify the projects before Congress by showing that benefits were indeed greater than costs. The situation changed, however, and more attention was given to economic analysis. A point was reached where considerations of economic efficiency began to dominate all other considerations. This tended to divert attention from other justifiable project objectives and to force decisions to be made solely on the basis of a favorable benefit-cost ratio. The difficulties that at times were encountered in identifying sufficient economic benefits led to a concern with the problem of secondary benefits and whether they should be included. The response of one critic to this was that

> There is no such thing as a secondary benefit...(It) is in fact a benefit in support of an objective other than efficiency. The word benefit (and the word cost, too) has no meaning by itself, but only in association with an objective; there are efficiency benefits, income redistribution benefits, and others. Thus, if the objective function for a public program involves more than economic efficiency-- and it will in most cases--there is no legitimate reason for holding that the efficiency benefits are primary and should be included in the benefit-cost analysis whereas benefits in support of other objectives are secondary and should be mentioned, if at all, in separate subsidiary paragraphs of the survey report.[29]

While there was something to be gained from the emphasis on economic efficiency, a redressing of the balance was also called for to recognize that other objectives were of equal validity. The identification of multiple objectives in both Senate Document 97 and the Water Resources Council's Principles and Standards for Planning are indicative of the growing sophistication in the process. The emphasis has been moved from the benefit-cost analysis to resources planning. The development of alternative plans is required by the latest document as is "a complete display or accounting of relevant beneficial and adverse effects." Also, it is recognized that an effect may be beneficial in the eyes of part of the public and detrimental as seen by others.

THE INTEREST RATE

We do not live in a timeless environment. Many of the purchases we make we expect will provide a flow of services over time. Yet, we have many needs that require immediate satisfaction and we have an understandable preference for that which is available now as against that for which we must wait.

The interest rate is the conventional measure of that time preference and theoretically, the interest rate should be a measure of the rate at which society as a whole expresses its preferences for the present over the future. But the interest rate also reflects the relative availability of money for investment which would cause no problems to the extent that the money market was free of outside influences. The Federal Reserve Board, however, through open market sales and purchases of federal bonds and through the reserve requirements imposed on national banks can cause the money supply to expand or contract, which in turn has an effect on the interest rate.

The interest rate is not a single rate either. The degree of risk is generally assessed by lenders and a premium which reflects that risk is added to the interest rate. The so-called "prime rate," the rate charged by banks to those customers who are considered the best risks and the rate which the federal government pays on loans are perhaps the most frequently cited to provide a benchmark against which change can be measured.

Discounting Future Values and Costs

The interest rate in project planning is not used to measure the annual return on the investment, that is, the internal rate of return. It is used to discount the sum of the annual benefits and the sum of the annual costs that reach out into the future to arrive at a present net worth or value of the proposed project. The higher the discount rate, the less significant are the future benefits and future costs in determining a present worth. While a project might show a positive benefit-cost ratio of a discount rate of 4 percent were used, as higher rates of discount are used, the present worth of the project declines and a point will eventually be reached at which benefit-cost ratio will be unfavorable.

431

Opportunity costs, that is, the difference in net return
from a chosen investment and the greater returns from a more
lucrative investment if it exists, are of importance in the
decision-making process, whether it be the private or the public sector. An investment decision may be made which does not
yield as great a return as an alternative investment simply because considerations other than monetary gain are given priority. How much should be given up to accomplish the other objectives is, of course, a matter of judgment, so it is important to be aware of the opportunity costs.

The Opportunity Costs of Public Investment

The rate of return on private enterprises in the economy
is the opportunity cost that is of most concern in decisions
relative to public project investments. Although the federal
budget takes more than 100 billion dollars a year from the private sector, these expenditures actually create investment opportunities. At the margin, however, there is no doubt competition for investment funds and taxes divert the funds that
might be used for private purposes to public purposes.

The interest rate indicates in an imperfect fashion the
investment opportunities in the private sector and the private
time preference. It has averaged approximately ten percent in
recent years. This would be in excess of the social rate of
discount, the appropriate rate for determining the feasibility
of public projects. But how the social rate of discount should
be determined is still a matter on which there is no agreement.
The discount rate presently used for federal projects is based
on the rate of interest the federal government must pay for a
particular type of bonds. The current rate is in the neighborhood of five percent. The Water Resources Council, with pressure from the Office of Management and Budget, is advocating
the use of a seven percent rate.

In addition to raising the threshold which projects must
meet in order to qualify for funding, an increase in the discount rate tends to give preference to projects with short useful life. A high discount rate also discriminates against
projects with a high initial investment but low operating costs
while it favors projects requiring a low initial investment but
which may have high operating costs. The annual high operating
costs are discounted whereas the initial investment is not discounted.

Although the discount rate, particularly at rates higher
than those that have been used in the past, is becoming increasingly critical in qualifying water projects, it is of no
consequence in highway projects because of the difference in
the procedures used. In principle, however, the procedure is
as applicable to highway projects as to any other public works
project.

ECONOMIC ANALYSIS OF HIGHWAY PROJECTS

The developments that have been recounted above do not apply to public works efforts other than those dealing with water and related land resources. That leaves a substantial amount of public works activities with much less effort being made to apply economic analysis. Highway construction, the largest single category of public works investments, is the most obvious case.

One reason the economic analysis of highway projects has never been as formalized as the process in the water resources field perhaps can be attributed to the fact that federal agencies carry out the formulation and execution of the water projects while the highway projects are planned and built by the various state highway departments. Federal standards and guidance are attempted, but the political situation does not permit close federal control over state operations. Standardization of procedures does occur to some extent through various non-federal sources. Important in this respect are the publications of the American Association of State Highway Officials.[30]

Distribution of Federal Funds

The allocation of funds is different, going directly to the states by a formula and matched by state monies. Although the formula gives weight to the population and area of the state, the existing road mileage and the estimated cost to complete the interstate system within the state, there are no direct measures of need, measures of urgency or priority, no indication of the volume of traffic now carried or expected, nor is there a measure of the magnitude of benefits relative to cost. The formula system for allocating the highway construction funds to the states is like so many other federal grant programs, frankly noneconomic. Some would call it political, for without general acceptance of the distribution pattern by the various states as meeting some standard of fairness or equity, such programs could not exist. It would be an accident, however, if the distribution of funds among the states was such as to provide the greatest total of transportation benefits for the nation from the monies available.

Until the Interstate Highway System was planned and funded, the division of financial responsibility for highways introduced problems where highway users from different states were involved. Maine provides one example of the difficulties encountered immediately after World War II before the Interstate System came into being. The situation is dealt with in detail by Owen and Dearing.[31]

The summer seasonal flow of automobile traffic along the coastline of southern Maine was particularly difficult to handle on Route 1 and a major need to provide new highway facilities

existed. Such an improvement, however, would be of major benefit to out-of-state visitors. The road system for the entire
state was also deficient which meant that the coastal highway
would have low priority.

The distribution of funds was biased against the coast
line route for state law required that highway funds be distributed in an equitable manner among the counties on the basis
of population and mileage eligible for federal aid. However, no
consideration was given to the volume of traffic served by the
different roads. The available funds were used in scattered
parts of the state. The local communities along the coast did
not press the state for assistance because of the probability
that a new highway would bypass the communities and might cost
them in terms of lost tourist business. Furthermore, because
the state gasoline tax in Maine was three cents a gallon higher
than in Massachusettes, weekend travelers from Massachusetts
were more likely to buy their gasoline before crossing into
Maine. Thus, it was felt that such users of Route 1 were not
paying their way.

To solve the problem, a toll road was built and opened to
traffic in 1949. It placed the burden of financing directly on
the users, regardless of their state of origin. The Interstate
System, once established, did much to relieve problems of out-
of-state use with 90 percent of the funding coming from federally imposed taxes.

Policy Problems in Finance

Highway financing policy also involves problems of how
costs of construction and maintenance are to be divided:[32]

1. between highway users as a group and nonusers,
2. among operators of different types of motor vehicles,
3. among users of different segments of road system within a
    given state, and
4. between present and future highway users.

The above mentioned problems are primarily policy matters
to which economics may make a contribution. Yet another policy
question should be included which deals with the allocation of
funds within the entire transportation system. The difficulty
is of course that the system involves both the private and public sectors which even with the federal Department of Transportation, lacks a focal point where transportation policy and the
relative efficiency of the overall system can be addressed.
These circumstances may seem to suggest that the highway planning team need not be concerned about such matters.

On the contrary, these are matters that *should* concern highway planners even if their area of authority and responsibility is limited and the larger policy problems have yet to be solved. Highway projects require an analysis of costs and benefits and appraisals of this sort are usually made by highway planning teams. The only question is whether their analyses include *all* the relevant data.

The prevailing method of project evaluation ordinarily considers the public investment in the highway facility and the cost to the highway user of operating a vehicle on the highway. Highway surfaces, grades, and alignments, and the relative density of traffic, the variability of traffic speed and the number of stops that are required per unit of distance traveled are the major variables which determine motor vehicle operating costs.

Transportation services are both consumer and producer oriented. To the extent that transportation services are an input in the production of other goods and services which are marketed, it is relatively simple to put a dollar value on them. The alternative of doing without the services helps to establish the value, a task which is somewhat easier for transportation services used for production than for those used to satisfy final consumer desires. The cost of depending on substitutes for transportation of some type may even give an upward bias to the value of transportation services. For example, without the required transportation services from home to work, either employment opportunities or housing opportunities would be greatly limited. A manufacturer might find that critical raw materials were unavailable, or available only at a premium price. Similarly, the market for his output would be greatly limited.

Because transportation services of some type are already available in nearly every part of the United States, transportation benefits are largely thought of in terms of cost savings, and time saved in moving materials or passengers between two points. In production processes time saved can be a significant benefit if the saving is great enough to bring about a reduction in other costs, in unproductive waiting time, etc., or if there is an opportunity to employ the time of the people involved productively. There are many instances in which communication could be used much more intensively as a substitute for travel. If the business can be transacted in that manner, unproductive travel time has been eliminated.

Convenience, comfort, and safety are also factors usually cited as benefits. Efforts to put dollar values on these benefits have taken literally man-years of effort. There is now a question of whether it might not be preferable to set these benefits up in a separate account, following the practice which will be used by the water resource agencies.

In addition, transportation services can be provided by a variety of systems: rail, water, air, highway, pipeline, etc. Furthermore, a highway may be used for a system of transportation which emphasizes public carriers as well as one in which private vehicles predominate. To date there is little examination of *all* transportation alternatives and a comparison of combined public and private costs relative to the value of services provided in an effort to find the most efficient transportation system.

Inadequate Consideration of External Costs

The decision making process then is one of determining the sum of public and private costs for alternative types of transportation systems and subtracting this sum from the benefits which have been determined. Present practice appears to go this far and no farther, although the environmental impact statements required by the National Environmental Protection Act are causing highway planners to take account of adverse effects.

Informed and articulate property owners have forced the relocation of highways but other groups, particularly low-income urban residents with little awareness of how their interests might be threatened by highway plans and little knowledge as to how the threat can be fought, have more often been victims of the highway. Park and open space, because it may already be in public ownership or because of the low cost to prepare the site, has also been extremely vulnerable to highway encroachment. The difficulty of putting a dollar value on such land suggests to some that its value is so low as to be inconsequential. At least, it can be obtained without too much difficulty.

Highway planning teams appear to have ignored entirely, or to have viewed in an extremely narrow, almost private way, the external or social costs associated with a highway project. While a broader interpretation of social costs would not have eliminated all the controversy and criticism that has been associated with highway projects in recent years, it seems likely that the furor would not have been as great as it has. Highway planning teams will find it necessary to examine the total social impact of highways in both their positive and negative aspects.

Some writers are critical of the method used to determine highway benefits of the conventional type. In fact, it has been asserted that

The problem of estimating benefits is avoided by the adoption of design standards. The underlying issues remain:

What are to be accepted as standards?   How is one to know
whether standards are too high or too low?   Uncertainty
about standards implies uncertainty about whether particular projects should be built?[33]

A Case Example

A number of the problems mentioned above which highway
planners face are illustrated in the case situation which follows.   The problems in the example are by no means economic in
their entirety but they do involve values of various types and
the basic concern of economics is with values.

The community in the case example lies not far from one of
the routes on the Interstate Highway System.   Paralleling that
highway at a distance of four miles is the highway which formerly carried the through traffic that now uses the Interstate
Highway.   Traffic on the older highway is still heavy.   The
highway serves local and regional needs but also is the only
major road from one direction to a community some distance beyond. Traffic on the interstate highway headed for that distant
community leaves the Interstate and comes in to the only town
on a new four-lane highway which terminates at the junction in
town with the older highway. The older highway passes directly
through the center of the business district and on beyond the
town to a highway junction where the direction changes by 90
degrees.   It continues in the new direction, passes through a
small village, and then gradually assumes its former direction.

Several years ago city officials and some of the businessmen in the town convinced the state highway department that a
"by-pass" was needed to carry the traffic coming off the Interstate by way of the connecting road and to take it on a route
paralleling the existing road after its direction changed.   The
proposed route lay in an undeveloped flood plain between the
city and the river, crossing the river upstream from the city
and rejoining the existing road beyond the village.

At a point near the beginning of the by-pass, the new road
would pass through a part of town where incomes are among the
lowest in the community.   The houses there are generally substandard and have extremely low market value. A high percentage
of the families, however, own their own homes. Compensation for
the loss of their homes however, would not go far in buying
better housing.   In addition, the community of low-income families would be disrupted.

The routing of the new road is also opposed by groups in
the community who desire to see the land along the river developed for park and open space purposes. The proposed route would
take this land in some places.   It would also impose a barrier
between the city and the river, limiting ready access to the

area. At one place the stream bed itself would have to be relocated, destroying for some time the natural appearances of
the stream as well as established trees growing in the vicinity
of the river.

The business community is divided in its support of the
route location. Businesses located along the present highway at
the point beyond the by-pass oppose it because it would reduce
the traffic past their businesses. Other businesses support the
location but insist it would not be located at a greater distance from the city or business would be lost as a result.

City and county government officials support the project.
They recognize that a true by-pass would require a location at
a greater distance from the city and particularly out of the
area on that side of the city most ready for development. They
also recognize the probability of having to provide a route in
a similar location in the near future to serve local traffic
unless the by-pass is built. The by-pass would be a "gift" to
the community, financed entirely from federal and state funds.
If the route were located in a more remote corridor and the
city and county were required to build another road, it would
probably have to be financed from local funds.

The residents of the village are not happy about the volume of traffic that passes directly through their community.
Their primary concerns are the hazards to safety and the noise
of the traffic. They support the new route.

The highway department has proposed two alternate routes
but has said nothing concerning the possibility of upgrading
the present road to overcome safety hazards that have been mentioned as one reason for building a new road. Cost estimates
presumably are available for the alternative routes. An origin
and destination study indicates that the largest part of the
traffic will be local.

Clearly, a situation such as just described, involves more
than the comparison of road construction costs and the cost to
the user of operating his vehicle on the road. This is perhaps
the place to start, however. The analyst would first wish to
determine whether all of the alternatives have been considered.
Is it possible to by-pass the village with the existing highway?
Can the low-income community be avoided?

When all alternatives have been set forth, cost comparisons can be made of them combining construction costs and the
cost of vehicle operation. Economic benefits to users should be
determined. A gross comparison of benefits relative to costs
can be determined. This, however, can be considered to be no
more than the first approximation.

The external costs and benefits should be tallied next. The cost of uprooting the homeowners and resettling them will have both monetary and psychological costs. The loss of access to park and open space land plus some loss of the land itself should be listed. It may be impossible to put a value of this adverse effect. The probable gain to merchants from the road relocation should be examined. Is it real or just an illusion? The fact that the city and county may avoid having to pay for a somewhat similar road in the future should also be considered. This does raise the question, however, as to whether the highway is now really a federal and state responsibility or a local one.

These externalities, both positive and negative, can be displayed with each of the alternatives and the estimates of the direct costs. Even if monetary values cannot be assigned to some of those costs and benefits, it is still possible to ask the question, for a difference of X dollars in direct benefits, is the city justified in imposing the burden of disrupting a part of the community and forcing some of the residents to find new housing. Will the difference saved in motor vehicle operating costs in the route and the difference in annual maintenance costs be great enough to justify the loss of easy access to open space and park land and the loss of some of the land. (It should be remembered that the state highway department, not the community, is bearing the construction costs.)

WEIGHING VALUES

This approach to solving the problem of weighing intangible values is not greatly different from one proposed for the water resource projects.[34] Alan Dickerman has proposed that basic values be identified and that project objectives be related to them as either making positive, neutral or negative effect on the particular values. He notes that project objectives can be expected to show differing relationships to some values. He suggests that value priorities be established.

When Dickerman speaks of goals, he is referring to such things as irrigation, navigation, flood control, transportation, the generation of electricity, etc. These goals in turn contribute to objectives such as development, environmental quality, efficiency, equitable income distribution and conservation. These in turn are the means to achieve such values as liberty and individual freedom of action; improvement in individual well-being; equality of persons and groups; and stability in all occurrences affecting the individual. Dickerman further believes that society ranks those values in the order listed.

Objectives, too, may be ranked. In a hypothetical evaluation of two conflicting plans, Dickerman has listed the objectives in descending order, from left to right in a matrix and

has considered the impact of two different goals on them as
shown below. He next determined what impact each project would
have on each objective.

GOALS	OBJECTIVES			
	Development	Environmental Quality	Conservation	Income Distribution
Dam Construction	Positive	Negative	Positive	Neutral or Positive
Preservation of Unique Feature	Neutral	Positive	Neutral	Neutral

In this example neither plan has a negative effect on the
development objective.   Moving on to the next objective, the
first project has a negative impact while the second has a positive impact and it is selected. He suggests that the decision
criterion be that of choosing the least negative effect.   He
further suggests that the efficiency criterion should be treated as a constraint function instead of the objective function
in policy planning.

SUMMARY AND CONCLUSIONS

The basic economic rationale for public works undertaken
by the government is that the services provided by such projects would not otherwise be available from the private sector.
The services produced by such projects are often of the sort
identified as public or collective in character. If available
for the benefit of one, others too may enjoy them at no additional cost. If the benefits to the private person or group are
less than the cost of the project, even though total benefits
to all would be more than adequate to cover costs, there is no
incentive for the private individual to undertake the effort.
Flood control is an example.

In many instances the cost of providing the service to one
person is the same, or nearly the same, as that of providing it
to many and the enjoyment of the service of such goods by one
does not limit the enjoyment of them by others. An example is
broadcasting. A radio or television signal is not reduced in
strength if the number of receiving sets is increased. A distant scenic view and water and air free of pollution are also

examples. To provide most of the services of this type usually requires public action because of the inadequacies or inability of the market to overcome these problems.

Public works projects are sometimes used to accomplish other objectives. They have been used in efforts to stimulate economic activity in lagging regions or to bring about a redistribution of income. Projects to accomplish these objectives may or may not be economically efficient.

All other things being equal, the well-being of society is greater if the resources of that society are used efficiently. This calls for a finely tuned balance at the margin in seven different production, exchange and consumption relationships. Economic analysis can determine in an approximate fashion the extent to which these are significant departures from these norms.

Increased attention in recent years has been given to public works projects involving water and related land resources to achieve a higher degree of economic efficiency in such projects. While this has been desriable activity, it is also one that must be handled with a sensitivity for objectives in such projects which may be in conflict with economic efficiency. It is the responsibility of economic analysts to indicate that these conflicts exist but it is ultimately the decision of the public, through the political processes, to determine the extent to which the various objectives are to be compromised.

The move toward multi-objective planning, set in motion in 1962 by Senate Document 97 and modified by the new "Principles and Standards for Planning" produced by the Water Resources Council, provides the framework in which the consequences of a project with respect to economic efficiency, regional development and environmental values can be displayed and reviewed by the relevant group of decision makers.

Highway projects, which are by far the most significant of all public works in terms of the total funds involved, lag behind the water related projects with respect to economic planning and evaluation. The National Environmental Protection Act, however, imposes the requirement that planners of highways financed with federal funds make the same environmental impact studies relative to highway proposals as are required of other projects supported by federal funding. Thus, the scope of the concerns of highway planners has been enlarged. As with other public works projects, highways have both positive and negative externalities, often involving nonmarket goods and services, but the procedures for taking such values into account do not appear to be highly developed or widely used.

While a public works project has limited objectives and is
not necessarily the most appropriate means for accomplishing
other societal objectives, its planners are under the obligation to take account of the full range of benefits and costs to
society that the project generates. The planner of private
enterprise projects is finding it increasingly necessary to
take account of externalities. Where the project may resemble
one of private enterprise, the public nature of the project and
its rationale requires far greater responsibility for these
impacts.

The traditional involvement and contribution of economics
has been in the area of market goods and services and market-
determined values because these have been the areas with which
society has been most concerned. Given the present level of
output of goods and services, many people in our society are in
effect saying that the value of additional quantities of things
to them is much less than the availability of increased opportunities for the enjoyment of what may be called environmental
goods and services. While such goods and services do not carry
monetary values, there has been some succees in arriving at
reasonably acceptable estimates of these values. Even in the
absence of such values, the determination of the value of those
alternative goods or services forgone to obtain the nonmarket
goods provides at least a minimal measure of their value to
consumers.

The usefulness of economics lies in the analytical tools
and approaches it offers and in this particular system and
order of approaching problems of value and human behavior to
achieve as high as possible a level of total satisfaction. One
needs only beware of a narrowness of concern that is limited to
those things with monetary values attached. The concern of
economics is rightfully with well-being or welfare, and while
the attainment of a welfare optimum for a society may require
the attainment of a high level of economic efficiency, it is
also related to the income distribution of that society, concepts of equity and justice, and other similar considerations.
Although economics may not provide the total answer, :t does
have an important contribution to make to public project planning, and as yet, it is not being fully utilized in all areas.

BIBLIOGRAPHY

1. R. Burnell Held.   Professor of Recreation Resources,
Colorado State University, Fort Collins, Colorado.

2. Kenneth E. Boulding.   "Economics as a Moral Science."
American Economic Review. 59:1-12. March 1969.

3. V. C. Walsh. *Introduction to Contemporary Microeconomics*. New York, McGraw-Hill Book Co. 1970, p. 3.

4. Ian McHarg. *Design with Nature*. New York, The Natural History Press, 1969. p. 25.

5. Joan Robinson. *Economic Philosophy*. Garden City, N.Y. Doubleday and Co., Inc., 1964. pp. 132-133.

6. Ibid. p. 133.

7. Ibid. pp. 136-140.

8. J. K. Galbraith. *The Affluent Society*. Boston, Houghton Mifflin Co., 1958, and *The New Industrial State*, Boston, Houghton Mifflin Co., 1967.

9. Mason Gaffney, "Applying Economic Controls." *Bulletin of the Atomic Scientists*. June 1965. p. 20.

10. Melville J. Herskovits. *Economic Anthropology*. New York, Alfred A. Knopf. 1952; Karl Polanyi, "Anthropology and Economic Theory," in Morton H. Fried (Ed) *Readings in Anthropology*, Vol. II, New York, Thomas Y. Crowell Co., 1959; Bronislaw Malinowski, *Argonauts of the Western Pacific*, New York, E. P. Dutton & Co., 1961, first published in 1922.

11. Karl Polanyi. *The Great Transformation*. New York, Rinehart, 1944. pp. 43-55.

12. Boulding. op. cit., p. 1.

13. See in particular, Marion Clawson, *Methods of Measuring the Demand for and Value of Outdoor Recreation*. Reprint No. 10, Resources for the Future, Inc., Washington, 1959; Lionel Lerner, "Quantitative Indices of Recreational Values", *Economics in Outdoor Recreation Policy*, Report No. 11, Committee on Water Resources and Economic Development of the West, Western Agricultural Economics Research Council, 1962; John V. Krutilla, "Evlauation of an Aspect of Environmental Quality: Hells Canyon Revisited," *Proceedings, Social Statistics Section*. 1970. American Statistical Society, Washington, D.C., 1971; and J. B. Stevens, "Recreation Benefits from Water Pollution Control," Water Resources Research, Vol. 2, 2nd Quarter 1966. pp. 167-182.

14. *Statistical Abstract of the United States*. 1971. Washington, Government Printing Office. 1972, p. 403.

15. Ibid. pp. 401, 403.

16. Ibid.

17. Ibid.

18. Lionel Robbins. "The Economic Function of the State in English Classical Political Economy," in Edmund S. Phelps (Ed) *Private Wants and Public Needs* (rev. ed.), New York. W. W. Morton & Co., Inc., 1962. p. 99.

19. See Francis M. Bator, "The Anatomy of Market Failure." *Quarterly Journal of Economics.* Vol. 72, No. 3, August, 1958; Paul A. Samuelson, "The Pure Theory of Public Expenditure," *Review of Economics and Statistics,* Vol. 36, No. 4, (November 1954).

20. Peter O. Steiner. *Public Expenditure Budgeting.* The Brookings Institution, Washington, D. C., 1969.

21. The Colorado-Big Thompson Project, which brings Colorado River water across the continental divide for use in areas east of the Rocky Mountains, was made possible by a precedent-setting statute. The Colorado legislature enacted a law which permitted a water conservancy district to be organized throughout the area which was to benefit from the use of the water. The innovative feature of the law was the provision which permitted all real property in the district, rural and urban, to be taxed to repay a portion of the project costs. This was justified on the basis that the community at large would be indirectly benefitted from the use of the water even though the primary use of the water was expected to be agricultural. Water delivery started in 1951. By 1970 the population of five of the six largest communities in the conservancy district has grown by a factor of 151 percent since 1950. Nonagricultural uses were taking 28 percent of the water compared to 10 percent in 1951 and the ten reservoirs of the system were providing opportunities for an estimated 1,825,000 visitor-days of recreation use.

22. Richard A. Musgrave. *The Theory of Public Finance.* New York. McGraw-Hill. 1959.

23. Milton Friedman argues against many functions now undertaken by the federal and state governments, basing his arguments on the loss of efficiency which he believes would be corrected if the activities were assumed by the private sector. His arguments appear to be based on the assumption that the programs are carried out for one purpose only, ignoring income redistribution and anti-cyclical objectives. Milton Friedman, *Capitalism and Freedom.* Chicago, University of Chicago Press, 1962.

24. Robert H. Haveman. *Water Resource Investment and the Public Interest.* Nashville, Tenn., Vanderbilt University Press. 1965. p. 67.

25. Kenneth E. Boulding.   "Welfare Economics" in *A Survey of Contemporary Economics*.   Bernard F. Haley, (Ed),   Homewood, Ill., Richard D. Irwin, Inc. 1952.

26. Charles W. Howe. *Benefit-Cost Analysis for Water System Planning*. Water Resources Monograph 2, American Geophysical Union, Washington, 1971. pp. 4-6.

27. R. G. Lipsey and R. K. Lancaster. "The General Theory of the Second Best," *Review of Economic Studies*. Vols. 38, 39, (1956-57) No. 1, p. 11.

28. Published with a notice of public review and hearing in the *Federal Register*. December 21, 1971, Vol. 36, No. 245, Part II.   This was subsequently revised and approved and published as, "Water and Related Land Resources, Establishment of Principles and Standards for Planning," *Federal Register*, September 10, 1973, Vo. 38, No. 174, Part III.

29. Arthur Maass.   "Benefit-Cost Analysis: Its Relevance to Public Investment Decisions." *The Quarterly Journal of Economics* LXXX (May, 1968) 208-226.

30. American Association of State Highway Officials, *Road User Benefit Analyses for Highway Improvements*, Washington. This report is widely used in the United States despite its theoretical and practical shortcomings.

31. Wilfred Owen and Charles L. Dearing.   *Toll Roads and the Problem of Highway Modernization*.   Washington, Brookings Institution, 1951.

32. Robert W. Harbeson.   "Some Allocation Problems in Highway Finance," in *Transportation Economics, A Conference,* National Bureau of Economic Research, New York, Columbia University Press, 1965, p. 140.

33. John B. Lansing. *Transportation and Economic Policy*. New York. The Free Press. 1966. p. 236.

34. Alan R. Dickerman. "A Value-Oriented Approach to Water Policy Objectives," Department of Economics, Colorado State University, Fort Collins, Colorado. Unpublished. 1971.

GLOSSARY

Average Cost Usually, total average unit cost, or the total cost of a given level of output divided by the units of output.

Cardinal Number	A number used to indicate quantity.
Cardinal Ordering	Arranged according to the quantity indicated.
Collective Goods	Goods of such a nature that all may use them and no one is able to establish a property right to the Public goods.
Consumer Sovereignty	The concept that all economic activity is ultimately traceable to the decisions of consumers and that production is initiated in response to an expression of consumer desires.
Demand	A schedule of the quantities of a good or service that can be sold for each of a series of prices at a particular place and at a particular time.
Diminishing Marginal Utility	The decline in the satisfaction from an additional unit of a good or service consumed. The first unit yields the greatest satisfaction and although additional units may yield satisfaction, each additional unit consumed adds less to total satisfaction than the previous unit consumed.
Discount Rate	A percentage figure used in a mathematical formula to reduce the numerical value of a probable future cost or benefit to the value it would have in the present. A rate of interest.
Economic Efficiency	The use of economic resources in such a way and for such a purpose that they generate the greatest net benefit or the least cost given the existing circumstances.
Economic Man	The concept of a decision maker whose single purpose which guides all decisions is the maximization of economic benefits and the minimization of economic loss with no consideration given to non-economic values or to other motives.
Opportunity Cost	The difference between the higher return available on an alternative investment and the return on the investment actually made. Similarly, money not currently invested, but held as cash or in a demand bank account has an opportunity cost represented by the returns it could be earning.
Pareto Optimum	A condition in the organization of economic activity in which it is impossible to improve the economic well being of any individual without decreasing it for someone else.

Pecuniary   Relating to money as opposed to economics which
pertains to production, management and use of
wealth.

Utility   The satisfaction which the consumer gains from
the consumption of a good or service.

Welfare   A field of economics devoted to the scientific
Economics criticism of public policies in terms of whether
they contribute to economic efficiency and an optimum state of well being for all.

Willingness The highest price a consumer would pay in order
to Pay   not to be deprived of a desired good or service.
The pricing strategy of public utilities is often
one that exploits the willingness of consumers to
pay. The cost to the consumer of the first units
of the service are priced at lower rates making
it more attractive for the consumer to purchase
additional units. This pricing policy contrasts
with a fixed price which does not change regardless of the number of units of the good or service purchased.

Effective   An expression of willingness to purchase a given
Demand   quantity at a given price backed by the ability
to pay the price.

Externality A loss or a gain to another as a consequence of
an action of one party or a transaction between
two parties which is not reflected back as a loss
or a gain to the parties whose actions are responsible for its occurring.

Extra-Market Beyond the market or outside of the market.
(Non-Market)

Indifference A geometric figure, usually drawn as a curve in
Curve   two dimensions, convex to the origin of two axes
of the diagram. A two-dimensional figure will
show combinations of two different commodities
which will yield equal satisfaction to the user.
Given any of the combinations mapped on the curve,
the consumer has no greater preference for one
set than another. He is indifferent to the various sets.

Internal   The annual net benefits generated by the project
Rate   as a percentage of annual investment and operating
of Return costs as opposed to the market rate of interest
or the discount rate.

Marginal   The additional or last unit. Marginal values are
                    often derived using calculus by taking the first
                    derivative of the particular function for which
                    the marginal value is to be determined.

Marginal   The added cost of producing one additional unit
Cost   of a good or service consumed.

Marginal   The added satisfaction yielded by an additional
Utility   unit of a good or service consumed.

Market   Instances in which the free market does not call
Failure   forth the quantities of particular goods for which
                    there is a demand.

Market   The value that a good or service can command in
Value   exchange which is dependent on the quantities of
                    that good or close substitutes that are available
                    for exchange and the relative desire of purchas-
                    ers for the good.

Ordinal   A number indicating sequence in a series, such as
Number   first, second, third.

Ordinal   Preferences indicated simply in terms of A is
Ranking of   preferred to B with no measure of how strong the
Preference   preference is.

AUTHOR NOTES

The author, Dr. R. Burnell Held, is Professor of Recreation Resources in the College of Forestry and Natural Resources,
Colorado State University.   Dr. Held is a natural resources
economist.   His professional career has involved experience in
four different universities, the federal government and with a
private research foundation over the past 25 years.

Dr. Held has been on the faculty of Colorado State University since 1967 and has taught and conducted research in the
Department of Recreation Resources as well as the Department of
Economics. He has served a total of six years of service in the
federal government, the most recent being assignments with the
Bureau of Outdoor Recreation of the Department of the Interior
where he was Chief of the Division of Research and Education
and, prior to that, Assistant Director for the Scenic Roads and
Parkways Study conducted for the President's Council on Recreation and Natural Beauty as an activity by the U. S. Department
of Commerce.   He was also a staff member of Resources for the
Future, Inc., for a ten-year period.

He is a member of the American Economics Association, the American Agricultural Economics Association, and the National Recreation and Parks Association. He is the co-author of three books on natural resources and land use subjects as well as other research publications and journal articles in that subject matter field. He received his doctorate as well as his M.S. and B.S. degrees from Iowa State University.

IO

SYSTEMS ANALYSIS

BY A. BRUCE BISHOP

SYSTEMS, ANALYSTS, AND PUBLIC PROJECTS

The foregoing chapters have treated a host of physical, environmental, and societal considerations related to public works projects. The NEPA imperative of environmental design and other supporting laws requires that these highly complex and interacting factors be incorporated into decision making among alternative project proposals. Systems analysis provides a way to undertake this formidable task, offering the decision maker a practical philosophy on how best to approach large and complex problems of choice. The holistic view of systems analysis, as opposed to a disciplinary approach, attempts to encompass all the significant elements and variables of a problem, while simultaneously excluding redundant or unimportant considerations. The aim of systems analysis is to evaluate alternative courses of action by examining the consequences of those actions and evaluating them with reference to a set of objectives. In doing so, it makes heavy use of advanced computer technology for processing the often massive quantities of data and for testing numerous alternative solutions in order to identify preferable choices.

The method of approach and scheme of logic used in developing analyses are highly influenced by the nature of the problem, the data available, and the groups involved in the decision. Thus, systems analysis is necessarily both a general approach for problem solving and a collection of specific methods and techniques. It provides a perspective on the use of available tools, combining whatever analytical techniques are appropriate for designing an acceptable solution. Within the hierarchy of management or government decision-making levels, systems analysis can be applied from the program policy level (where alternative policies are identified and evaluated using fairly subjective and qualitative approaches) to project levels

451

(where it most often utilizes quantitative methods such as benefit-cost analyses, operations research algorithms, or other mathematical techniques generally associated with systems evaluation).

Environmental design of public projects is inherently a systems approach. The systems relations of the other chapters have been illustrated by depicting the interactions of man-made public works systems with existing natural and social systems (Figure 10-1).

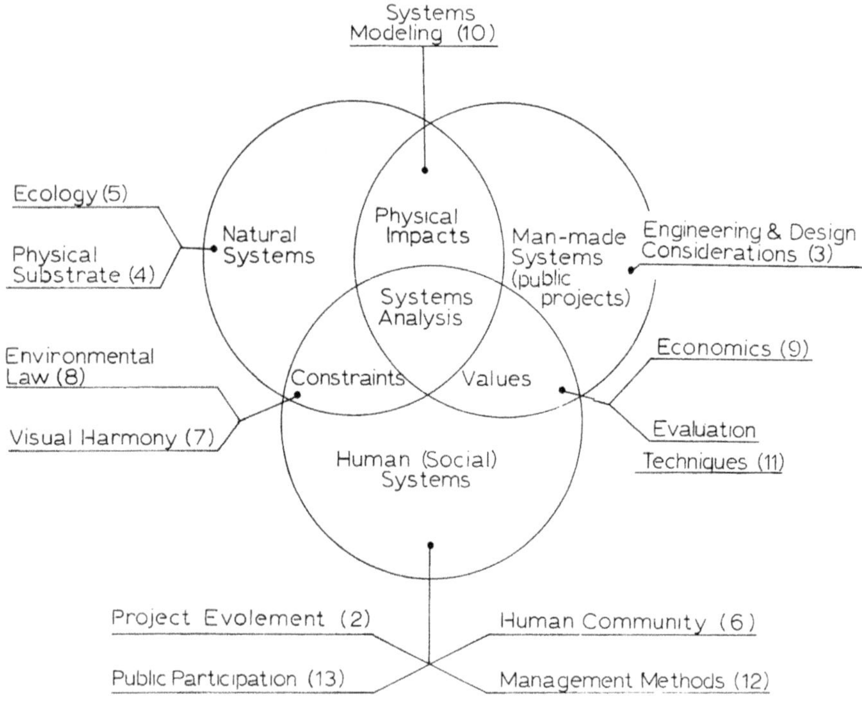

Figure 10-1. Systems relations of treatise chapters.

Concepts of systems analysis provide the framework for integrating the disciplinary knowledge of other chapters. As the diagrammatic (Figure 10-2) shows, the interdisciplinary systems design team synthesizes the essential data on resources, objectives, variable relations, and constraints of real systems into an adequate system model. The model is then used to analyze and evaluate the consequences of alternatives subsequent to selecting a preferred course of action. At the same time key inputs on objectives, alternatives and values must be incorporated from the public and their representative decision makers.

452

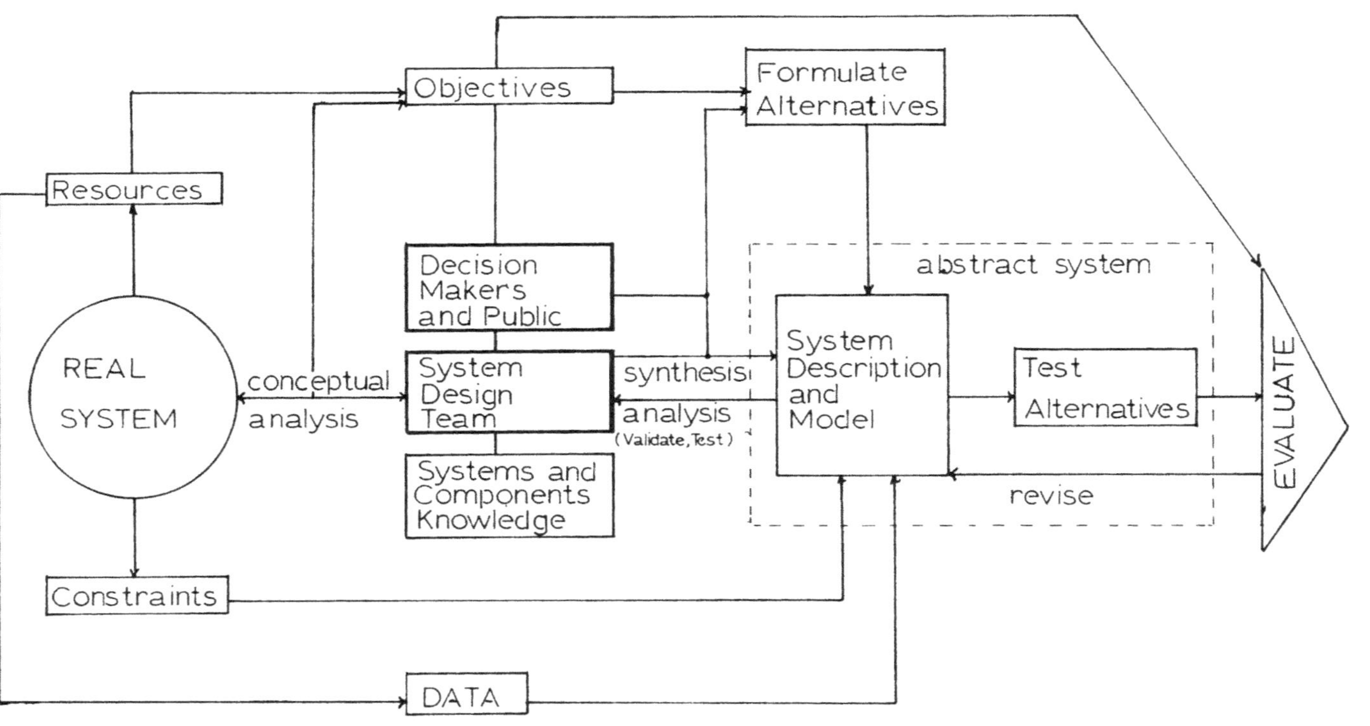

Figure 10-2. Role of systems design team.

The purpose of this chapter in relation to the entire book,
then, is to provide integrating concepts as well as insight into analytical tools for handling complex decision problems. The
extent of the literature precludes a comprehensive description
of the analytical techniques available to the system's analyst.
Theoretical and mathematical development is therefore employed
only as necessary to identify the conceptual structure of the
methods applicable to public projects planning and decision
making.   A convenient summary of references is provided for
those who desire to fully familiarize themselves with specific
techniques. Case study examples are used to highlight the general systems approach and to place the various analytical techniques within a context of application to public projects.

Describing System Structure

A system conceived as a collection of interdependent elements or components that operate on inputs is to produce outputs.   In Figure 10-3 the system is defined by a boundary drawn
to separate those components that are interrelated as part of
the system from those existing outside the system as part of
the "environment." The boundary depends on the physical system
itself, plus the sets of assumptions used and the purposes for
which the analysis is being conducted.   For example, if one is
dealing with the mass transportation problems of Chicago, highway, rail, water, and other transportation modes must be considered as well as strong regional connections between Chicago,
its suburbs, and other parts of the mid-west.   The question is,
which of these technological and spatial elements should be included within the system boundary.

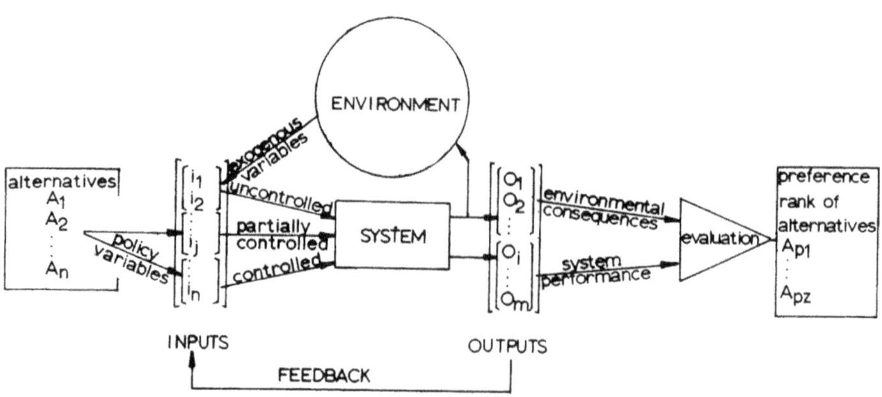

Figure 10-3. Description of a system.

Once this point is settled by the analysts, the inputs and outputs, acting as the bonds between the system and its environment can be identified. The inputs include controllable or decision variables which represent design choices that are open to the analyst. Assigning values to controllable variables establishes a policy alternative. In the mass transit example, if the demand function for transit ridership is known, and prices can be set by the transit authority, then price is a controllable variable, and the demand can be at least partially modified through setting fares. Similarly, routes, schedules and other service features can be controlled, and can thus be used to partially shape demand. On the other hand, constraints imposed on the system by the environment such as land use patterns, development, population growth and distribution, and other spatial and demographic variables may be uncontrollable. These would enter the system as state descriptions or forecasts.

The outputs describe the performance of the system or its consequences upon the environment. They indicate the effects of applying design and planning decisions via the input variables and are evaluated against system objectives and criteria in order to assess the worth of the respective policy alternatives in terms of time, reliability, costs or other appropriate units.

If the inputs and outputs are independent, the system is closed; but when outputs affect inputs through a feedback process, the system is open. State variables represent the condition of the system itself at any given time and place. In a static system inputs, outputs, and state variables are not a function of time, whereas in a dynamic system they are specified as rates of flow or rates of change. The representation which is most appropriate depends on the nature of the prototype system, the analytical objectives and accumulated data.

A system may be composed of several subsystems varying in their levels of refinement. The subsystems comprise a hierarchy of system structure (Figure 10-4). The concept of a hierarchical ordering of system parts is important to the analyst, because it helps in deciding how to simplify or expand the system boundaries as might be necessary for problem analysis.

For example a basin-wide network of water sources, users, and polluters is highly complex, but by decomposing the system at successive hierarchical levels into appropriate subbasins, regions, and interconnected systems serving various users, conceptual simplifications can often be obtained. Thus, the systems approach often can be most efficiently implemented by separating of large complex systems into multilevel components. Studying and modeling the systems at the lower hierarchical level and then coordinating these systems at higher levels ultimately yield an analysis of the entire system. The analytical

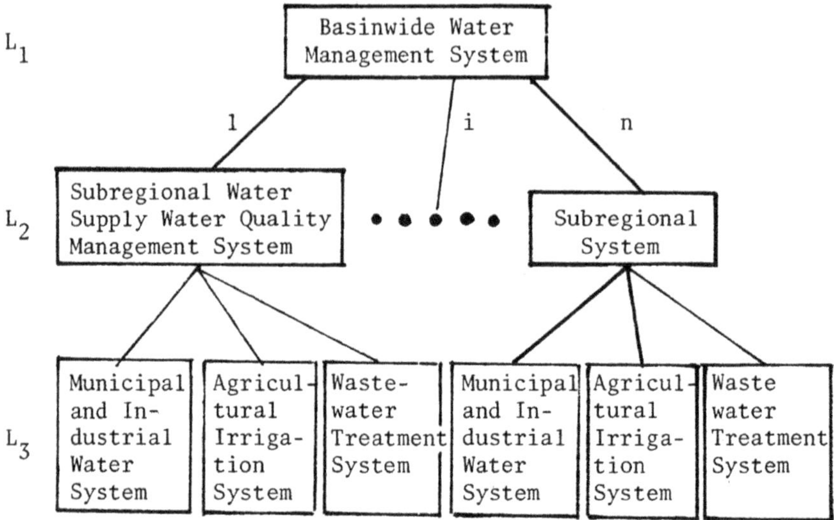

Figure 10-4.   Hierarchical structure of water resources system.   Levels are: L_1, Coordination of subbasin systems and subbasin water transfers, operation of mainstem facilities; L_2, Operation of subbasin systems of dams, aqueducts, pumping stations, and treatment plants, system supplies; and L_3, Operation of individual local systems, distribution networks, treatment plants, user demands.

advantages of the hierarchical approach include conceptual simplification of complex systems, reduction of dimensions and computational effort, and the ability to treat each subsystem with analytical techniques specifically suited to its nature.

The Systems Analysis Process

The feedback loops, reviews, and checks shown in Figure 10-5 illustrate that systems analysis is not simply a step by step procedure. Rather, it is a dynamic process which, after an initial start up period, has work proceeding concurrently on several tasks.   The results and assessments of each stage of analysis are continually being fed back to alter or refine ongoing work on other tasks. The analysis process continues until model outputs are deemed sufficiently refined within the limitations of data and the accuracy requirements, and the evaluation of alternatives completed.

The balance of this chapter describes the specific stages within the process (Figure 10-5) and the related ysstems analysis methods and techniques applicable to public projects design.

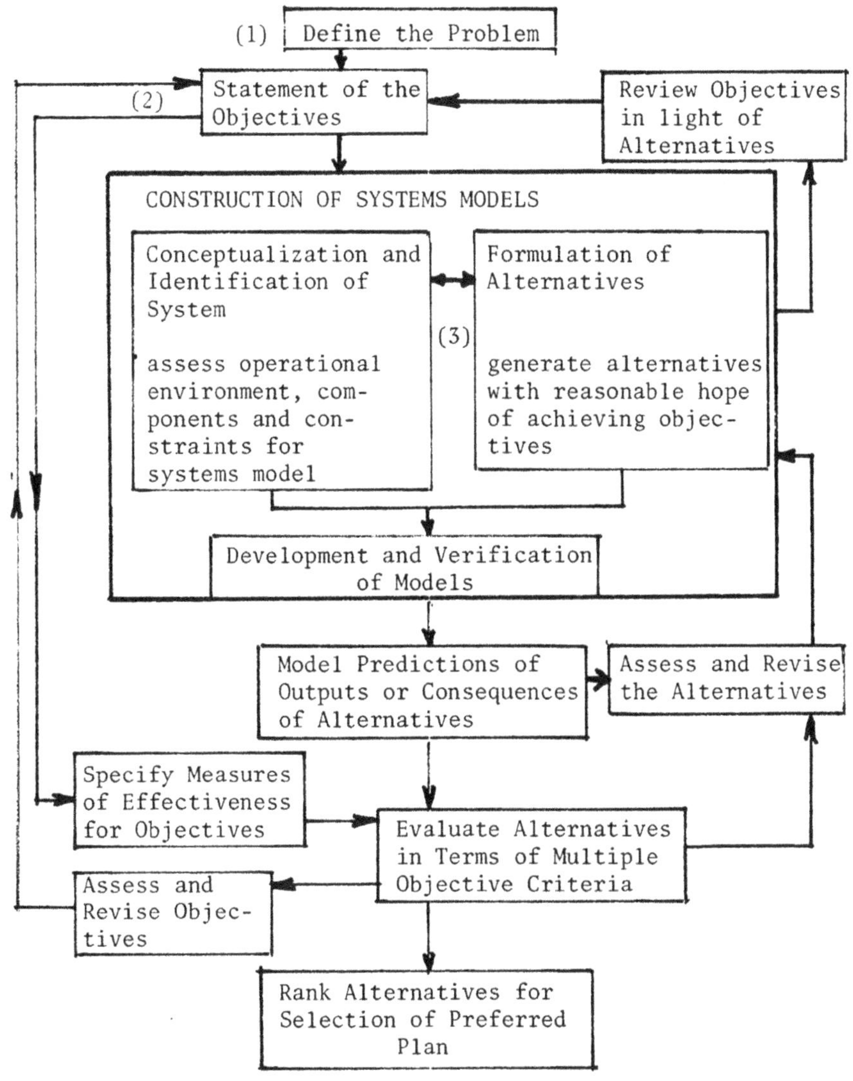

Figure 10-5.   Systems analysis process.

DEFINING AND LIMITING THE PROBLEM

A good systems study begins with an identification of the problem(s) to be analyzed. Public projects are generally planned and designed in response to the desires and aspirations of society that require the provision of public services or facilities.

The general specification of these public wants is usually
a function of the political process at the federal and state
levels. To cite examples, Congress establishes a national system of interstate highways, creates a TVA, or directs that a
study of the water resources of a particular basin be made in
order to recommend projects for flood control, water supply,
etc.

Such political expressions, however, only cover the broad
scope of the problem. To focus a study during the planning and
analysis phase, the range of socio-economic and environmental
factors which may be either improved or disrupted by a public
project, and magnitudes, locations, and durations or time periods should be made as specific as possible. Often these items
will be highly significant to the planning and evaluating of
proposed public projects.

Problem Description: A Case Example

A specific description of any problem requires the inclusion of the various aspects (Table 10-1) of resource capacity
for primary and related uses, their geographical scope and distribution, the time period of analysis and the magnitudes of
demand.

The following case is presented to illustrate and integrate concepts discussed throughout the chapter. The situation
is real, although the presentation here is modified in the interest of brevity.

Description of the study area. Pleasant Valley, a mountain
valley apprxoimately 12 miles long and eight miles wide, is six
miles from the urbanized Thriving City area (see the sketch map
of Figure 10-6). The valley is devoted to rural agriculture
which supports a community of 500 people. The one access road
to the valley is a two lane winding road through Narrow Canyon.
The valley and the city are in the Bluewater River drainage
basin. A number of years ago a dam was constructed at the head
of Narrow Canyon. The resultant reservoir supplies municipal
and industrial needs of the city as well as irrigation water
for agricultural lands surrounding the city. The reservoir has
also become a major watershed recreational area for city residents during the summer. The Bluewater River drains the valley,
feeds the reservoir, and (below the dam) runs through the canyon and the city. The river is one of the best streams for
sport fishing in the area and the canyon itself provides recreational opportunities, such as picnicking, hiking and sight-
seeing. Following is an overview and summary of the major
problem areas.

1. Transportation. The recreational opportunities in Pleasant
   Valley act as a major traffic generator drawing ever larger

Table 10-1. Dimensions of problem definition.

	Areas of Public Projects		
Dimensions	Water Resources	Transportation	Power
Resource Capacities Primary Uses	Water management: water supply for various uses, flood control, recreation, water quality, navigation, etc.	Transportation service: highway, rail, mass transit, air, navigation for people and goods.	Electrical energy supplies: base loads, peak load from hydro, steam, nuclear
Related Uses	Land use, urban and regional development, recreation resources, fish and wildlife resources, economic, environmental and social impacts.	Land use, urban and regional development, community services and facilities, parks and recreation, economy, environmental and social impacts.	Land use, urban and region plans, economic and industrial development, resource extraction, environmental impact, social and community services.
Scope (Spatial Area)	River basin, subbasin, contiguous urban area, water districts or service areas.	Service areas, internal and thru-trip generation, inter and intra urban areas.	Service area, distribution network.
Time	Useful project life, forecast periods, planning horizons.	Useful project life, forecast periods, planning horizons.	Useful project life, forecast periods, planning horizons.
Magnitudes	Demand forecasts, alternative futures.	Demand forecasts, alternative futures.	Demand forecasts, alternative futures.

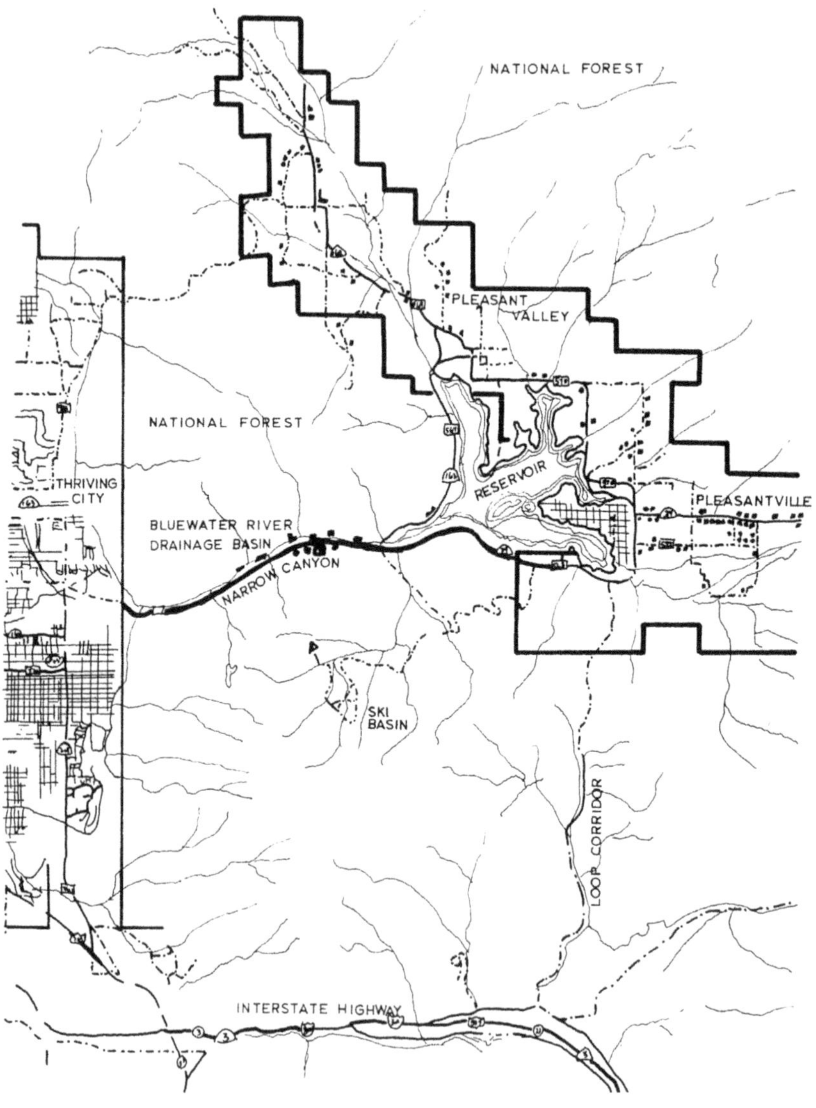

Figure 10-6. Map of Pleasant Valley and surrounding area.

traffic volumes. Traffic congestion during weekend peak periods is a serious problem, with vehicles moving at low speeds, with frequent stoppages. Accident rates have been increasing.

2. Water management. The water supply sources are being used at near capacity to satisfy present needs, and the growing city will require an expansion in available supply in the immediate future. In addition, water quality in the reservoir is deteriorating due to the increasing recreational use, while the quality of the river below the city is affected adversely by municipal and industrial effluents.

3. Community Quality. The above mentioned problems also interact with the desires of Pleasant Valley and Thriving City citizens for more jobs, higher incomes, and increased recreational opportunities for leisure time.

Problem assessment and possible futures. The major issue in future resource use is related to improvement of transportation access into the Valley. The weekend congestion suggests that access should be improved, especially since expanding population and increasing development of Thriving City would intensify the problem. On the other hand, little or no road improvement might limit growth and development in the valley if that is desired. Thus, an important consideration in assessing the level of future traffic demand for the valley is the question: "What ought to be the future of the valley with respect to the carrying capacity of the resources?" The kinds of changes imposed on the valley will be defined by such factors as land use planning and zoning, availability of utilities and other services, the general investment climate, and recreation opportunities. Four possible futures can be logically predicted for the valley: each implying a different forecast of traffic demand and impact on use of resources.

Future 1. At one extreme is the "no development, return to nature" alternative. Under this possibility, further development would be precluded, and efforts would be made to buy private lands and businesses and to convert them to public use as park areas, natural areas, wildlife sanctuaries, etc.

Future 2. This alternative would involve maintaining the present character of the valley by preserving current land use patterns for agriculture, residential, and limited commercial use. Housing or recreational development beyond that already under construction would be precluded, and efforts would be toward maintaining a stable population in the valley.

Future 3. Controlled development would be encouraged. This would include an expansion of both summer and winter recreational potential. Summer home and other residential development would be allowed limited to low density.

Future 4.   At the other extreme, the valley's recreational and residential potential would be fully developed with attendant commercial expansion to satisfy the need of recreationalists as well as permanent and temporary residents. Residential and vacation home development would be encouraged as both high density condominiums and lower density single unit dwellings. Resort areas with varying recreational facilities would also be encouraged.

Each future implies a different forecast of traffic demand and impact on use of resources.

Forecasting Futures

Implicit in the identification and specification of problems is the idea of planning for the future. No public project or policy should be designed just to meet current conditions; hence, it is incumbent on the planner to forecast or perhaps even hypothesize a set of future conditions as a basis for analysis.   This is no easy task, and certainly a simple extrapolation of past trends into the future does not provide an appropriate basis for system planning. Such projections too often tend to become regarded as a set of "requirements" which must then be met by the specified point in time to satisfy the future needs of society. This "requirements" approach to planning incorporates one fallacy and generates another: first, the implication that present growth rates and current practices will continue; and second, its projected requirements assume the nature of self-fulfilling prophecies. Freeway planners ought to be well aware of this fact since they have repeatedly discovered that if freeways are built, vehicles will use them. Opening a new freeway causes changes in traffic patterns, population distribution, and the economy of a region, which, in turn, are at least partially responsible for creating the traffic that planners had projected.   Thus, in order to adequately define and delimit the problem, the analyst will have to delve into areas of economic, social, technological, and ecological forecasting. Two interrelated concepts are particularly indispensable: the economic concept of demand considered in conjunction with supply (or the capacity of a system) and the idea of alternative futures.

Demand and supply relations. The question that is directly related to deciding the desirable size or magnitude of a public project is how to forecast the likely demand for the goods and services to be supplied by the project.   The anticipated need for a public work may be expressed, for example, in terms of volume of water to be delivered, number of vehicles on a road link, visitor-days of recreation, or amounts of energy supplied.

A demand function depicts the relationships between the quantity of a project's output that would be consumed as a

function of price and other influencing factors such as quality
of output, consumer tastes, and prices of substitute or complementary goods or services. In Figure 10-7 and the demand function, 'D', cast in terms of willingness to pay shows the relationship between the prices of a good and the quantities that
would be used at those prices.   Applied to a transportation
situation, such a demand curve would indicate the volume of
travel that could be expected at various trip prices, when
price includes travel time, comfort, convenience, etc.

On the supply or output side, the curve "S",   represents
the quantities of goods or services that will be supplied at
different prices.   In the case of public projects, the concept
varies depending on the "goods" in question. For those items
where a market demand exists, such as water or electrical energy, the supply curve, S_p , is interpreted as the amount that
suppliers are willing to provide at various prices.   However,
when the "product" is available to the public at large (public
goods), as for example a road system, then the curve, P_o, represents the user-cost or price-output function faced by users
as they jointly consume various quantities of project output.
In this case P_o tends to become asymptotic to the vertical line
C, which represents the operating capacity of the system.   For
example as usage of a highway approaches capacity, each additional user is faced with higher and higher "trip prices" in
terms of vehicle operating costs, time, and in convenience until the road becomes completely clogged.   For public projects,

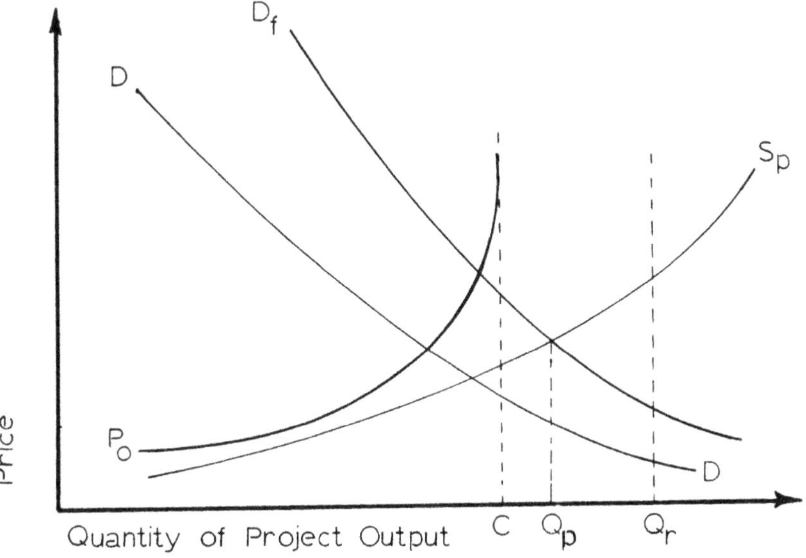

Figure 10-7.   Demand curves and project output or supply:
D_f - willingness to pay (market and non-market costs); S_p =
user cost (market and non-market).

the economic concepts of supply and demand are explored for traffic transportation systems by Wohl and Martin (1967), and by Thompson et al. (1971) and Howe et al. (1971) for water resources.

If the supply and demand schedules for future planning periods were definable, then information developed from the relationship of these curves would help in estimating the actual level of service or output to be provided by the alternatives. To contrast this with the requirements approach (Figure 10-7), the vertical dashed line R represents a requirement level of output, Q_r, that "must be provided" to meet future needs. However, if D_f is the demand curve for the output, its intersection with the supply curve S_p indicates that the project size should be set at Q_p , which is less than Q_r .

The significant concept illustrated by this example is that the ultimate demand for a public good depends on the price of the output, and the price users are willing to pay in terms of both monetary and non-monetary costs, and not just on projected needs. Thus, a public project proposed to augment the capacity or increase the output of an existing system should be responsive to the eventual relations between the costs to the consumers of levels of supply and the quantities that will be taken at various prices. Such forecasts, which depend on many socio-economic and environmental factors, are, of course, difficult to make. Therefore, "forecasting" levels of project outputs requires an understanding of the interrelation and influence of these variables at future planning horizons.

Alternative futures. Any demand schedule is probably valid only for a specific time and particular consumer behavior, as affected by income, in population, life style and other factors. More specifically, a change in demand determinants variables, e.g., growth in population or income, may result in a shift in the demand curve, as denoted by the dashed curve D_f. Without the aid of a crystal ball, the problem of forecasting demand curves for some planning horizon is indeed a difficult task. However, demand analysis and sets of alternative predictions for the futures can provide a better basis for systems planning than simply extrapolating current trends. Alternative futures, then, are brief descriptions of possible sets of future conditions, and should provide insight into likely levels or magnitudes of "demand" for system outputs. Since shifts in demand are expected in response to such factors as changes in income, population, and leisure time, alternative descriptions of possible future levels of various demand determinants are essential when estimating the probable total magnitudes of change. In this respect, a statement of alternative futures can serve as input when formulating program objectives, because the objectives actually help generate the alternative future that is deemed most desirable by society.

FORMULATING OBJECTIVES

In effect, objectives translate desired futures into operational terms by stating the conditions to be met in designing
a satisfactory solution. At the same time, objectives provide
a basis for, or at least an insight into, the criteria against
which the viability of alternatives should be judged. Generally
speaking, two types of goals or objectives should be sought
when public projects are introduced into the natural and social
environment.

1. Harmony: Harmonizing in the planning, design, construction
    and operation of the public work means attempting to blend
    it with its surroundings, including both the natural and
    social environment. In other words, this objective insures
    that systems alternatives are responsive to the question
    "Does it fit/" from the standpoint of ecological, human,
    and physical impact. Objectives designed to achieve harmony
    attempt to account for the overall system and subsystem re-
    lationships, the compatibility causal linkages and their
    environmental consequences.

2. Optimality: Objectives concerned with optimality aim for
    effective operation of the system within the environment
    with respect to predetermined performance criteria. When a
    public work incorporates multiple objectives, optimization
    in one area may preclude optimization in another. This
    common situation implies the necessity of making tradeoffs
    among objectives on system performance, and hence the need
    for an implicit or explicit weighting of objectives in or-
    der to determine the "preferable" or optimal system config-
    uration. "Optimal solutions" therefore depend directly on
    the criteria and values selected. Selecting a set of ob-
    jectives for planning public projects is not a simple or
    routine task. The differing viewpoints of specific interest
    groups involved in public works planning generally produce
    a multiplicity of objectives that often conflict or are mu-
    tually exclusive. Thus it may be necessary to work with
    several alternative sets of objectives in order to avoid
    the problems created by constructing studies on a single
    narrow set of goals.

    This in turn requires some investigation of the relation-
ships among the objectives themselves. In this respect   two
considerations are important: (1) hierarchical relationships
in objective sets and (2) compatibility within sets of objectives.

    The hierarchical ordering of objectives is illustrated in
Figure 10-8. Broad program objectives are generally contained,
either implicitly or explicitly, in the federal and state legislation that empowers public works planning and construction

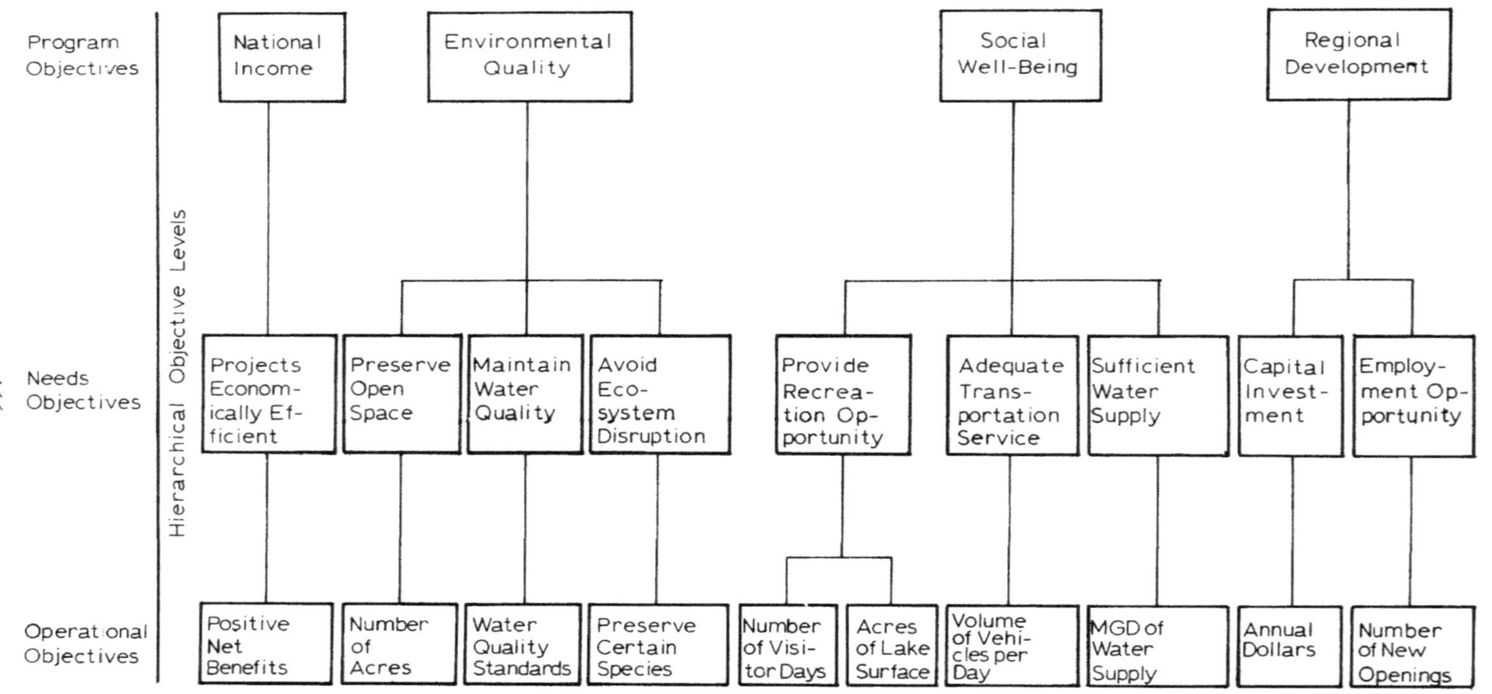

Figure 10-8. Hierarchical levels of objectives.

agencies. Such broad statements, however, have little practical
value when attempting to provide a proper scope and focus for
analysis of a specific problem at the regional or local level.
Thus a second level of objectives that identifies more specifically the goals and desires of a region, such as open space,
recreation opportunity, or employment opportunity, is required.
At the third hierarchical level, objectives are specified, so
far as possible, in operational terms - facilities for so many
visitor days of recreation, so many acres of open space, numbers of vehicles to be accommodated by a highway link, and so
on.

At this level of specification, objectives will closely
parallel or perhaps mirror the areas delimited by detailed
problem definition. It can also be seen that the operational
objectives are a function of the alternative future(s) being
considered. Consequently, these objectives will significantly
affect the description of the system, the configuration of alternative solutions, and hence the results of the analysis. For
this reason, it is important to carefully analyze the compatibility of objectives with alternative futures, and the internal
or cross-compatibility of the objective set itself in order to
determine a mutually consistent set of objectives for planning
a public project.

An analysis of the compatibility of objectives with a set
of alternative futures is illustrated in Table 10-3. The alternative futures refer to those stipulated in the description
of Pleasant Valley in the previous section. The objectives are
from the hierarchical set shown in Figure 10-8. In the Table,
a (+) indicates the objective is compatible with a given future,
a (-) incompatible, and (0) has either no effect or an indeter-
minant effect at this time. Such an analysis is necessary to
find what objectives can be reasonably pursued within the scope
of some desired future state. For example, the objectives of
maintaining open space, preserving certain species in the ecosystem, and maintaining the valley water supply sources are incompatible with the future calling for full development of the
Valley. But they are compatible with the status quo or limited
development described in Futures 1 and 2.

Additional investigation of the internal or cross-
compatibility of the objective set is shown in Table 10-4. The
attempt here is to identify those objectives within a set that
are mutually exclusive, i.e., where the accomplishment of one
particular objective would preclude the achievement of another.
Again in terms of application to the Valley case, the water
quality objective is not compatible with providing unlimited
recreation, additional vehicular access, or the development implied by more capital investment and jobs. Where two objectives
are in conflict, as indicated by cross-compatibility analysis,
choices on tradeoffs between the level of achievement of the

Table 10-3. Compatibility of objectives with policy futures.

	Future 1	Future 2	Future 3	Future 4
Positive Net Benefits	0	0	+	+
Number Acres Open Space	+	+	-	-
Water Quality Standards	+	+	0	-
Preserve Species in Ecosystem	+	+	-	-
Number of Recreation Visitor Days	-	0	+	+
Acres of Lake Surface	0	0	0	0
Volume of Vehicles Per Day	-	0	+	+
Mgd of Water Supply	+	+	0	-
Annual & Capital Investment	-	0	+	+
Number of New Jobs	-	0	+	+

two objectives will need to be made, or perhaps one objective
will have to be abandoned in favor of the other.   Efforts to
devise definitive sets of objectives for public projects have
been made by the Water Resources Council (1970), planning agencies themselves, and the Technical Committee (1971).   Such efforts can be useful in developing lists of appropriate objectives for a particular study.   Then the most fruitful approach
for most public work projects is to simply concentrate on deriving compatible sets of objectives for various possible "futures."

     Finally, goals or objectives are not necessarily static.
Since setting objectives is a social and political function,
they will often be vaguely expressed at first and, only as
planning translates them into physical and functional form, do

Table 10-4. Internal or cross-compatibility of objectives.

	Economic Benefits Exceed Costs	Acres Open Space	Water Quality Standards	Preserve Species	Number Recreation Visitor Days	Areas of Lake Surface	Volume of Vehicles	Mgd Water Supply	Annual Capital Investment	Number of New Jobs
Positive Net Benefits	x									
Number Acres Open Space	0	x								
Water Quality Standards	0	+	x							
Preserve Species in Ecosystem	0	+	+	x						
Number Recreation Visitor Days	+	-	-	-	x					
Areas Lake Surface (Flat Water Recreation)	+	0	0	-	+	x				
Volume of Vehicles Per Day	+	-	-	-	+	+	x			
Mgd of Water Supply	+	+	+	+	-	+	+	x		
Annual & Capital Investment	+	-	-	-	+	+	+	0	x	
Number New Jobs	+	-	-	-	+	+	+	+	+	x

they become crystallized as alternatives are put forth and impacts are perceived. The setting of objectives for public
projects is a process of feedback and adjustment as changes in
objectives in turn call for a reformulation of alternatives.
Ultimately, objectives and criteria emerge on which the merits
of project alternatives can be judged.

The construction of a systems model is essentially a process of abstracting from reality; hence, every model is an analog of reality.   As symbolic representations of real systems, models are a central part of systems analysis.   In fact, the critical phase of systems analysis is mapping from the real world to the abstract world of the model, and then translating the results of the model back to real world terms. The mapping requires careful assessment of the significant variables in the system and their interactions. The translating calls for careful interpretation of results in terms of the assumptions and approximations on which the model is based.

A systems model may be defined as a description of the functional relations of the variables relevant to a specified problem.   A model is constructed in order to ask questions of it with the hope of getting some hints or guides, or perhaps even answers, about the consequences of alternative policies. It should be recognized at the outset that the structure of any model will depend on the kinds of questions being asked of it, and on the structure of the real system.

The use of a model as an analytical device, of course, is not new.   Engineers have long used models for determining the relationships and effects of physical variables that influence whether a given design can fulfill its primary physical functions (e.g., the construction of a dam to store water for power generation, municipal and industrial uses).   The accompanying analyses, however, have often not been adequate in terms of the secondary and tertiary effects of implementing solutions (e.g., the impact of the dam on streamflow regimen, aquatic life, recreation and cultural patterns, development and land use, and so on).

A primary concern in systems modeling of public works is understanding the controllability and relative efficiency of different variables in producing given changes.   The model facilitates the use of such powerful system analytic techniques as operations research methods, and the computational capabilities of digital and analog computers. Systems analysis and the modeling of the interrelation of the project and the environment can thereby provide information to planners and decision makers in three important ways:

1. A system model may be able to forewarn the planners of the possibility that a change in one part of the system may yield otherwise unforeseen and undesirable consequences in another part of the system.

2. System models can demonstrate that changes may be secured in one element, not only by a frontal attack upon it, but

also by a circumspect and indirect manipulation of more distantly removed variables. These, because of system interdependence, may ultimately produce the desired changes in the target variables.

3. Systems analysis by the use of models, directs attention to the multiple possibilities of intervention with respect to a single problem.

Elements of Systems Modeling

The construction of systems models incorporates two parallel tasks: (1) conceptualizing the real system and formulating it as a systems model, and (2) developing a set of alternatives by which the system can be manipulated to achieve the desired objectives. These tasks are largely interactive since the identification of subsystems, components, and constraints within the "real" system will help generate appropriate sets of alternatives. On the other hand, the specification of alternatives may suggest portions of the system structure that need a more adequate description.

Conceptualizing and identifying system structure. The usual procedure for conceptualizing the real system is to work at building up a framework of the interaction of system parts. This is done by separating the system into its subsystems and identifying components or variables at a level of detail that is compatible with the specificity of the objectives and goals.

Some very useful descriptions of model structure also provide the needed conceptual tools to move from prototype to an abstract system description. These techniques are described briefly as follows.

Networks, matrices, and graph theory. Considering a system as a collection of components, a network diagram conveniently represents relationships among components. Transactions, transmissions, and cause and effect linkages among components can be expressed by connecting them with an arrow indicating the direction of flow or interaction. For example, in Figure 10-9 an arrow directed from component A to component B might be variously interpreted as A affects B, A's output is an input to B, etc., while the components themselves--A, B, and so on--represent system attributes or variables.

Depending on the nature or meaning of the variables and their connecting links, various tools are available for further analysis. If the network is a sequence and time duration of a set of tasks or operations, then PERT and CPM can be used to analyze the critical sequencing of the operations. If the network represents the flow of some physical quantity to or through various points, then algorithms are available to find

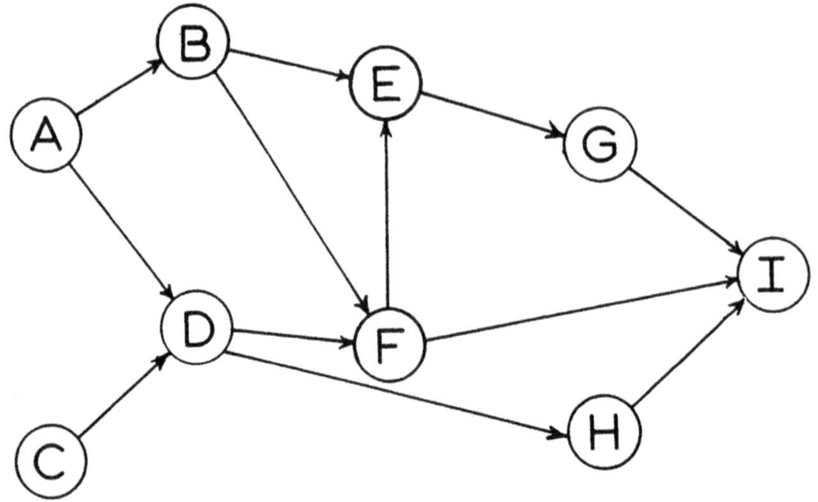

Figure 10-9.   Sample network diagram.

maximum possible flow or an optimum network design given cost
or other constraints (de Neufville and Stafford, 1971).   If the
diagram represents a causal structure of variables, then the
network provides a useful "blueprint" for constructing analytical, mathematical, or simulation models.

Networks effectively characterize many problems encountered in public works systems, e.g., the links in a highway or
freeway system, water supply distribution systems or wastewater
collections systems, or configurations of energy distribution
nets.   To develop an example from the Bluewater Basin in the
case study, Figure 10-10 shows the network flow for the water
resources system as water is diverted and used, and effluent
water is returned to the system.   Each node represents a water
using activity as well as a source. An arrow into the activity
represents a flow of water from other sources or activities
into that one; conversely, an arrow leaving an activity represents the effluent flow from that activity to another.

Many system descriptions formulated as networks can be
conveniently represented in a matrix format.   For example, if
the nodes in a network diagram are listed as row and column
headings, then the existence of a connecting arrow can be denoted by an entry in the appropriate cell of the matrix. Figure
10-11 is a matrix representation of the Bluewater Basin network
diagrammed in Figure 10-10.   The matrix representation has the
advantage of easy storage and manipulation of system data by
computer.   Network diagrams often serve as an initial description of a system, which then can be investigated by some other
analytical means.

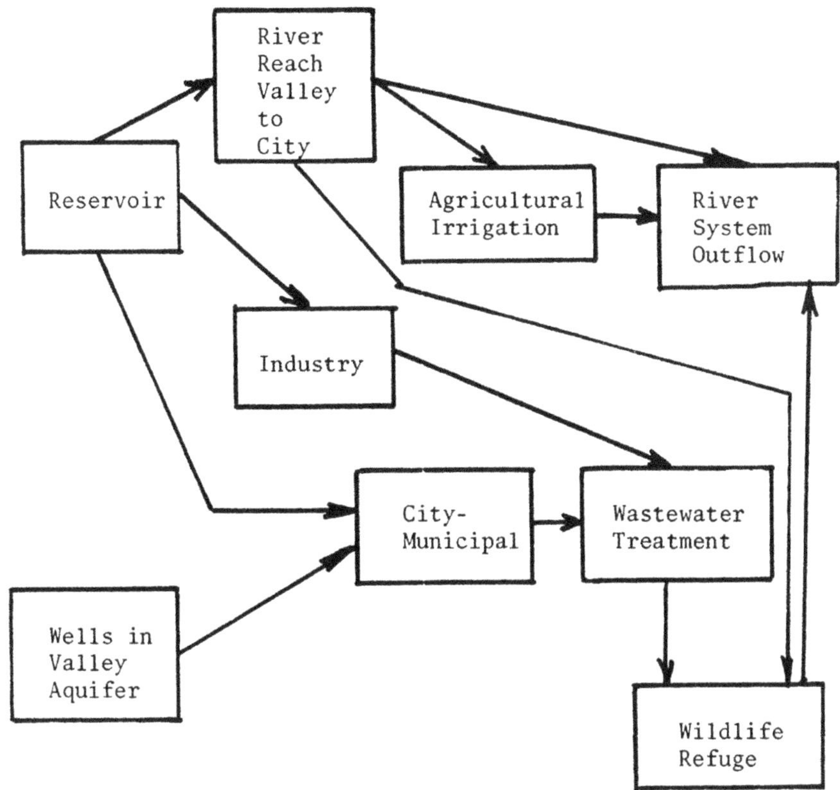

Figure 10-10. Network diagram for Bluewater River Basin.

A network diagram that shows the connection between components in a system can also be considered as a geometrical diagram called a graph.   A directed graph, one in which the direction of flow between two components is specified, can be conveniently represented as a matrix for analysis. There exists a mathematical theory of graphs (Gordon, 1966: Deutsch, 1969) which offers very useful analyses, such as the lengths of paths between components, their relative importance or hierarchical ordering, and simplifications of apparently complex system structures.   By assigning weights to connecting links, value relations can also be studied.

Block diagrams and signal flow graphs. ₁ Block and logic operation diagrams facilitate examining the connecting operations or transformations between system components.   These operations may be arithmetic (addition, multiplication) and functional (integration, differentiation) operations, or logical operations and conditions.   The former are most useful in mathematical system theory, while the latter are familiar in computer programming.

Source or Activity \ Destination Beneficial Use	River Reach: Valley to City	Municipal	Industrial	Agricultural (Irrigation)	Wastewater Treatment Plant	Wildlife Refuge	River System Outflow	Water Supply at Source or Activity
Reservoir	X	X	X					Annual Inflow
Wells in Valley Aquifer		X						Groundwater Pumpage
River Reach: Valley to City				X		X	X	River Flow
Municipal Effluent					X			Municipal Wastewater
Industrial Wastewater					X			Industrial Wastewater
Agricultural Return Flow							X	Irrigation Return Flow
Wastewater Treatment Plant						X		Treatment Plant Effluent
Wildlife Refuge							X	Refuge Outflow
Demand Levels for Beneficial Use	Reservoir Releases to River	Municipal Demand	Industrial Demand	Agricultural Demand	Plant Inflow	Refuge Demand	Downstream Outflow	

Figure 10-11. Matrix description of Bluewater Basin water resource system.

Figure 10-12 is a block diagram representing a simple system with feedback for converting the input 'x' to the output 'y'.   The blocks represent a multiplication operation and the circle a summation operation.   The combination of operations within the system boundary which converts x to y is the system "transfer function."   The function can be written as a mathematical expression (model) of the following form

$$T = \frac{y}{x} = \frac{k}{1-bk} \tag{10-1}$$

The dual of the block diagram is the signal flow graph; i.e., in the flow graph, the variables become nodes and the block transformations become the arrows.   The flow graph is usually more appropriate where detailed algebraic analysis is required.

In most cases network and block diagrams are used as a basis for modeling the behavior of large scale complex systems. Such diagrams provide both a substantive analytical insight and a framework for identifying functional relations of variables that indicate what other types of analytical models, mathematical models, probability models, etc., can describe the system functions.

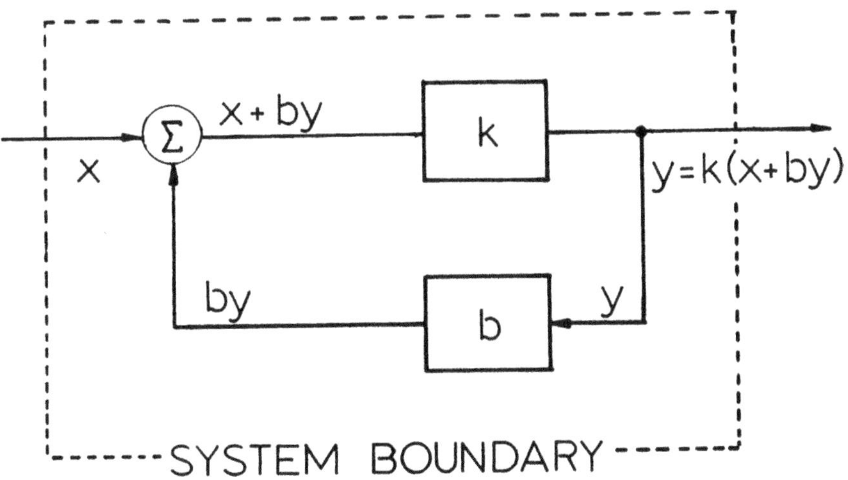

Figure 10-12. Sample block diagram.

475

Specifying constraints. As subsystems and their components
are investigated, attention should also be given to identifying
'system constraints that will control and influence interactions
among the variables and subsystems. Sets of constraints as related to public projects may include technological and environmental factors that directly affect the physical and operational capability of the systems involved, as well as social,
legal, economic, and political factors that arise from the behavioral patterns, mores, and traditions of society.

An illustration of how various kinds of constraints might
affect a system is presented in Figure 10-13. Again referring
to the Bluewater River Basin example, the diagram focuses on
constraints that affect a particular water using activity, and
the relations between the influent water supply and the disposition of the effluent or return flow in terms of the possibilities for salvage and reuse, or treatment and disposal. For
example, several social or economic constraints may control
flows in the system, i.e., the water influent to the activity
and the quantities of water that could be salvaged for sequential reuse by another activity or recycled to the same activity.
At the same time, technological constraints on water use or
wastewater disposal processes will affect the level of pollutants added to or removed from the system.

Formulating a description for such a system requires,
therefore, an assessment of the types of constraints that may
govern its operation, and at what points they will exert their
influence.

Constraints may set minimum levels for certain systems
operations, maximum levels for particular resource usages, or
threshold levels for certain critical parameters. Constraints
should always be investigated as to their rigidity to see if
they can be altered, and within what limits, before arbitrarily
injecting them into the analysis as fixed or absolute.

Developing alternatives. Formulating alternatives is essentially a creative process of resource allocation. Particularly in respect to public projects, where multiple objectives
are present and it is necessary to strike a balance between objectives of harmony and optimality in project outputs, large
numbers of alternatives must be generated and tested in the
course of analyzing a problem.

Broadly speaking the kinds of alternatives to be considered in planning civil works projects involve structural (construction) and/or nonstructural (policy) aspects. Table 10-5
contains a brief summary of the dimensions which describe an
alternative's physical characteristics for structural solutions,
or policy configurations for nonstructural solutions. The descriptive dimensions include staging overtime, spatial features,

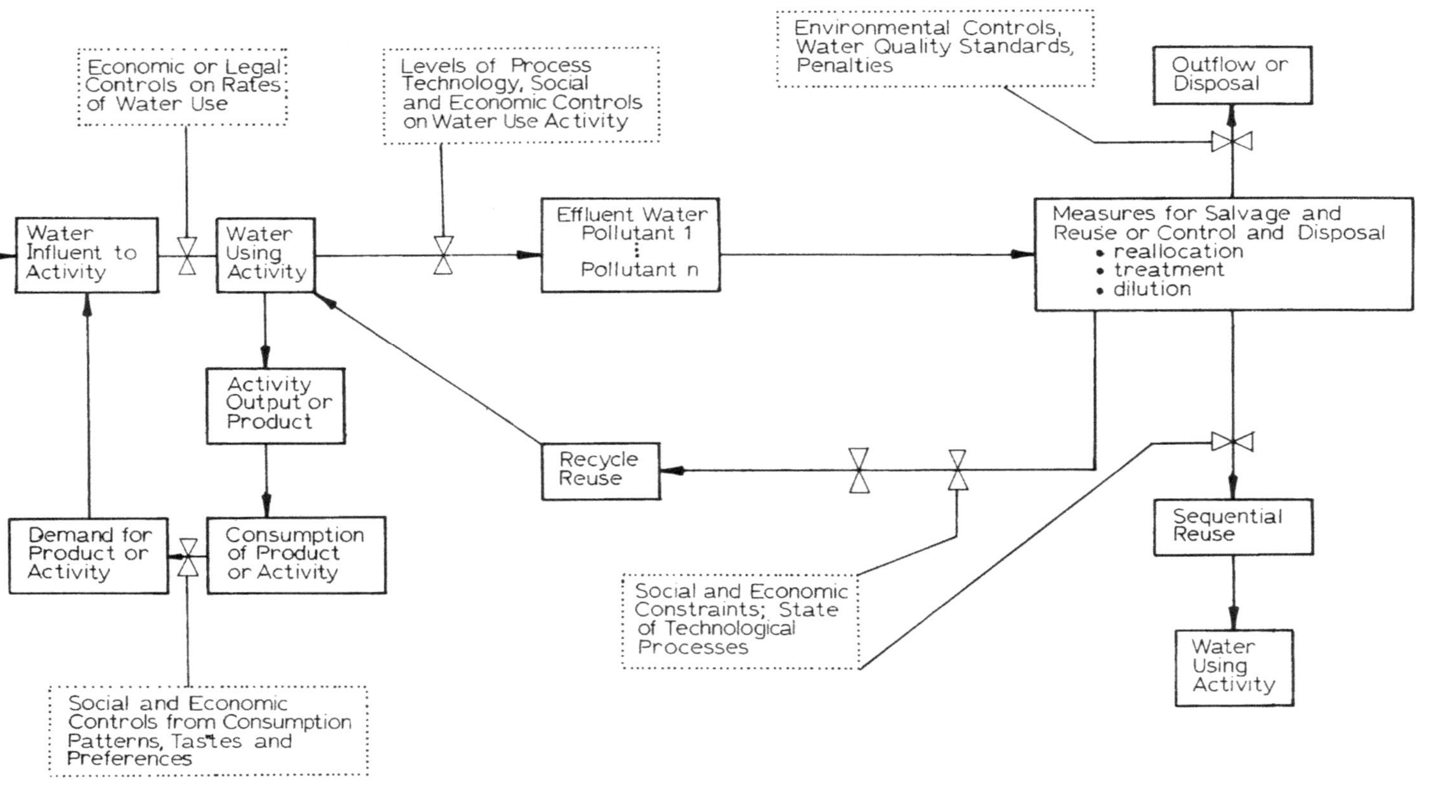

Figure 10-13. Relation of social, economic, environmental, and technological constraints on water using activities in the Bluewater Drainage Basin.

Table 10-5. Considerations in the formulation of alternatives.

Descriptive Dimensions	Configuration of Alternatives	
	Structural (Physical)	Nonstructural (Policy)
Staging over time	Construction/ operation	Phasing and implementation
Spatial features	Location and scale	Points of intervention in social, political or legal system
Operational arrangements	Operating rules	Administrative or organizational arrangements

and arrangements for implementation such as operating rules for physical alternatives or administrative and organization arrangements for policy alternatives. These dimensions also provide a framework for describing policy input variables to systems models for simulating or predicting the consequences of the alternatives.

To return to the example of transportation access to Pleasant Valley, a wide range of operational alternatives to complement the possible future could be conceived. Nonstructural policy alternatives might include:

1. Close the canyon to normal traffic and turn it into a park for walking, bicycling, picnicking, fishing, and other recreational pursuits.

2. Regulate traffic into the valley by charging a toll with a variable rate structure for cars, campers, boats, etc.

3. Undertake no development or construction on the existing route other than normal maintenance; in other words the status quo or do-nothing alternative.

These fairly simple statements of alternative policy plans already contain information about implementation, intervention in the socio-political system, and the administrative arrangements necessary for carrying out the policy. From this point, of course, much additional information would be developed to adequately describe the operational aspects of the alternatives.

Structural alternatives for transportation access could include:

4. Develop a full four-lane highway in the canyon with a 50 mph design speed to accommodate the high traffic volumes expected with future valley development.

5. Develop a loop road as a four-lane facility at a 60 mph design speed which will connect with the Interstate system to the south. The present canyon road will be maintained as is.

6. Limit development of the canyon road to providing left turn and passing lanes, some road and bridge widening, and straightening of curves.

7. Close the canyon to private vehicular traffic and provide access only via public mass transit facilities.

Detailed information on requirements for construction and operation including location and scale, would be needed as input information for systems modeling and evaluation.

In addition to these basic alternatives other combinations are conceivable and possible. Through systems analysis and modeling a comprehensive range of alternatives can be explored efficiently. In this respect, each of the possible alternatives cited above would have definite implications for the futures discussed previously for the Pleasant Valley example. These implications (as predicted by systems models), once understood, can help in assessing the degree to which a particular future will be facilitated by the implementation of an alternative.

Synthesizing a systems description. Systems models are a synthesis of statements delimiting the specific aspects of problems under study, statements of objectives, descriptions of alternative solutions, and theory and observation of the activities or resources involved. A useful heuristic device which helps to integrate this information in systems conceptualization is the use of questions of the form "If alternative x (or future condition y) occurs, then what are the effects on A, B, and C."

To draw again from the Pleasant Valley case, the following two questions--the first relating to a transportation alternative and the second to the water resources problem--provide an illustrative example:

"If alternative 4 (or 5) were implemented, how would this change in the existing transportation system affect (a) peak traffic volumes between the city and the valley, (b) cost and time of travel of the motorist, (c) use of land

and recreation resources of the valley, (d) the environment of the canyon (or loop road corridor), and (e) the quality of existing water supply sources?"

"If the city reaches a population of N (number), what will be the effect on (a) the adequacy of existing water supplies, (b) pollutant levels in the Bluewater River Basin, and (c) the management of wastewater, including salvage and reuse, in meeting water quality requirements?"

This questioning process is not an end in itself, but facilitates identification of affected variables, the exploration of major subsystems and the important components within each subsystem, and the time scale and spatial boundaries. The questioning process also leads to the formulation of a causal diagram of the system under study. The diagram then serves as a basis for systems modeling. Figure 10-14 is a causal diagram developed for the first question posed. As the questions are analyzed, links in a chain of cause and effect relationships are identified, along with various subsystems indicated in Figure 10-14 by the dotted boundaries. By testing temporal and spatial domains from the problem statements against the structure of the causal diagram, those endogenous variables which should be included within the system boundary are separated from the variables that are exogenous or external to the system. It should be re-emphasized that one significant purpose of the earlier chapters of this volume is to provide knowledge about basic subsystems so they can be integrated into an adequate system description.

Systems modeling: methods and techniques. Broadly speaking the purpose of modeling may be either predictive or prescriptive. Predictive models of systems are constructed to clarify the internal structure of a system and predict its behavior under various manipulations. On the other hand prescriptive models strive not only to reproduce the results of the system itself, but also to evaluate the consequences of alternatives using a predetermined measure of performance.

Identifying the model structure for predictive or prescriptive models must be based either on formal theory or at least some very strong plausibility arguments. Systems models cannot be devised by simply using statistical manipulations of data and information to determine variable interactions. Moreover, all the information relevant to the system may not be quantifiable as numerical data. Hence, systems modeling techniques for both quantitative and nonquantitative models will be reviewed.

Table 10-6 provides a general classification of modeling methods and techniques useful in systems analysis. The entries in the Table are classified under the heading of predictive or

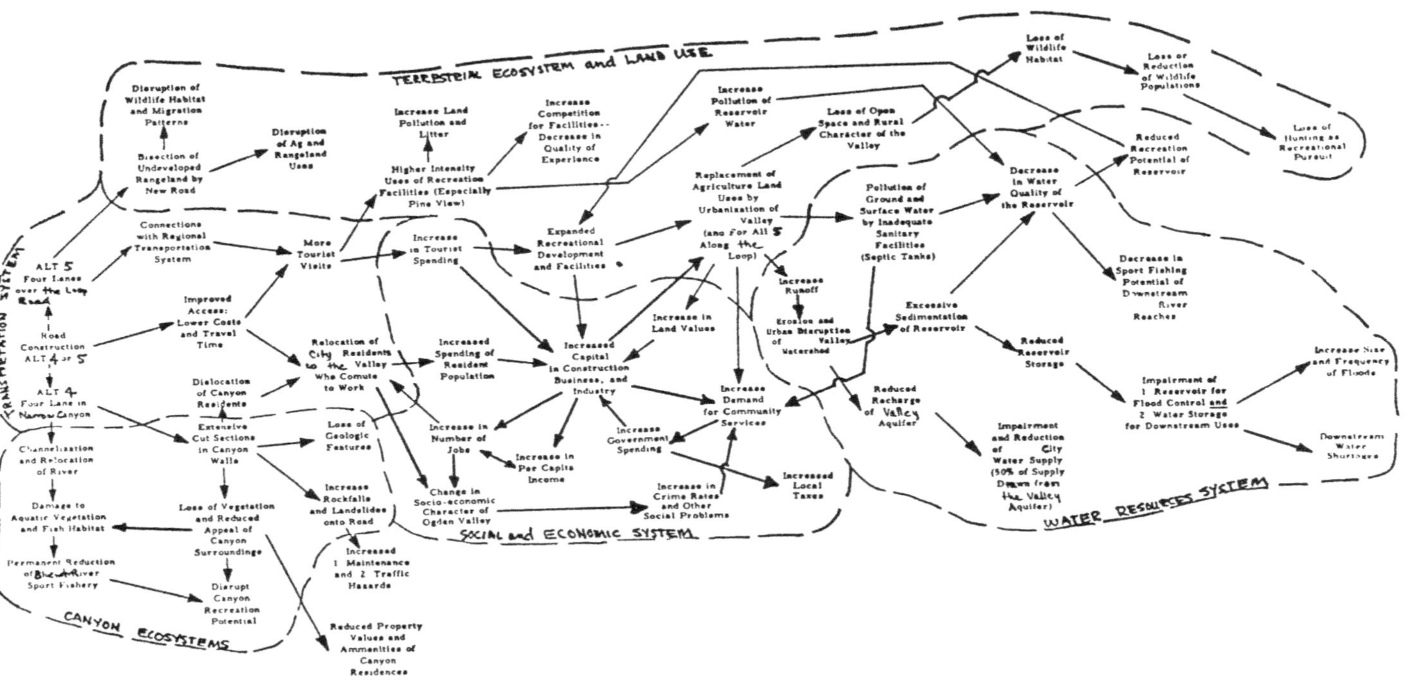

481

Figure 10-14. Cause-effect diagram for consequences of highway alternatives.

Table 10-6. An overview of approaches to systems modeling.

PROBLEM TYPE	APPLICATION OR USE	
	Predictive	Prescriptive
Quantitative	(See Figure 10-15 for further classification)	
	Systems Transformation algebraic equations, differential equations, state variable formulations, input-output analysis	*Optimization Procedures* *Classical Optimization* Theory; differential calculus, lagrangians, optimal control theory
		Mathematical Programming: linear, non-linear, dynamic, integer and stochastic programming
	Networks graph theory	*Networks* CPM and PERT
Stochastic	*Stochastic Processes* Inventory theory, queueing theory, Markov processes	*Decision Analysis* Statistical (Bayesian) sion theory, game theory
	Statistical Models Regression analysis, component and factor analysis, stepwise multiple regression, discriminant analysis, econometric analysis	
Simulation	Deterministic and stochastic model components and models	Monte Carlo methods, search techniques for dominant solutions
Non-quantitative	*Verbal Models* Scenarios, survey research	*Verbal Models* Delphi inquires
	People Models Role playing	*People Models* Operational gaming

prescriptive models and according to whether the theoretical
basis for model construction is quantitative or nonquantitative.
The remainder of this section is organized along the lines indicated in Table 10-6. It is of course impossible, and unnecessary, to provide detailed descriptions of the various modeling approaches. Whole textbooks are devoted to these subjects.
Instead, this discussion tries to provide a basic understanding
of the techniques and some idea of their applicability to public projects.

Quantitative Models

The network delineation of the model structure, as discussed earlier, will aid in determining whether a quantitative
model is appropriate. Figure 10-15 is a pictorial representation of a "modeling space," with each axis representing an important dimension for quantitative types of models. Depending
on whether variable relationships are probabilistic or deterministic, static or dynamic, and linear or nonlinear (as represented by the corners of the cube) various analytical techniques are required to handle them. Table 10-7 summarizes the
attributes of quantitative models as a guide to identifying
suitable analytical methods. Early in the systems analysis
process, gross models will likely be sufficient; later more refinement will be needed.

It should be pointed out before examining specific modeling techniques that is is unlikely that any one model will be
able to answer all questions associated with a particular systems study. Often, in fact, efficiency is achieved by using
different models for different subsystems and hierarchical
levels, and to answer different questions. These problem-
specific models may subsequently be linked to provide analysis
of broader systems questions, such that the outputs of one
model are the inputs to another. This technique of "nesting"
models usually follows the lines of the hierarchical relationships of subsystems and systems. Referring to the hierarchical
systems arrangement presented in Figure 10-4, separate models
might be developed to analyze the operations of the municipal
and of the agricultural water supply systems. At the basin
level, however, these models would be linked together in modeling the total water resources system.

To summarize, quantitative models may either describe and
predict the behavior of the system, or provide prescriptive
solutions for decision makers by identifying preferred alternatives based on predetermined criteria.

For simplification, the interface and overlap between
models and evaluation of alternatives will not be treated here.
Instead models and their basic solution outputs will be treated
in this section, while the optimality of prescriptive solutions

Table 10-7. Attributes of quantitative models.

Attributes of Quantitative Models	Time Variance		Random Variance		State or Domain Variance		Functional Variance	
	Static	Dynamic	Deterministic	Probabilistic	Single Variate	Multivariate	Linear	Nonlinear
Systems Transformation								
Linear, nonlinear systems								
1st order diff. eq.		X	X		X			
State variables		X	X			X		
Input-output analysis	X		X			X		
Optimization								
Classical optimization								
Differential calc.								
(no constraints)	X		X			X		X
Lagrange Mult.								
(w/constraints)	X		X			X		X
Optimal control theory								
Mathematical programming								
Linear prog.	X		X			X	X	
Nonlinear prog.	X		X			X		X
Integer prog.	X					X	X	
Dynamic	X	X				X		X
Stochastic	X			X		X		X
Stochastic								
Inventory				X				
Queuing		X		X	X			
Markov		X		X		X		
Multivariate	X			X		X		
Networks								
Graph Theory								
CPM & PERT								
Simulation								

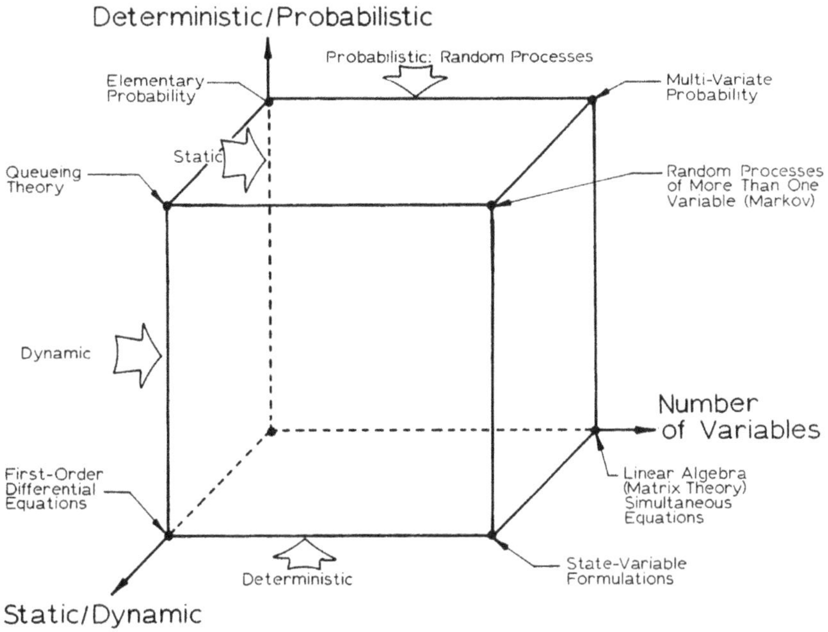

Figure 10-15. Conceptualization of a "modeling space."

will be treated in the subsequent section in the context of
basic considerations in the evaluating of alternatives.

Systems transformations: deterministic equations in systems modeling. An equation or set of equations can constitute
a mathematical model of the relations among system components
or variables, and describes the transformation of inputs to
outputs. For deterministic problems, such equations take various mathematical forms as represented by the bottom corners of
the modeling space in Figure 10-15, viz. algebraic equations,
sets of simultaneous equations, differential equations, and
state-variable formulations. These mathematical models have
been useful in describing and predicting various aspects of
systems transformations important to the design of public projects. For example, a conceptual aspect of virtually all public
projects is that of the production function. A production
function is a mathematical model that represents the transformation of resources into products or project outputs. The general form for the production function is:

$$0 = f(r_1, \ldots r_i \ldots r_n) \tag{10-2}$$

in which 0 is the project output of goods and services and r_i's
are the input resources or production requirements. The function can be represented by the simple block diagram of Figure
10-16.

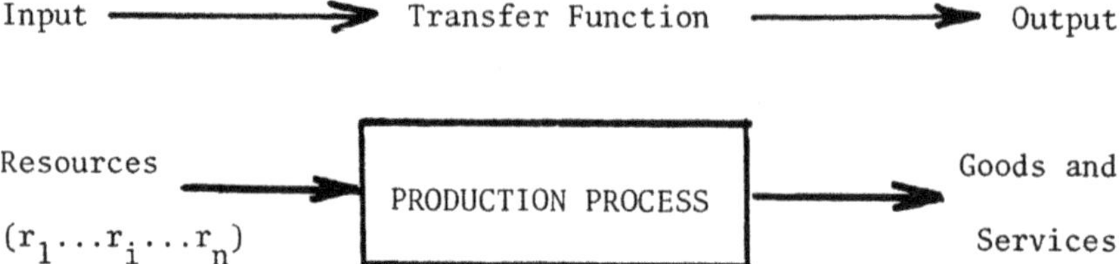

Figure 10-16. Block diagram of production function.

The relationship represented by equation 10-2 and the block diagram in terms of purely physical quantities--units of output for given combinations of input resources. Some simple examples of production functions that can be represented as an algebraic equation are the power output from a hydro-generating station as a function of the head and flowrate, or streamflow as a function of rainfall intensity and duration over a drainage basin, or the amount of BOD removal in a wastewater treatment system as a function of influent BOD concentration and detention time in treatment.

Systems with several interrelated variables requires sets of simultaneous equations for their description. An example, somewhat analogous to the production function, is the economic input-output model which describes how inputs are combined to produce several outputs in order to meet the demand for goods and services by society. The input-output approach has been further extended for use in forecasting water demands and waste residuals for production processes, two important facets in planning many public projects.

For dynamic systems, where a number of the parameters vary with time, differential equations and state variable formulations provide the basic mathematical model. For example, differential equations can be used to model water quality parameters that change with time, and to model ecological relationships where the chemical and biological behavior of the ecosystem varies with time. State variable formulations find useful applications in describing dynamic systems of several variables. This concept can be illustrated by referring again to Figure 10-16. Assume that the internal state of the system (the transfer function or production process) can be measured by a set of state variables $\bar{q}$. The time rate of change of the state variables then is

$$\frac{d}{dt} q(t) = f(\bar{q}(t), \bar{r}(t)), \qquad (10\text{-}3)$$

a function of the input vector $\bar{r}$ and the level of the state variables $\bar{q}$. The system output 0, then is given by

$$\bar{0}_{(t)} = g(q(t), \bar{x}(t)) \qquad (10\text{-}4)$$

where $\bar{x}$ is the vector of inputs. These two equations thus relate the system input to the output through the state variables $\bar{q}$. State variable formulation have also found applications in water quality modeling (Thomann, 1971).

Stochastic models of systems. Many of the phenomena that require quantitative treatment in planning public projects are stochastic or random in nature. A few examples include precipitation and streamflow, arrival of vehicles at a tollgate or the left turn lane of an intersection, or the fluctuation of water quality parameters in a stream. The theory of probability, statistics, and stochastic processes, as denoted by the corners of the top face of the modeling space in Figure 10-15, provides the basis for modeling such random events in systems analysis.

Statistical methods. Effective modeling in many system studies requires the ability to manipulate and analyze large quantities of data. Statistical methods in such cases help establish or estimate from data the fundamental relationships among variables to serve as descriptive models. They are also used to test the reliability of models and the representativeness of the data. The following briefly summarizes some key techniques of statistical analysis which might find useful application in systems modeling or as systems models in their own right.

Regression analysis is used to establish a relationship or model that will allow the estimation of a dependent variable as a function of a set of independent variables. This amounts to fitting an equation of the form

$$Y = f(X_1, X_2, \ldots X_n) \qquad (10\text{-}5)$$

to a set of observation of the variables $(Y, X_1, X_2, \ldots X_n)$. As is implied in the specific case of linear regression analysis, an effort is made to fit a linear equation to the data of the general form

$$Y = a_o + \Sigma a_n X_n, \; n = 1, 2, \ldots N \qquad (10\text{-}6)$$

in which a_o is a constant and a_n's are the coefficients of the independent variables. The function implies a causal relationship among the variables which can be diagrammed as

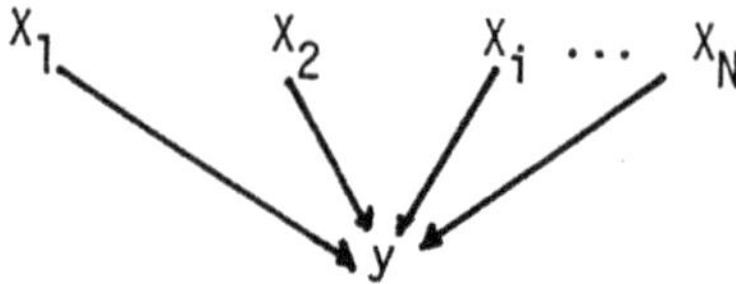

indicating that each variable, X_i, affects Y independently and linearly.

Conversely, regression analysis provides a test to determine if sets of observations are related. To isolate an example from Figure 10-14, suppose it is desired to know if a decline water quality levels of the reservoir is related to the intensity of recreation use of the reservoir and the level of development in the valley as measured by the number and/or size of septic tank disposal sites. The causal diagram adapted from Figure 10-14 is shown as,

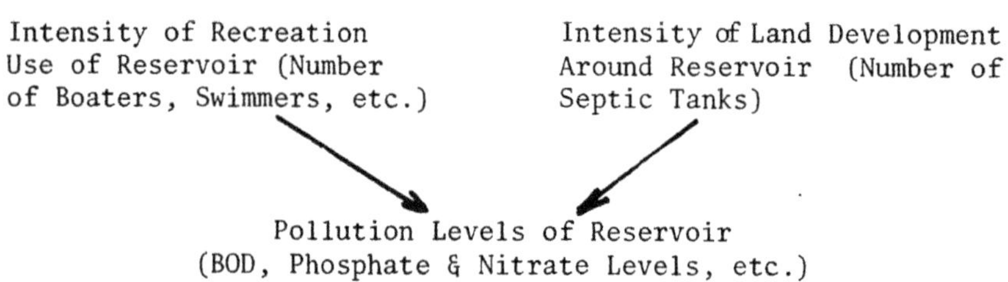

and after a sufficient number of observations are made on the variables indicated, then a regression analysis can be performed to find if in fact a relationship does exist and what it is.

Multivariate techniques refer to several statistical modeling tools used for describing the interdependencies between system variables. Variables to be related in a systems model are often highly dependent on one another (although often referred to as "independent" variables), and the objective of multivariate analysis is to determine the individual contribution of one variable in an entire set of interdependent observations involving many variables. The purpose of *component analysis* is to reduce the number of variables needed to represent or model a phenomena, and as an added bonus, to rank the importance of the several variables in the model description. As equation 10-7 shows, the model is based on the assumption that variables are additive. If k is the number of original variables, y, the procedure constructs a new variable, Z_i, as a linear combination of k (or fewer) of the original variables.

$$Z_i = a_1Y_1 + a_2Y_2 + \ldots + a_kY_K \tag{10-7}$$

Further Z_i's are constructed until as much of the variance is accounted for as feasible. Referring to the previous example, if several variables have been observed to relate to "recreation intensity" and "land development" component analysis might be used to collapse these into simpler diagrams of a few new key variables, and then a regression analysis could be performed on these components to develop estimating functions for models. While component analysis tries to reduce the complexity of a problem, *factor analysis* attempts to identify cause and effect relationships in a set of variables. In very gross terms, an equation of the form $X_1 = a_1f_1 + a_2f_2$ is estimated for a variable X_1 using certain "factor" components, f_1 and f_2, which are combinations of the variables y in the set. However, the dependent variables appearing on the left of the equation is not specified, so the coefficients, a_1 and a_2, derived in the model represent influence values, or the percent change of the variable, X_i, per unit change of a factor on the right-hand side. Finally, discriminant analysis tries to decide to which of several alternative populations an individual should be assigned where characteristics of the populations overlap. Once the discriminant function is estimated, the sample characteristics of the individual can be entered in the function and a value calculated for comparison with population means to determine where it should be assigned.

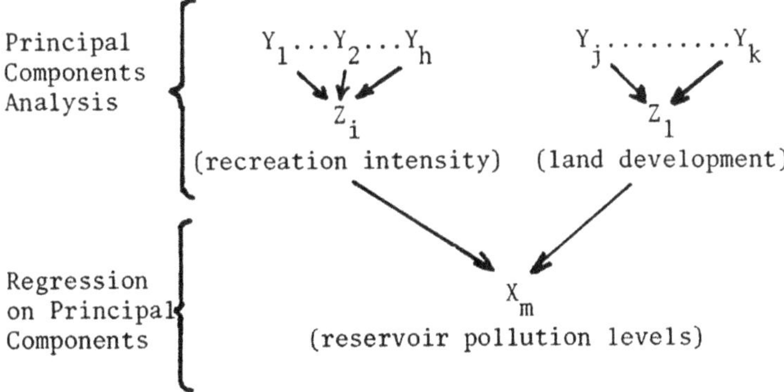

Econometric analysis. Econometrics is the statistical branch of economics which in recent years has given considerable attention to the problem of simultaneous interactions of variables in a system and how to statistically estimate their parameters. Because the interactions of variables, usual regression techniques are not adequate. Thus, econometric analysis goes beyond the correlation models of multivariate techniques to complex causal modeling with sets of simultaneous equations. Without the detailed theory and application of econometric techniques, an example from the case study provides some further insight into this modeling approach. A causal model consists of a set of independent equations, relating several interdependent variables, which can be solved simultaneously. Each independent equation is a structural equation which represents some technological or behavioral aspect of the system. In other words independent equation must have a sound theoretical basis of its own. The economic subsystem in Figure 10-14 illustrates the basic approach. The four variables outside or exogenous to the subsystem and the seven variables endogenous or internal to the subsystem would be related by eleven independent equations which are theoretically sound and statistically significant, as estimated using econometric methods. For example one equation in the estimated set might be:

> Capital Construction = f (tourist spending, resident spending, and government spending),

and so on.

Stochastic processes. A stochastic process is a random phenomenon which is developing in time in a manner controlled by probabilistic laws. Hence, the key element differentiating stochastic processes from other statistical models is the parameter of time. While a number of stochastic processes have been studied in nature, two classes of problems common to business management have particular application to the public projects. These are inventory theory and queuing theory.

Inventory theory relates to stochastic analysis in the management of inventories or stocks in business. The kinds of applications of inventory models to public projects is illustrated by the following example. Without resorting to a mathematical description of the problem, suppose that the recreational use of the reservoir in Pleasant Valley depends on maintaining the lake level within a given range. Holding reservoir levels within specified limits amounts to maintaining water storage "inventories" over time which are sufficient to meet the contingencies of randomly fluctuating inflows and random draft requirements arising from water demand for downstream uses. The sketch diagram in Figure 10-17 depicts the situation.

The stochastic inflow to the reservoir is i , and the draft to meet sotchastic water demand for various uses is d . The initial storage is denoted by S_i and final storage at the end of a time period by S_f . M is the maximum storage level possible. For any time period then the change in storage ΔS is:

$$i - d - \Delta S = S_f - S_i, \quad S_f \leq M .$$

With this basic relationship and the appropriate probability distributions for i and d , the "inventory" problem to

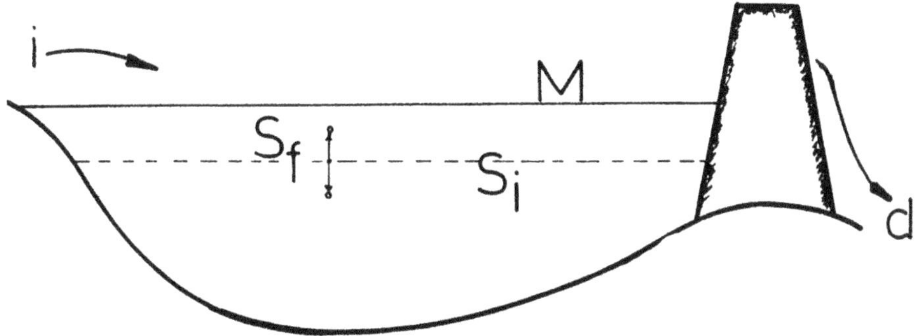

Figure 10-17.   Pleasant Valley Reservoir inflows, storage and releases.

be analyzed is how to obtain a stationary distribution for reservoir storage (i.e., within the levels required for recreational use). Other analyses might also consider costs of maintaining certain storage levels versus the cost of unsatisfied
demand for downstream uses.

Queuing theory provides a mathematical description of random processes which lead to the formation of queues or waiting
lines. Queuing theory also provides a useful model for studying
problems arising in public works planning. Applications are
often found, for example, in study of traffic flow.   To illustrate briefly, consider the situation of the present two lane
road in Narrow Canyon, as described in the case example.   At
several points in the canyon, access to recreational areas requires a left turn.   If there is oncoming traffic the turning
vehicle is obliged to wait, along with other vehicles which arrive in the interim, thus forming a queue.   Some of the questions that can be analyzed by queuing theory include: What is
the probability that a left-turning vehicle will have to wait,
or that other arrivals will be faced with a queue? and also,
What would be the expected waiting time for the left turning
car, or the waiting time for other vehicles arriving in the
queue?   A study of these questions could be very important in
the analysis of Alternative 6, which calls for providing left
turn and passing lanes at certain points in Narrow Canyon, and
what difference it might actually make in the traffic flow.

Optimization models and procedures. The problem of optimal
policy analysis, stated in general terms, is to find the values
for a set of decision variables $(x_1, x_2, \ldots x_n)$ which maximizes
(or minimizes) an objective function V relating the decision

variables, the system states $(s_1, s_2, \ldots s_i)$ and parameters $(p_1, p_2, \ldots p_j)$, subject to any constraints on the decision variables, states and parameters. In symbolic form:

$$V = \max \nu \; (x_n, \; s_i, \; p_j) \tag{10-8}$$

subject to constraints

$$f(x_n, \; s_i, \; p_j) \leq 0 \tag{10-9}$$

This general conception of prescriptive optimization models circumscribes a number of analytical procedures which have direct application to public project analysis.

Classical optimization methods. The classical optimization methods are familiar tools of differential calculus. If the functions specified by equations 10-8 and 10-9 are continuously differentiable, extreme points or optima are found by taking the partial derivatives of the function with respect to the decision variables, setting them equal to zero and solving the resulting set of simultaneous equations. This procedure is modified somewhat for the constrained system by introducing an unknown function called the Lagrange multiplier. While calculus methods may find application in optimization of smaller subsystems, the development of satisfactory objective functions and computational difficulties limit their use. They do, however, provide the basis for other more powerful techniques of optimal control theory, gradient search procedures, and steepest-ascent methods for finding optima. For example, in water pollution control studies Goodman and Dobbins (1966) proposed a simultaneously equation model with optimal cost plans identified by steepest-ascent algorithm, and Fan et al. (1971) examined cooling water discharge from power plants as a nonlinear distributed parameter system using optimal control analysis to determine the most efficient operating policy for the plant without violating stream water quality standards.

Programming models. A large group problems which cannot be analyzed with the classical optimization techniques are solvable through mathematical programming. Programming models have the same general format as equations 10-8 and 10-9, where the problem is to determine the values for a set of decision variables which maximize or minimize a given function of the decision variables while, at the same time, satisfying a set of constraints on how the resources can be used or combined. The functions, however, need not be continuous, the constraint equations do not have to be equalities, and the number of variables treated need not agree with the number of equations in the system. This format is the basis for many resource allocation problems, and there are numerous applications of programming models in analyzing public projects as a mechanism for resource management and allocation.

In *linear programming* models, both the objective function
(relating the set of decision variables) and the constraints
(specifying how resources may be combined) must be linear. An
example of the structuring of a linear programming model is
drawn from the water resources management aspects of the Bluewater Basin in the case example. The identification of water
management problems pointed out the need for additional water
supply for Thriving City and the contiguous agricultural land,
and at the same time cited the deterioration of water quality
in the Bluewater River. The network flow diagram for the water
use and wastewater flow system appeared in Figure 10-10, and a
matrix description of the system in Figure 10-11. By recasting
Figure 10-11, the components of the water resources system can
be interpreted as supply sources and demand requirements (Figure 10-18) illustrating that water from several origins of supply can be allocated to satisfy the demand of various water
using sectors. Each row heading in the matrix represents a
possible origin of supply. The systems of water users, indicated by the column headings, are grouped by sectors of municipal, industrial, agriculture demand, and other uses. Note that
the unconsumed effluent from each demand sector is available for
reuse in the system. For example, the agricultural sector is
both a destination of supplies from various sources, and the
origin of irrigation return flow which can be reused in supplying another sector. Hence, destinations of use (columns) are
also origins of secondary supply (rows). Effluent from any demand sector can be allocated to another sector for sequential
reuse, or to the same sector for recycle reuse, thus interfacing both water supply and water quality considerations in the
formulation. In the context of broad system planning, the matrix of water supply sources and sector demands depict all possible combinations for satisfying the aggregate system demand
with the aggregate available supply. Thus, each element in the
matrix represents a possible means of satisfying all or part of
the demand requirements of a sector with all or part of the
water from a given source.

The matrix of Figure 10-15, illustrating the components of
the water supply system, is analogous to the transportation
problem from linear programming. In the general transportation
problem (Gass, 1964, pp. 193-214), a homogeneous product, water
in this case, is supplied in the amounts a_1, a_2,...a_m from each
of m origins and is demanded in amounts b_1, b_2,...b_n by each
of n shipping destinations. The structure of the problem is
set out in Figure 10-19, where each supply origin-demand destination pair has a cost and an associated decision variable. The
decision variable x_{ij} represents the amount shipped from the
i^{th} origin to the jth destination. The cost of shipping a
unit amount from the ith origin to the jth destination is
c_{ij}.

Supply Origins	Demand Destinations	Municipal	Industrial	Agricultural	Wildlife Refuge	Treatment and/or System Outflow	Supply Availabilities
Primary Supply	Surface Water			initial allocation of primary supply			annual reservoir storage
	Groundwater						annual recharge
Secondary Supply	Municipal Effluent	recycle reuse	sequential reuse	sequential reuse	sequential reuse	sequential reuse	municipal waste system outflow
	Industrial Waste	sequential reuse	recycle reuse	sequential reuse	sequential reuse	sequential reuse	industrial wastewaters
	Agricultural Waste	sequential reuse	sequential reuse	recycle reuse	sequential reuse	sequential reuse	irrigation return flows
Treatment Operations	Wastewater Treatment	sequential reuse	sequential reuse	sequential reuse	sequential reuse		plant capacity
Supplementary Supply	Imported Water			allocation of supplementary supply			annual importation
	Use Sector Demand	municipal demand	industrial demand	agricultural demand	refuge demand	treatment on down-stream outflow	Totals

Figure 10-18. Allocation alternatives for the water resource system.

Destinatons

i\\j	(1)	(2)	·	·	(j)	·	·	(n)	
(1)	c_{11}	c_{12}			c_{1j}			x_{1j}	a_1
·									
·									
(i)	x_{i1}	x_{i2}			x_{ij}			x_{in}	a_i
·									
·									
(m)	x_{m1}	x_{m2}			x_{mj}			c_{mn}	a_m
	b_1	b_2			b_j			b_n	

Figure 10-19. The transportation problem tableau.

The costs incurred in allocating water from any origin to
any destination depend on the water quality of the source, the
quality requirement for the use, and the facilities required to
treat, transport, and deliver the water.   The objective is to
determine the amounts X_{ij} allocated from the supply sources
to meet use sector demands so as to minimize costs. The mathematical statement of the linear programming problem is to find
values for the variables X_{ij} which minimize the total cost,
TC:

$$\text{Min TC} = \sum_{i=1}^{m} \sum_{j=1}^{n} c_{ij} \, x_{ij} \tag{10-10}$$

subject to the constraints:

$$\sum_{j=1}^{n} x_{ij} \leq a_i \quad i = 1, 2, \ldots, m \tag{10-11}$$

$$\sum_{i=1}^{n} x_{ij} = b_j \qquad j = 1, 2, \ldots, n \qquad\qquad (10\text{-}12)$$

and $\quad x_{ij} \geq 0$

The constraint equation 10-11 states that the sum of the water allocations x_{ij} over category i must be less than or equal to the available supply a_i from that category, while 10-12 states that the amount of water x_{ij} allocated to sector j must equal the use sector demand b_j .

The use of the linear programming model for the assessment of water supply and wastewater system alternatives seeks prescriptive answers, in terms of minimizing costs, to the questions: Which supplies should be allocated to satisfy which demands? In what amounts? and, At which points in time? A similar set of questions would apply to wastewater management. These questions regarding system alternatives are explored through optimal solutions to the model obtained through linear programming procedures. An example solution for the Bluewater River Basin problem for a forecast demand for the year 2000 is given in Figure 10-20. The entries in the matrix represent the optimal levels of the decision variables x_{ij} indicating the quantities of various supplies allocated to satisfy the sector demands. Converting the matrix information to the network flow diagram of Figure 10-21 gives a picture of how the optimal solution looks when implemented in the system. By solving the model for forecast demand levels of future periods, a number of alternative system configurations can be examined including time staging and redistribution of supplies, timing and capacity of wastewater treatment operations, and the effects of various policy decisions on system operation.

Applications of linear programming in public projects analysis are found throughout resource management literature. The example presented here is discussed in detail by Bishop and Hendricks (1971a,b). Examples of other applications include optimum design of a reservoir system (Stephenson, 1970) sewage treatment plants, wastewater collection systems, river basin water quality systems and water supply systems (Deninger, 1970).

Nonlinear programming models are of the same general format as other optimization models, but the objective function and the constraints have to be expressed as nonlinear functions. No general efficient solution algorithm exists for the nonlinear programming problem although some are available for special cases. For example, Carter et al. (1971) developed a separable convex programming model to determine optimal low flow augmentation releases for maintenance of stream quality.

Supply Origins \ Demand Destinations	Municipal	Industrial	Agricultural	Wastewater Treatment (Including tertiary)	Wildlife Refuge	River System Outflow	Available Supply (1000 F)
Surface Water and Reservoir Storage	83						83
Wells in Valley Aquifer	48						48
River Reach Valley to City		81		48	141		270
Municipal Effluent		209				4	213
Industrial Wastewater			203				203
Agriculture Return Flow			6	74			80
Wastewater Treatment Plant							122
Wildlife Refuge	122					65	65
Demand Level	253	290	209	122	141	69	Total Demand = 893,000 AF Avg. Cost/AF = $26.90

Figure 10-20. Year 2000 allocation matrix for Bluewater Basin.

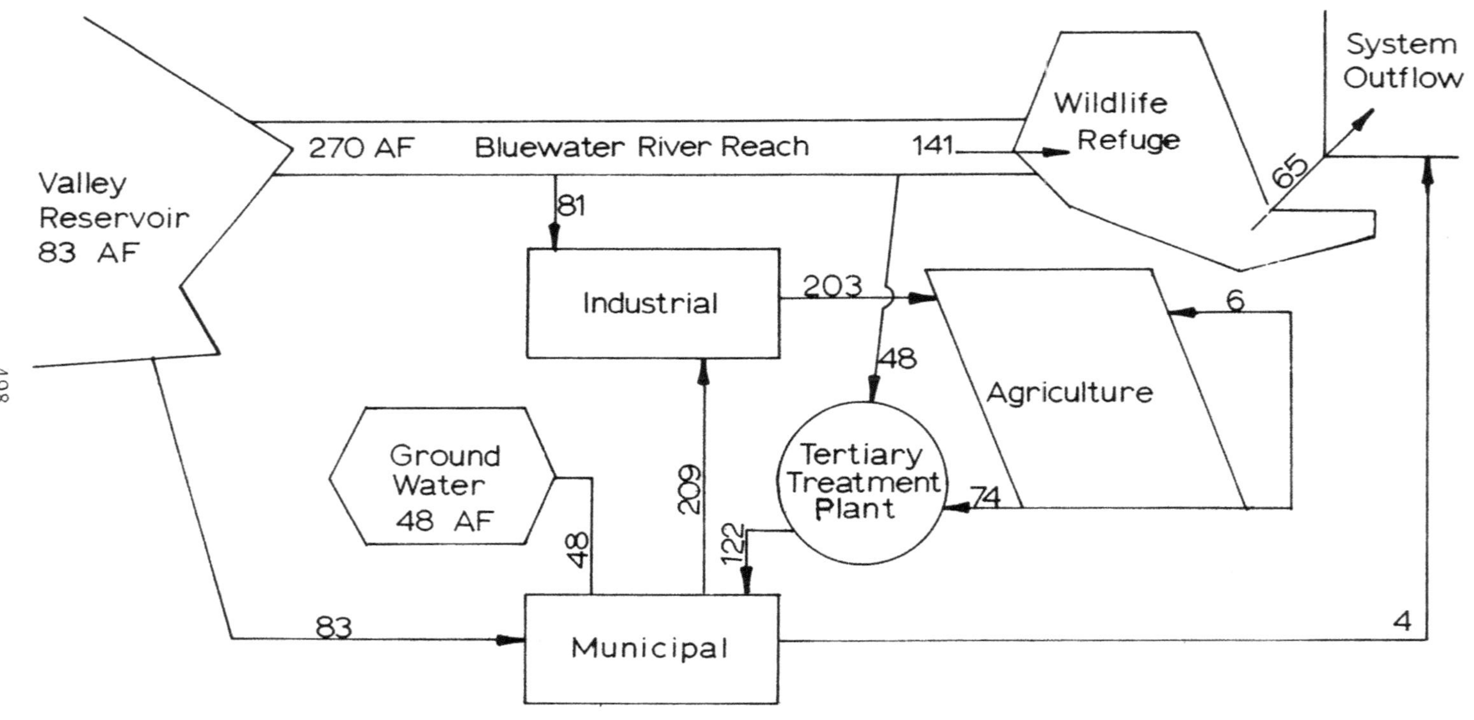

Figure 10-21. Schematic of year 2000 allocation pattern for Bluewater Basin.

Another variation of the basic programming model is *integer programming* where the decision variables in the model are required to be whole numbers. A typical example of such cases is where capacity or production processes, such as electrical generators, must be added in whole units. Hughes (1972) used a mixed integer programming model for determining the operation and time sequencing of sources for a city water supply system.

Dynamic programming is a mathematical technique useful for modeling problems where a number of interrelated decisions can be made in sequence. In effect, dynamic programming is an efficient trial and error procedure for selecting an optimal combination of decision variables. The structure of problems which can be cast in a dynamic programming model is diagramed in Figure 10-22. The problem is composed of several "stages" and within each stage there are a number of "states" representing the possible conditions of the system at that stage. The connecting arrows signify those states in each successive stage which can be attained from a preceding one. The problem is to determine the sequences of states (decisions) to go from stage to stage that will render an optimal solution according to some objective criterion such as minimizing costs. This can be seen clearly by reworking the model structure into the "tree" structure also diagramed in Figure 10-22. Seen in this light the problem is to select the optimal branch. To mathematical formulation of the problem requires an objective function and a recursive equation which is used to calculate the optimal decision at the last stage. By working back through the decision stages the set of optimal decisions for the entire problem is determined. The objective function and the constraints need not be linear or continuous. Hall and Dracup (1970, p. 83) point out that water resources systems have properties which often make them more amenable to dynamic programming models than other modeling approaches. A number of articles in the public works literature illustrate the use of dynamic programming, including optimizing conjunctive use of ground and surface water (Aron, 1971), optimizing reservoir releases for water quality control (Jaworski et al., 1970), waste treatment plant design (Shih and DeFilippi, 1970), new links in a transportation system (Tillman, 1970), and investment in solid waste disposal facilities (Clark, 1970).

Decision analysis. Decision analysis, as it fits within the framework of systems models, provides a logical and quantitative structure for considering the elements of risk and uncertainty in the outcomes or consequences of alternative decisions. Risk or uncertainty in determining model parameters or project consequences may arise from many sources, e.g., future demand for facilities, availability of resources, changes in technology, changes in goals and values of society, and actions taken by other decision makers. Decision models to deal with these problems are of two basic types--statistical or Bayesian

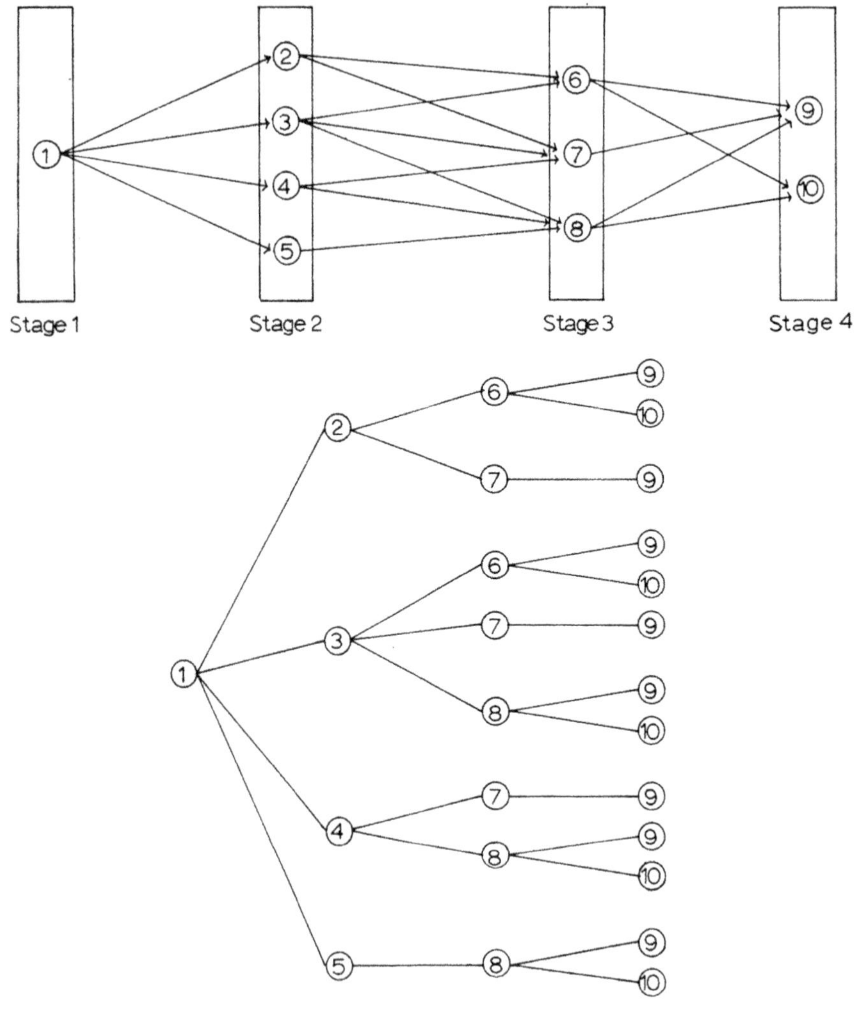

Figure 10-22.   The dynamic programming problem.

decision theory and game theory. The former is applicable to
cases where the probability of outcome is due to the state of
nature, and the latter to cases where the outcomes are affected
by other decision makers under circumstances where their joint
decisions determine the outcomes for each. These are prescriptive models since they will identify a preferable course of action for a decision maker based on the criterion or decision
rule which the decision maker specifies.

Statistical (Bayesian) decision theory. Statistical or
Bayesian decision theory is a useful method of evaluation for
dealing with situations of risk and uncertainty. The basic
analytical model is the decision tree, illustrated in Figure
10-23. The "X" represents the decision point with alternatives
$A_1 \ldots A_n$ open to the decision maker. Given the decision, there
is a set of possible outcomes $0_1 \ldots 0_n$ with the respective probabilities of occurrence P_1 through P_n. Each outcome can be
assigned some value of Payoff (V_i) in monetary or some other
unit that measures the relative utility of the outcome to the
decision maker. One of several criteria for decision making is
to select the alternative which maximizes the expected value or
utility of the outcome. Expected value for each alternative can
easily be calculated from the tree as:

$$E(A_1) = \sum_{i=1}^{n} P_i V_i (0_i) \qquad\qquad (10\text{-}13)$$

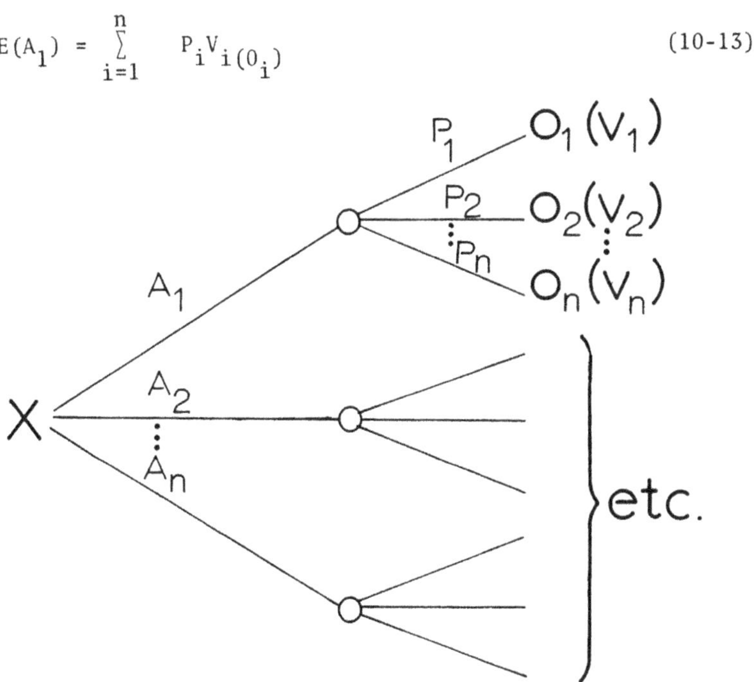

Figure 10-23. Decision tree.

To apply statistical decision theory, the decision maker constructs a decision tree with the alternatives or actions that are open to him, and the possible outcomes given each of those alternatives. The next two steps are the critical ones insofar as the prescriptive value of the answer is concerned. First, a value or utility for each of the outcomes is assigned. These can be dollar values such as total annual cost, or some other units that are value related. Second, a probability for each combination of action and outcome is assigned. Probability assignments based on data would be considered more objective, but if data are not available, then subjective probabilities are delivered by judgment of the engineer or planner and the decision maker. Using the probability and the value for each action-outcome-payoff branch, the expected value for each action is calculated. The action achieving the highest expected value would be selected as the preferred alternative.

Game Theory

Game theory models describe the interaction of two or more decision makers in a decision situation. The models are mathematical abstractions relating the players choices to the outcomes or payoffs that can be mutually expected as each player adopts his course of action. The "game" situation is conveniently represented by a payoff matrix, such as the one for two players in Figure 10-24. The row headings A_1, A_2, ...,A_n designate the decision options open to player A , and similarly the column headings B_1, B_2,...,B_m for player B. The value of the payoff V_{ij} to player A , or conversely the loss to B , for each combination of decisions by A and B is entered in the matrix. Thus if A chooses A_1 and B chooses B_1 , then A gets payoff V_{11} or B loses V_{11} . Several different criteria have been proposed as a basis for decision evaluation in such a game. The maximum criterion says to choose the alternative that maximizes the minimum payoff. The maximax criterion suggests maximizing the maximum value of the outcome. Other criteria have been suggested which lie between these extremes of pessimism (maximin) and optimism (maximax).

Therefore, the outcomes depend not only on the structure and dynamics (one-shot as compared to an iterative) of the situation, but also on the strategic intent of the players. It is interesting to consider the possibility of abstracting some aspects of a planning problem to a gaming situation. For example, in the alternative to construct a 4-lane facility in Narrow Canyon where conflicts of interest exist, three different strategic attitudes may be encountered in dealing with the public and interest groups:

1. Self-interest--other groups are not hostile to your desires; they are merely interested in their welfare and not directly concerned with your desires.

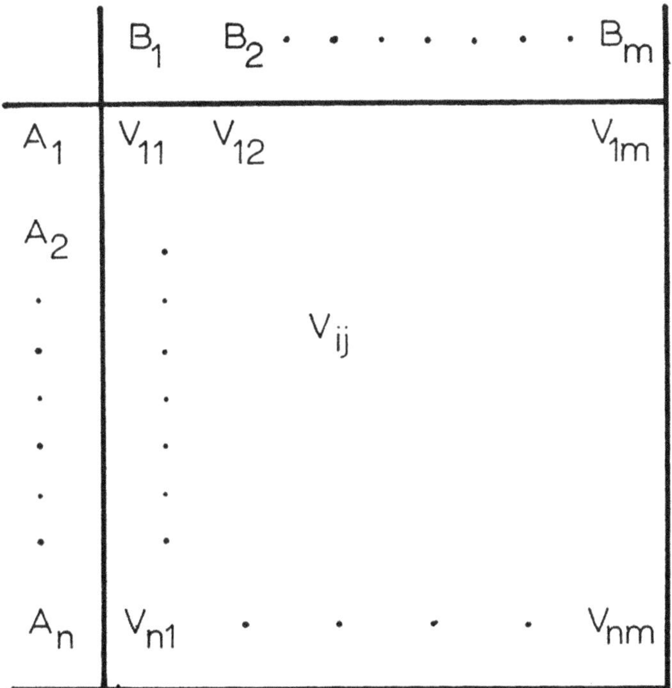

Figure 10-24.   Payoff matrix for two person game.

2.   Competitive--the opponent is hostile to your desires.

3.   Cooperative--The cooperative solution is essentially the
     spirit behind welfare economics. The idea is that there are
     gains to be had from cooperation and rational individuals
     will realize these gains. They will argue only about their
     share of the joint product. Cooperative solutions are
     normative. They recommend that even though some of the
     goals of individuals may be opposed, any parallelism of in-
     terests should be exploited to the utmost.

     In planning studies efforts to portray planning conflicts
in a game model may offer worthwhile insights as to the structure of the problem and the payoffs to various groups under
various alternatives.

Simulation.   Simulation is a technique for modeling and
studying large scale problems with complexities that do not fit
the normal structure of quantitative analytical models. Where
an analytical model is likely to be expressed as a set of equations, simulation models will usually contain smaller analytical models, equations, algebraic descriptions, and a set of

rules in the form of a computer program for specifying what
happens under various conditions. Figure 10-25 is a generalized
illustration of elements in a simulation model. The model has
parts analogous to those in the real system--state variables,
exogenous variables (non-controllable data inputs), policy variables (controllable inputs), functional relations, and outputs.
Combining these elements into a simulation model requires an a
priori understanding of the system structure, plus experience
and intuition. Most often simulation models are set up for a
digital or analog computer. A number of computer languages
which facilitate programming of simulation operations   are
available. The model is operated by generating input data and
then recording the system output behavior for various alternative configurations. The outputs can then be examined comparatively. Since an infinite number of model runs to test alternatives could be performed, examining solutions is an experimental process which requires sampling and analysis of the generated output in accordance with sound statistical theory.

Nonquantitative process models. For systems or parts of
systems which are not amenable to quantitative descriptions,
verbal and human interactive modeling techniques have been developed to provide decision makers with useful information
about system behavior. These nonquantitative analyses may serve
either as models or as useful devices for conceptualizing system structure and developing missing data for quantitative
models. In this way, some of the quantitative-nonquantitative
dichotomies in system models may be overcome.

Scenarios. A scenario is a verbal model of the system being analyzed and the conditions under which it must perform.
Within this general definition, scenarios may take various
forms: an accounting of a sequence of hypothetical events, a
plan of action under a set of given or assumed conditions, or a
set of parametric values selected for empirical comparison and
analysis. A well-documented scenario, logically constructed to
show system interactions or series of events, is essentially a
nonquantitative systems analysis. It inquires into the circumstances under which a system must operate, the tasks it must
perform, the natural and physical conditions, and the political,
social, and legal constraints. The scenario can be a useful
beginning point for a systems study, incorporating and integrating the features of the first four phases of the systems
analysis process. From a statement of the problem and antici-
pations of the state and of the environment (e.g., those contained in the Pleasant Valley case), objectives and criteria
for alternative plans and a conceptualization of system structure eventually emerges. A scenario format may consist of language, numbers, and symbols depending on the analysis desired.
For public projects scenarios could well-treat the issues of
problems and needs, provide insight into possible futures, elaborate the major policy alternatives to be considered and their

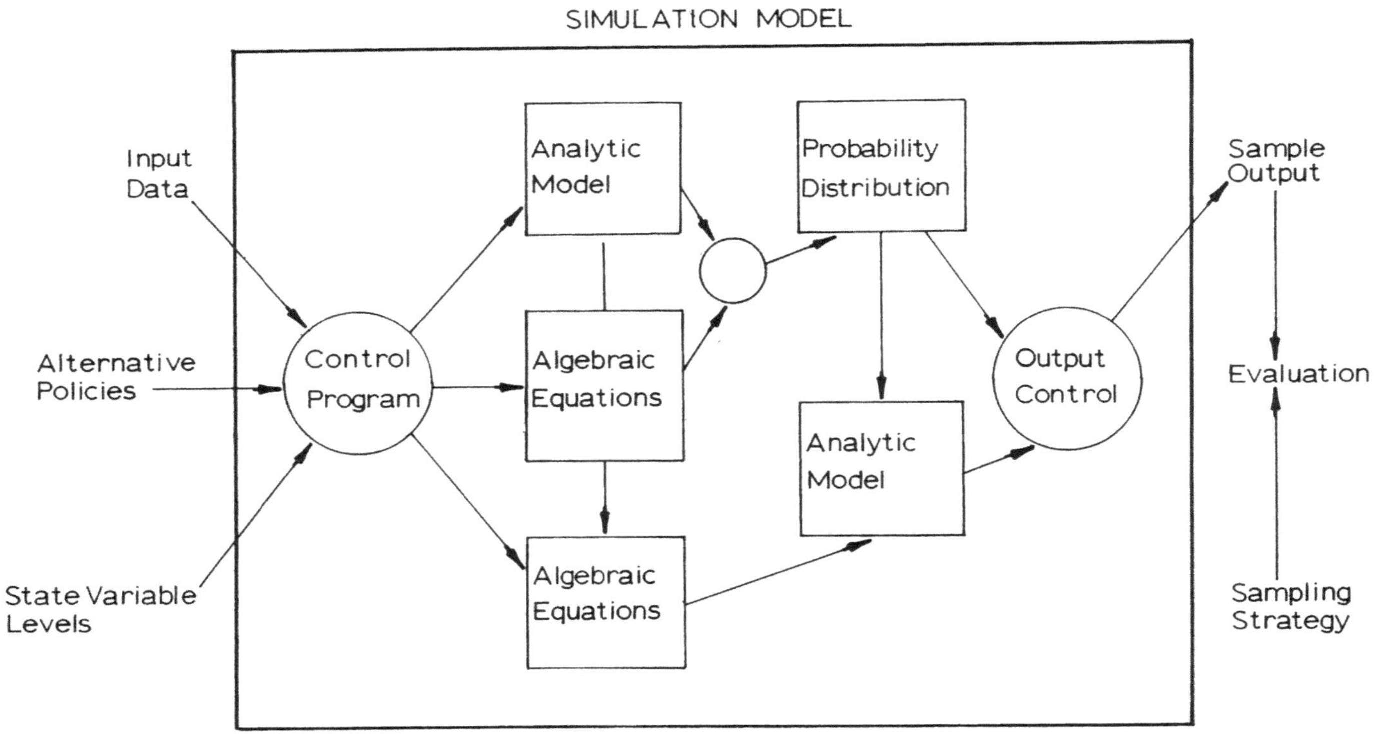

Figure 10-25. Simulation modeing concept.

potential consequences, inquire into the logic and assumptions behind systems design, and explore objectives, criteria, and measures of effectiveness.

Delphi. The Delphi method is a means of generating a consensus of expert opinion on a particular subject. If properly structured, the Delphi inquiry itself could stand as a systems model or serve as a technique for generating parts of systems models for which quantitative data are unavailable, e.g., generating input data, defining system structure, and specifying functional relations. The Delphi technique (Quade, 1968) produces information or forecasts by subjecting the views of individual experts to each other's criticisms in ways that avoid face-to-face confrontation. Direct debate is replaced by interchange of information through carefully designed questionnaires thus providing anonymity of opinions and of arguments in defense of opinions. Participants are asked to give both their opinions and the reasons for them. These responses are edited by a steering committee and fed back to participants as new and refined information. The process continues through several rounds of response, feedback, and soliciting additional information until further progress towards consensus appears negligible. The Delphi method has been usefully employed in forecasting and policy analysis in the defense area, but only recently have efforts been made to apply the methodology to problems of public works planning and evaluation (Crawford and Bishop, 1972). In these studies, the Delphi process has been used to generate three kinds of information required in the analysis and evaluation of public projects: (1) using an interdisciplinary panel of experts to assess the structure of the system and the likely consequences of implementing the proposed alternatives, (2) using a random sample of citizens and a panel of community leaders in the study area to identify and weigh sets of criteria by which alternatives shall be evaluated, and (3) using an expert panel and a community leaders' panel to assess the impact of consequences of alternatives on the sets of value criteria.

Operational gaming. Operational gaming or role playing is a human interactive modeling technique. The application of the technique to public projects design has been little explored, but it could be a potentially useful device in situations where the viewpoints of many interest groups must be represented. Players would be assigned roles with instructions either to optimize his individual position or to achieve some compatible solution with others given his value set. Insofar as players can truly behave as if the situation were actual, the results would be a direct analog of real individual and group reactions to alternative plans or changes in the environment. The basic steps in setting up operational gaming are (a) select the roles or viewpoints to be represented, (b) specify the options or moves available to the players, and (c) specify the rules which produce consequences for taking certain actions. An additional benefit of human interaction is the opportunity for feedback by which to modify objectives or alternatives, and to check the assumptions in analytical models.

EVALUATING ALTERNATIVES

The final task in the systems analysis process is the evaluation of alternatives. The primary purpose of systems evaluation is to present information on project outputs in a way that will aid decision makers in choosing among alternatives. To achieve this objective, an evaluation method must provide the decision maker with a rational means for assessing the tradeoffs among the many attributes of alternatives in making preference decisions.

In a systems analysis, evaluation of alternatives may have to be conducted at several different levels of decision making. At the broadest policy level, the concern is with overall allocation of resources. At the planning level it is the selection of an appropriate system to most effectively meet the desired use, conservation and management of the resources. Finally, at the design level, it is the evaluation of project features or components and their optimal operation within the system.

Description of Project Outputs

As discussed earlier, the objectives to be achieved in the planning and design of public works projects fall into two broad classes, harmony and optimality. Consequences or outputs of public projects are likewise twofold (recall Figure 10-1): (1) those outputs (consequences) directly related to the primary project objectives in terms of system performance (optimality), and (2) those outputs which have impacts on the natural and social environment in terms of system compatibility (harmony).

The nature of the output itself may be evaluated as either a benefit or a cost, where cost might also be thought of as a negative benefit or a disbenefit. Figure 10-26 is a schematic classification of benefits and costs incorporating several of the terms for category descriptions which appear in the literature. A public project introduced into the normal framework of a community has some effects which can be measured in the market place (termed market costs or benefits), and others for which no market mechanism exists (referred to as non-market costs or benefits). The user effects result from direct use of the facility, while non-user effects result from the physical existence of the facility and its associated activities. Practically all firms, institutions, and households find themselves in both positions, instead of one or the other. This means

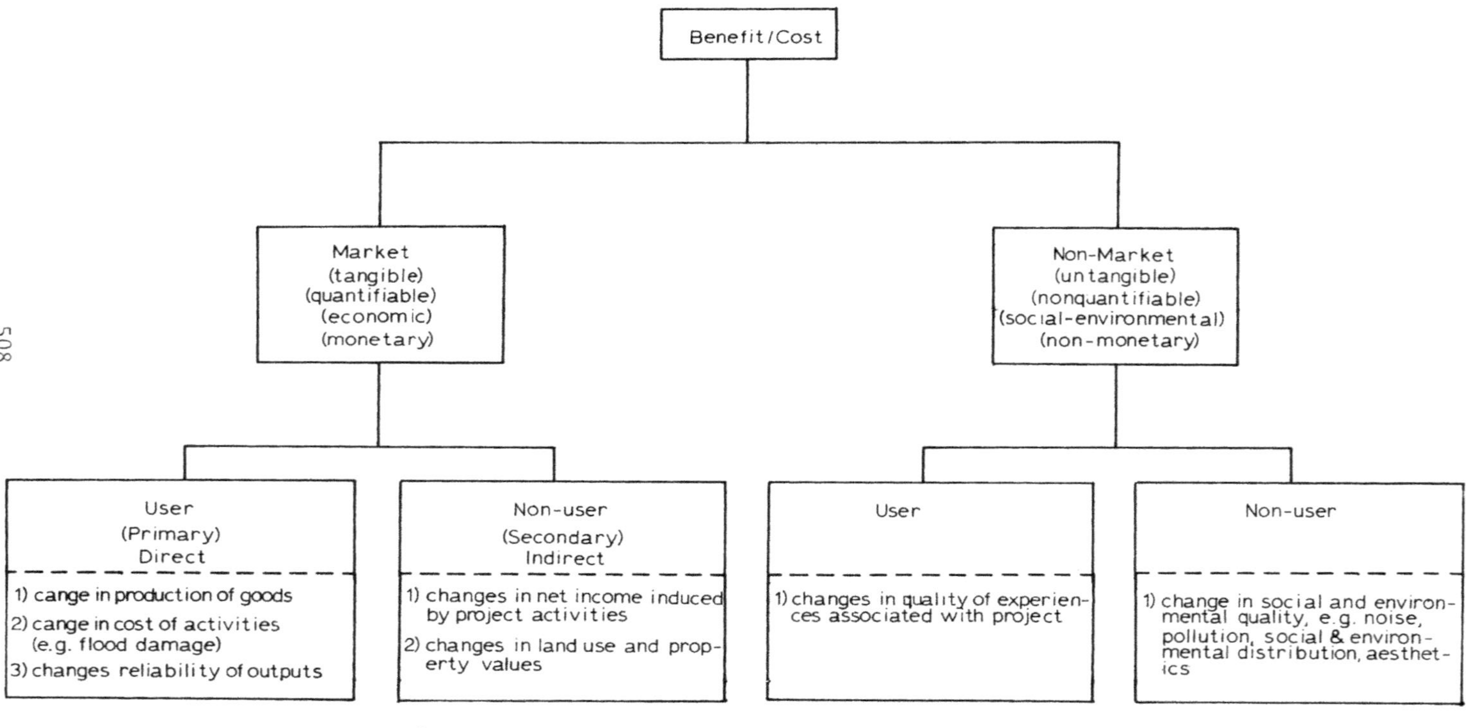

Figure 10-26. Classification of benefits and costs.

"user" and "non-user" do not necessarily identify two different
groups of individuals which receive the costs and benefits of
a project. Rather, the terms denote a separation of kinds of
effects, and the decision maker must weigh the importance of
the consequences in both areas. For example, in the case study
a major objective in changing the present highway might be to
reduce the transportation costs to the users. Assuming this is
accomplished, savings in vehicle operating costs accrue to the
user as direct monetary benefit, while businesses in the community may realize indirect benefits through increased sales
because of the improved highway system. At the same time, both
users and non-users may experience changes in their social or
personal well-being which are not measurable in monetary terms
through the market place.

Objectives, Outputs, and Methods for Evaluation

Large scale projects to promote the public welfare are
multi-objective and multi-purpose in scope. Likewise there are
multiple project outputs to be evaluated. The existence of
multi-dimensional outputs and criteria for evaluation is clarified by the concept of a response surface. A response surface
is generated by plotting a point in N-dimensional space representing the level (value) of N different project outputs for an
alternative. The collection of such points for all possible
alternatives forms a response surface. A pictorial example of a
response surface (Figure 10-27) developed by the North Atlantic
Regional Water Supply Study shows the aggregation of project
outputs into three dimensions - national income, regional benefits, and environmental quality benefits. The points plotted
are for three different plans, where the coordinates of the
points indicate the magnitude of the project outputs along the
three evaluative dimensions.

Choosing among alternative plans as represented by the
points on the response surface, requires an environmental, engineering, economic, and social analysis of the effects of public projects. To do this, a basis must be established   for
evaluating both the market and non-market outputs of project or
policy alternatives. The basic principle for evaluation and
decision making in a setting of noncommensurate project outputs
and multiple evaluative criteria is that decision must be based
on the differences among alternatives.

Making comprehensive comparisons of the differences among
alternatives requires: (1) identifying and quantifying of market and nonmarket consequences, (2) determining the relevant
time period of analysis for the consequences, and (3) using
evaluative criteria for assignments of values of consequences.
Elaborating on these three points, evaluation of the differences among alternatives depends on identifying and quantifying
the outputs of system models which measure the relative merits

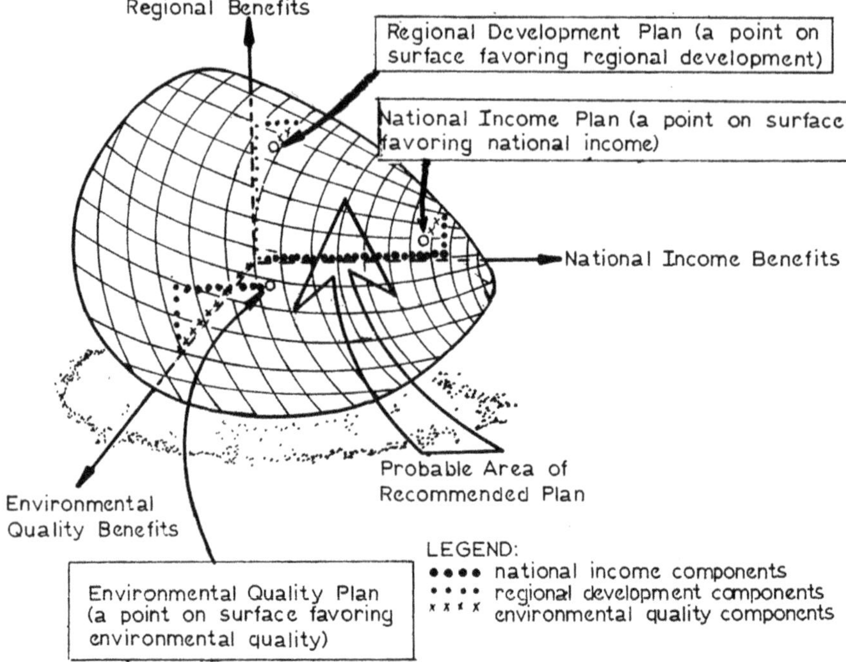

Figure 10-27.   Three dimensional response surface.

of the alternatives.   The viewpoint taken in the evaluation should also be specified since this will affect the accounting of benefits and costs, and comparative evaluations from several viewpoints may be necessary.   The time period over which the consequences of the various alternatives are to be evaluated is also important. Otherwise short run consequences might be given more weight in the decision as compared to the long run effects, or vice versa.   With a proper perspective of the time horizon, the decision maker may be able to "discount" the value of a consequence conceivably in ways similar to the application of compound interest formulas in economy studies.

Selection of a preferred alternative is a difficult but necessary task; one which requires making value choices and tradeoffs.   The evaluative criteria applicable to public proj- ects cover a broad range, but at the same time the human mind can encompass and select among only a limited number of such value relationships.   Thus a primary aim of any evaluative scheme is to eliminate as many irrelevant or dominated alter- natives as possible and to provide a means for clearly focusing the remaining value choices. Computer models and optimization techniques can be of considerable assistance to the decision maker in this regard.   In short, systems evaluation procedures

should display the appropriate information to decision makers
in as understandable and manageable a form as possible.

Single criterion evaluation. If consequences of alternatives can be valued on the same comparative scale, then the
preferred alternative is the one which produces the largest net
benefit (benefits minus costs). When both benefits and cost can
be measured in monetary units, then evaluations can be performed using the tools of engineering economic analysis (Grant
and Ireson, 1964). Since investment in project facilities and
the returns from project operation occur over long periods of
time, to correctly evaluate both present and future benefits
and costs they must be compared at the same point in time. This
is accomplished by "discounting" the costs and benefits by an
appropriate interest rate applied over the useful project life
to obtain the net present value of project outputs. This procedure stated in equational form is:

$$NPV = \sum_{t=0}^{n} \frac{B_t}{(1+r)^t} - [K + \sum_{t=0}^{n} \frac{C_t}{(1+r)^t}] \qquad (10\text{-}14)$$

in which

    NPV = net present value
    $(1+r)^t$ = discount factor for period t at interest rate r
    B^t = benefit in time period t
    K = initial capital cost of project
    C_t = cost of project operation and maintenance in time
        period t

Besides comparison of present worth of benefits minus
present worth of costs, engineering economic analysis may also
be formulated as a benefit cost ratio, equivalent uniform annual costs (benefits), and rate of return including incremental
rates of return. These methods are presented and discussed by
various writers (Grant and Ireson, 1964; Winfrey, 1969;
DeGarmon, 1967; Howes, 1971; and others). All of the methods
when correctly applied will give equivalent answers. The principal difficulties in benefit cost studies are the selection of
an appropriate time period and discount rate, since the results
of the analysis are often sensitive to these factors.

In cases where project costs are monetary but benefits are
measured in some other unit, then cost-effectiveness can be
used for single criterion evaluation. The procedure is best
illustrated by a Bluewater Basin example. Assume that the objective is to reduce pollution of the reservoir resulting from
the present recreation development and facilities. A systems
model is constructed and three possible alternatives are tested
to determine the effectiveness in reduction of the pollution
level as a function of the process or program cost. The results

of the analysis are given in Figure 10-28. However, the information given by just the cost-effectiveness curves is still not sufficient to make a decision. Either a level of effectiveness must be specified and then the cost minimized for that level, or the limit on cost specified and the effectiveness maximized. For example, if the cost cannot exceed C_2 and reduction of pollution to level E_1 is all that is required for the reservoir uses then Alternative 2 operated at cost C_1 is the best choice. On the other hand, if the goal is to reach some minimum level of effectiveness E_2 regardless of the cost then Alternative 3 at cost C_2 would be the choice. However, some flexibility should be allowed in the analysis for if C_2 is a reasonable cost to pay, then by only a slight increase in costs to C_3 large gains in effectiveness can be achieved with Alternative 1. In fact, the approach of setting costs at the place where slope of the cost-effectiveness curve flattens is a judicious one since little is gained by further expenditures past that point.

Multiple criteria evaluation. Project alternatives which have several noncommensurate outputs involving both market and non-market values require multiple criteria comparisons for evaluation. Insight into this very important problem of trade-offs between the pluses and minuses of noncommensurate project consequences is provided by the theory of welfare economics, the subject of Chapter 9. Other evaluation techniques which attempt to incorporate several decision factors, such as environmental mapping, land use modeling, and matrix techniques are discussed in detail in Chapter 11. The discussion here focuses on approaches to multiple criteria evaluation within the context of systems analysis and operations research. To illustrate the problem, consider the joint optimization of two objectives corresponding to the outputs, $0_1 = f(x)$ and $0_2 = g(x)$, where x is a set of input levels associated with a range of alternatives. 0_1 and 0_2 are plotted in Figure 10-29 as a function of input levels for the alternatives. These constitute a pair of objective functions to be "maximized" simultaneously:

$$\text{MAX} \begin{cases} 0_1 \\ \\ 0_2 \end{cases} \tag{10-15}$$

subject to any applicable constraints: $h(x) \leq 0$.

Examination of Figure 10-28 indicates that some alternatives (input levels) can be immediately eliminated from further consideration because they are dominated by better combinations. This includes the area to the left of "a" and to the right of "b", since in these regions the functions 0_1 and 0_2 are both decreasing. This reduces the range of alternatives to those

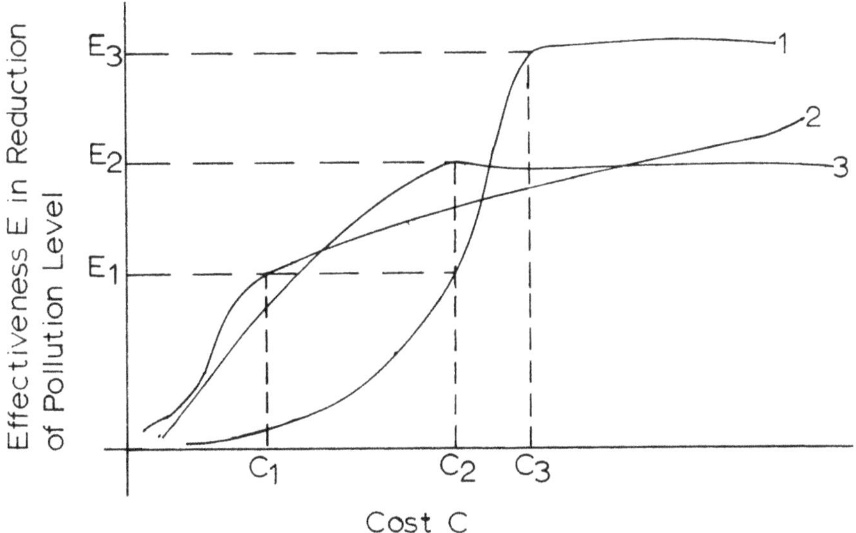

Figure 10-28. Cost-effectiveness curves of reservoir pollution control alternatives.

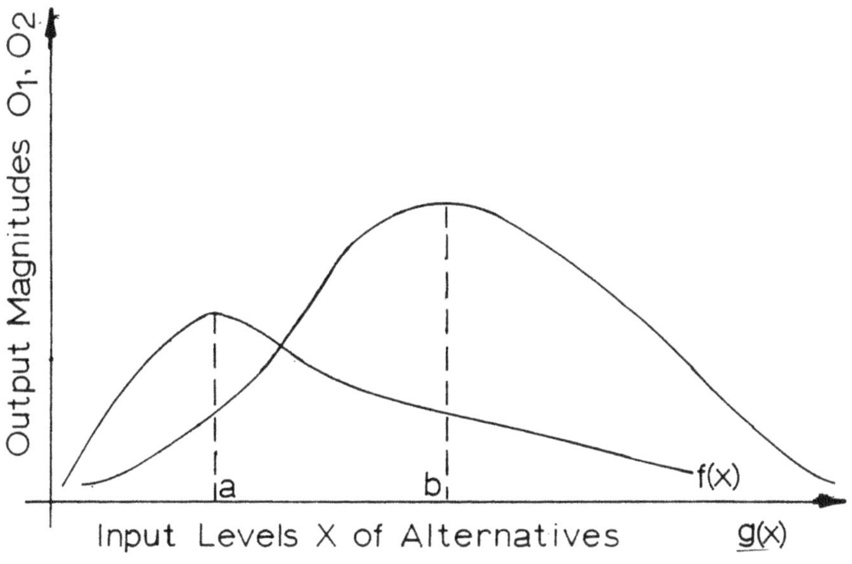

Figure 10-29. Joint optimization of multiple outputs.

between "a" and "b", called the efficient region. However, to select the joint optimum point within the efficient region depends on the tradeoffs or relative weights for the outputs O_1 and O_2. The tradeoffs or weightings of the objectives cannot be deduced analytically, but are value judgments that must be supplied by the decision maker. Three approaches that can be used in making these values decisions are the following:

1. Weighting functions. The weighting function combines the two objective functions through a parameter θ that indicates the relative weight that each shall have in the combined objective function. Symbolically this is expressed as:

MAX $\theta [O_1] + (1-\theta) [O_2]$

subject to $h(x) \leq 0$

$$0 \leq \theta \leq 1$$

Optimization is then performed on the single function which has built-in the decision makers assessment of the relative value of each objective.

2. Constraints on minimum acceptable levels. The constraint method optimizes one of the objective functions subject to the other acting as a constraint which specifies some minimum acceptable level of its output. Stated in symbolic form:

MAX $[O_1]$

subject to: $O_2 \geq v_2$

$$h(x) \leq 0$$

Here the decision maker must select the minimum acceptable value, v_2 for O_2 before the optimization is performed.

3. Direct tradeoff analysis. Evaluation through direct tradeoffs rests on the concept that values are manifest in preference choices among alternatives. Evaluative procedures provide information as to the impact of consequences on evaluative criteria, which the decision maker can use in making direct tradeoffs through preference decisions among alternatives. Bishop (1969) proposes the use of graphical means for presenting the criteria and impact information, and outlines the logical procedure to be used in arriving at a preference ranking of alternatives.

Sensitivity and Reliability Analysis in Evaluation

Quantitative and analytical models in systems analysis are employed to evaluate the optimal arrangements and operation of

components or the use and manipulation of resources in a system.
However, the parameters, data inputs and variable relations are
seldom, if ever, known with complete accuracy due to random or
stochastic variations arising from elements of risk and uncertainty. Hence, the actual levels of output consequences are
also subject to error. Thus, any comparative evaluations of
alternatives should be further investigated to determine the
sensitivity and reliability of decisions to possible changes
in data or parameter values. In short the analysis is not complete even when the sets of consequences from alternatives have
been evaluated and a preference ranking of alternatives obtained. Decisions among alternatives must also be tested to see
how they stand up under questions of uncertainty and sensitivity, reliability and controllability.

Sensitivity analysis. To account for possible change or
variation in data or parameter values, model outputs and evaluations should be subjected to a thorough sensitivity analysis.
Some of the specific purposes of sensitivity analysis are to
investigate the effects on system outputs due to (1) changes in
or relaxation of constraints, e.g., amounts of resources available, (2) changes in prices of resources or costs in production,
and (3) changes in the structure of the system, e.g., through
improved technology. Essentially, sensitivity analysis consists
of varying data and parameters over their range of feasible
values and noting the changes, if any, in model outputs. Variables to which the output is very sensitive are indicative of
areas where assumptions should be carefully checked or where
data should be refined before making a decision. Similarly,
varying parameters with time provides insight into how optimal
solutions change over time. By examining temporal changes of
outputs, phasing of projects can be evaluated as well as optimal combinations of tradeoffs between short run versus long run
solutions. Techniques for sensitivity analysis are usually
built-in features of mathematical programming and simulation
models, and the analysis can be performed with very little additional time and cost.

Uncertainty in data. Outcomes or consequences of alternatives are often connected with data and parameters which are
uncertain. Problems of uncertainty may be handled separately or
in conjunction with Bayesian analysis by: (1) buying more time
in which to assess the outcomes and perhaps determine the actual probabilities, (2) buying more information and data with
which to assess the problem and alter probabilities, (3) buying
flexibility in the decision as a hedge, (4) using "a fortiori"
analysis, i.e., an analysis to determine if a dominant alternative exists for all possible outocmes, and (5) using sensitivity analysis as discussed previously.

Reliability and controllability of alternatives. The choice of an alternative should not be made without consideration of its reliability in producing the desired outputs and its controllability insofar as keeping it operating at the optimal point. This is illustrated by the graph of Figure 10-30 comparing the value outputs for two alternative programs or projects. Alternative 1 promises a very high return and, based on considerations of optimality alone, is the obvious choice between the two alternatives. However, the range of controllable inputs for which this return can be obtained is very narrow. If the reliability of the alternative were such that inputs could not be maintained within this range then the return would drop off drastically. On the other hand, Alternative 2 does not produce as high a return, but the range of input values for which this return can be obtained is much broader. Hence, Alternative 2 can be more easily controlled with greater reliability over the range of input values, and for this reason may be the better choice. The final evaluation and decision among alternatives should consider, then, not only optimal returns or outputs, but also the shape of the output function and its implications for system reliability and controllability.

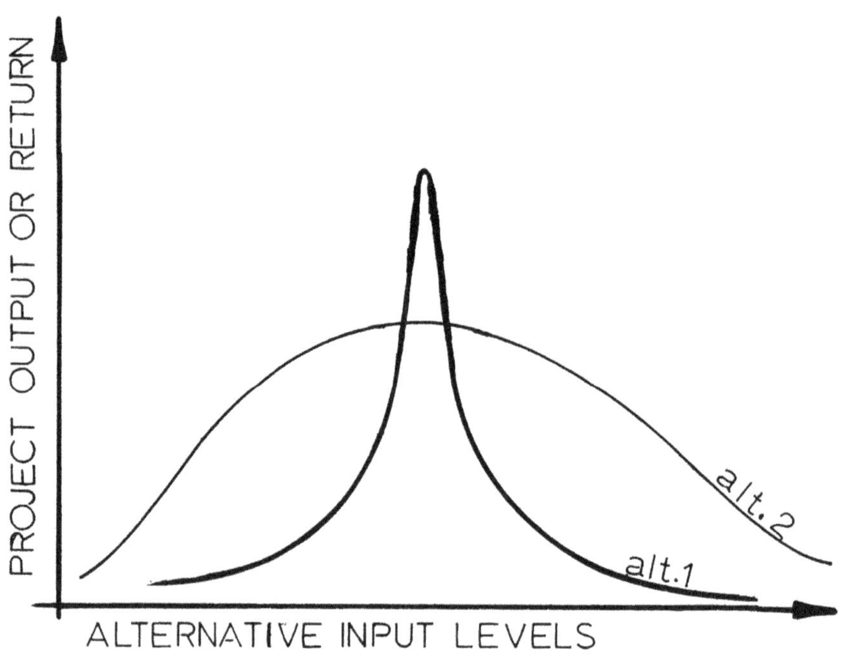

Figure 10-30. Output levels illustrating characteristics of reliability for alternatives.

In the end, systems analysis always boils down to a decision among alternatives. Systems evaluation seeks to measure the performance of various alternatives as analyzed by systems models, and to organize the information in such a way that decision makers can make rational judgments as to which alternative should be implemented. Hence, interaction between analyst and decision maker is required in order for the decision maker to input his values and decision criteria into the process. Comparison of alternatives for simple problems may require only a single objective and measure of effectiveness, but virtually all public works programs have many objectives and as many measures of effectiveness. The evaluation of the multi-objective public works systems is both an art and a science, and the analyst must exercise ingenuity in developing realistic models and in generating information along all dimensions appropriate to the decision. The challenge for the systems analyst is to devise appropriate evaluation techniques for assessing the trade-offs among many attributes of alternatives in order to assist the decision maker in choosing among alternatives.

CAVEATS AND GUIDEPOSTS

While systems analysis provides a logical method of approach and important insights into combining powerful quantitative tools and the computational capabilities of computers for solving the large complex problems of public works, planners and decision makers should also be alerted as to its limitations and pitfalls. Hence, it is appropriate to conclude with a few basic caveats and guideposts for undertaking each of the tasks in the systems analysis process.

Definition of the Problem: Let the problems and questions about the public project shape the analysis rather than the particular phenomena involved, and as a corollary emphasize the questions to be answered and not the analysis itself.

Statement of Objectives: Statement of appropriate sets of objectives is far more important than finding a perfect optimization procedure.

Construction of Systems Models: Avoid overconcentration on the model to the extent that relationships in the model become more important than the system being studied, and as a corollary avoid excessive attention to detail that is not warranted either by the level of analysis required for decision making or the accuracy of the data available. Recognize the assumptions underlying the model when performing the analysis.

Evaluation of Alternatives: Detailed evaluation of alternatives is not nearly so important as the creative formulation of alternatives--a good idea is worth a thousand detailed evaluations. The subjective elements in evaluation should not be neglected. Realize that techniques of systems analysis and evaluation of alternatives are an aid to and not a substitute for the decision makers.

REFERENCES

Biderman, Albert D. 1966. Social Indicators and Goals. *In* Bauer, Raymond A. (Ed.) *Social Indicators*. The M.I.T. Press, Cambridge, Mass.

Bishop, A. B. 1972. An Approach to Evaluating Environmental, Social, and Economic Factors in Water Resources Planning. *Water Resources Bulletin*. Vol. 8, No. 4. August.

Bishop, A. B. 1969. Socio-Economic and Community Factors in Planning Urban Freeways. EEP Report No. 33, Stanford Program of Engineering Economic Planning, Stanford University, Stanford, Calif.

Bishop, A. B. and D. W. Hendricks. 1971. Water Reuse Systems Analysis. *Journal of the Sanitary Engineering Division*, ASCE, Vol. 97, No. SA1, Proc. Paper 7898, February, pp. 41-57.

Bishop, A. B., D. W. Hendricks, and J. H. Milligan. 1971. Assessment Analysis for Water Supply Alternatives. *Water Resources Bulletin*, Vol. 7, No. 3. June.

Carter, B. J., Jr., J. P. Heaney, and E. E. Pyatt. 1971. Valuation of Flow Augmentation Releases. *Journal of the Sanitary Engineering Division*, ASCE, Vol. 97, No. SA3, Proc. Paper 8203, June, pp. 345-359.

Clark, Robert M. 1970. Economics of Solid Waste Investment Decisions. *Journal of the Urban Planning and Development Division*, ASCE, Vol. 96, No. UPI, Proc. Paper 7151, March, pp. 65-79.

Crawford, A. B., and A. B. Bishop, et al. 1973a. A Methodology for Evaluating the Environmental and Social Impacts of Project Alternatives: A Case Study. Report to Army Corps of Engineers Institute for Water Resources. Feb. Utah State University.

Crawford, A. B. and A. B. Bishop. 1973b. A Preliminary Analysis of the Environmental and Community Impacts of 4 Alternatives for Improving Highway Access to Ogden Valley.

Crawford, A. B. and A. B. Bishop. 1972. Use of Delphi Techniques in Assessing Value Impacts of Highway Alternatives. Mimeograph, Utah State University, Logan, Utah.

de Neufville, R. and D. H. Marks, eds. 1974. *Systems Planning and Design*. Prentice-Hall, Englewood Cliffs, N.J.

de Neufville, R. and J. H. Stafford. 1971. *Systems Analysis for Engineers and Managers*. McGraw-Hill, New York.

Deutsch, Ralph. 1969. *Systems Analysis Techniques*. Prentice-Hall, Inc., Englewood Cliffs, N.J.

Gass, S. I. 1958. *Linear Programming Methods and Applications*. McGraw-Hill Book Company, New York.

Gordon, C. K. 1966. The Mathematics of Structure. *International Science and Technology*. May, pp. 61-70.

Grant, E. L., and W. G. Ireson. 1964. *Principles of Engineering Economy*. The Ronald Press. New York. 5th Edition.

Hall, W. A. and J. Dracup. 1970. *Water Resources Systems Engineering*. McGraw-Hill. New York.

Hare, Van Court, Jr. 1967. *Systems Analysis: A Diagnostic Approach*. Harcourt Brace & World. New York.

Howe, C. W., C. S. Russel, R. A. Young, and W. J. Vaughan. 1971. Future Water Demands. Resources for the Future, Inc., National Technical Information Service, PB 197877, Springfield, Va. March.

Hughes, T. C. 1972. A Mixed Integer Programming Approach to the Use of Multiple Water Sources for Municipal Water Supply. Utah Water Research Laboratory Report PRWG 73-3. Utah State University, Logan, Utah.

Jaworski, N. A., W. J. Weber, and R. A. Deininger. 1970. Optimal Reservoir Releases for Water Quality. *Journal of the Sanitary Engineering Division*, ASCE, Vol. 96, No. SA3, Proc. Paper 7361, June, pp. 727-742.

Manheim, M. L., and F. Hall. 1967. Abstract Representation of Goals - A Method for Making Decisions in Complex Problems. Proceedings, 1967 Transportation Engineering Conference, New York Academy of Sciences.

Oglesby, C. H., A. B. Bishop, and G. E. Willeke. A Method for Decisions Among Freeway Location Alternatives Based on User and Community Consequences. Highway Research Board Record No. 305. Washington, D.C., p. 1-15.

Shih, Chia Shun, and John A. DeFilippi. 1970. System Optimization of Waste Treatment Plant Process Design. *Journal of the Sanitary Engineering Division*, ASCE, Vol. 96, No. SA2, Proc. Paper 7230, April, pp. 409-421.

Stephenson, D. 1970. Optimum Design of Complex Water Resources Projects. *Journal of the Hydraulics Division*, ASCE, Vol. 96, No. HY6, Proc. Paper 7320, June, pp. 1229-1246.

Thompson, R. G., M. L. Hyatt, J. W. McFarland, and P. Young. 1971. Forecasting Water Demands. National Water Commission, Report NWC-F-72-030, Arlington, Va., November.

Tillman, Frank A. 1970. Model for Planning a Transportation System. *Transportation Engineering Journal of ASCE*, Vol. 96, No. TE2, Proc. Paper 7289, May, pp. 229-238.

AUTHOR NOTES

Dr. Bruce A. Bishop, Associate Professor of Civil and Environmental Engineering at Utah State University, is engaged both in teaching graduate and undergraduate courses, and in research through the Utah Water Research Laboratory and the Environment and Man Program. He is principal investigator on a number of projects involving systems analysis, multiple criteria evaluation and public participation related to environmental and public works planning. He has had a wide variety of professional experience ranging from the Idaho Water Resources Board, the California Division of Highways, and the Institute for Advanced Planning, U. S. Army Corps of Engineers. Since joining Utah State in 1972 he has acted as a consultant to a number of private and government engineering organizations in public participation, planning and analysis, and decision making.

II

EVALUATION AND PLANNING TECHNIQUES

BY DAVID SEADER

"To build or not to build?" becomes the ultimate question of any public project design work. This question that once seemed relatively simple to answer is now fraught with thorny issues and subject to touchy, emotional tempests. Projects that were thought to enhance system-wide efficiency and promote community growth have stopped cold against the barriers of environmentalism, political power brokerage, and economic and social justice. Project implementation has taken on a new meaning over and above that of physical construction.

As has been repeatedly emphasized in this work, the ground rules for designing and implementing large scale public projects have changed. The newest fact of life for engineers and designers is that they must incorporate into their design decisions a vastly enlarged set of design parameters, including ecological, economic, social and political. The information input needed to assess the decision of whether or not to build a project has increased to a point where the seams of the traditional evaluation methodologies have burst and a search for new ones has begun.

The twin prods of environmental and political awareness have shattered forever the simple, investment-oriented ways of evaluating projects, and have left in their place a morass of problems and issues. New expanded techniques for analysis which cross disciplines and coordinate total evaluation have only begun to respond to the demands of the new ground rules.

THE EVALUATION FUNCTION IN THE DESIGN PROCESS

The evaluation process can serve several important objectives in the design of projects. These range from simple impact assessment to overall program structuring. The major purposes in ascending order of magnitude are:

1. assessment of need for a project through the tracing of effects and comparison with no project at all,

2. assessment of impact and framework for developing the environmental impact statement,

3. analysis and selection of alternative solutions and proposals,

4. synthesizing basis for compatible or complementary schemes, and

5. organizing framework and rationale for the entire design process.

The last purpose is most significant, especially as evaluation methods are expanded to be more comprehensive and flexible. The structure of the final evaluation may grow to encompass the structure of the entire study, generating development processes consonant with the assessment system. Indeed, the evaluation function is made much simpler if it is fully understood before design begins and its requirements are met throughout the design process.

The remainder of this chapter is devoted to the exploration of techniques which are slowly evolving to serve this structuring purpose. They are large-scale, wide ranging, flexible techniques capable of exerting positive and sensitive influence on the evolution of public projects.

COMPREHENSIVE EVALUATION TECHNIQUES

To borrow from musicological terminology, the state of the evaluative art is in a romantic era. This type of development phase is characterized by great experimentation, breaks with past traditions, grinding ideological conflicts, and bold new initiatives. This current state reflects the new demands placed on designers by public officials, pressure groups, other professionals and general citizenry to know more about the impact of proposed projects.

The response by the designers has been to search far and wide for any possible assistance in satisfying the seemingly endless required flow of justification for implementation.

Professionals are reaching for every opportunity to reconstruct
the evaluative function in this present state of flux.   Little
direction is being given by policy makers, and certainly no one
has yet found a universal approach.   The result has been a pas-
tiche of concepts and mechanisms assembled to generate evaluation techniques.   Such is the legacy of a romantic era.

Yet there have been emerging better and healthier evaluation techniques in a survival-of-the-fittest process.   Some
methods have been successful initially, others have adapted and
grown.   Some regularity in the field has emerged since 1970,
giving five general types of methodology: (1) mapping; (2)
modeling; (3) scoring; (4) matrices; and (5) social accounting.
This taxonomy will become clearer through the following sections of this chapter.   In each section, the general method is
introduced and a typical example of the type is presented.
Where possible, an example of an actual application is described, followed by a critical summary of the process.

The samples are only a cross-section of techniques and
each one could easily have had substituted for it other fine
examples. They were selected only to illustrate one of the five
methodology types, not to represent the last word on the subject.   The cross section will, however, demonstrate the many
different approaches taken to the same problem of evaluation.
A summary section at the end of the chapter traces similarities
between the techniques and might be used as the most skeletal
of outlines for any new technique. This selective presentation
must suffice at present until that skeleton is eventually
filled with codified, standardized techniques.   Until the time
that the classical period of retrenchment is entered, any
practitioner is free to devise that evaluation system which
best satisfies his or her needs. This chapter is, then, work in
progress.

One final note about evaluation: the evaluative function
is distinctly a human activity and by its nature will ever remain so.   Any methodology devised can only assist a decision
maker by posing, categorizing and structuring the options
available and the consequences of action. One should not search
these pages for simple solutions to current dilemmas.   The reliance must rest upon human sensibilities, intuition and judgment.   There will never be an adequate replacement for these in
the planning and design process.

Environmental Mapping

One of the techniques used to overcome the limitations of
traditional evaluation analysis is a method that can be called
environmental mapping.   This technique is typified by the work
of Ian L. McHarg in his *Design with Nature* (1969).   His method
is a reaction to the highway planning methodologies employed by

highway engineers. His attitudes about them are revealed by the
following analogy:

> A plumber is a most important member of society--our civ-
> ilization could not endure long without his services; but
> we do not ask plumbers to design our cities or buildings.
> So too with highways:...[the engineers'] competence is not
> the design of highways, merely of the structures that com-
> pose them--but only after they have been designed by per-
> sons more knowing of man and the land. (p. 32)

McHarg's frustration with highway planning was rooted in the
route selection process which included only what he calls price
costs and benefits: land prices, construction costs, relocation
costs, travel time saved, network efficiency, etc. Missing from
the analysis were so-called non-price items, such as neighborhood destruction, scenic quality, land use impacts, wildlife
and resource impairment. McHarg would widen the measurement
criteria to embrace social as well as economic values. Figure
11-1 is a table of criteria to be used in route selection analysis. The left-hand column lists benefits and savings, both
price (traditional) and non-price. For each category, there is
on the right-side a cost category which completes the other
half of the expanded benefit/cost framework. According to
McHarg, the route selected should provide "maximum social benefit at the least social cost." Disregarding the logical paradox
of simultaneously minimizing one thing while maximizing another,
one aspect of the decision rule stands clear: the measurement
of *social* values in route selection. His proposition is that
social, natural and economic processes can be given some social
value, with "value" here meant as in "valuation." As can be
seen from the list in Figure 11-1, the disparate elements to be
valued do not have common measures, thus limiting summary by
dollar value or the like. The environmental mapping technique
is an attempt to overcome the measurement problem while still
arriving at a best route.

Description: The method is relatively simple and straightforward. It is totally graphic and requires only visual analysis. Each variable (or "value") affected by highway construction, such as wildlife habitats or neighborhood cohesiveness,
and each variable affecting highway construction, such as difficult topography or soils, are mapped separately on a uniform
base map. Areas of the study area where the values are the
highest or where the costs are the greatest are given the darkest tone (gray). Other areas are given progressively less
shading as their values and costs decrease. The lightest area
on each map (white) is thus the area where values would least
be destroyed or where the least cost would be incurred. The
ranking and measuring system varies from variable to variable
at the discretion of the user. The measuring system for each

BENEFITS AND SAVINGS	COSTS
Price Benefits	**Price Costs**
Reduced time distance	Survey
Reduced gasoline costs	Engineering
Reduced oil costs	Land and building acquisition
Reduced tire costs	Construction costs
Reduced vehicle depreciation	Financing costs
Increased traffic volume	Administrative costs, Operation and maintenance costs
Increase in Value (Land & Bldgs.)	Reduction in Value (Land & Bldgs.):
Industrial values	Industrial values
Commercial values	Commercial values
Residential values	Residential values
Recreational values	Recreational values
Institutional values	Institutional values
Agricultural land values	Agricultural land values
Non-price Benefits	**Non-price Costs**
Increased convenience	Reduced convenience to adjacent properties
Increased safety	Reduced safety to adjacent populations
Increased pleasure	Reduced pleasure to adjacent populations
	Health hazard and nuisance from toxic fumes, noise, glare, dust
Price Savings	**Price Costs**
Non-limiting topography	Difficult topography
Adequate foundation conditions present	Poor foundations
Adequate drainage conditions present	Poor drainage
Available sands, gravels, etc.	Absence of construction materials
Minimum bridge crossings, culverts, and other structures required	Abundant structures required
Non-price Savings	**Non-price Costs**
Community values maintained	Community values lost
Institutional values maintained	Institutional values lost
Residential quality maintained	Residential values lost
Scenic quality maintained	Scenic values lost
Historic values maintained	Historic values lost
Recreational values maintained	Recreational values lost
Surface water system unimpaired	Surface water resources impaired
Groundwater resources unimpaired	Groundwater resources impaired
Forest sources maintained	Forest resources impaired
Wildlife resources maintained	Wildlife resources impaired

Figure 11-1. Criteria for highway route selection (McHarg, 1969, p. 33).

variable must always be clearly understood in order to be meaningful to others. Usually, three ranks will suffice: high, intermediate, and low. For example, the map of residential values would have the following zones: market value over $50,000, market value from $25,000 to $50,000 and market value less than $25,000.

After all maps are prepared, they are made into transparencies and superimposed to create a composite map. This composite map has a series of shaded areas of various tones, depending on the shades of all the component maps. The lightest tone areas are the least social cost areas within the study area. Routes can then be tested as they transect the study area, or routes can be generated by connecting the lightest areas on the map.

Example: The method was employed in route selection for the Richmond Parkway on New York City's Staten Island. One group of factors included those criteria normally employed by highway engineers--geology, slopes, soils, drainage. Other categories used in mapping include historic, wildlife and scenic values. The following is a list of all the categories McHarg used, together with the grades of values.

SLOPE

Zone 1   Areas with slopes in excess of 10%.
Zone 2   Areas with slopes less than 10% but in excess of 2-1/2%.
Zone 3   Areas with sloeps less than 2-1/2%.

SURFACE DRAINAGE

Zone 1   Surface water features--streams, lakes and ponds.
Zone 2   Natural drainage channels and areas of constricted drainage.
Zone 3   Absence of surface water or pronounced drainage channels.

SOIL DRAINAGE

Zone 1   Salt marshes, brackish marshes, swamps, and other low-lying areas with poor drainage.
Zone 2   Areas with high water table.
Zone 3   Areas with good internal drainage.

BEDROCK FOUNDATION

Zone 1   Areas identified as marshlands are the most obstructive to the highway; they have an extremely low compressive strength.
Zone 2   The Cretaceous sediments, sands, clays, gravels, and shale.
Zone 3   The most suitable foundation conditions are available on crystalline rocks, serpentine and diabase.

SOIL FOUNDATION

Zone 1   Silts and clays are a major obstruction to the highway; they have poor stability and low compressive strength.
Zone 2   Sandy loams and gravelly sandy to fine sandy loams.
Zone 3   Gravelly sand or silt loams and gravelly to stony sand loams.

SUSCEPTIBILITY TO EROSION

Zone 1   All slopes in excess of 10% and gravelly sandy to fine sandy loam soils.
Zone 2   Gravelly sand or silt loams soils and areas with slopes in excess of 2-1/2% on grave-ly to stony sandy loams.

Zone 3   Other soils with finer texture and flat topography.

LAND VALUES

Zone 1 $350 a square foot and over.
Zone 2 $2.50-$3.50 a square foot.
Zone 3   Less than $2.50 a square foot.

TIDAL INUNDATION

Zone 1   Inundation during 1962 hurricane.
Zone 2   Area of hurricane surge.
Zone 3   Areas above flood line.

HISTORIC VALUES

Zone 1   Richmondtown Historic Area.
Zone 2   Historic landmarks.
Zone 3   Absence of historic sites.

SCENIC VALUES

Zone 1   Scenic elements.
Zone 2   Open areas of high scenic value.
Zone 3   Urbanized areas with low scenic value.

RECREATION VALUES

Zone 1   Public open space and institutions.
Zone 2   Non-urbanized areas with high potential.
Zone 3   Area with low recreation potential.

WATER VALUES

Zone 1   Lakes, ponds, streams and marshes.
Zone 2   Major aquifer and watersheds of important streams.
Zone 3   Secondary aquifers and urbanized streams.

FOREST VALUES

Zone 1   Forests and marshes of high quality.
Zone 2   All other existing forests and marshes.
Zone 3   Unforested lands.

WILDLIFE VALUES

Zone 1   Best quality habitats.
Zone 2   Second quality habitats.
Zone 3   Poor habitat areas.

RESIDENTIAL VALUES

Zone 1   Market value of $50,000.
Zone 2   Market value $25,000 - $50,000.
Zone 3   Market value less than $25,000.

INSTITUTIONAL VALUES

Zone 1   Highest value.
Zone 2   Intermediate value.
Zone 3   Least value.

In all, sixteen categories were used to summarize the
study area; the first six relating to physiographic obstructions, and the remaining ten outlining areas of potential impacts on values. Figure 11-2 shows some of those values mapped,
with tones of gray and white representing the three zones outlined in each of the separate categories described. All zones 1
are darkest; all zones 3 are white.

After all categories were mapped, converted into transparencies and superimposed, a composite map was produced (Figure
11-3). The lightest areas are the least social cost zones from
which a least social cost corridor was selected. Other proposed
alignments were shown to impose more social cost (i.e. net loss
of social value) simply by overlaying a plot of the corridor on
the composite map.

Discussion: The technique of environmental mapping is easy
to apply given the availability of data. The selection and description of the actual data categories is a major problem in
using this method, as it is in most methods. McHarg emphasizes
natural processes to the detriment of political, social and
cultural processes and values, especially accessibility characteristics, but that can be corrected by proper selection of
variables. Much harder to rectify are the inherent problems of
comparing two or more value categories which have no common
measure and weighting the relative importance between them.
Mechanically, certain variables can be weighted more than
others by superimposing more than one copy of its category map
but that does not solve the problem of weighting justification.
The variable weighting problem is generic to evaluation techniques, however, and not limited to mapping.

Another problem stems from the precision of measurement of
variables. The simple three-fold division of shading suffices
for visual analysis and indeed makes the comparisons between
categories easier. But, because it is a graphic technique,
mapping soon becomes cumbersome if too many variables are included or too fine a gradation system is employed. Particularly
difficult is trade-off analysis between alternatives, because
the data components get blurred in a single shade or color.
What does one do, for example, if no least social cost solution
is apparent? Detailed comparisons, while not impossible, are
extremely difficult.

Other limitations of environmental mapping are its static
nature and its inability to relate one potential public improvement to others. The maps are a picture of the present, and
as such do not incorporate any selected route or public project
into the study area's infrastructure for further analysis. The
time dimension is missing. Other influences on values that
might occur simultaneously with the one under study cannot be
easily incorporated. These limitations can be overcome, but

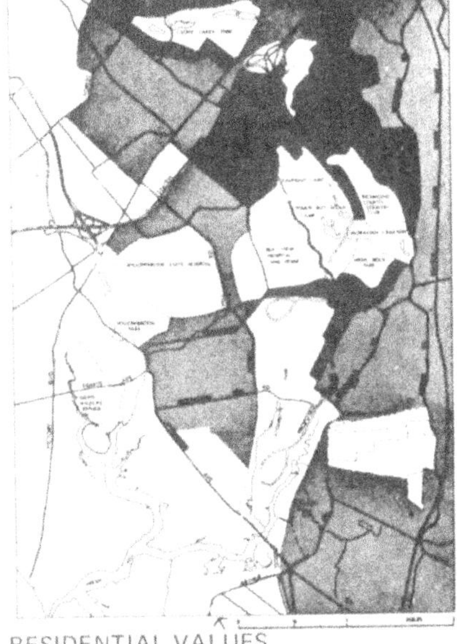

Figure 11-2.   Examples of value category maps (McHarg, 1969, p. 39).

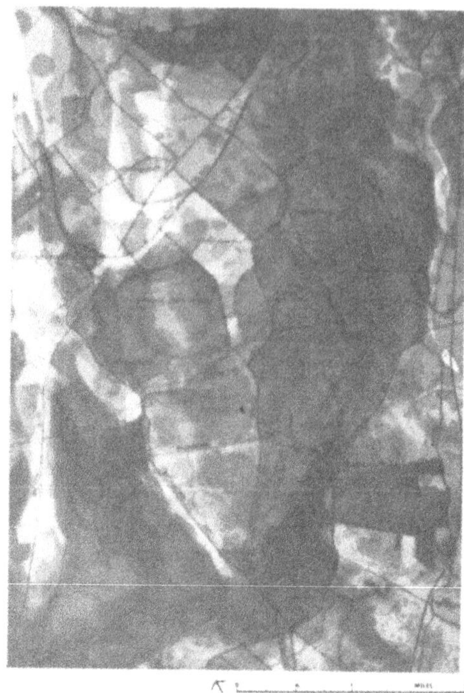

Figure 11-3.   Composite social   value map   (McHarg, 1969,
p. 40).

only in ways that involve so much effort as to negate the value
of the method itself.

The advantages of the method lie in its innate simplicity,
allowing easy if only preliminary analysis, and providing for
the comparison of diverse values if only on a qualitative and
imprecise level.   This latter advantage begins to solve the
dilemma of including non-price considerations into the evaluation process in a manageable way.   The data requirements for
this method are the same as with traditional regional or urban
planning.

Because of its simplicity, the method can easily be applied in a variety of situations and problems, from site selection to general land use and land capacity analysis. For example, McHarg reports in his book on an elaborate Staten Island
land use planning exercise using many more variables and introducing colors.   Additionally, it is easily understood by professionals, policy makers and laymen.   The method forms a logical basis repeated by other variations of geographic techniques.

By employing the capabilities of digital computers to the method of environmental mapping, an extended version of the technique is created--digitized mapping.   The theory or method is no different, but computerization overcomes some of the limitations of environmental mapping. Data is not only mapped and displayed but coded in computer compatible format.   Thus, each individual category map is stored as a computer map or digitized map.   The composite map produced is then made automatically by the computer rather than manually by graphic overlays.

Description: One of the most successful digitized mapping programs is GRID, created by the Harvard University Laboratory for Computer Graphics and Spatial Analysis in Cambridge, Massachusetts (1971).   GRID takes a matrix of values and produces as output a graphic representation of those values.   It uses a standard line printer and standard keyboard characters to produce print symbols of varying shades of gray.   Figure 11-4 is an example of the print characters.   This is similar to the McHarg technique.

The symbols are assigned according to the value associated with each of the print character locations or grid cells.   For example, a base map will be overlaid with a rectangular grid. For each grid cell, a value for a certain variable, say land value, will be assigned. The final matrix of values is encoded and inputted into the GRID program. GRID then assigns a graphic symbol to the grid cell depending on its value. The program does not interpolate values, nor does it contour values into iso-value maps. It produces an exact rperesentation of the data given to it.

The symbol used is at the option of the user.   The user specifies which symbol will be associated with which range of values.   The user also assigns the number of different graphic symbols that will be used, the minimum and maximum of the data range and the various subdivisions of the data range.   The result of the coding is a map with shades of gray corresponding to data values on the original map. All other information, like rivers, roads and boundaries, must be added manually to the output. Reference numbers are provided around the edges of the map.

This is quite an elementary version of digital mapping, yet it contains all the essentials. The following example will demonstrate more clearly the potential of the method and its relation to McHarg-type analysis.

Example: The Four Corners Commission of the southern Rocky Mountain states (Utah, Colorad, Arizona, New Mexico) have begun using a digitizing mapping system similar to GRID (Gibler,

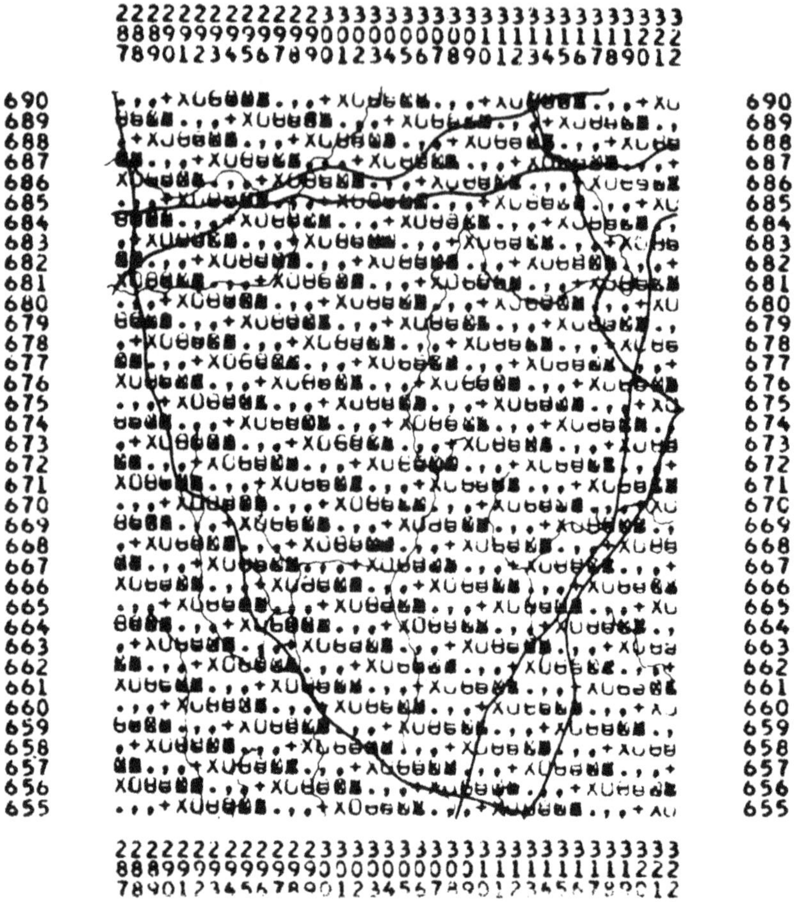

Figure 11-4. Examples of GRID map output symbols.

DATA LEVEL=	0	1	2	3	4	5	6	7	8	9
SYMBOL=	.	,	•	+	X	0	Θ	⊖	⊠	

Figure 11-4. Examples of GRID map output symbols.

1970). It is the Composite Mapping System (CMS) developed in
the Economic Development Administration.   The CMS is described
as follows:

> The new process can readily handle a great variety of geo-
> graphic data, local economic data, transportation and so-
> cial service phenomena; and can accept inputs of any form,
> tabular, printed maps, or sketch maps.   Its outputs are
> cartographic and statistical.   It permits any combination
> of spatial data to be overlaid, with various weights, and
> readily composed into simulated patterns, called "compos-
> ite maps." (Olson and Nez, 1972, p. 21)

The strength of CMS is its ability to take a combination
and weighting of input maps from individual variables and to
produce a composite map. While the problem of justifying a
weighting scheme still holds, it is mitigated by the case with
which various weightings may be tested through the production
of different composites. The sensitivity of the analysis to the
weightings can then be inferred. The compositing of values,
conditions or variables is still left to the sophistication of
the user.

The State of Utah's use of CMS in finding efficient industry locations well illustrates the power of CMS. In finding
efficient location patterns for furniture production, Utah made
a composite of relevant location factors with appropriate
weightings. The variables and weightings used were:

Lumber and Wood Products	.12
Federal and State Highway Accessibility	.15
Railroad Accessibility	.05
Local Settlement Density	.08
Regional Market Zones	.25
Vocational Training Accessibility	.10
Labor Availability Rating	.25
	1.00

The listed factors each had its own factor map prepared.
The computer then combined the individual values at each point
location, according to the weightings to produce the composite
value. The composite values, in turn, were converted into
graphic point symbols and assembled into a map. Figure 11-5
summarizes this process. The darker the area, the better the
location. (Note: the large blank patch on the maps is Provo,
Utah.)

Of course, a different weighting scheme will produce a
different "optimum" location map and many should be tested. The
weighting illustrated here comes from collected technical literature. One interesting possibility raised with CMS is the
determination of weightings through multiple linear regression
analysis. If the *actual* industry location pattern is compared
with the individual factor maps, grid cell by grid cell, the
statistical techniques of regression analysis can be used to
elicit an appropriate weighting of variables. This use of CMS
has inherent statistical problems, but may be of use in handling the complex problem of variable weighting in environmental impact studies.

As a land use locator CMS is limited, however, Land uses
and public improvements compete for sites, and an industry or
project might not be able to obtain the locations it most desires. The single CMS does not reflect this competitive, interactive process. If several trials are used sequentially,

SEVEN-FACTOR COMPOSITE MAPPING OF OPTIMUM LOCATION (OPLOC) FURNITURE PRODUCTION

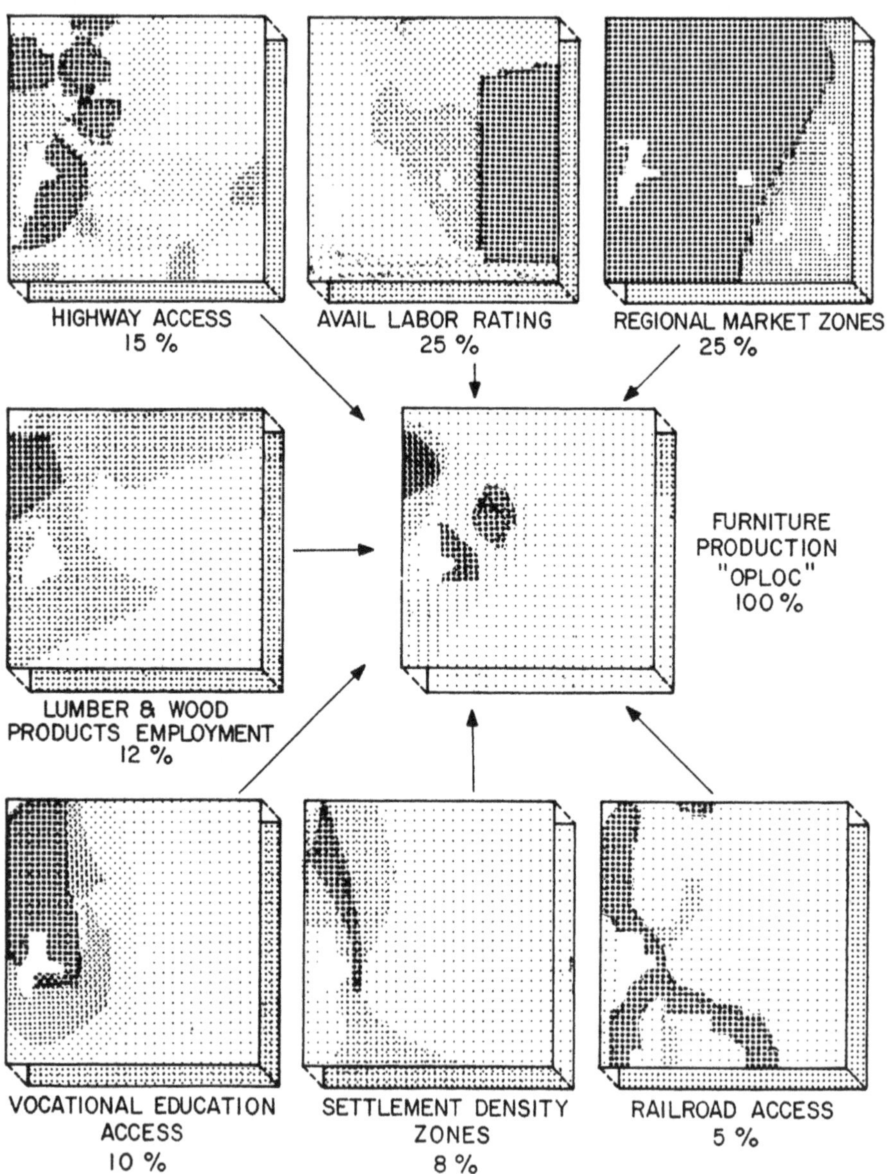

Figure 11-5.   CMS composite map based upon   seven   input factor maps.

however, the problem may be overcome. The more effective technique of modeling, described in the next section, aids in the
analysis of the dymamic location process.

Discussion: It can be seen that the above example of location analysis is analogous to the problems faced in route selection, corridor analysis, site selection, impact and alternative analyses. As data becomes more available and accessible
the computerized form, CMS, GRID and other forms of digitized
mapping will become highly effective and convenient analysis
systems.

The digital storage and retrieval of data in mapped form
allows for several extensions of environmental mapping, such as:

1. inclusion of many variables without loss of comprehensibil-
    ity,

2. east of weighting and recombining variables at little ex-
    penditure of cost and effort,

3. east of decomposition of composite map into components for
    detailed analysis, and

4. ability to use more precise measuring systems for variables.

Because digitizing is no more than a mechanical extension
of environmental mapping, it cannot assist in eradicating the
more fundamental limitations of mapping, namely justification
of weighting, cross-measurement of variables, lack of interaction with land uses and static analysis. It will always be
limited if used by itself, however, as it does not embody any
analytic model in and of itself and relies solely on the art of
variable selection and weighting.

While the introduction of digitizing and computer utilization relieves much of the computational and data burden from
the method, it imposes some of its own limitations. Firstly,
there is computer availability, still a problem even in this
cybernetic age.   Second is the availability of computer expertise.   The programs presented in this section are relatively
easy to use but, to be of maximum utility to a user, one must
know how to adapt them to one's own needs. For example, the
coupling of mapping techniques to computer data banks and information systems will no doubt fully exploit the technique,
yet it also requires specialized knowledge. Lastly, and most
importantly, there is the tendency toward mis-believing computer output, either by deifying it or damning it. Although it
does no more than its graphic counterpart, digitized mapping
may be imbued with all the trappings of spaceage technology for
no reason.

If used correctly, digitized mapping cannot only do the
job of environmental mapping, but can serve as the introduction
to a large variety of computer aided techniques which will amp-
lify its basic value. Simulation, modeling, statistical analysis, and information systems all become powerful design and
planning tools when coupled to digitized mapping techniques.

Lane Use Modeling

One of the responses to the limitations of pure mapping
techniques has been the advancement of land use modeling. In a
land use/evaluation model, the interactions between land uses,
public improvements, economic activities, policies and projections are studied and structured in an attempt to reproduce the
interworkings of reality. If a model is structured correctly,
experiments, called simulations, can be made to test changes in
the system. If such changes are new public projects or policies, then the impacts of the projects can be traced. If the
model has a time dimension, then the dynamics of change may
also be observed. The model becomes a laboratory for the investigation of alternative infrastructure developments and
their attendant impacts upon the landscape.

DYLAM II is a land use model that has been developed and
applied in achieving two major objectives:

1. the allocation of future increments in land use consumption
    or development throughout a study area to create probable
    future land use patterns; and

2. the testing of the impact of alternative public policies
    and improvements upon the land use pattern of a study area.

The former objective allows for a control case, that is, how
the landscape would look in the absence of further infrastructure implementation, and the latter objective becomes an evaluation mechanism.

DYLAM II was created jointly by Columbia University's Division of Urban Planning and the firm of Parsons, Brinckerhoff,
Quade & Douglas, Inc. (Seader and Grava, 1971).

Description: The DYLAM model (Dynamic Land Use Allocation
Model) was developed to locate units of physical development in
two dimensional spaces according to the relative attractiveness
of various sites. In this way, the model approximates the future location decisions of land users--developers, industrial-
ists, governments--based upon the premise that each user has a
list of characteristics it desires (or wishes to avoid) in a
location and that users select those locations which best satisfy those desires. The model is a matching model which rates
all locations in a study area and selects the best location for

each type of land use, depending upon that use's locational desires.

The DYLAM model does not rely on sophisticated, obscure and complex theoretical concepts. Its basis is very pragmatic--something that has been observed by planners and developers over a long period of time and is generally accepted. That observation is that land uses have specific needs or priorities of desires in terms of locational factors, and they seek those sites that can best satisfy those relative requirements. Industrial districts, for example, seek flat land with railroad and highway access, and they are limited in their choices by restrictive zoning, among other things.

The basic unit of land is the grid square or lot. A study area is overlayed with a rectangular system of lots, like a checkerboard. This grid system can be of any scale, to include problems from neighborhood planning to regional planning. All data is collected, coded and displayed by grid square. Figure 11-6 (four parts) shows some examples of coded data.

The data base is much like that of the digitized mapping systems. Previously projected future amounts of land needed to support increases in population and economic activity are also expressed in units of lots or parts of lots. Units of land uses are placed in a grid square like checkers on a checkerboard according to the location algorithm.

The model rank-orders every available lot according to how well it satisfies the needs and desires of a land use type. After scoring each lot and eliminating those having critical conflicts with the needs of a land use, the model then assigns the number of "lots-worth" of the land use needing location to the best available grid squares until all the land required is found. A lot-worth of land use looking for a location may not find an ideal site, but will take the best available site if enough location criteria are satisfied.

One of the basic features of the DYLAM model is the idea that below a certain level of satisfaction with the available lots (the level set by the model's user) a land use simply will not locate, and there will be an incomplete distribution. The assumption is that the land user would rather settle elsewhere if available locations within the study area do not meet its minimum acceptable requirements.

The DYLAM model itself has no internal projection capabilities. Thus, before the model can make future patterns, it must have period by period projectsion of the new amounts of each land use to be located within the study area throughout the entire study period. Any method may be used to arrive at these

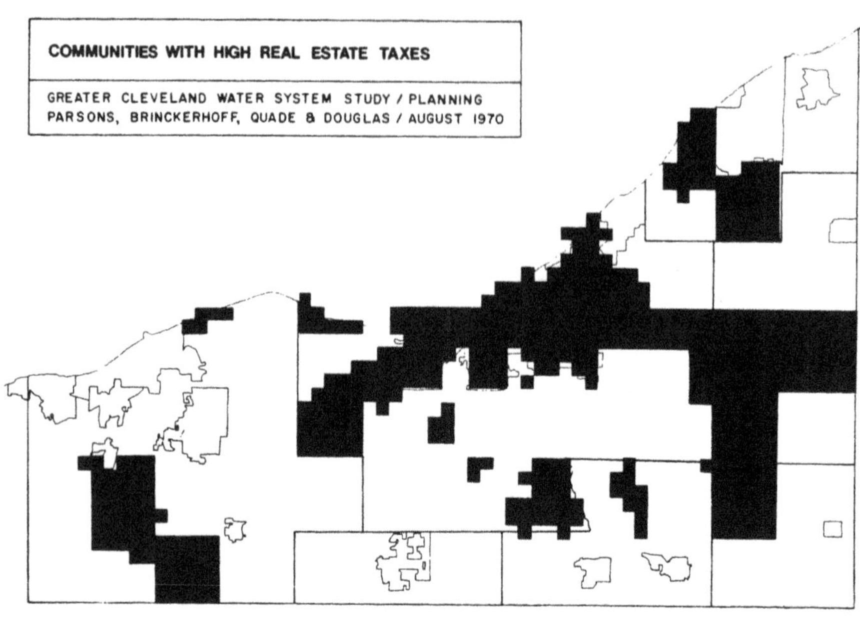

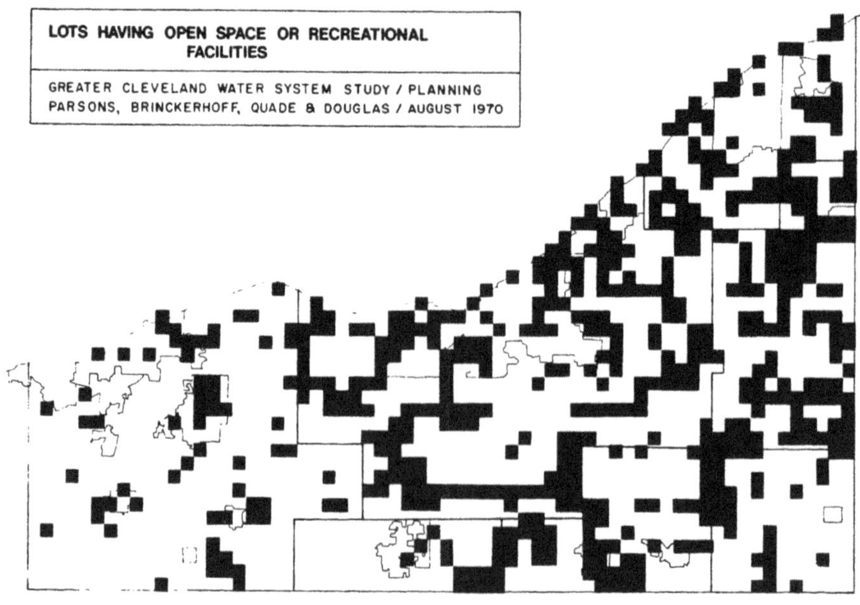

Figures 11-6A and 11-6B.   Typical Dylam II Data Maps.

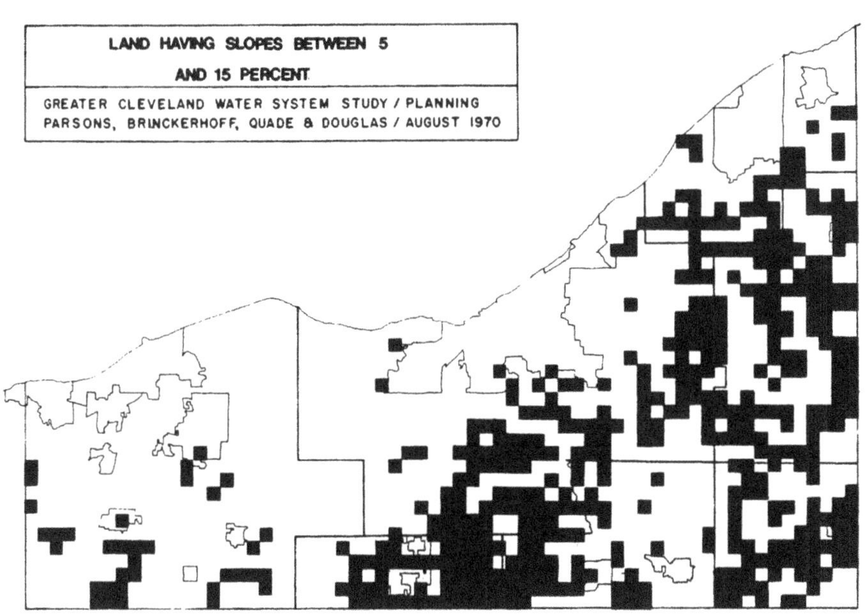

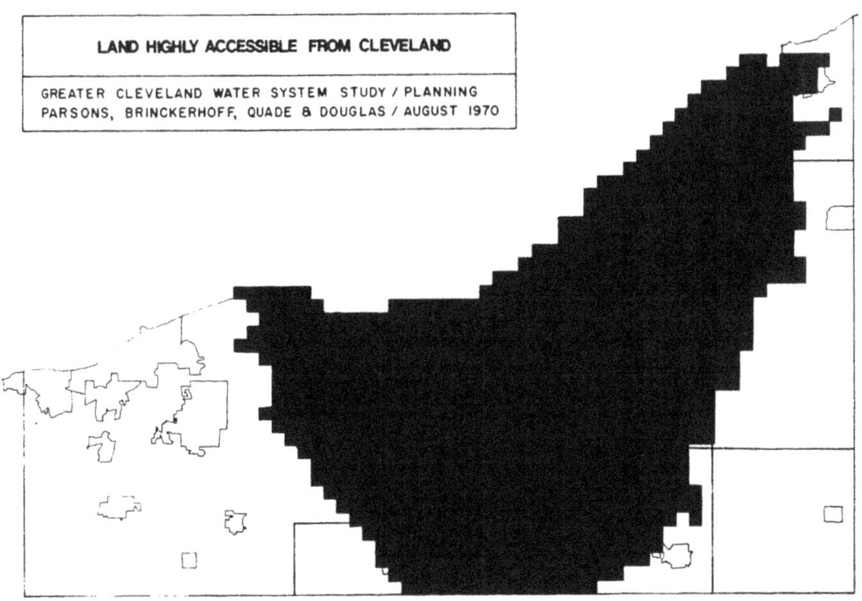

Figures 11-6C and 11-6D.   Typical Dylam II Data Maps.

increments, but the model does not perform this task. Population projections, economic indicators and the like can be used,
but the final expression of the future must be in the form of
x-number of units of each use for each incremental time period.
The basic work of the model is to allocate the new units of
land uses according to a rational pattern or, to a certain degree, to redistribute existing activities if the conditions affecting their original location change appreciably.

The model makes its distribution patterns in discrete time
increments, simulating years or, perhaps, decades. Each new
time period builds upon the pattern of the previous time period
and adds the projected new development, in terms of the number
of new lots of each land use. The model produces a land use
pattern as its output, reflecting the placement of all development as distributed by the matching procedure outlined above.

Figure 11-7 is a typical DYLAM II output, showing maps of
two successive time periods. In addition to the maps, a tabular
summary of changes in the land use pattern is produced for each
time period.

The model itself does not derive any secondary subject information from the allocated activities, but can very easily be
coupled with otherwise separate procedures of estimates or calculations to obtain other sets of usable data. For example, in
the study below, the land use data was converted directly into
water consumption figures by a subroutine that was not an integral part of the DYLAM II model. The secondary information
generated in this way has a direct interrelationship with land
use information, and a direct utility to more specialized design tasks.

The land use pattern obtained is not intended to be "optimal." It is an approximation the sum of the multitudinous private and public location decisions made over a given time period which determine the shape of the future of a study area. The
location algorithm reflects the rational human location decision process and, to the extent that location dynamics are rational, models that process well. It can even be argued that if
the study area and the required amounts of new land are large
enough, irrational decisions will balance and the pattern will
still be valid.

Beyond the simple extention of land use patterns, DYLAM II
possesses features which make it more sensitive to the complexities land use dynamics:

1. there is a lot aging mechanism which accounts for decay of
    uses over time, and simulates the depreciation of land im-
    provements. This feature is particularly helpful in older
    areas where redevelopment is the dominant factor of the fu-
    ture.

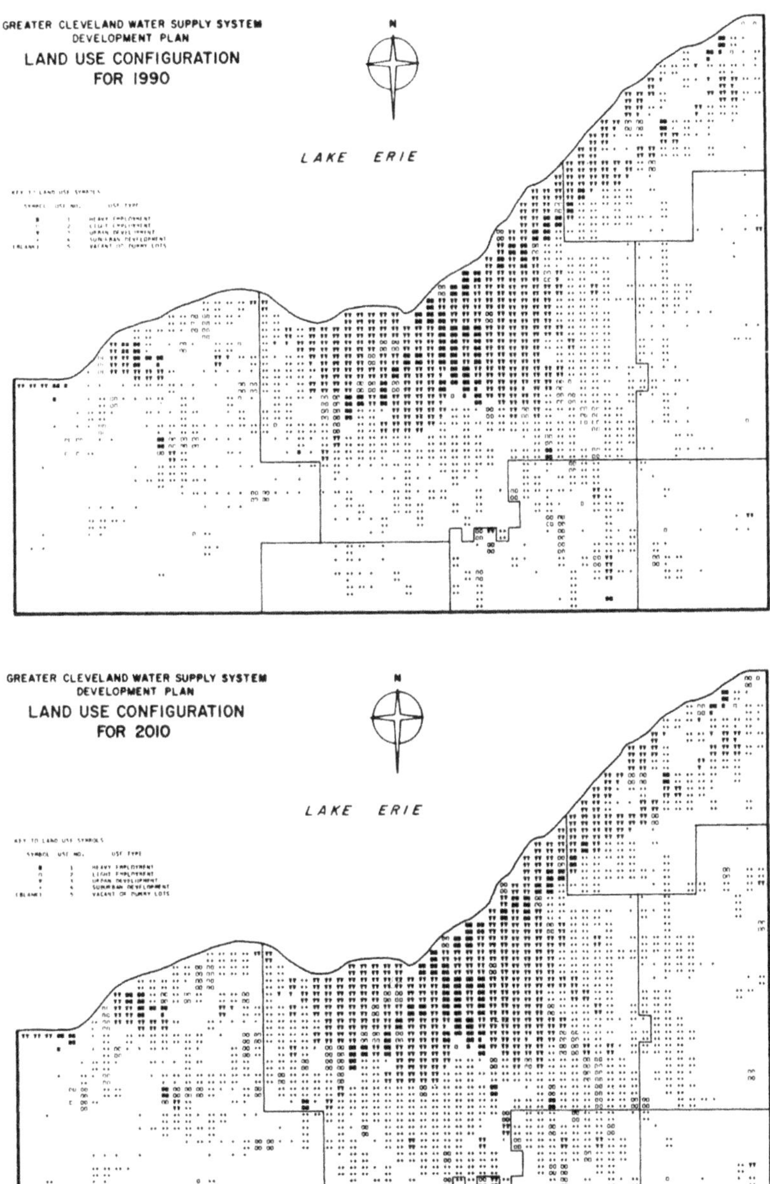

Figures 11-7A and 11-7B. DYLAM II land use output may be
time period.

2. there is a land intensification feature which allows one land use to usurp another, allowing the simulation of increasing urbanization or land use intensification.

3. there is a surrounding lot feature which changes the attributes of a lot depending upon the land uses which locate on its borders. This is useful for simulating nuisance factors, economies of scale and interactions of uses.

4. there is a use clustering factor by which lots already partially occupied with a certain land use become more attractive for further occupation by that use than lots of otherwise equal attraction. This allows simulation of use clustering, in-filling and agglomeration.

5. most important to the changing of an area's characteristics, however, are the public improvements and policies which determine the attractiveness of a site. For example, if a new highway is built through a grid square, the accessibility of that lot increases, making it more desirable for location. The model accounts for these changes with external public policy inputs through which the user can either: (1) reflect probable changes in the study area's services and facilities, or (2) test the effects of various proposed policies and improvements on the study area's development pattern.

This last feature is the one which allows DYLAM II to test marginal land use realignments as functions of policy or infrastructure changes. It is here that the model becomes an active evaluation tool for public projects.

Example: DYLAM II was employed in the greater Cleveland Water Supply System Study undertaken by Parsons, Brinckerhoff, and partially funded by a HUD Urban Systems Engineering Demonstration grant. The objective of the study was the design of a water supply system for the greater Cleveland area, which encompasses approximately 2000 square miles, to be staged over the next forty years to the year 2010.

The DYLAM II model was used to derive the demand pattern for water by projecting the land use pattern for residential and industrial land use. These patterns could then be used by water resources engineers to plan the extensions of the water supply system, and to design facilities with the foresight granted by the knowledge of the future distributions of demand. Additionally, DYLAM II was used to test sensitivity to changes in the future patterns by tracing alternative infrastructure patterns.

Because of the scale of the project, grid squares of about one square mile were selected, necessitating a redefinition of land uses into four broad categories: heavy (water use) industry, light (water use) industry, urban (density) development and suburban development. Each of these uses was composed of varying percentages of traditional land use types, reflecting the general types of broad land uses existing in Cleveland.

To accommodate the locational desires of the composite land use types, a data base of 41 lot characteristics was assembled. The list included a wide variety of locational factors from physiographic, infrastructure and access to community facilities, prestige and wealth. The lot characteristics are as follows:

A. Physical data (nine characteristics)
   1. Slopes greater than 15%
   2. Slopes between 5% and 15%
   3. Soils not suitable for heavy structures
   4. Land reserved for public use
   5. Flood plains
   6. Marshy or swampy land
   7. Land with poor drainage characteristics
   8. Industrial watercourses
   9. Recreational watercourses

B. Utilities (two characteristics)
   1. Water service
   2. Sewer service

C. Site amenities (eight characteristics)
   1. Open space and recreation facilities
   2. Regional medical facilities
   3. Regional cultural or educational facilities
   4. Regional commercial facilities
   5. Outstanding school systems
   6. High schools
   7. Government facilities
   8. Incompatible uses

D. Institutional (five characteristics)
   1. Large aggregations of land
   2. Real estate taxes
   3. Income taxes
   4. Favorable development attitudes
   5. Unfavorable development attitudes

E. Access (twelve characteristics)
   1. Rail lines
   2. Transit lines
   3. Highways
   4. Arterial roads
   5. Airports
   6. Road network nodes

    7. Port accessibility
    8. Access to major employment areas
    9. Access to Cleveland
  10. Access to Akron
  11. Access to secondary centers
  12. Access to tertiary centers

F. Socio-economic (five characteristics)
   Communities that:
    1. have outstanding community facilities
    2. have outstanding housing
    3. are wealthy
    4. are prestigious
    5. land of high agricultural value

Throughout the study time period, land uses searched the landscape as summarized by the 41 variables and selected the most appropriate sites in competition with one another. The maps in Figure 11-7 reflect the results of that selection process.

As a test of the sensitivity of those patterns, a separate run of the model was undertaken with one significant change: the creation of an airport in Lake Erie. This controversial project was injected into the model to trace the realignment of land uses in the region. Figure 11-8 shows the airport clearly as a box of five lots in Lake Erie. These lots were made extremely attractive to all land uses to test where development would be drawn from in the study area if the project were built. These maps may be compared with those in Figure 11-7 to highlight the marginal development areas in the original projections. This sensitivity analysis alerted system designers to the marginally important parts of future service districts. This increased the flexibility of the system designs.

As a final output, the land use patterns were converted into water demand patterns and mapped to help determine new service areas. Figure 11-9 displays a map of water demand patterns. Alternative demand patterns could have been generated by varying the parameters of the model, the tested public improvements, the projected new amounts of land and the development mechanisms of the region. Either land use or some secondary variable (like water demand) may be used as an output measure for various combinations of inputs. Thus the model can be used by policy makers to test consequences of various assumptions about and plans for the future of a study area.

Discussion: A model like DYLAM II can provide a comprehensive and flexible framework for most land use impact oriented studies. It allows for wide ranging experimentation in a systematic manner. The use of land use models does have limits, however.

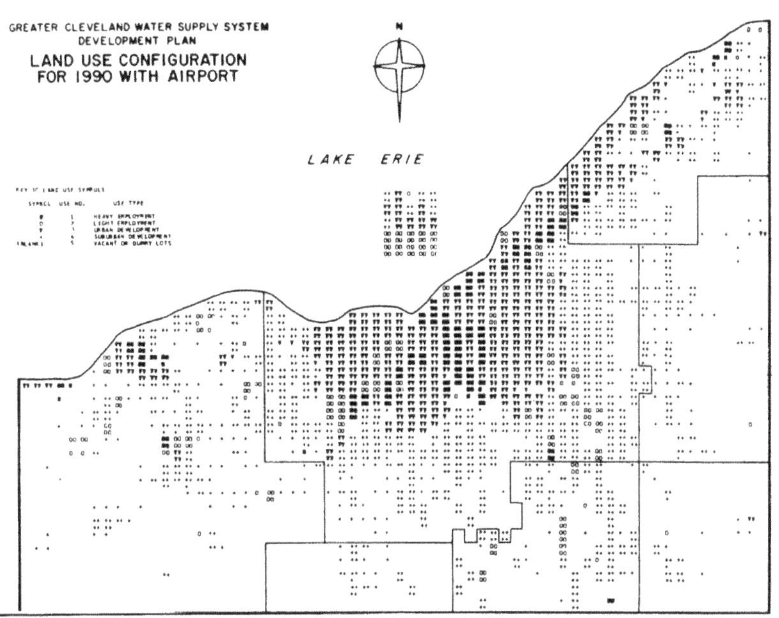

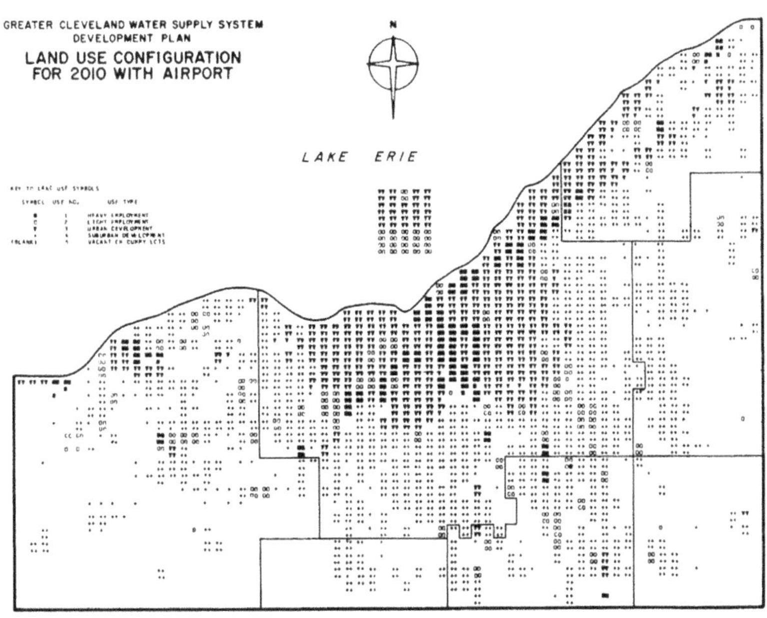

Figure 11-8. Cleveland land use maps with proposed airport-in-lake.

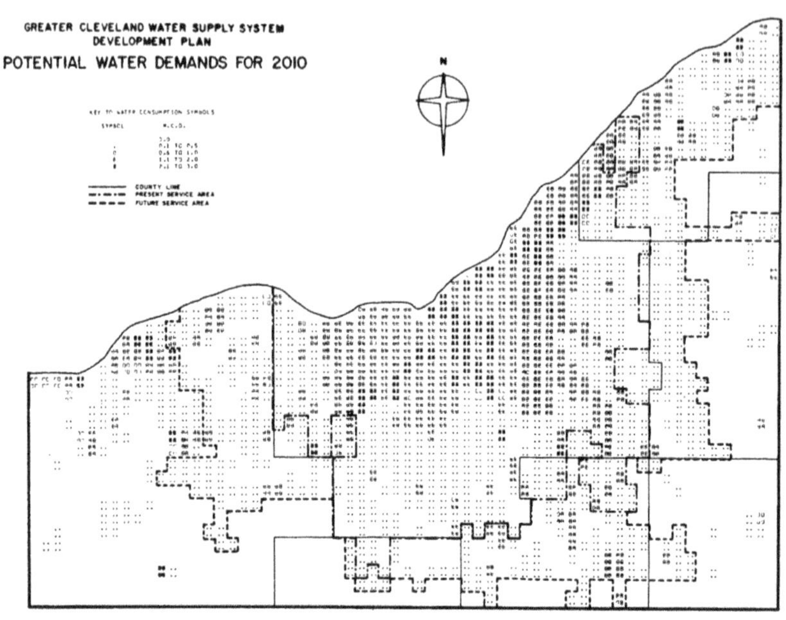

Figure 11-9. Water demand pattern map derived from DYLAM II.

One of the severest limitations is difficulty of adapting a model from application to application. Models are either structured to respond to one singular situation or are so flexible as to demand detailed validation exercises. Some of the better known urban planning models (e.g., Goldner, 1968; HRB, 1968; Center for Real Estate, 1968), while competent are so complex and cumbersome as to prevent their being used efficiently in any case but the one originally intended. On the other hand, a model like DYLAM II, which was designed to be highly adaptable, requires painstaking calibration to insure confidence in its projections. In the Cleveland study, over thirty five separate computer runs were needed to determine the parameters, compared with ten runs of the finalized model itself. Either type of model, rigid or flexible, requires a great deal of knowledge about the workings of the study area before results can be certified. The demanding nature of models, though, provides an opportunity for model users to learn the details of the location dynamics of their study areas. The tradeoff between learning about the system and the need to obtain results is one that must be made each time a model is considered for use in a study. This holds true for individually tailored models or "off the shelf" models like DYLAM II.

A complementary limitation to the previous one is the need
for a great deal of data for the operation of models. Even in
models like DYLAM II where the data base is chosen by the user
rather than being specified *a priori*, the data requirements are
large. Certainly data must be available comprehensively, completely and compatibly for most models. Also many interrelationships between sectors, land users, variables and parameters
must be stated and, therefore, must be known. This places a
large burden on the model user. On the other hand, the model
imposes a discipline and organizing structure to a study which
is invaluable.

Even with these limitations, computer-aided models will
take on a growing importance in the future because of their
great utility to the planning and design fields. Models are
being conceived and developed to investigate impacts of specific types of projects, like flood plain development (INTASA,
1972). Special computer languages for modeling, like DYNAMO
have been written leading to ambitious investigations of the
future of the city (Forrester, 1969) and the future of the
world (Meadows, et al., 1972), stressing the relationships between the growth rates and capacities of various elements of
the society. Work is accelerating in all facets of modeling.

As applications are amassed, it will become increasingly
easier to use models quickly and cheaply to study impacts and
analyze alternatives in the design and implementation of public
projects. This is especially true of models which have a flexible input and analysis mechanism to account for "soft", qualitative data in evaluation, and those which have a geographic
base that can be used both for analysis and display. The coupling of comprehensive analysis techniques and digitized, composite mapping techniques produces an impressive evaluation
tool--the land use simulation model.

Environmental Impact Matrix

One of the consequences of the Environmental Policy Act of
1969 was the need for establishing the criteria for the preparation of the environmental impact statement required in Section
102(2)(c) of the Act. Many techniques have been proposed using
an evaluation matrix of some kind (e.g., Committee on Power
Plant Citing).   One of these methods was developed by the Department of Interior, U. S. Geological Survey (Leopold, et al.,
1971). The method encompasses more than just the evaluation of
impact.   It relates to the entire process of generating an environmental impact statement.

The heart of the system, however, is a generalized matrix
that can serve as a checklist or reminder of the entire spectrum of possible environmental impacts that can result from
proposed projects. The matrix can be used as a guide for those

preparing or even evaluating impact statements. If more than one alternative project is under consideration, then this technique may be helpful in selecting or synthesizing them into a final scheme.

Description: The technique is basically a cross reference system for keeping account of most possible impacts of most possible forms of intervention. A generalized matrix is created by listing the actions which cause environmental impacts vertically (columns), and listing exisitng environmental conditions that might be affected horizontally (rows). Thus, each grid cell of the resulting matrix is the intersection of a possible action (e.g., harbor dredging) and a condition that the action might affect (e.g., fishes and shellfishes). The evaluator must then assess that interaction in terms of impact.

To be generally useful, the listed number of actions totals 100. Any project might be composed of several of those actions, not just one. A highway, for example, includes: alteration of ground cover, paving, alteration of drainage, bridges, cut and fill, etc. Similarly, 88 environmental conditions are listed which might be affected. Each of the actions to be undertaken may affect each condition, resulting in 8800 possible interactions. Practically, many fewer are usually uncombined, especially if unnecessary replication is avoided by concentrating on first-order effects of specific actions and not general or second order effects. The entire range of actions and conditions is as follows:

I. Existing Characteristics and Conditions of the Environment
   (Rows)

   A. Physical and Chemical
      Characteristics

1. Earth	3. Atmosphere
a. Mineral resources	a. Quality (gases, particulates)
b. Construction material	b. Climate (micro, macro)
c. Soils	c. Temperature
d. Land form	4. Processes
e. Force fields and background radiation	a. Floods
2. Water	b. Erosion
a. Surface	c. Deposition (sedimentation, precipitation)
b. Ocean	d. Solution
c. Underground	e. Sorption (ion exchange, complexing)
d. Quality	f. Compaction and settling
e. Temperature	g. Stability (slides, slumps)
f. Recharge	
g. Snow, ice, and permafrost	

h. Stress-strain (earth-
   quake)
   i. Air movements

B. Biological Conditions
   1. Flora
      a. Trees
      b. Shrubs
      c. Grass
      d. Crops
      e. Microflora
      f. Aquatic plants
      g. Endangered species
      h. Barriers
      i. Corridors
   2. Fauna
      a. Birds
      b. Land animals includ-
         ing reptiles
      c. Fish a nd shellfish
      d. Benthic organisms
      e. Insects
      f. Microfauna
      g. Endangered species
      h. Barriers
      i. Corridors

C. Cultural Factors
   1. Land Use
      a. Wilderness and open
         spaces
      b. Wetlands
      c. Forestry
      d. Grazing
      e. Agriculture
      f. Residential
      g. Commercial
      h. Industrial
      i. Mining and quarrying
   2. Recreation
      a. Hunting
      b. Fishing
      c. Boating
      d. Camping and hiking
      e. Picnicking
      f. Resorts

3. Aesthetics and Human
   Interest
   a. Scenic views and
      vistas
   b. Wilderness qualities
   c. Open space qualities
   d. Landscape design
   e. Unique physical fea-
      tures
   f. Parks and reserves
   g. Monuments
   h. Rare and unique spe-
      cies or ecosystems
   i. Historical or arch-
      aeological sites and
      objects
   j. Presence of misfits
4. Cultural Status
   a. Cultural patterns
      (life style)
   b. Health and safety
   c. Employment
   d. Population density
5. Man-Made Facilities and
   Activities
   a. Structures
   b. Transportation net-
      work (movement, ac-
      cess)
   c. Utility networks
   d. Waste Disposal
   e. Barriers
   f. Corridors

D. Ecological Relationships
   such as:
   a. Salinization of water
      resources
   b. Eutrophication
   c. Disease-insect vec-
      tors
   d. Food chains
   e. Salinization of sur-
      ficial material
   f. Brush encroachment
   g. Other

II. Proposed Actions Which May Cause Environment Impact (Columns)

A. Modification of Regime
   a. Exotic flora or fauna introduction
   b. Biological controls
   c. Modification of habitat
   d. Alternation of ground cover
   e. Alternation
   f. Alternation of drainage
   g. River control and flow modification
   h. Canalization
   i. Irrigation
   j. Weather modification
   k. Burning
   l. Surface or paving
   m. Noise and vibration

B. Land Transformation and Construction
   a. Urbanization
   b. Industrial sites and buildings
   c. Airports
   d. Highways and bridges
   e. Roads and trails
   f. Railraods
   g. Cables and lifts
   h. Transmission lines, pipelines and corridors
   i. Barriers including fencing
   j. Channel dredging and straightening
   k. Channel revetments
   l. Canals
   m. Dams and impoundments
   n. Piers, seawalls, marinas, and sea terminals
   o. Offshore structures
   p. Recreational structures
   q. Blasting and drilling
   r. Cut and fill
   s. Tunnels and underground structures

C. Resource Extraction
   a. Blasting and drilling
   b. Surface excavation
   c. Subsurface excavation and retorting
   d. Well drilling and fluid removal
   e. Dredging
   f. Clear cutting and other lumbering
   g. Commercial fishing and hunting

D. Processing
   a. Farming
   b. Ranching and grazing
   c. Feedlots
   d. Dairying
   e. Energy generation
   f. Mineral processing
   g. Metallurgical industry
   h. Chemical industry
   i. Textile industry
   j. Automobile and aircraft
   k. Oil refining
   l. Food
   m. Lumbering
   n. Pulp and paper
   o. Product storage

E. Land Alternation
   a. Erosion control and terracing
   b. Mine sealing and waste control
   c. Strip mining rehabilitation
   d. Landscaping
   e. Harbor dredging
   f. Marsh fill and drainage

F. Resource Renewal
   a. Reforestation
   b. Wildlife stocking and management
   c. Ground water recharge

        d. Fertilization appli-
           cation
        e. Waste recycling

     G. Changes in Traffic
        a. Railway
        b. Automobile
        c. Trucking
        d. Shipping
        e. Aircraft
        f. River and canal traf-
           fic
        g. Pleasure boating
        h. Trails
        i. Cables and lifts
        j. Communication
        k. Pipeline

     H. Waste   Emplacement   and
        Treatment
        a. Ocean dumping
        b. Landfill
        c. Emplacement of tail-
           ings, spoil and over-
           burden
        d. Underground storage
        e. Junk disposal
        f. Oil well flooding
        g. Deep well emplacement

        h. Cooling   water   dis-
           charge
        i. Municipal waste dis-
           charge   including
           spray irrigation
        j. Liquid effluent dis-
           charge
        k. Stabilization   and
           oxidation ponds
        l. Septic   tanks,   com-
           mercial and domestic
        m. Stack   and   exhaust
           emission
        n. Spent lubricants

     I. Chemical Treatment
        a. Fertilization
        b. Chemical deicing of
           highways, etc.
        c. Chemical stabiliza-
           tion of soil
        d. Weed control
        e. Insect control (pes-
           ticides)

     J. Accidents
        a. Explosions
        b. Spills and leaks
        c. Operation failure

In tracing effects through this method, two important measures of impact are used for each action: magnitude and importance.   Magnitude is used in the sense of size, degree, extent
or scale.   A highway may make an impact of great magnitude on
"drainage" but affect "landmarks" by only a small magnitude.
The importance of an interaction is the significance of its
consequences, that is, the results of changing that particular
condition and other factors in the environment.   Thus, the previously mentioned highway which affected landmarks with a small
magnitude, might have an effect of great importance if the one
landmark touched was the Washington Monument.   In the course of
the assessment, it is crucial to keep the two concepts separated.

It is generally easier to assess magnitude than importance
(though neither may be simple to measure).   More quantitative
measures apply to magnitude assessment than importance.   Importance rating also includes highly subjective judgments based on
values, biases and vested interests.   This method does nothing
to untangle the problem of importance weighting though it is
cognizant of the difficulties. Both measures are given a score
from 1 to 10 for each interaction. A low score indicates small

magnitude or little importance. A high score denotes the opposite. A "+" may be used to mark positive or beneficial effects
and a "-" to denote detrimental or negative effects.   Thus,
each action gets two numbers from -10 to +10 for each of the
conditions it affects.

The matrix is completed by identifying (and marking) each
proposed action. Going down the column for each action, a slash
is placed in each cell at an intersection with a possible impact.   The slash is placed from the upper right corner to the
lower left corner.   At this point the matrix may be reduced in
size to show only slashed boxes.   Then, in each box with a
slash, the MAGNITUDE rating is placed in the upper left half
and the IMPORTANCE rating in the lower right.   Figure 11-10
shows a sample interaction.   When all boxes are completed, the
entire range of impacts is laid bare for analysis and evaluation.   Discussion of each impact can lead to completion of the
impact statement, project reassessment, redesign or modification. The important result is the denouement of the complex set
of interactions and impacts.

Example: A hypothetical example was generated using data
from a mine leasing application.   The mine was for a phosphate
deposit estimated at 80 million tons of crude ore located in
Los Padres National Forest, California. To develop the area, a

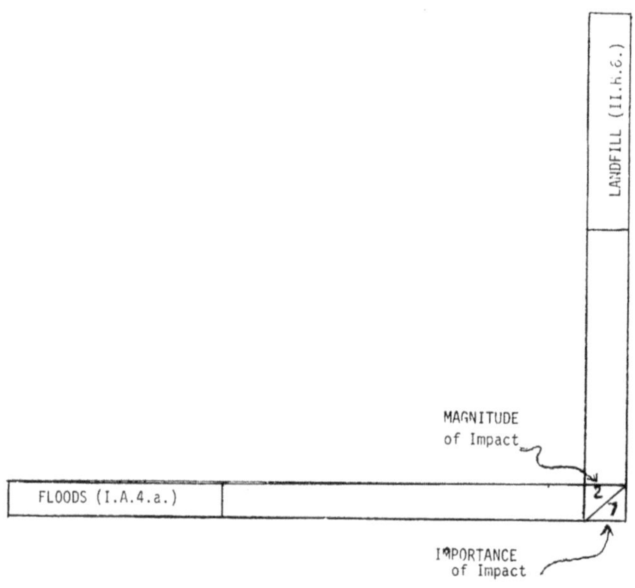

Figure 11-10.   Sample interaction using the environmental
impact matrix.

new access road would be needed. Additionally, plans called for open-pit mining, intersecting local stream channels and watersheds, an ore-processing plant and trucking of product. Water from a deep well would be used for ore-processing. Thus, the following activities (actions) were anticipated:

> Industrial sites and buildings
> Highways and bridges
> Transmission lines
> Blasting and drilling
> Surface excavation
> Mineral processing
> Trucking
> Emplacement of tailings
> Spills and leaks

Some other actions were considered from the matrix but were discarded as being of minor importance when compared with the nine above. Some of these were: "alteration of drainage," "sedimentation" and "erosion control."

Overriding environmental concerns were the placement of a mine in a national forest, and the effect upon the California Condor, a rare endangered species present in the region. In addition, water pollution, air pollution, vegetation and wild life impact, noise, power requirements and soils received attention.

The analysis is summarized in the reduced matrix in Figure 11-11. From that matrix, one can see that the activity making the most impacts is "surface excavation," affecting nearly all the listed conditions. The most important condition affected is "rare and unique species," reflecting the importance of the Condor's resting habitats. The other greatly affected condition is "wilderness qualities" of the forest area, as well as its "scenic views and vistas." Although no minus signs were used, all listed impacts were negative.

This, in essence, is the process. Accompanying text was used to complete the assessment. Alterations in mine plans could be requested to reduce impacts based upon the results of the matrix. Evaluators can use the matrix to determine where effort can be made to result in the least negative impact.

Discussion: As a checklist, the environmental evaluation matrix is excellent. It provides a comprehensive, rigorous look at all possible effects of proposed public projects. The matrix is also quite a flexible framework that can be further detailed where necessary to get at finer grain impacts, or can be expanded to allow for simultaneous consideration of alternative solutions. For example, each of the proposed actions could be given two columns instead of one, with one set labeled "A" and

		II B.b. Industrial sites and buildings	II B.d. Highways and bridges	II B.h. Transmission lines	II C.a. Blasting and drilling	II C.b. Surface excavation	II D.f. Mineral processing	II G.c. Trucking	II H.c. Emplacement of tailings	II J.b. Spills and leaks
I A. 2. d.	Water quality					2/2	1/1		2/2	1/4
I A. 3. a.	Atmospheric quality						2/3			
I A. 4. b.	Erosion		2/2			1/1			2/2	
I A. 4. c.	Deposition, Sedimentation		2/2			2/2			2/2	
I B. 1. b.	Shrubs					1/1				
I B. 1. c.	Grasses					1/1				
I B. 1. f.	Aquatic Plants					2/2			2/3	1/4
I B. 2. c.	Fish					2/2			2/2	1/4
I C. 2. e.	Camping and hiking					2/4				
I C. 3. a.	Scenic views and vistas	2/3	2/1	2/3		3/3		2/1	3/3	
I C. 3. b.	Wilderness qualities	4/4	4/4	2/2	1/1	3/3	2/5	3/5	3/5	
I C. 3. h.	Rare and unique species		2/5		5/10	2/4	5/10	5/10		
I C. 4. b.	Health and safety							3/3		

Figure 11-11. Environmental impact matrix for ore mining example.

one "B". Then the evaluation could proceed for two alternatives and comparisons can quickly and easily be made. Another quality of the matrix is that it is operationally easier to use. Even the mechanical use of the matrix still insures a sensitive evaluation, especially if it is used as the abstract of a full-blown, detailed statement explaining all the component decisions made in deriving the matrix.

These qualities belie the environmental evaluation matrix's sizable shortcomings. For fine scale impact analysis the two numbers generated for each impact fall short of being suitable. Even though any row or column may be expanded, the impact of "airport" on "residential land use" cannot appropriately be enlarged and still be useful within the matrix framework. Yet this is precisely the crux of many project controversies. Some interactions are too subtle to be captured by sets of paired numbers. Also, expansion of the matrix to too fine a scale or to include too many interactions ruins its elegance and efficacy.

The method requires large amounts of work to be done outside of its own framework, thus limiting its ability to organize the necessary efforts. A fair amount must already be known about the impacts before the method can be used, limiting it to an after-the-fact summary. Also ignored but vital are the interactions between various boxes of the matrix, the causalities and impact chains which haunt the designers of most public projects. The matrix itself provides little assistance in sorting things out. On the other hand, the matrix method does little to help one grasp the overall impact in some summary manner, leaving a sense of incompleteness to the analysis.

Before it can be used to any extent, an actual physical design must already have been made, limiting the matrix's aid in the planning stages.

The matrix technique is non-geographic, making it a secondary form of communication. In planning and design, impacts must usually be related to the landscape if evaluation is going to be useful as a tool.

The most important limitation in the method, though, is the derivation of the importance numbers. The method pivots upon these ratings for analysis emphasis, yet provides no assistance in deriving them. In fact, the representation of importance by one number is a severe constriction of the evaluation process. This is not to say that other techniques do not also suffer from the same myopia in importance rating, only that this one makes the rating too glib. In the above example, the importance of saving Condors ranks high, but perhaps only to some interests. More important than the individual value judgment itself, however, is the trade-off between values. Just how important *are* Condors relative to recovering those phosphates? This is the type of question that first generated the entire controversy over impact statements, yet this avowed (though admittedly preliminary) solution does not actually address the issue that gave it birth.

These shortcomings must be judged only in the light of the newness of evaluation techniques. Overall, the environmental evaluation matrix, if used creatively and non-mechanically can provide a sophisticated assessment of environmental impact, especially as a first cut at the problem. Its overwhelming advantages of comprehensiveness and ease of application outweigh

the many smaller shortcomings of which this technique is not
the only perpetrator.   The future of this type of evaluation
mechanism is great if combined or developed jointly with mapping techniques or expanded to account for differences in values among interest groups. The authors of the method admit that
the method is preliminary, yet it is a vital and well taken
first step.

Environmental Scoring

One of the searches of evaluation methodologies is for
some universal measurement · or currency to substitute for money
in assessing benefits, costs, and impacts. If some quantitative
measure could be devised to encompass social, institutional and
ecological values as well as economic ones, much of the environmental evaluation problem would be solved.   Precise trade-
offs and comparisons could be made with little of the agony
currently attendant with such analyses.   The entire evaluation
system would possess an order and structure which would render
it highly tractable to the demands made of it. It is the promise of this type of pay-off which keeps the quest alive, though
no universal currency has yet been found.

To approximate an actual universal measure, the notion of
an abstract score has been devised as a surrogate. All measured
and derived variables are related to a dimensionless, independent ordinal numbering system which represents relative value.
Once a score has been associated with each condition, impact or
action, detailed analysis is possible (e.g., Stover, 1972). Alternatives can be compared both in summary and by component to
make quite finely scaled trade-off and syntheses.

One such scoring system is the Environmental Evaluation
System (EES) devised by Battelle-Memorial Institute Columbus
Laboratories for the U.S.Bureau of Reclamation (Whitman, et al.,
1971) to aid in the Bureau's assessment of the environmental
impact of its proposed projects. Battelle-Columbus took a different approach to the subject than the Department of the Interior's environmental evaluation matrix.   Battelle-Columbus
assembled a checklist of impacts to be measured for each project, and created a weighting system to arrive at a composite
score for the proposal, whereas the matrix left the rating as a
magnitude/importance pair of numbers and made no attempt at uniformity of measurement or summary of proposal.

Although much of the method was tailored specifically to
the Bureau's needs and is still just proposed, the system itself has wide-ranging possibilities in its basic form.

The essence of the method is to measure all environmental
impacts of a project, convert each measure to a score, add the
score to arrive at a composite and, finally, to compare the

score with the score of conditions without the project.   The
final comparison for a net score is made on a with-without
basis rather than a before-after basis.   The environment is
scored once as if no project were to be built, so that inevitable degradation of the environment not attributable to the
project is eliminated.   The environment is scored again after
completion of the project to summarize changes in all the environmental factors due to the project.   Thus, each project or
project alternative has associated with it a net score relating
solely to the projects immediate impact.

The first step in the EES was the identification and in-
ventorying of all possible impact parameters.   A list of 66
parameters was assembled by Battelle-Columbus ranging from
stream temperatures and food chains to cultural significance
and "mystery."   The 66 parameters were sorted into seventeen
environmental components, which were in turn placed in four
general environmental categories: ecology, environmental pollution, esthetics, human interest. A hierarch of generality was
thus created so that projects could be discussed at various
levels of specificity.   The entire hierarchy is as follows:

I. ECOLOGY

A. Species & Population
  1. Rare and endangered plant and animal species
  2. Productive plant species
  3. Game animals
  4. Other animals
  5. Resident and migratory birds
  6. Sport fisheries
  7. Commercial fisheries
  8. Pestilent plant and animal species
  9. Parasites

B. Habitats and Communities
  10. Species diversity
  11. Food chains
  12. Land use for habitats and communities

C. Ecosystems
  13. Productivity rate
  14. Hydrologic budget
  15. Nutrient budget

II. ENVIRONMENTAL POLLUTION

D. Water Pollution
  16. Algal blooms
  17. Dissolved oxygen
  18. Evaporation
  19. Fecal coliforms
  20. Nutrients
  21. Pesticides,herbicides, defoliants
  22. pH
  23. Physical river characteristics
  24. Sediment load
  25. Streamflow
  26. Temperature
  27. Total dissolved solids
  28. Toxic substances
  29. Turbidity

E. Air Pollution
   30. Carbon monoxide
   31. Hydrocarbons
   32. Particulate matter
   33. Photochemical oxidants
   34. Sulfur Oxides

F. Land Pollution
   35. Land use and misuse
   36. Soil erosion
   37. Soil pollution

G. Noise Pollution
   38. Noise

III. ESTHETICS

H. Land
   39. Land forms
   40. Geologic surface material

I. Air
   41. Pleasantness of sounds

J. Water
   42. Surface characteristics
   43. Water-land interface characteristics

K. Biota
   44. Vegetation
   45. Fauna

L. Man-Made Objects
   46. Visual
   47. Condition
   48. Consonance with environment

M. Composition
   49. Interaction of land, air, water, and man-made objects
   50. Color

IV. HUMAN INTEREST

N. Educational-Scientific Significance
   51. Geological significance
   52. Ecological significance
   53. Archeological significance
   54. Unusual water phenomenon

O. Historical Significance
   55. Related to persons
   56. Related to events
   57. Related to religions and cultures
   58. Related to architecture and styles
   59. Related to the "western frontier"

P. Cultural Significance
   60. Related to Indians
   61. Related to religious groups
   62. Related to ethnic groups

Q. Mood-Atmosphere Significance
   63. Isolation-solitude
   64. Awe-inspiration
   65. "Oneness" with nature
   66. Mystery

Each chosen parameter is intended to have significance worthy of separate consideration, and to be mutually exclusive of the others. Major changes in any parameter can be noted with "red-flag" status as a dnager signal to designers and evaluators.

Associated with each parameter is a measure by which a
score can eventually be given to that parameter. Some of the
measures are straightforward and others are as elusive as the
parameters they measure. The following is a small sample of the
detailed measure provided:

(3) Game animals - Number of species of this type, quantity
per unit per species, and a real distribution of species.

(11) Food chains - Linkages between higher and lower order food
chain species - actual compared to desirable.

(17) Dissolved oxygen - Monthly average in mg/l and daily variation - coefficient of variation.

(32) Particulate matter - Concentration in ppm, $\mu g/m^3$.

(38) Noise - Level of sound in decibels (db) and duration and
variety of sound.

(43) Water-land interface characteristics - Proportion of
shoreline in vegetation, rip rap, man-made structures, eroding
soils; visibility of drawdown effects on shoreline; and distance of mud flats and silt deposits extending into water body.

(46) Visual design - No unit of measurement.

(51-66) Human interest parameters - Each of these parameters
is measured by something called a "package of significance,"
sort of the gestalt of significant changes in each of the specific parameter areas. The measure is the change of significance of all sites and objects affected by the project in relation to the small significance of the package of the specific
values impacted.

It was recognized that the parameters were not of equal
importance and that the relative value of each affected the
overall impact of the project. As noted previously, no common
measure of value exists in the field, and so the Environmental
Quality Unit was invented. A total of 1000 units is distributed
among the parameters according to their importance. Thus the
units of any parameter represent a maximum value that the given
parameter can be worth relative to the others. If it is determined that a project has less than maximum environmental quality as described by a parameter, the units assigned would naturally be less than the maximum for that parameter. Figure 11-12
is a representation of the distribution made of the 1000 EQU's
on the component level. As can be seen the greatest value is
associated with environmental pollution (321 units) and the
smallest to esthetics (159 units). The single most important
component is water pollution (160 units).

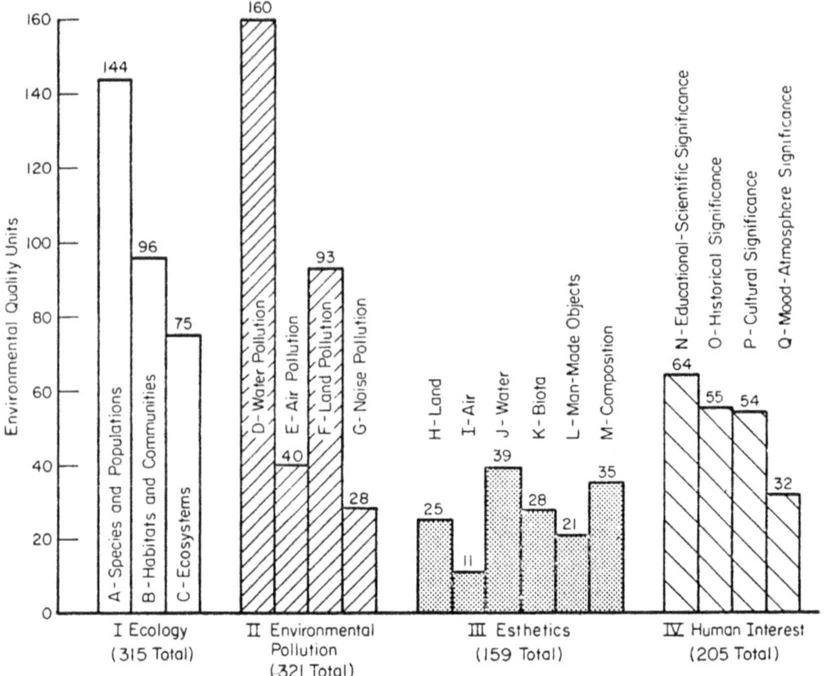

Figure 11-12. Relative significance of environmental evaluation system components.

The values given the parameters were derived to reflect the needs of the Bureau of Reclamation. Other weighting systems can and must be derived for other purposes. The weighting methodology itself is a vital and revealing process in the course of developing a project. The weights associated with the parameters are highly dependent upon the biases and judgment of any individual charged with the responsibility for selection. Any weighting system must be subject to careful review.

The process of weighting has two steps: (1) valuating all parameters (or components) relative to one another, and (2) placing numeric EQU value on various quality levels within any given parameter. Various methods exist to undertake the first step; the one used by Battelle-Columbus in that of paired comparisons. In the paired comparison technique, categories of indicators are first listed in descending order of importance.

560

The first item on the list is given the value of "1". Then each succeeding list item is compared with the one above it and assigned a portion of the value of the item above. For example, the second list item might be three quarters the importance of the first and therefore, would have a value of .75. The third item, when compared to the second might be considered one third as important, and given a value, therefore, of .25 (1/3 x .75). When all succeeding pairs have been valued, the numbers are totaled and each item's value is expressed as a percentage of the total for the category. The EQU's are then distributed according to those percentages.

The valuation should be repeated a few times until a consistent set of weights is derived. Many separate estimates may be made or averaged for a final weighting system. The valuation is done in three stages, firstly among categories, secondly within each category among components, and thirdly within each component among parameters for the final distribution.

The second step involves a determination of how scores will be assigned within each parameter, depending upon the value of the parameter measure. The score can be assigned by subjective judgment and probably must for the "softer" parameters. Or, a value function can be created which directly transforms the quantitative measure of parameter into an appropriate number of EQU's.

The weighting system completes the environmental scoring system. Impacts are theorized or measured, parameters are given values, values translated into EQU's and EQU's combined into scores both at various levels of specificity and as a total. The EES provides a score-sheet to aid in the process. As mentioned earlier, impacts are similarly composited for the area without the project and a net difference is taken as a final score. Thus, every aspect of every proposed project alternative, including no project, is precisely scored and is, therefore, comparable with every other aspect. The completed score sheet is an expression of the anticipated impacts and provides a basis for further design, trade-off or synthesis analysis, and completion of impact statements.

Discussion: Scoring systems provide a valuable shorthand in developing commensurable impact indicators and summary expressions of value. They overcome in some respects the difficulty of equating diverse parameters and trading off things like species lost against jobs created. They do, however, tend to obscure more fundamental measurement problems behind a cloak of false precision. In the EES, those parameters which have always been measured accurately are still measured that way. The elusive parameters, though, are still veiled in subjectivity and obscure scoring techniques, even though the weighted score looks as precise as the carefully assayed quantitative parameters. Once again, it is not this method alone which suffers from this problem, but here the problem is aggravated by the need for eventual summary of complex variables by single numbers. The measurement problem is ameliorated but by no means solved by the imposition of a score.

The sensitivity of the scoring to the weighting system cannot be overemphasized. One project may be an industrialist's dream, an ecologist's disaster, a landscape architect's eyesore or a politician's salvation although all may agree on the measured values of all the parameters. The difference is in the eye of the evaluator.   The EES is acutely aware of the value judgment problem.   It takes great pains not only to emphasize that the specific weighting system is intended to fulfill a single need (the Bureau's) but also to expose the weighting derivation process to observation and review. This latter quality is perhaps the most important aspect of the system. After all criticisms are made of weightings, the problem remains: how is it to be done? The EES makes an attempt to solve that dilemma.

The other significant sensitivity besides weighting is to the listing of impacts themselves.   Duplications or omissions may skew results undeservedly. Great care must be taken to obtain a list which is comprehensive, detailed, but not overlapping or unduly biased in any direction.   The EES emphasis impacts to the natural environment to the detriment of social, economic and institutional impacts.   While the system was tailored for work in mostly rural, natural areas, it cannot long survive without addition of the crucial impacts to suburban and urban environmental quality.   The creators of EES eliminate consideration of these impacts as being secondary, indirect and, therefore, beyond the range of the Bureau of Reclamation. They claim the system should account only for direct and indirect physical impacts, and direct use impacts, not indirect use impacts. The distinction between direct use and indirect use becomes hazy too soon for comfort to exclude any reasonable impact that can be measured, however remote.   If it is attributable to the project in the with-without sense then it should be added and used for evaluation. How else is evaluation methodology to grow more rigorous and effective?

The use of scoring systems, especially in conjunction with other techniques, adds a powerful evaluation weapon in the designer's absence.   The greatest aid comes from the measurement assistance and the delineation of values and vested interest through weighting. Other systems, like the balance sheet method following emphasize just those aspects of evaluation, and are useful if they are extensions of scoring. The scoring technique by itself, is self-contained, relatively straightforward and easy to apply depending, of course, on data availability.   But it should not be forgotten that the assignment of cardinal numbers to a wide variety of values is an untenable approach   to

environmental analysis.   If used in isolation of this understanding the approach would be unsound.

Very few evaluation techniques respond to the pressures
brought to bear on projects by interest groups and concerned
communities.   Most techniques adhere to a single viewpoint in
evaluating impacts and effects, usually those of the client or
professional. In fact, none of the preceding techniques account
explicitly for the conflicting value judgments inherent in the
assessment process, although the scoring system is cognizant
of the problem.

Any method that would account for the varying values,
goals and objectives of the various interests surrounding the
project would be a large improvement over others, especially in
urbanized areas.   As the density of development in the project
area increases, social forces begin to outweigh natural and
physical forces in determining the success and shape of the
public project.   Evaluation must also shift gears if it is to
respond to changing priorities of interest and remain effective.

Urban planners have recognized this shortcoming in even
expanded methodologies, and have tried to develop techniques to
overcome the limitations (Lichfield, 1968).   Much work in the
field has been done by Nathanial Lichfield in transforming
cost-benefit analysis to meet the needs of inter-disciplinary
urban evaluation. The method evolved by Lichfield is the Planning Balance Sheet (Lichfield, 1964).   It is similar to an accounting technique. It is the only type of expanded methodology
that retains the basic framework of cost-benefit analysis and
the only type that explicitly retains price costs and benefits.
The method does, however, broaden the concept of costs and benefits to include non-physical, social and intangible items as
well as a cross reference to users, non-users and other groups.

Description: The planning balance sheet first groups the
various interests into various homogeneous groups distinguished
by the kinds of operations they perform. Groups of individuals
are classed generally into producers and consumers.   Producers
are identified as those groups creating or controlling economic
and social goods and services.   The definition is broad enough
to include employers, legislators, planning agencies, consultants, developers, businessmen's groups and landlords.   The
consumer groups would include residents of various types, system users, employees, as well as some of the producer groups in
a different role.   For example, a local government may be a
producer by creating an urban renewal area, but may also be a
consumer by desiring office space or public housing in that renewal area. Thus when groups are represented by function rather
than title, the accrual of costs and benefits are easier to

trace and clearer in their implications. These groups are listed as one dimension of a matrix, usually the rows.

The other dimension is the array of alternatives, be they system configurations, sites, or plans. Each alternative is subdivided into a costs (disadvantage) column and a benefits (advantage) column. These columns are further subdivided into capital (once for all) items and annual (continuing) items. Figure 11-13 presents a typical matrix for a two plan analysis. Groups X, Y, Z are producers and group W, Y' (group Y in a different role) and Z' are consumers.

| | PLAN A | | | | PLAN B | | | |
| | Benefits | | Costs | | Benefits | | Costs | |
	Cap.	Ann.	Cap.	Ann.	Cap.	Ann.	Cap.	Ann.
Producers								
X	$a	$6	----	$d	----	----	$b	$d
Y	i_1	i_2	----	----	i_3	i_4	----	----
Z	M_1	----	M_2	----	M_3	----	M_4	----
Consumers								
W	----	$e	----	$f	----	$g	----	$h
Y'	i_5	i_6						
Z'	M_1	----	M_2	----	M_3	----	M_4	----

Figure 11-13. Schematic planning balance sheet.

The table is completed by measuring the impact of the plans or projects on the groups. The measures may be monetary as represented by $ in Figure 11-13, they may be non-monetary yet quantitative as shown by the M's in the figure, or they may remain intangibles, as shown by the i's. Problems of measurement still remain in this system, although the problem is somewhat mitigated because relative difference between alternatives are stressed, not absolute amounts. Differences are often easier to measure and control than absolutes. Scoring techniques may be employed to make all table entries comparable and to allow summary tabulations to be made for selection and/or plan modification.

Great advantages may be obtained if the method allows for
some participation by the various groups involved. Elements of
self-scoring can be added to minimize evaluation biases, and a
system of weights added to emphasize important groups or sets
of impacts.   In this way, the planning balance sheet becomes a
project generation, project synthesis, liaison, communication
and feedback tool as well as an evaluation tool.   The method,
if used correctly, becomes an integral part of the design process at an early stage.

Example: The planning balance sheet has been applied successfully by Parsons, Brinckerhoff, Quade & Douglas, Inc.
(PBQ&D, 1970) in the development of a model neighborhood plan
under the HUD Model Cities program in San Juan, Puerto Rico.
The area is a *barrio,* dilapidated land use area along the
waterfront, valuable because of its proximity to the central
business district.   The objective was to develop a plan for
housing, industry, schools, offices, commerce, etc. which would
accommodate the various interests of the city.

Eight major groups with strong, valid, interests in the
area were identified. They were (1) the dwellers of the *barrio*,
(2) the residents of metropolitan San Juan, (3) the local government, (4) a group of businessmen organizing and developing
an office/financial complex near the area, (5) the business
community in general, (6) the Puerto Rico Planning Board, (7)
the City Demonstration Agency (Model Cities agency), and (8)
the local housing authority.

Groups were not divided into producers and consumers, but
left only as identifiable interests. The groups were contacted
and asked to participate in the evaluation process. In the case
of the metropolitan residents, the obvious problem of representativeness was resolved by having the other seven groups serve
together *in toto* as the citizens' representatives.   The barrio
residents were represented by a tenants committee, eliminating
the need to organize a representative group.   The eight groups
were brought into the process early enough to make significant
contributions to the development plans.

While the interest groups were being assembled, the consultant prepared a preliminary array of twelve alternative development schemes, ranging from megastructures to marina oriented bedroom neighborhoods.   This was later reduced to four
viable schemes.   Each scheme was just a sketch, rather than a
fully detailed plan, built around one organizing principle.
These four alternatives were placed before the residents for
consideration: a neighborhood oriented plan, a marina/waterway
cluster plan, a mass-transit/commutation oriented plan, and a
mixed use/counter-node plan.   In this way, mutual feedback between groups and plans could be obtained without locking into
one preferred alternative and alienating the participants. This
engendered an air of mutual trust and appreciation of opponents.

The organizing framework for this interactive design and evaluation process was the planning balance sheet. Several modifications of the basic format and technique were made. Figure 11-14 displays the final resultant balance sheet as well as its basic structure. Alternatives and factors were arrayed across the top as columns; the groups and their concerns formed the rows of the balance sheet.

The balance sheet was divided into two sections, intangible factors and tangible factors, and data was accumulated and analyzed separately. An intergroup weighting system was utilized, wherein 100 points were distributed among the various groups by the consultant, based upon the immediacy of their interests. A scoring system was used to ease tabulations within each category and each alternative was given a summary total score for tangibles and intangibles.

For scores to be assigned, each group was first asked to identify the factors which were important to it. Groups selected factors ranging from "access" and "tax base" to "image" and "tourism." This allowed groups to set the terms of their own evaluation, thus limiting evaluation biases and creating implicit goals and objectives. The use of the self-enumeration overcomes what would otherwise be a limit of the process, that of not responding to specific group needs and goals through ignorance, bias or design. The absence of goal-orientation had been noted by critics of the balance sheet (e.g., Hill, 1968) who have devised similar techniques to the illustrated Parsons, Brinckerhoff method.

Groups then distributed the importance weightings allotted them among their factors by any method they desired. For each alternative a rank was given for every factor, reflecting the decision for the group about the relative effect of each alternative on the category of concern. Ranks of 1 to 5 were used for intangibles and 1 to 3 for tangible items. The ranking was made by the consultant with group participation because technical considerations limited the ability of groups to adequately measure differences. The "value" for each factor was derived by multiplying the assigned ranking by the factor weight. The "values" of each alternative were summed from each factor score to arrive at a bottom line or total score for the alternative. Totals were made for each alternative by: net intangible gains, tangible losses and costs, and tangible gains. Tangibles and intangibles were sensibly kept apart during scoring and analysis. The results clearly show the mass-transit oriented alternative the clear favorite. It was highest in gains and next to lowest in cost.

The result of this selection process also contains the vital citizen involvement crucial to the implementation of controversial projects. The importance of public participation is

also stressed in Chapter 12.   Properly used, the balance sheet
can provide the inputs for a better design process.

Discussion: The emphasis placed on interest groups and
their values provides the social accounting method unique inputs in the evaluation field.   The technique minimizes natural
and environmental parameters, but does not need to if properly
modified.   The retention of cost-benefit methodology in part
might overcome resistance to its use by those familiar only
with traditional techniques.

This method is most useful in the design stage of a project.   Its emphasis on group participation and alternative selection minimizes its utility in impact assessment as currently
conceived.   It also lacks the geographic dimension crucial to
effect assessment, although some preliminary attempts have been
made to incorporate spatial concerns (Gebert, 1971).

The planning balance sheet is highly dependent upon the
selection of participants, making judicious and representative
selections imperative. This selection process is sometimes as
difficult as the evaluation itself. The product of this evaluation is a concensus alternative, "best" in the sense of acceptability.   The question of environmental impact still remains
in the pure sense, and must be evaluated concurrently with the
balance sheet.   Social, political and environmental evaluation
can be woven together within this framework, however.

Social accounting does not totally solve the weighting and
scoring problems.   Yet the process is open and the opportunity
for group self-measurement greatly enhances the prospects of
eliminating unnecessary value judgments in evaluation.   The
planning balance sheet also allows for a degree of active participation which is unavailable in other techniques, which by
itself is an advantage of the technique although the orchestra-
tion of such an effort is demanding.   The ability to have a
common ground for adversary groups to discuss and debate may
overcome some of the blind opposition to project implementation
and may provide for project design which is assured of compatibility with social and political priorities.

COMPARISON OF METHODS

The criticisms discussed under each of the six techniques
should not be taken as shortcomings peculiar to a method, but
rather the general frustration with the state-of-the-art of interdisciplinary evaluation.   The six methods presented as pro-
totypes and the similar methods now in use in the field represent great strides forward in evaluation technology and should
not be underestimated.   Each in its own way comes to grip with
some of the painful realities of the field.

Balancing the Pluses and Minuses of

Eight Interest Groups Involved in the Area

INTEREST GROUP INTANGIBLE FACTORS	WEIGHT	Neighborhood Oriented		Waterway Clusters		Mass-Transit Oriented		Mixed Land Use	
		Rank	Value	Rank	Value	Rank	Value	Rank	Value
City Demonstration Agency									
Possible immediate action	7	3	21	4	28	5	35	4	28
Desirable urban environment	5	5	25	5	25	5	25	4	20
Five-year implementation	3	3	9	4	12	4	12	4	12
Total for Group	15		55		65		72		60
Slum Residents									
Relocation housing choice within area	4	2	8	3	12	5	20	3	12
Density low as possible	4	5	20	3	12	−2	−8	3	12
Jobs	4	2	8	3	12	5	20	4	16
Access and mobility	2	2	4	2	4	3	6	5	10
Community facilities	2	5	10	3	6	3	6	3	6
Recreational park	3	1	3	3	9	5	15	4	12
Total for Group	19		53		55		59		68
San Juan Residents									
Housing, all incomes	4	3	12	4	16	5	20	4	16
Job sources	4	2	8	3	12	5	20	4	16
Eliminate social problems	3	2	6	4	12	3	9	4	12
More and better streets and highways	2	3	6	2	4	4	8	3	6
Improve waterways potential	2	1	2	5	10	3	6	3	6
Total for Group	15		34		54		63		56
Planning Board									
Better quality of urban environment	4	3	12	5	20	4	16	3	12
Develop waterways for city recreation	3	1	3	5	15	3	9	3	9
Implementation feasibility	2	3	6	4	8	5	10	2	4
Permit future proposal implementation	2	5	10	4	8	5	10	5	10
Master plan compliance	3	4	12	4	12	5	15	2	6
Total for Group	14		43		63		60		41
Housing Authority									
Provide relocation housing	2	2	4	3	6	5	10	3	6
Build housing quickly	3	3	9	4	12	5	15	3	9
New concepts for development	4	2	8	4	16	4	16	3	12
Total for Group	9		21		34		41		27
Financial Interests									
Usable commercial space	1	2	2	4	4	5	5	4	4
Image	1	2	2	4	4	5	5	4	4
Access to "New Center"	1	3	3	2	2	5	5	2	2
Total for Group	3		7		10		15		10
General Requirements Equally Satisfied by All Alternatives	25								
ALL GROUPS GRAND TOTAL	100		213		281		310		262

Figure 11-14 A. Planning balance sheet for a neighborhood design in San Juan.

Balancing the Pluses and Minuses of

Eight Interest Groups Involved in the Area

INTEREST GROUP / TANGIBLE FACTORS	WEIGHT	LOSSES AND COSTS — Alternatives								GAINS — Alternatives							
		Neighborhood Oriented Rank	Value	Waterway Clusters Rank	Value	Mass-Transit Oriented Rank	Value	Mixed Land Use Rank	Value	Neighborhood Oriented Rank	Value	Waterway Clusters Rank	Value	Mass-Transit Oriented Rank	Value	Mixed Land Use Rank	Value
Slum Residents																	
Rents not now paid	5																
Equity required for condominium ownership	5																
Compensation for owned structures	4																
Personal income	7									1	7	2	14	3	21	3	21
Increased property value	4									1	4	2	8	3	12	2	8
Increased taxes	6	1	6	2	12	3	18	1	6								
Total for Group	31		6		12		18		6		11		22		33		29
San Juan Residents																	
Increased property values	9									1	9	3	27	3	27	2	18
Travel to major recreation	4	3	12	1	4	1	4	3	12								
Increased taxes	7	1	7	2	14	3	21	1	7								
Total for Group	20		19		18		25		19		9		27		27		18
Housing Authority																	
Soil stabilization	13	3	39	3	39	1	13	2	26								
Public housing construction	13	2	26	3	39	2	26	2	26								
Total for Group	26		65		78		39		52								
Local Government																	
Utilities & city services	7	3	21	2	14	3	21	2	14								
Tax base	2									1	2	2	4	3	6	2	4
Tourism	1									1	1	2	2	2	2	1	1
Channel development & pollution control	2	2	4	3	6	1	2	2	4								
Expressways	1	3	3	3	3	3	3	3	3								
Major parks & recreation	1			2	2	3	3	1	1			1	1	3	3		
Total for Group	14		28		25		29		22		3		7		11		5
San Juan Business Interests																	
Income from land & commercial space	3									1	3	2	6	3	9	2	6
Increased taxes	1	2	2	2	2	3	3	1	1								
Increased tourism	2											2	4	3	6		
Total for Group	6		2		2		3		1		3		10		15		6
ALL GROUPS GRAND TOTAL	100		120		135		114		100		26		66		96		58

These factors applied to a portion of the area that had already been built up and contained a mixture of substandard and acceptable housing. In the absence of firm information as to what proportions of housing should be cleared, rehabilitated, or left alone, meaningful ranking values could not be assigned.

Figure 11-14 B. Planning balance sheet for a neighborhood design in San Juan.

Two important considerations of real world complexity should be made note of specifically. One is the scale of project and the other the time horizon of impacts.

1. Scale--In carrying through an evaluation, limits must be set as to what part of the universe must be considered for study. Certainly the project site is too small an area, but the economic region may be too large. Similarly, the order of magnitude of effects considered must be limited practically, but not so constrained as to exclude important interactions remote from the site yet due to the proposed project. A suitable way of delineating the scale might be to construct impact chains spatially and quantitatively into a hierarch and cutting off analysis at some reasonable point.

2. Time Horizon--Most methods (except the land use model) consider impacts upon completion of the proposed project and so exclude a vital link to the future. It is possible that projects with short term benefits may prove deleterious in the long run. A way to overcome the problem may be to expand each interaction to a long- and short-term pair and have each considered separately. A discounting method (similar to compounding) could be employed to insure the proper perspective between short and long range effects. If the two are treated as one, however, an explicit statement of the time consideration should be included in the method.

Although each method has been presented as a pure type, they are not mutually exclusive and each freely draws upon the mechanisms of the others. While each method confronts various aspects of the hydra of evaluation, some generic characteristics seem to run through them all:

1. Selection of appropriate variables--Whether sixteen or sixty-six, a list of parameters is a requisite; things must be chosen for measurement. The methods herein provide a large menu from which to choose, yet the final selection must depend upon the actual application or proposed project.

2. Measurement of parameters--The selected variables must be given values either as sum totals for a project as in the matrix and scoring techniques or as a spatially oriented value, as in mapping and land use modeling. The maps used shades of gray, the model a binary code, the matrices a relative magnitude and the scoring an abstract score.

3. Weighting of parameters--Each method recognized or had the ability to recognize differences in the importance of various parameters. Parameters were emphasized through weights, either numeric, graphic or coded. Each weighting reflected a value judgment of the method's user.

4. Comparison of actions with changes in parameters--This step
    is pivotal to each method. At some point, the proposal is
    compared to the environment, and the interaction is gen-
    erally noted as a change in parameter. The interaction was
    limited in the mapping cases to an overlay, but appeared as
    a change in a land use code in the land use model, a change
    in the net score in the scoring system and a magnitude in
    the matrices. The interaction was generally compared to the
    no-project case (the *ceteris paribus* condition).

5. Compositing of interactions--Each method had some mode of
    presenting the summation of its techniques, that is, an
    output. This output generally was a tying together of the
    threads of analysis into a finished fabric. The mapping
    techniques used a composite map, the model a land use map,
    the environmental matrix the matrix itself (actually no
    composite) and the balance sheet and scoring methods a com-
    posite score.

These five items compose the essence of any interdisciplinary technique for evaluation. Each facet of the analysis
has its own intricacies and pitfalls. Each aspect deserves
careful further research, as there is much room for improvement.
From these five items, however, a complete methodology can be
generated, limited only by the imagination and sophistication
of the practitioner.

REFERENCES

Bigler, Craig, "Using Composite Mapping in Optimum Location, A Preliminary Test with Nine Industries in Utah," Office
of State Planning Coordination, State of Utah, Salt Lake City,
Utah, Nov. 6, 1970.

Center for Real Estate and Urban Economics, *Jobs, People
and Land: Bay Area Simulation Study (Bass)*, Institute of Urban
and Regional Development, University of California, Berkeley,
1968.

Committee on Power Plant Siting, *Engineering for the Resolution of the Energy-Environment Dilemma*, National Academy of
Engineering, Washington, D.C., 1972.

Gebert, Gordon A., "COMPOSITE: ANALYSIS OF MULTIPLE IN-
TEREST GROUP VALUE SYSTEMS RELATIVE TO SPATIAL ATTRIBUTES,"
School of Architecture and Environmental Studies, City College
of the City University, New York, 1971

Goldner, William, *Projective Land Use Model (PLUM)* (Berkeley, California: Bay Area Transportation Study Commission,
1968).

Hill, Morris, "A Goals-Achievement Matris for Evaluating Alternative Plans," *Journal of the American Institute of Planners,* January, 1968.

INTASA, *A Computer Simulation Model for Flood Plain Development,* Institute for Water Resources, Corps of Engineers, Alexandria, Virginia, 1972.

Leopold, Luna, et al., *A Procedure for Evaluating Environmental Impact,* Geological Survey Circular 645, U. S. Geological Survey, Department of the Interior, Washington, D.C., 1971.

Lichfield, Nathaniel, "Cost Benefit Analysis in Plan Evaluation," *Town Planning Review,* Volume 35, 1964.

Lichfield, Nathaniel, "Evaluation Methodology of Urban and Regional Plans: "A Review," *Regional Studies,* Volume 4, Pergamon Press, 1968.

McHarg, Ian L., *Design with Nature,* (Garden City, N. Y.: Natural History Press, 1969).

Meadows, Donella, et al., *The Limits to Growth* (New York: Universe Books, 1972).

Michel, Henry and Sigurd, Grava, "The Planning Balance Sheet and the Barrio," *Worldwide P & I Planning,* Nov.-Dec., 1970.

Olson, Kenneth C. and George Nez, "Land Use Planning & Economic Environmental Considerations," *State Planning Issues '72,* The Council of State Planning Agencies, 1972.

Parsons, Brinckerhoff, Quade & Douglas, *Model Neighborhood Area Land Use and Development Plan* (San Juan, Puerto Rico: Municipality of San Juan, 1970)

"Program Description and Availability Memorandum for GRID," Laboratory for Computer Graphics and Spatial Analysis, Harvard University, Cambridge, Mass., 1971.

Seader, David and Sigurd Grava, "A Demonstration of the Use of Computer-Aided Land-Use Modeling for Regional Service System Design," *Proceedings of the 6th Annual Urban Symposium,* Association of Computing Machinery, New York, 1971.

Stover, Lloyd, *Environmental Impact Assessment: A Procedure,* Sanders and Thomas, Miami, Florida, 1972.

Urban Development Models, Special Report 97, Highway Research Board, Washington, D.C., 1968.

Whitman, Ira L., et al., *Design of an Environmental Evaluation System*, Battelle-Columbus Laboratories, Columbus, Ohio, 1971.

AUTHOR NOTES

This chapter was authored by David Seader, New York, N.Y. Mr. Seader is currently a senior urban planner for urban systems for the firm of Parsons, Brinckerhoff, Quade & Douglas, Inc. He has managed a number of projects for the firm involving urban and regional land use modeling and analysis. Projects under his supervision have included studies of Cleveland, St. Louis, Fairfax County, Virginia, Lakewood, Colorado and the Republic of Singapore.

Mr. Seader is also an Assistant Professor of Urban Planning at Columbia University. He teaches graduate courses in computer applications in urban planning and architecture, as well as a variety of basic subjects in the urban planning Comprehensive Workshop. His academic interests include computer-aided analysis, regional planning, effect assessment and computer mapping and graphics.

PUBLIC PARTICIPATION

BY DESMOND M. CONNOR

Public participation has become the unbidden companion of
the planner and engineer in a growing number of projects across
the country. This upsurge of involved citizens stems from a
number of causes, for instance, a disillusionment by many with
previous values on unlimited growth, dissatisfaction with some
of the negative side effects of major projects, a lack of confidence in the willingness and ability of politicians and administrators to interpret and accept citizens' goals.

The recent increase in the awareness and active involvement of people with planning and decision-making in major projects has deep roots in our society. From the well-springs of
democracy through to the continuing relationship of politicians
with their constituents, there is a tradition which is demonstrated more specifically in the long-standing custom of public
hearings on highways and the responsive stance taken by such
agencies as the U. S. Bureau of Reclamation and the U. S. Army
Corps of Engineers. The period of effective confrontation of
major works programs is difficult to place, but the results of
the Sierra Club's opposition to the Echo Park Dam in 1957 encouraged many to support similar battles subsequently.

In this context, this chapter is a response to several
core questions: what is public participation, why involve the
public and how does one design and manage the public participation component in planning a major public project?

Public Participation - A Part of the Whole

In the opening exploration of the traditions and assumptions underlying this book, the salience of the public's stake
in both the environment and in major public projects is obvious.
Similarly, the exposition of the project evolvement process
specifies many components, such as goals, leadership, influence,

attitudes and understanding, which can hardly be addressed without constructive relationships with the persons who possess them.

The illustrative treatments of transportation, water and power provide further reminders of the place of the public, not only as consumers of services but, increasingly, as a source of preferences concerning how, when and where such services will be provided.

The importance of the public's values as data for disciplinary models used by ecologists, sociologists, architects, economists, lawyers and systems analysts are evident in the respective chapters. In one case study after another, illustrations are provided that public participation is indeed a factor to be reckoned with; if left unmanaged it can readily destroy an orderly planning project.

Public participation is thus an important consideration in more than half the chapters of this book, as it addresses the question of planning and designing public projects in a context of the total environment, human and non-human. Some of the reasons for this will be reviewed shortly.

Is public participation a passing fad or a legitimate component of the comprehensive planning process? For a perspective, see how various forms of participative management have growing acceptance in the field of administration, how public participation has become part of a broad range of government programs and how consumerism shows no signs of fading from the business arena.

In Britain, a Royal Commission has produced a report "People and Planning" to provide practical ways in which local authorities could best implement the new Town and Country Planning Act. In Canada, the Canada Water Act and other federal legislation make specific provision for public participation.

Recent national surveys of city administrations and planning agencies at the state, regional and local levels indicate substantial satisfaction with the results of public participation and a general expectation that it will continue because of its usefulness and because legislation makes it unavoidable.

From this evidence, it appears that public participation is already a powerful component in our national value system and in related fields such as administration, business and planning, both here and abroad. There is no indication that citizen involvement will decline in the future. The reasons for this will become clearer in the next section, which examines the nature of public participation, its background and some terms useful to describe and direct it.

WHAT IS PUBLIC PARTICIPATION?

The two words "public" and "participation" are in such common usage that their meaning seems clear to most of those who use them. Even when employed together, there is a strong tendency to assume that not only does everyone know what "public participation" means, but that everyone shares the same meaning. Experience disproves the assumption all too often.

This section will therefore provide a working definition of public participation as a field of activity, an indication of its disciplinary origins, and some major concepts useful in understanding and managing public participation. Some short vignettes are included to illustrate major areas.

A Definition

For working purposes in this discussion, public participation in planning a public project is *a systematic process of mutual education and cooperation which provides an opportunity for those affected, their representatives and technical specialists, to work together to create a plan which combines and reflects their values, knowledge, experience and best judgement at the time in a democratic manner.* (Citizen involvement will be used synonymously with public participation to relieve repetition for the reader.)

This definition assumes that:

1. Public participation is neither a single unitary act, such as a public hearing, nor a haphazard set of occurrences, but a planned process, responsive to the unforeseen but guided by a general concept.

2. The process of public participation is largely a learning experience by which each participant acquires a more complete understanding of both the issues and how other parties see the issues. Each participant is potentially both a learner and a teacher in a series of events which foster a shared understanding of the situation and ways to deal with it. A growing mutual trust and confidence between the parties is, of course, an essential foundation for learning and creative cooperation. This process of public participation rejects those approaches which focus on conflicts or negotiations between protagonists from initially fixed and polarized positions.

3. Public participation encourages the acceptance of civic responsibility in ways meaningful for the people concerned, but it does not compel attention.

4. Since "those affected" often include the unborn, future migrants and the currently unaware, the representatives of these persons are included in the process, e.g., elected representatives from appropriate levels and agencies.

5. The responsibility for the technical aspects of the project remains with its professional staff; public participation does not remove their responsibilities, nor those of the elected representatives under whom the project is carried out.

6. There is a focus on creating a plan which represents the best contributions of all the participants, within the constraints of the situation, rather than minimally adapting a standard practice to one more occasion.

7. Since technology and human values change, this concept recognizes that the plan must preserve flexibility for future needs, problems and opportunities.

8. The democratic manner of operation recognizes the use of both an open process to gather information, ideas and preferences as directly as possible from citizens and to respond to them, and yet also a representative process of legislative democracy by which the political system functions in making final decisions on matters of public policy.

    More specifically, public participation is happening when:

    - planners listen to residents concerning their attitudes, goals, fears and factual suggestions;

    - citizens find early and convenient opportunities to make positive contributions ("citizens" may include visitors as well as residents):

    - citizens learn from planners and others a broader and deeper knowledge and understanding of their environment, its potential and its fragility;

    - individuals, interest groups and agencies are identifying their own positions, recognizing those of others and working towards a cooperative solution through which all the participants can win, rather than becoming locked into a destructive win-lost syndrome, or a position in which all are worse off.

    - relationships between planners, politicians and other people are strengthened so that communication barriers are breached, and mutual trust increases as a foundation for communities to function more effectively in every way.

Public participation is NOT:

- selling a pre-determined solution by public relations
  techniques;

- planning behind closed doors when information could be
  shared;

- one-way communication, e.g., planners telling people what
  is best for them;

- public confrontations between "people power" versus the
  bureaucracy;

- by-passing elected representatives or impairing their
  freedom to exercise their decision-making responsibilities.

Manheim (p. 2) has identified the persistent "myth of rationality: the belief by the professional and the public that,
because of his education and training, he was uniquely qualified to adjudge what was best for society in his domain of competence."

From this outline, the nature of public participation as
used in this context should be clearer. Further insights into
the concept can be obtained by considering its roots in various
disciplines.

Disciplinary Background

Public participation in planning as a field of applied
work draws upon theory, concepts and experience from political
science, adult education and various forms of applied sociology.

Political science has perhaps the longest tradition of research and analysis into the nature of public policy formation,
political decision-making, voting behavior and related subjects.
Specific studies have explored how citizens form opinions on
issues, the nature of interest groups and the identification of
publics for specific topics. Adult education as a sub-
discipline is distinguished from the general field by its focus
on the processes by which adults learn, the characteristics of
the adult learner and by its development of methods to foster
learning through techniques not usually employed in the classroom. Applied sociology is concerned with the practical applications of how people behave in groups, communities and organizations, including how people adapt new ideas and practices.

Each field, of course, has viewed public participation
from its own perspective, e.g., in terms of power, influence,
communications, learning, decision-making etc. For our purposes

it is essential to identify useful concepts from each of these
disciplines and use them to examine the public for a project,
select the staff required, design its operational phases and
implement them.

*The Anatomy of Public Participation**

In any description of public participation, "Who is the
public?" remains one of the most central yet elusive concepts
to pin down. *The public* for a major project may be viewed in
all-inclusive terms as all persons who are likely to be affected by the project, directly or indirectly, positively or
negatively, severely or in a minor way, whether currently present, aware and interested or not. More narrowly, the public may
be seen as those resident in the area directly and substantially affected by the project who care to use the means available
to learn about it and register their views.

Both positions, and points between them, raise many ethical issues and questions of technical practice.

To begin with, there is no single public, but rather a
whole series of publics which crystallize around particular issues. *A public* may be seen as any loose association of individuals who are held together by common interests and objectives and by various means of communication. Many of the publics for a given project have no formal organization, e.g., the
potential recreational users of a proposed multipurpose water
storage dam.

Within the many publics for a particular project, *interest
groups* may be identified which are organizations of persons
with common interests and objectives. Such associations readily
become *pressure groups* or *special interest groups* when they
focus their attention and efforts on a particular subject.
Frequently, several of these special interest groups will
emerge, each claiming to speak for their entire public and
often claiming that their public is somehow more important than
other publics for the project under consideration. How to
weight the often conflicting desires of various publics and
special interest groups is a matter which has to be worked out
in the context of the particular project.

The persons who constitute these publics and interest
groups are not simply lone individuals, but are knit together
by a complex linkage of *social systems* and sub-systems. (See

---

*This section is intended to provide the user with a set of
 basic terms useful in describing the phenomena of public participation and in dealing with staff and consultants in this
 field.

Chapter 6, "The Human Community.") There are typically a number of social systems with a project area, often with widely different values, goals, attitudes and standards of behavior. Some social systems significant for the project area may lie beyond its boundaries. To understand the citizens and their individual and collective responses to the project, it is vital to grasp the essentials of their way of life, the context of their community and their views of their society. Methods appropriate for gathering social data are outlined later under "Operational Techniques."

Understanding the *types of participants* can provide a firm basis for developing effective strategies of public participation. Citizens are likely to become active participants only when they become aware of the issues, believe they can influence the decisions involved and see some benefit, tangible or otherwise, in investing the effort in this activity rather than in some other. Active participants who emerge early in a program are likely to be existing leaders.

Among the many kinds of leaders in a given community, some are followed in many fields, while others are looked upon as credible in one. It is useful to discover who are the *influentials* for a given population and project, as demonstrated in the Susquehanna Communication-Participation Study, where water influentials were defined as "those people who have the greatest demonstrated or perceived ability to make or affect policy decisions about water resources in their area of the Susquehanna River Basin" (Borton and Warner, p. 83). Nearly 200 water influentials were identified in a five-county area and were involved in an intensive process of communication and consultation.

One effect of different types of participants becoming active over time is the *"cumulative curve of citizen involvement."* Initially, only a very small proportion of the population likely to be affected by a planning project will recognize that (a) their interests are affected and (b) the agency will respond to their actions. This results from the very general statement of the issues at the outset, and a combination of a low information flow and low level of credibility of many planning agencies.

As the project proceeds, issues become more clearly defined, more people recognize that they have a direct or indirect stake in the outcome, information flows (from agencies, media, interest groups, grapevines, etc.) increase, and the credibility of at least some of these sources of information rises.

Given a decision which the active public generally understands and accepts, the level of citizen involvement generally

declines as other issues compete for public attention. However,
if the decision is not understood or is seen as unacceptable by
many, a further escalation of citizen involvement, usually in
the form of protest, can be expected.

The concept of a cumulative curve of citizen involvement
has many implications for participative practice:

1. Only a relative small response should be expected to any
    citizen involvement program at the start of a project, un-
    less the project is set up in response to public outcry.

2. When a quiet start occurs, active solicitation of interest
    may be required to identify those concerned with the issues
    involved.

3. People will be discovering the project for the first time
    right up to the eleventh hour, so they will need to be ac-
    quainted with what has already occurred, as well as given
    the opportunity to participate from that period onwards,
    i.e., introductory orientation material and procedures are
    needed throughout the life of the project.

4. People with different backgrounds, values and goals may
    identify themselves with the project at different times, so
    opportunities to respond with full flexibility to their in-
    terests must be provided.

5. The cumulative curve of citizen involvement has very direct
    implications for project budgets. Recently, over 50 percent
    of the public participation budget in a four month study
    was expended in the final month.

6. The nature of the curve underlines the fact that the citi-
    zen participation process must be rapidly responsive to un-
    expected developments in the planning process, e.g., reori-
    enting the interested public to some new alternatives aris-
    ing from technical studies completed in the early part of
    the planning project may be accomplished fairly easily,
    compared with the situation which arises when new opportu-
    nities become apparent towards the end of the project.

The public participation process consists of several
phases which develop through the period of the planning study.
While different writers (Bishop, p. 61; Warner, p.39; Manheim,
p.7) have identified various formulations, each has essentially
outlined two parallel and mutually interacting chains of activities in the technical planning process and in public participation, which should commence at the outset of the project and
last until its completion.   The basic activities start with a
study of the people and the area, and continue through information exchanges to develop a larger shared perspective by all

involved, to the development and testing of alternatives until
a concluding decision is reached (see Table 12-1).

Table 12-1. Public participation in the planning process.*

Public Participation Process	Planning Process	Time Estimate** (months)
1. Introduction and Start-up	Introduction and Start-up	1
2. Initial Data Collection	Preliminary Design and Review	3
3. Mutual Education of Issues	Technical Studies and Development of Objectives, Criteria, Priorities, etc.	6
4. Public Response to Project; Evaluation of Alternatives	Project Alternatives	1
5. Present Decision	Decision by Representatives	1
		12

*Adapted from Connor (a., p. 16).
**Based on a one-year planning process.

For these activities and interrelated phases to occur
productively, public participation has clear implications for
staff selection and training, both for all members of the study
team and especially for persons hired specifically to foster
public participation in the project. All must be willing and
able to review their activities with non-specialist members of
the public and those working full time with citizens must be
selected for their relevant knowledge, attitudes and social
skills, and be prepared to extend these as the project requires.

The *organizational climate* of the planning office is important for public participation. If it is directive, cold and
repressive, these characteristics are likely to pervade the
public participation activities with disastrous results. On the
other hand, a cooperative, stimulating atmosphere is likely to
generate a similar response amongst members of the public.
Mutual trust and confidence are essential ingredients of a
sound organizational climate; unless these pervade relationships within and between the various agencies involved, the

planning process will be substantially less productive and efficient.   The free flow of information and responses are vital
in this regard.

Feedback, or the systematic gathering of responses from
individuals and groups to reports by a planning team, distin-
guishes public participation from traditional, one-way public
relations programs. Responses must be solicited in ways appropriate for the prospective respondents and then processed into
a form useful for the planners. The simple pooling of ignorance
is hardly a sound basis for planning, so the education of all
participants is fundamental.

Adult learning processes, to be effective, require that
many traditional assumptions pervading school and university
classrooms be discarded. For instance, adult learners always
have something to contribute to the issues at hand; those who
would teach cannot assume initially that they are credible
sources of information; adults may resist learning something
which, if accepted, would threaten their present and accepted
way of living.

The preceding pages have outlined a descriptive definition
of public participation, some underlying assumptions, some prevailing myths, its disciplinary background and some core concepts associated with it.   In the next section, the objectives
and reasons for public participation will be outlined, together
with some limitations likely to be incurred.

FUNCTION IN PLANNING

There must be compelling advantages for a planner to include members of the public in the planning process.   The objective sought by adding citizen involvement are reviewed next,
followed by some reasons for the present attention given to
this subject.   Finally, some of the limitations of public participation are identified so that a balanced perspective may
result.

Objectives

There are a number of objectives which can lead a planning
team to engage in public participation, ranging from technical
considerations to political factors:

Mutual education.   This is a basic objective of public
participation to ensure that the goals, technical information
and working assumptions of the people are appreciated by the
planners, and vice versa.   More specifically, the various publics, agencies and political figures identified at the start of
a project typically have little understanding of the aims, data

and beliefs possessed by the other parties; often they do not
fully grasp their own positions concerning the project. Through
the use of appropriate communication methods, each participant
should acquire a better grasp of the others' positions, and
often a new sense of his own. A broader, more balanced and
shared perspective is the result, which assists cooperation in
generating and selecting alternative solutions and other planning decisions. This provides a broader base of understanding
and support for the recommendations of a planning project and
thus improves its prospects for implementation.

Enlarged planning space. Through the mutual education
process, a broader arena for planning may result, as when an
initially narrow interpretation of the terms of reference for a
study are seen to frustrate its underlying intent.

Additional data. Planners can obtain important information
from persons who often have decades of year-round experience of
the environment. When official records are often recent and
especially when the project does not permit years of original
data collection, the systematic recruitment of local observations by laymen and professionals can supplement other data
sources.

Technical expertise. Residents of the area often possess
technical expertise in the key subjects of the project. They
can contribute this valuable resource in support of the project;
if alienated, they can conspire powerfully against it. Retired
professionals, especially, have the training, the time and the
contacts; constructively involving these people is a survival
skill essential to project success!

Creative capacity. Perceiving solutions to problems is not
a prerogative of technical experts. Indeed, their training
often equips them with as many blinkers as insights. Concerned
laymen can often see sound alternatives which experts do not,
e.g., a technically sound route for a major urban freeway was
identified by a citizens' group; many teams of specialists had
considered the problem for years without noticing this technically sound and widely acceptable solution.

Social data. Information on goals, attitudes, values,
preferences and priorities are a crucial input to the planning
process; the citizens are one of the most important valid
sources of this knowledge. Attempts to give people what planners think is best for them, or what planners think they want,
have led to one debacle after another. Such disasters leave
behind the original problem, a heavy financial expenditure with
little to show for it and a corrosive residue of ill-will.

Involvement.   Increasing numbers of citizens   want to experience the process of creative planning as well as its product. Often they have a substantial sense of ownership in their
part of the environment - to ignore this is dangerous.

Managerial solutions.   This type of solution for environmental problems, as opposed to purely structural solutions,
seeks to initiate changes in people's behavior. The likelihood
and east of changing behavior is greatly increased if people
systematically become aware, interested, informed and thus convinced that new behavior is needed e.g. reducing domestic water
use.   This participative approach minimizes the need for regulatory legislation, enforcement procedures and a needless proliferation of a law-and-order society.

Acceptance. Public participation fosters the acceptance of
the planning agency as a legitimate and trustworthy vehicle for
the competent and sensitive solution of problems. Without such
confidence, the reception and interpretation of the agency's
recommendations may well be hostile.

Conflict resolution.   A strong mutual education component
in a public participation program stimulates the resolution of
conflicts between the parties. However, merely making them more
aware of their differences may heighten conflict unless this
process is handled with professional competence and personal
concern. Opportunities for the many publics and interest groups
to identify and express their positions is a fundamental first
step,   following which methods already developed in inter-group
relations and management may be employed.

Reasons

In the small, simple and relatively stable community,
human relationships are typically close and direct. Under these
conditions, there may be little need to foster the existing
processes of communication and decision-making.

However, in the large, complex and fast-changing community,
the relationship of one person to most others is usually remote,
fragmentary and indirect.   The full impact of this situation
today is presented by Alvin Toffler.

At the commencement of a major planning study today, a
situation may exist like that schematically shown in Figure
12-1. Developers or other entrepreneurs have established a
relatively close and direct relationship with a number of key
elected representatives.   These in turn have a similar relationship with their planners.   Through the politicians and directly, the developer-entrepreneurs have also created a strong
linkage with the planners.   Reverse flows of information and
influence are, by comparison, weak and sporadic.   For the most

part, other citizens are not part of the triangular relationship of planners, politicians and developer-entrepreneurs. The
bond between other citizens and elected representatives waxes
and wanes with the rise and fall of specific issues, but is
typically tenuous except during election campaigns. Regardless
of the specific relationships, the main point is that communications between these four major parties are far from balanced.

Under traditional practices, a major project was generally
planned with little or no opportunity for contributions from
the public in its early stages. During this period many citizens assumed that, in most cases: "growth is good; the new is
better; experts know best; the main criteria are cost and technical effectiveness." When planning was completed, public
hearings might be held, but essentially to air grievances
rather than to permit basic modifications of the plan.

Recently, many more people have called these former assumptions into question and have placed high priorities on
rather open-ended values such as the quality of life and the
preservation of the environment. As a result, and during the
last five years especially, many technical plans have been completed only to encounter a massive delayed reaction.

The record is clear. *When opportunities for constructive
citizen participation in major planning events are too little
and too late, an overwhelming negative response is likely to
the extent that the public believes that (a) vital issues are
at stake and (b) change is possible in the plan.*

Rising levels of formal education and the informal spread
of knowledge through the media, especially television, is a
major contributing factor to a more responsible public. Another
element is dissatisfaction with traditional forms of voting and
representation by which a politician is elected usually on a
broad and sometimes contradictory platform for a period of
years, often without specific direction from his constituents
on particular issues. The issues in question may not have even
existed when the politican was running for office.

Given this analysis, one approach is to try to create a
situation in which there is effective two-way communication between all of the parties involved - politicians, citizens,
planners and entrepreneurs, i.e., to bring a balanced flow of
technical and social data in Figure 12-1. General and specific
means to accomplish this will be outlined later; however, it is
first useful to recognize some of the limitations of citizen
involvement programs.

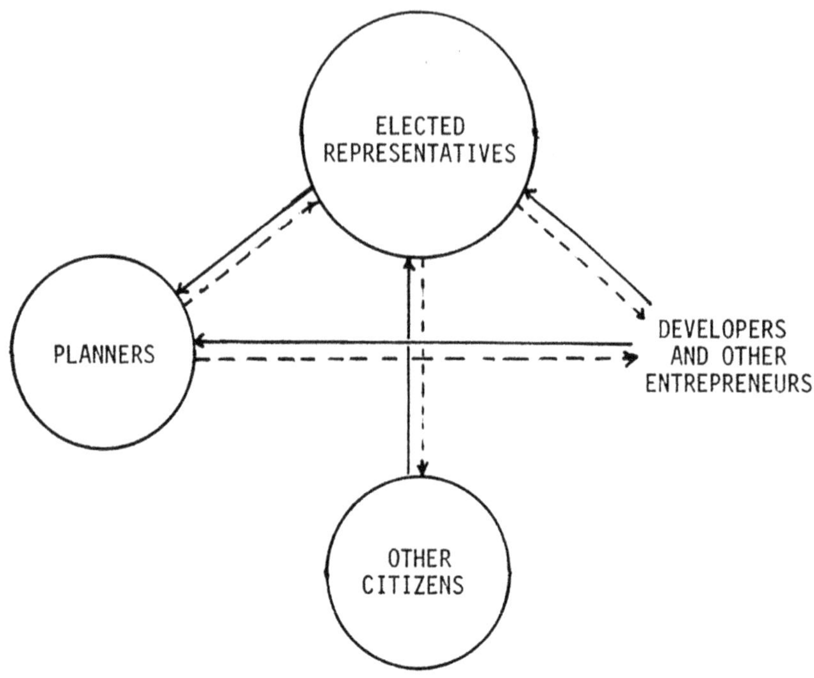

Figure 12-1. Typical communication and influence patterns in planning projects.

Limitations

Citizen participation contains some inherent restrictions and potential dangers. As a *responsive* process, the public involvement component is never fully predictable, since it is impossible to forecast accurately just how many persons will respond to what issues with what level of intensity.

The *cost* of a given study will certainly increase through the addition of a citizen participation component. However, this may be regarded as a cost-effective investment for a better plan which, being also politically more viable, is thus less likely to end up on the shelf, where it represents a major expenditure yielding minor returns.

Public participation programs may be used to serve a variety of *vested interests*; neither equity, justice nor rational decision-making are necessarily advanced by citizen involvement, which may in fact foment goal conflicts into a noisy stalemate.

These are serious limitations and dangers; they reflect the volatile nature of the issues, persons and groups which become involved in these programs. These negative aspects of

citizen involvement can be reduced to a considerable degree by
the creation of an appropriately designed program to fit the
project, people and agencies involved. This is the theme of the
section which follows.

DESIGNING A PROGRAM

Before examining specific operational techniques, it is
appropriate to review some design requirements of any effective
public participation program, some general strategies which
have been employed and some budgeting and other management considerations.

All too often, the lack of initial design time and the absence of opportunities to evaluate and adjust an ongoing program result in massive waste and failures which could have been
successes. A sound analysis of the program situation at the
outset can return much more value for the resources invested
than later spending on operational expenses.

Some Design Requirements

There are certain characteristics which appear to be essential if a public involvement program of the type defined
earlier is to be developed and managed effectively. If any one
of these elements is absent, the probability of ineffective
operations or outright failures is increased.

A sound public participation program should be designed so
that it is:

Service-oriented. The citizen involvement sequence is not
a continuing counter-attack of the planning process, but must
be integrated into it in a supportive manner. The planning
process has a scheduled start, certain intermediate stages and
a completion date. Public participation must be designed at the
outset to fit and support these activities. For instance, to
contribute successively to the identification of goals and priorities, to the development of alternatives and to selection
amongst them.

Connected directly and continuously. In any major public
works project, citizen participation is not an intermittent activity which can be done on a part-time basis--at least one
person will need to work with the public and the planning group
at all times. This person will function as an honest broker
between the various parties.

Rapidly responsive. There must be a capacity to reply
quickly to changing issues amongst the public and to unforeseen
developments in the planning process by contacting individual

citizens, the fundamental units of both the planning and the
political processes, quickly and directly as well as through
their leaders and various groups in a multi-step information
flow. While the latter will be often useful and sometimes sufficient, there will be occasions to contact as many as possible
immediately and with a minimum of distortion and selective interpretation.

Inherent in this responsiveness is a *respect* for the citizens and *acceptance* of them as valid sources of data, priorities, preferences, some technical expertise and potential alternative solutions. *Cooperation* as the pervading process in
relationships with groups and agencies is also implied in this
responsiveness.

Mutually educational. The public should become more aware
of planning matters, planners should become more understanding
of community concerns, and all parties ought to develop a
broader and shared perspective of the opportunities, the problems, the alternative solutions and the criteria for a satisfactory decision.

Multimedia. The program must use not only various means of
mass communication as appropriate, but also techniques appropriate for individuals and groups of diverse cultural backgrounds as these occur in the project area. This is particularly important in ensuring that as many persons as possible
become aware of the project, its implications for them and the
opportunities they have for participating at different stages
of the project.

For example, in developing a regional official plan for
half a million people in an area of 1,100 square miles, urban
residents were reached through newspaper supplements, public
affairs programs on television, hot-line interviews on radio,
community associations, direct mail and personal contacts.
Rural people were approached through local organizations, identified influentials and other leaders; a second language was
widely used by many residents, so most program material was
distributed in both languages by bilingual staff.

Multidirectional. Information flows must be provided upward, downward and across the population considered. One-way
communication is anathema in a participatory program. Realistically, this often calls for a translation from the language
comfortable for one group into that of others involved; area-
wide forums, for instance, are often needed to foster lateral
communication between groups affected by a project.

Encouraging responsibility. Individuals and groups involved must develop a sense of personal commitment to the outcome. Frequently, this can take the form of working to convert

groups which are protesting so as to provide instead constructive suggestions for resolving issues. (See "Constructive Citizen Participation.")

Politically acceptable. This is necessary so neither the
project nor its public participation component will be curtailed, but will be supported. The program's viability usually
rests on continuing to keep political leaders informed of project activities and involved in them as appropriate. They must
see its benefits directly rather than hear indirectly of its
risks or else, for example, they are likely to misinterpret the
gathering of public preference data as a binding referendum.
When kept continuously appraised of the situation, many politicians are more open and willing to take risks with the public
than their agency administrators.

In a recent project, elected representatives were invited
to confidential briefing sessions before each major phase of
the project unfolded - they were never caught unawares by unexpected developments. At the same time, they were led to see
that political intervention in the early stages of the work
would be self-defeating, in contrast to providing for full disclosure of information and an unrestricted discussion of alternative solutions.

Participatively managed. The spirit and style of the administration of the public participation element will be reflected in its field operations. The quality and quantity of
public participation is unlikely to exceed that within the
project group; if the latter is directive, traditional and defensive, these same characteristics will probably pervade the
citizen involvement phase of the project. If an innovative
spirit, mutual trust and cooperation prevail in the project
group, they will be more easily fostered amongst public participants.

In one case, a directive and insecure project manager had
a public participation program wished upon him by the government responsible. Though a semblance of public participation
was carried on, this manager and his organization had a very
limited capacity to respond to contributions from the public,
which became very dissatisfied. This constraint could have been
reduced or overcome by working to improve the quality of management and the organizational climate of the agency.

Adequately funded. Data gathering, analysis, output and
response must be appropriate for the project. This involves
both the allocation of sufficient funds and accessibility to
the money as it is needed. Some specific factors in budgeting
for public participation are addressed later in this section.

Suitably staffed. Enough competent persons must be avail-able as required. This involves experience and training in applied social science, adult education and the media appropriate for the project situation. One or more of these persons will work as specialists in public participation. Other technical staff who also deal with the public will need to be selected with this factor in mind and given some on-the-job opportunities to increase their interpersonal skills. (These will be valuable within the project team as well as in working with citizens.)

These characteristics, then, are seen as design requirements for an effective public participation program as outlined earlier in this chapter within the context of this treatment of public projects. They must be built in to the initial design of the program, and must then be monitored during the operational phases to see that all continue to be met appropriately.

These design requirements and other program elements can be woven into several general strategies which are outlined next.

Some Participative Strategies

A *strategy* has been defined by Bishop (p. 36) as "a procedure, established in advance, which determines how, when and in what depth various parties will participate in planning, evaluation and decisions." Developing such a strategy requires answers to such questions as:

a. Who *should* participate, i.e., who has a legitimate interest?

b. Who *will* (is likely to) participate?

c. *How much* participation is *desirable?* How much is *possible?*

d. On *what issues* should there be citizen participation? What is the appropriate *timing* for participation? At what *stages?*

e. How should expressed views be *weighted?*

f. Does *residence* in a particular area, at a particular place, increase the weight to be attached to views on issues that have a geographic impact? How are areas or communities defined?

g. How does the *lack of interest,* lack of *time,* lack of *knowledge,* or *apathy* of substantial portions of the people affected by almost any public action influence the evaluation of participation?

h. What weight should be given to the expression of *organized and articulate interest groups,* when it is recognized that many who may be affected by public action are not represented and do not express their interests? (Wengert, p. 9)

Replies to these questions depend partly on the project situation and partly on the values of the respondent. While there is no simple, general correct answer, the fact that we can now recognize and work on these issues is itself a milestone in the planning process.

One approach to some of these questions, for example, is to recognize that there is an informal division of labor in society on subjects of public concern, as there is in other aspects of social life. A public participation program for a major project should provide a wide range of locally appropriate opportunities for people, as individuals and groups, to obtain information, ask questions and respond to issues and alternatives. As citizens, all have both a responsibility and an opportunity to participate in their society; those who wish to will do so. The remainder pass up the opportunity on their own responsibility. By the end of a comprehensive program to distribute information and generate alternatives, the responding individuals and groups typically include a majority of the interested, the concerned and the influential. The remainder typically follow their leadership.

Conflict. Among the general strategies for public participation, one of the best known is that developed by Alinsky which focuses on self interest, neighborhood identify and confrontation tactics with the establishment. It hardly seems appropriate for the type of programs envisaged in the context of this treatment.

Advocacy. Another well-known approach, termed advocacy planning, seeks to provide skilled representation for low-income groups which would otherwise be unlikely to receive a hearing in traditional planning practices. This approach assumes that other groups would be adequately served and takes for granted many Alinsky premises; it is therefore of dubious validity in terms of the position taken in this chapter.

Collaboration. The collaborative approach has a focus on understanding human activities, fostering two-way communication and working in a cooperative manner.

Education. The concept of planning as education is also consistent with the foregoing design requirements, but when education becomes the primary objective and planning goals are set to one side, this strategy is better suited to citizenship development.

Behavioral change. In this strategy, direct efforts are made to change the behavior of persons and groups who have assumed polarized positions on key issues. It may frequently become part of a more encompassing approach to citizen involvement in public works planning.

Staff supplement. This approach relies on citizen volunteers to assist agency staff to accomplish their objectives. It, too, may be used as part of a total public participation program, but is hardly sufficient in itself.

Co-optation. This strategy attempts to overcome obstacles by absorbing dissident leaders into the project organization, whether formally or informally. The compromises involved may levy a high price on the organization's objectives, the goodwill of citizens drawn into committees or both; it is therefore dubious approach to use substantially.

Law. A legal strategy for defending the environment by Sax has been widely distributed at a time when many environmental groups are changing their approach from participation and education to a power orientation.

In any given project, many of these strategies would be employed to some extent in some phase of the program. Just when, where and how would depend on the project situation; the program would be tentatively described in the preliminary design and developed more specifically as the operation occurs. One of the real problems in designing such a program concerns the budget for public participation.

Budgeting

As indicated earlier, there is, unfortunately, no way to predict what level of effort will be needed, how many citizens will take the opportunities offered, how many issues will develop, etc. in a responsive program of public participation. In these circumstances, a budget estimate is the best that can be provided.

Some factors to consider in budgeting include:

a. the technical nature of the project, i.e., the relative difficulty expected in transmitting information about it to the various publics. For example, informing a community about its transportation needs seems easier than explaining high voltage transmission line requirements and alternatives.

b. the type of public participation desired, i.e. the relative emphasis on each of the design requirements. For instance, the higher the educational input, the higher the immediate cost, but the greater the prospect of longer term public cooperation on the project.

c. the duration of the program.

d. the dimensions of the project area and the distribution of the population likely to be affected within and beyond this area.

e. the cultural homogeneity or diversity of the population, e.g., language, ethnic groupings, occupations, etc.

f. the degree of local awareness and support or opposition concerning the project by the population and various levels of government.

g. the competence, coverage and costs of the mass media available to reach the population under consideration.

h. the number of public participation staff required, which depends on the above factors plus the level of staff competence required, the support available to them and the project from related agencies, and the more specific goals of the project management group. Some local standards for comparison are the areas and populations assigned to public health staff, school inspectors, clergy and resource agency staff, e.g. county agents. A working maximum is about one staff person per 100,000, but this ratio really depends on a full appreciation of the project as a whole.

The major items of cost in most public participation budgets are staff salary and associated administrative overheads; taking administrative overheads at 100 percent of staff costs, these two often account for two thirds of the budget. A quarter of what remains may go to printing and media expenses to cover the cost of brochures, information kits, advertising meetings, questionnaire expenses etc. The balance provides for training, consultation and miscellaneous expenses.

The size of the population considered is probably the factor exerting the greatest leverage when developing a public participation budget, since it directly affects both staffing and overhead and media costs.

In a recent four-month highway relocation study in a city of 300,000, the budget for the total project was $150,000, of which public participation accounted for $35,000. Some $10,000 of this was spent on printing and media expenses, while most of the balance went to staff salaries and overheads. In another 2-1/2 year comprehensive study of a 450-mile-long river basin with a population of 200,000, the budget for public participation is over $1 million and exceeds the cost of the technical studies, which amount to $775,000. In a two-year metropolitan transportation review for a city of 2-1/2 million persons, an estimate of $500,000 has been developed for citizen involvement out of a total cost of $1,500,000. A 10-month conceptual study of 70 miles of lakefront with an urbanized population of 600,000 generated a public participation budget of $200,000 compared with a technical budget of $194,000.

These examples, taken from recent professional practice, all employed the same basic model presented in this chapter. The proportion allocated to public participation varies from 25 percent to 56 percent, according to the factors reviewed above. There is no ready rule of thumb yet discernible to estimate these budgets accurately on a per capita or other basis.

The U. S. Army Corps of Engineers is reported by Warner (p. 94) to have allocated 25-35 percent of the total budget to public involvement in its Upper Rock River Basin comprehensive study in Illinois and Wisconsin. The importance of adequate financial and manpower resources is emphasized in reports on the Susquehanna study.

Another vital factor in designing a public participation program is project management. Although touched on already as a design requirement, some further aspects are developed below; these should be read in conjunction with the chapter on project management.

Management Considerations

Since the public participation program is an integral part of the total planning process, its design requirements to be rapidly responsive and participatively managed set standards for the management of the project as a whole. A planning agency with a public participation component can hardly be managed in the classic, traditional style characterized by rigid authority, formal communications and little delegation. Instead, a flat organizational structure should replace the pyramid and the climate of the operation should be typically open, flexible, trusting and focused more on effective and creative contributions to the task at hand rather than the status of the participants. This kind of management can be achieved if its importance is fully recognized and reflected in organizational design, staff selection and continuing attention to the character of the agency by the application of organization development techniques.

Preliminary Design

Before a public project is authorized, an essential pre-planning phase is to develop from given terms of reference for the project as a whole, an appropriate statement of objectives, goals, methods, phasing, staffing and budget for the public participation program. The topics discussed in the preceding section are, of course, especially important.

To create this preliminary design requires the services of a senior consultant, or team of persons with specialized public participation competence, who will work with the staff of the agency responsible for the project. A reconnaissance visit to the project site is essential. Depending on the project situation, this preliminary design phase may last from a week to more than a month.

Staffing

The recruitment, selection, orientation and training of staff make up one of the most crucial elements in a public participation program. This applies both to full-time staff for this program and to other members of the project team who will have dealings with the public. For the latter, technical competence alone is not sufficient when an open planning process implies that project staff must meet other criteria such as their capacity and willingness to understand other kinds of people and their points of view, and to communicate with them in terms they understand.

In a small project, the participation staff may consist of one such person with the periodic assistance of the design consultant and occasional help from a media expert. Larger projects will require more than one person, perhaps one with greater media skills and another, for example, with particular ability to deal with an important ethnic group. Frequently, two part-time persons can provide better value for the same cost than a single full-time worker. Some on-the-job training of all staff will be required.

Gathering Data

Through its lifetime, the project must be continually aware of its environment, especially its volatile human environment. A number of methods can be employed together to ensure that the project team is aware of key issues in the various communities, and the reasons for them.

Before the project becomes established in the area, a good deal of information can be obtained concerning the people and their communities from previous studies, reports, maps, aerial photographs, census data, theses, former residents of the area,

analysis of back issues of local newspapers, etc. The analysis
of past examples of public participation in the area can provide many insights, and may include school boards and bond issues, as well as local, state and national political participation.

During the first three months of a project lasting more
than a year, the use of qualitative data-gathering techniques
can provide a rapid understanding of the main aspects of life
in the area with a minimum of disturbance. These techniques,
derived largely from anthropological field work methods, include:

    a. participant observation - living in the community, join-
       ing in its activities and gaining a view of their life
       and circumstances from the inside;

    b. life histories - the selective recollection and inter-
       pretation of past events by a small number of different
       kinds of people in the project area can provide many in-
       sights into values, goals, attitudes, cliques, factions
       etc.;

    c. key informants - identifying and cultivating a number of
       persons who have long residence in the area, secure po-
       sitions (at any level) and a high level of contact with
       others (e.g., personal service occupations) can provide
       a rapidly responsive network of contacts.

Once a qualitative appreciation of the people and their
area is developed in ways such as these, specific items on
which quantitative data are required by the project team can be
formulated into a sample survey instrument, if such is required.
Often questionnaires are poorly designed because of lack of
first-hand experience of the people and their environment.   On
other occasions, qualitative research can provide a sufficient
level of information more rapidly and cheaply than thorough
surveys, and without generating "survey fatigue" in many of the
respondents.

During the remainder of the project, the staff must constantly test and revise their understanding of the people as
conditions warrant.   In later stages, more specific means will
be required to solicit alternative solutions and responses to
them; these are reviewed subsequently.

Distributing Information

The methods used to distribute information should reflect
an understanding of the communications patterns currently used
by the people of the project area. These will probably vary
with the different publics identified for the project. Sensitivity to accepted and effective local means of distributing

information is not only learned but believed. A "person-centered" approach will often suggest ways to reduce the alienation fostered by exclusive reliance on the traditional use of the mass media. For example, a discussion of project issues between respected persons of the area and technical staff, reported in the press and carried by radio and television, can help to both communicate and project activities in a way that a simple news release to the media could not.

The distribution of technical information can occur through the use of a variety of techniques from personal meetings with decision-makers and key influentials, through workshops with organizational leaders to booklets, information kits, mailings, film, video-tape recording, radio, press (daily, weekly, ethnic), television, reference centers open to the public, mobile information centers visiting selected locations, community workers, advisory groups and public meetings. As noted earlier, which of these methods to use with what organizations and publics, for what purpose and at what time, is a matter to be projected initially in the preliminary design for public participation and developed in detail as its phases are implemented. (See Table 12-2)

The opportunity for feedback should be provided in every instance by the use of response sheets at meetings, short questionnaires at the end of publications or simply the request for letters on key issues. The use of existing organizations, e.g., meetings held under their auspices, is likely to create a more constructive environment, and to foster continued interest and work on the project, compared with project-sponsored public meetings. Formal public hearings are even less likely to bring about constructive results when, in the experience of many, they are typically a legal requirement convened when core project decisions have already been made.

Although the distribution of information occurs throughout the project, it is usually a primary activity during the middle of the sequence. One important element is to distribute sufficient technical information and descriptions of currently proposed alternatives that interested citizens and organizations may develop and present any alternative solutions which occur to them. All proposed alternatives can then be evaluated by the project team, using criteria which reflect both technical and community values. Those which are judged to be viable can then be rated by the public for their social acceptability.

Fostering Decision-Making

All of the previous activities - stimulating people to develop an understanding of their own goals and those of others, encouraging them to become better informed on the issues and

Table 12-2. Descriptive Dimensions Characterizing Various Public Involvement Mechanisms

INVOLVEMENT MECHANISM	DESCRIPTIVE DIMENSIONS*				
	FOCUS		Degree of Two-way Communication	Level of Citizen Activity Required	Agency Staff Time Requirements
	Scope (No. people)	Specificity			
Workshops	L	H	H	H	H
Public Hearings	M	L	L	H	M
Mass Media (Including use of news-papers, radio and TV)	H	L	L	L	L
Task Forces	L	H	H	H	H
Agency Publications	H	M	L	L	M
Speeches and Presentations	M	M	M	L	M
Survey Questionnaire	M	H	L	M	M
Advisory Boards	L	H	H	H	H
Informal Contacts	L	H	H	M	M

Source: Warner, p. 52

*H = High degree
M = Medium degree
L = Low degree

the project, and inviting their suggestions for alternative solutions - facilitate the decision-making process. This process
is complicated by the fact that at least three groups have a
stake in it - planners, elected officials and citizens. Bishop
outlines seven ways by which recommendations may be obtained:

   a. planners recommend,
   b. planners recommend, advised by citizens,
   c. elected officials recommend,
   d. elected officials recommend after public hearings,
   e. citizens review, board recommends
   f. referendum and majority decides,
   g. state commission recommends after public hearings.
      (Bishop, pp. 45-48)

    Methods which use only one of the three groups are clearly
of limited value. In only three of these ways is widespread
citizen participation possible, but in two of them formal public hearings are the main vehicle. Its limitations are outlined
in a recent article by a member of the U. S. Corps of Army Engineers:

> There is a potential for bias because those present may
> not be representative of all interests. Meeting places and
> times are not conducive to large attendance. In addition,
> the problem may not be clearly understood, if it is the
> first time that the issue has been brought up. Also, there
> are certain natural inhibitions in people which preclude
> adequate expression of opinion in formalized meetings.
> There may not be enough meetings or they may not be of
> sufficient duration to satisfy all points of view.
> (Tabita, p. 14-15)

    An approach based on value analysis was presented by
Fielding who employed several sample surveys and mathematical
techniques to indicate the preferences of the public and those
of the decision-making Transportation Committee. This raises
the question of the scientific strength of sample surveys,
which nevertheless limit participation to those selected at
random and are sometimes decried by non-participants, versus
the political viability of a more open referendum-like approach.
There are times when sample surveys are necessary as a form of
quantitative data collection, but according to the position
taken early in this chapter, they are no substitute for a more
open opportunity in which all of those interested in the outcome, and none of those uninterested, will contribute their
preferences.

Supporting Implementation

    Between planning and implementation there is usually a
period of hiatus which may last from months to years. During

this time the good relationships developed between the project
team and the community may deteriorate and local conditions may
change in ways important for the project. A continuing liaison
with key people is therefore essential.

Unfortunately, many projects are funded in ways which preclude continuity, even of the most token kind.

One challenge, immediately after the decision on the project has been taken, is to ensure that as many people as possible not only know the decision but the reasoning of those who
made it. People who feel they have been heard early in the
planning process, and whose preferred alternative was not
chosen, can often accept the decision of its rationale as presented to them. They may still not like it, but they are more
likely to accept it. Failure to carry out this follow-up step
is to invite continued misunderstanding and hostility to the
decision.

Another important step in the transition from planning to
implementation is to evaluate the public participation methods
employed throughout the project, involving people, politicians
and planners in the process. This activity is valuable for the
guidance it produces for the implementation phase, for its useful insights for other projects and as a personal de-escalation
device for the participants.

Depending on the nature of the project and the decision to
be implemented, support for implementation may require a public
participation component similar in type and scale to that used
in the planning phase. This will need to be designed according
to the project situation and the guidelines provided by the
evaluation of the participation component of the preceding
planning process.

PROSPECT

This chapter has attempted to outline the nature, rationale, design and operation of citizen participation in public
projects. In a field which is developing so rapidly, it is impossible to acknowledge, much less include, many other approaches and experiences.

It is evident, according to Tabita, that planning for public projects in the future:

> ...must at least encompass (1) a definition of problems
> and needs from all parts of the public sector at the out-
> set; (2) a wider range of alternatives; (3) a broader
> evaluation of the impacts in areas not fully considered
> before; and (4) a partnership with the public in arriving

at and acceptance of the current solution to the problem.
(p. 1)

This means that we are moving away from what may be called
disciplinary determinism. Economics once shaped the futures of
hundreds of thousands of people through the dominance in major
public projects of cost-benefit analysis, whose weakness is now
acknowledged. Architects designed physical structures which
largely determine social behavior; their influence often dominated urban renewal programs, for instance, with less than
ideal results. Engineering considerations of technical efficiency have held sway in many areas for decades, but are now
being challenged as the dominant considerations. In the field
of planning, the physical has long dominated the social, but
now the tables are turning as public values change and new criteria emerge.

The American public, which has experienced difficulty in
specifying its goals, appears to be struggling towards becoming
to a greater extent the master in its own house. While expertise is required, the public will not accept easily the dicta-
torship of any one discipline or any consortium of them.   Instead, a growing number of people are demanding an opportunity
to directly affect decisions which they feel directly affect
them. Participative planning, based on shared knowledge and
mutual understanding, must now become the rule rather than the
exception. No one said it was easy, but it seems possible.

REFERENCES

S. Alinsky (a), *Rules for Radicals*, (New York: Random
House, 1971).

S. Alinsky (b), The Professional Radical,(New York: Harper
and Row, 1970).

A. B. Bishop, "Public Participation in Water Resource
Planning," Institute for Water Resources Report 70-7 (Alexandria, Va.: 1970).

T. E. Borton and K. P. Warner, "The Susquehanna
Communication-Participation Study," Institute for Water Resources Report 70-6 (Alexandria, Va.: 1970).

· D. M. Connor (a),"From Partisans to Partners" in *Community
Planning Review*, Vol. 22, No. 1. Ottawa, Spring 1972.

D. M. Connor (d), "A Typology of Residents for Community
Development," *Anthropologica*, Vol. 9, No. 2, 1967.

Engineering Circular, U. S. Army Corps of Engineers No.
11965-2-100, June 1971.

E. A. France, "Effects of Citizen Participation in Governmental Decision-Making," in *Highway Research Record*, No. 356, (Washington: Highway Research Board, 1971).

E. Katz and P. F. Lazarfeld, *Personal Influence: The Part Played by People in the Flow of Mass Communications*, (Glencoe, Ill.: The Free Press, 1955).

M. L. Manheim, "Reaching Decisions about Technological Projects with Social Consequences: A Normative Model," Professional Paper P70-78, presented to the Regional Science Association, November 1970, M.I.T. Cambridge, Mass.

People and Planning - Report of the Committee on Public Participation in Planning (London, 1969).

E. M. Rogers, *Diffusion of Innovations*, (New York: Free Press of Glencoe, 1962).

J. L. Sax, *Defending the Environment: A Strategy for Citizen Action*, (New York: Alfred A. Knopf, 1971).

W. B. Shore, *Public Participation in Regional Planning*, A Report of the Second Regional Plan, Regional Plan Association, October 1967.

J. W. Simmie, "Public Participation - A Case Study from Oxfordshire," in *Journal of the Town Planning Institute*, Vol. 57, No. 4, April 1971.

A. F. Tabita, "Implications of Public Involvement in Water Resources Planning," Meeting Reprint 1627, ASCE Water Resources Engineering Meeting, Atlanta, Ga., January 1972.

A. Toffler, *Future Shock*, (New York: Random House, 1970).

K. P. Warner, *Public Participation in Water Resources Planning*, (Ann Arbor, Mich.: University of Michigan, July 1971).

N. Wengert, "Where Can We Go with the Public Participation in the Planning Process?" (Paper to National Symposium on Social and Economic Aspects of Water Resources Development) in *Proceedings*, Cornell, June 1971.

J. Q. Wilson, "Planning and Politics: Citizen Participation in Urban Renewal" in H.B.C. Spiegel (ed.), *Citizen Participation in Urban Development*, Vol. 1 (Washington: NTL Institute, 1968).

AUTHOR NOTES

Dr. Desmond M. Connor is a consulting sociologist and President of D. M. Connor Development Services Ltd. of Ottawa. He holds degrees from the University of Toronto and Cornell University and has taught at St. Francis Xavier University in Nova Scotia and Carleton University in Ottawa. He has conducted a graduate seminar in the subject of public participation in the School of Urban and Regional Planning at the University of Waterloo, Ontario. As a consulting sociologist, he is one of the few in his profession offering professional services as a full time consultant.

Since 1968 he has specialized in the design and management of citizen participation programs. Examples of the many projects he has undertaken for various public agencies in Canada include: highway relocation studies in Ontario; comprehensive river basin planning such as the federal-provincial Saint John River Basin Study in New Brunswick and similar projects in western Canada; and regional and park planning in Ontario.

13

MANAGEMENT METHODS

BY DOUGLAS A. BENTON AND HENRY L. MICHEL

*Most firms manufacture materials which can be pointed to as an
inventory representing some tangible asset. A service organization, however, offers human talent, and its inventory, its
people, goes down on the elevator every night.*

The purpose of this chapter is to describe the management
process required to effectively incorporate varied talent into
the conception, planning, design and implementation of public
works projects. Though the mobilization of varied professional
persons is the central question to which the chapter is addressed, this cannot be done by enumerating a point by point
procedure. Situations are too diverse for this. Rather, the
chapter will aid the reader in synthesizing an approach unique
to the specific needs of each management situation. Included in
the chapter are (1) an introduction to the nature of management,
(2) decision making as it relates to planning and control,
(3) the management of human resources, (4) the management of
conflict and change, (5) the design and operation of organizations and (6) implementation of the management tools available.

THE NATURE OF MANAGEMENT

Management is a separate discipline just like any of the
other disciplinary inputs discussed in this book. Management is
not a capability that is unique to any group or class of individuals who manage by virtue of a position in an organization.
Rather it is, to a varying extent, an activity in which all individuals engage when working on public works projects. This
widespread use of management does not mean that everyone is a
good manager, nor that management is "common sense."

There are several important elements that must be defined
if management is to be understood properly. One definition of

management is that it is an integrated hierarchy of people whose activities must be coordinated to achieve specific objectives. The aspects of *integration* and *coordination*, and the emphasis upon people and the accomplishment of specific objectives are important fundamentals to grasp if the true meaning of management is to be understood. The manager is a catalyst and integrator--especially in multidisciplinary projects where he must bring together human resources from various disciplines and truly manage them for the accomplishment of specific objectives.

Traditional Principles of Management

Individuals primarily responsible for the evolution or the beginning of principles of management include Frederick W. Taylor, an industrial engineer working in the United States with the Midvale Steel Company, and Henri Fayol, a geologist and managing director of a large mining company in France. Both Taylor and Fayol wrote at approximately the same time, around the early 1900's and postulated many of the principles which today are still in use--although modified by more recent research and application.

Taylor sought to replace rule-of-thumb management with a science, and hence his works became known as scientific management. He sought to select, teach and develop workmen scientifically, where in the past the worker had to rely upon his own ingenuity or, in many cases, the lack of it. Taylor believed that all work could be brought under principles of science and sought to divide that work and the responsibility between management and workmen. The emphasis on planning and the use of time study as a tool for studying managerial and workman behavior were Taylor's outstanding contributions.

Henri Fayol went further in taking a more macro approach to the organization. Fayol suggested that in addition to planning there were four other specific functions: organization, coordination, control and command. Fayol enumerated fourteen specific principles of management that he had studied and applied in this work. Included in these principles are the necessity for: (1) division of work, (2) parity of authority and responsibility, (3) unity of command whereby any individual, whether manager or laborer should have only one boss to whom he reports, and (4) subordination of individual interest to the general interest of an organization. The other principles and a further elaboration of Fayol's work can be found in most management references as well as Fayol's original work, which incidentally was not published in the United States until after the Second World War--a factor which further hindered the advancement of management theory.

Much of what Taylor and Fayol did has been modified in recent years by the development and application of other theories and explanations of the management concepts. Some of the outstanding contributions have been those of Max Weber and his theory of bureaucracy that suggests that the bureaucratic structure is not totally inappropriate for government and other large organizations in that it follows a scheme and delineates specific responsibilities for individuals within an organization. Similarly, Chester Barnard has contributed to the growing body of management thought by his suggestion that the informal group and its focus of authority are important prerequisites to effective formal authority.

More recently, behavioral research has made a substantial contribution to the field of management thought. Among others, psychologists such as Abraham Maslow, Frederick Herzberg, Rensis Likert, and Saul Gellerman, and their concern with motivation and productivity, have made notable contributions to the field. These behavioral science and other contributions will be examined further in the management of human resources section later in this chapter.

The Role of "The Professional Manager" in Multidisciplinary Projects

The manager of a large multidisciplinary project must be more of a manager than anything else. That means that the project manager must relegate any former discipline and enlarge his cognitive map to include the subject matter of management.

A natural scientist demands that progress on public projects which affect the environment be curtailed until determinations can be made as to the effect of those projects. Yet, all people make demands on our environment to provide us with accommodations, recreation areas such as ski slopes and golf courses, and transportation media capable of hurrying us to and from these areas. How do we reconcile these divergent points of view? Who makes, or should make, decisions on what public works projects should be funded? The political scientist? The economist? The engineer? The biologist? The sociologist? The professional manager? The answer has to be that it is a collective, coordinated determination of objectives and the necessity for professional managerial skills in achieving this coordination becomes evident.

By definition, the professional manager is a catalyst, coordinator and integrator. On one individual must rest the integrative burden of reconciling differences between divergent points of view. The manager may be dealing with people who have little or no experience in team effort. Many team members may be individually science-oriented and not trained to think in terms of team problem-solving. Effective team-building and the management of a multidisciplinary team are the subjects of a succeeding section.

The manager is subject to self-interest pressure groups and other external forces including laws and public opinion. He may be constrained by his own cognitive map or former discipline. Whoever serves as manager, regardless of background, has ultimate responsibility for all decisions and the various associated processes used to make them. Decisions are involved in each of the managerial functions: planning, control, organizing, and human resources management.

ELEMENTS OF DECISION MAKING: PLANNING AND CONTROL

Most works on decision making begin with an enumeration of steps involved in the decision making process. These include "defining the problem," "getting the facts," "examining alternatives," and "choosing the best alternative." These are all important steps in the decision-making process, but equally important--particularly in the design of public works projects-- are the necessity for "increasing understanding" on the part of all parties to a decision, "acquiring and processing information" necessary to make a rational decision within the limits of efficiency, and the necessity for "identifying pertinent or controlling known facts and/or unknown information" that is critical to further analysis. Identifying these variables of decision making as well as the more traditional steps, will serve useful purposes in making ultimate decisions more effective, efficient, and efficacious, as discussed in the social systems chapter.

All problems are disguised opportunities. "Definition of problem" need not have a negative connotation; in fact, this step might better be described as "definition of opportunities." Even the coordination "problems" of bringing together a multidisciplinary team have the opportunities of making a more complete project and the opportunity of creating a better solution. *The importance of weighing alternative remedies and examining each of the alternative courses of action cannot be overemphasized.* There is a necessity for rational analysis in identifying all or as many as practical of the significant consequences of the various courses of action. If alternatives to the design of projects or alternatives to any other decision are not examined, then a decision has not really been made.

Another important factor which must be recognized in the decision-making process with respect to public works projects is that: *every project has to be managed within certain constraints--the basic ones being time and money.* Most public

works projects are a result of multiple input and a conglomerate situation so that deadlines and budget limitations can seldom be altered.

There are five principal evolutionary steps in the evolvement of public works projects: (1) pre-feasibility (needs)
studies, (2) feasibility studies, (3) design, (4) construction,
and (5) maintenance and operation. Each of these steps are part
of the evolvement process discussed earlier in Chapter Two.
The determination of needs and the alternative ways of meeting
the needs elaborate upon the decision-making process discussed
above. The design and actual construction of a project as well
as the operation and maintenance of the facility are types of
implementation and follow-up of decision making.

Establishment of Objectives

Many projects and ventures lack clearly defined objectives.
The setting of objectives is closely related to planning, but
is sufficiently important that it deserves separate attention.

*The objectives of a business, or any institution, are mul-
tiple rather than singular.* Profit is an important objective of
business, but the myopic pursuit of immediate financial gain
without consideration of prerequisites to continued profit
(service, reputation, competitiveness, and others) foretell the
demise of individual businesses. To emphasize any one single
objective misleads not only the management, but other employees
and clients of an organization.

In the *Practice of Management* (1954) Peter Drucker has
suggested that a credo for the conduct of business, and by inference other institutions, should be that: "Objectives are
needed in every area where performance and results directly and
vitally affect the survival and prosperity of the business."
According to Drucker, the rigorous preselection of specific objectives in the key areas of performance and measurement of the
results will assist managers to:

1. organize and explain the whole range of business phenomenon
    in a small number of general statements;

2. test these statements in actual experience;

3. predict behavior;

4. appraise the soundness of decisions when they are still
    being made; and

5. enable practicing businessmen to analyze their own experi-
    ence and, as a result, improve their performance.

Objectives tailored solely to produce profit cannot meet
these criteria. The objectives defined to provide service (social responsibility; engineering functionality, quality, utility, among others) meet most of these criteria and can serve as
the central focus of all organizations--public and quasi-public
institutions as well as businesses. In short, accomplishing the
service objectives provides a means to meet the profit objective. One writer has suggested that:

> The right attitude of mind for the competent manager re-
> cognizes and accepts that there is no easy road to success,
> that service to customers is a matter of major concern to
> him, that productivity and economy are worthwhile objec-
> tives, that he must have the persistent intention of hav-
> ing things right and a readiness to improve whenever op-
> portunity occurs, that he must be ready for change and ac-
> quire the human skills to see change successfully accom-
> plished. (From E. F. L. Breck, *Construction Management in
> Principle and Practice*, 1971, p. 61).

*Within overall objectives such as service, profit, client
and employee relations, it is important that each unit and each
project have specific objectives which are clearly defined and
communicated.* There is a necessity for creating involvements
and obtaining commitments from all parties to a project. The
preceding chapter on public participation elaborated upon the
obtaining of commitments from the public. It is equally important that commitments to the objectives be obtained from individuals working on a project.

This process, more formally called "Management by Objectives" (MBO) is one means of creating involvement and obtaining
commitments from personnel working on a project. Management by
objectives means that every manager must develop and set the
objectives of his own unit or task force. The manager places
these goals in written form and then arranges for concurrence
of higher level management. As simple as the concept may sound,
it is not without its pitfalls and must be approached in any
organization with knowledge of the philosophy of employee involvement. Failure to give the goal-setter some guidelines and
flexibility in determining the objectives may spell failure of
the MBO system.

Communication of objectives and other plans is another important aspect of objective management. The objectives of a
public works project must be communicated not only to those
seeking to implement the project but to the public whom it affects as discussed in the previous chapter. For example, if the
objective of a project requires the shooting of charges underground at certain times of the day, it is important for safety
and involvement purposes that the times and reasons for the

tests be communicated to all employees and the surrounding community. Another example would be that if it is necessary to close an entryway to a place of business or a residence that the public be informed of the consequences during the *design* stage of the project.

Another aspect of objective determination is present when objectives appear to be in conflict. Economic objectives may and usually conflict with personal objectives in a project. A more serious situation is present when economic objectives conflict with social objectives. The problem here becomes which set of objectives to pursue or preferably how to reconcile conflicting objectives.

The director of the Environmental System Department at Westinghouse stresses the need for clear and reconcilable environmental objectives:

> If the total preservation of our natural environment is the objective, then the nation becomes poor because of the vast amounts of resources within our nation which cannot be used. If the objective is to maximize production without regard to environmental effects, then the nation will become poor when our quality of life reaches an undesirable level. (Dr. James H. Wright, "The Environmental Challenge--Values in Conflict," *ASEE--North Central Section Meeting*, University of Pittsburgh, April 16-17, 1971).

Obviously, an objective between these two extremes is necessary. It is part of the manager's job to reconcile the differences between these conflicting objectives by securing information, weighing alternatives, creating involvements and obtaining commitments of all parties.

Strategic Planning

Strategies are ways of accomplishing objectives. These strategies take into account the leadership skills, knowledge and ability of the manager and also his belief of what his competitors are planning and doing. This means that he takes into account various strategies such as:

1. one's own actions in attaining the goals.

2. probable actions by others in support of the goals.

3. probable actions by those opposed to the goals.

4. identification of major turning points within the course of the project.

5. general guidelines for continuation after setbacks or successes.

6. inherent flexibility to deal with variables and unforeseen events.

Specific objectives developed, during the course of the project, can then be tested against the overall strategy to determine first whether an individual objective furthers the project, second, what modifications to the objective could be made to enhance its usefulness, and third, which portions of conflicting objectives should be retained.

A prudent strategy permits the choice by managers of a wide range of tactical maneuvers in support of the overall goal and the interim objectives.

The competence of a public works project manager is tested severely in the establishing of an overall strategy towards achieving the goals of the project, and his wisdom becomes evident in his ability to modify strategy and rethink objectives in response to changing situations and new information.

Other Planning and Controlling Techniques

"Management by exception" is a general management principle that permits planning and control. The principle implies that managerial efficiency is greatly increased when managers concentrate their attention on those matters which are variations from routine, plan, or standard. That is, the more a manager concentrates his planning and control on exceptions, the more efficient will be the end results of managerial activity.

Some caution is urged on practitioners of the management by exception approach. Confusing activity with productivity occurs most frequently when several crises, or exceptions, must be dealt with simultaneously. This is also generally the most important period for arriving at prudent decisions. Management by exception works reasonably well in a smooth flowing operation with occasional minor anomalies that require the manager's attention. In a chaotic or fast changing situation, management by exception tends to increase the level of chaos. There is no substitute for competent human resources capable of making decisions and taking action.

Most engineers are very familiar with the techniques that have arisen within the past decade or two that permit management to have the ability to do a better job of planning the use of resources within time and cost limitations. For those practitioners in the environmental design and public projects areas who have not been exposed to these techniques, a brief synopsis is offered here.

*Time-event network analyses such as PERT (Program Evalua-
tion and Review Technique) and milestone budgeting provide management with an opportunity to approach the manpower, material,
and capital requirements from a sophisticated planning basis.*
The individual task to be completed in any given project must
be put into a network which is comprised of events and activities. The events represent a specified accomplishment at a
particular instant in time. An activity represents the time and
resources which are necessary to progress from one event to the
next. The logical sequence of events and activities in the
network facilitates making time estimates such as the optimistic
(the minimum time an activity will take), most likely (the normal time an activity will take), and pessimistic (the maximum
time an activity will take). Thus, this technique can define a
critical path since it identifies the sequence of activities
that will determine the minimum time in which the project can
be completed. Network analysis techniques can also be utilized
in determining program costs and the commitment of resources to
a project. Examples of time and cost network analyses are illustrated in the implementation section of this chapter.

*Information itself is an important variable in planning
and controlling.* Statistical decision theory, based on Bayesian
statistics, and management information systems can provide the
data necessary to make intelligent, more rational decisions in
all of the phases previously discussed such as objective determination and strategic planning as well as the commitment of
other resources to a project. An important aspect of any management information system is that it be able to provide the
information in summary form without inundating the receiver
with data.

Other systems analyses, including logistics systems and
industrial dynamics, as tools of management are included in
Chapter 10 and are not repeated here.

*Program Budgeting or PPB (Programming, Planning, and
Budgeting) is another useful technique in planning and control
of projects.* Program budgeting provides a systematic method for
allocating resources of an institution in an effective way to
meet its objectives. By putting the emphasis on the objectives,
the risk of becoming overly concerned with day-to-day activities
and budgetary limitations is overcome.

*Similarly, cost effectiveness or cost-benefit analyses are
of assistance to the manager in determining the effectiveness
of his project.* One difficulty is that accounting data are
seldom consistent with the program budgeting data and this lack
of information in these areas makes effective or useful cost-
benefit analyses difficult and costly to obtain.

Despite the sophistication of some quantifying techniques, there are still some critical intangibles about which decision makers, including project managers and legislators, should have knowledge. This is particularly true in the case of public work projects with respect to human resources management, eco-future needs, costs and technological developments.

HUMAN RESOURCES MANAGEMENT

Human resources management involves identification and programming of available manpower, their skills, abilities, and aptitudes.   It involves job analyses, job evaluation, and job design to ensure that each job is filled by the "right" person. Human resources management includes employee training and development, evaluation, and remuneration. Most important of all, human resources management includes the elements of motivation, leadership, and communications. Every manager must be aware of these facets of human resources management if he is to contribute toward the accomplishment of organization objectives.

The human resources management of public works projects begins with the recruiting and assembling of a multidisciplinary team.   Many members of a project staff may serve as part-time contributors or consultants on a project, but their role and contribution should be identified early in the project.

Procedures must be developed for creating ad hoc design teams whose composition will include all the disciplines required to solve design and environmental problems on a particular public works project. One of the fundamental considerations of who should be involved in a project is the factor of "interest"--a major factor of motivation.

Models of Motivation and Organizational Behavior

Some of the fundamental considerations in handling human resources are that the human resources are entirely different from the other resources involved in a project.   Fundamental considerations include human dignity, individual differences, motivated or caused behavior and hierarchies of needs.

Motivation is personal and self-inspired concept with differences among persons and within any individual over a period of time. On the other hand, motivation is also a social process guided by interaction with others including superior-subordinate-peer relationships. Familiarity with the various types of motivation will help the manager in developing his style of management and leadership.

Historically, deficiency motivation has been the most predominant explanation of the motivation process. Representatives

of the deficiency theory hold that behavior results from a need
which is termed a "deficit."   In short, psychologists such as
Freud and Lewin are suggesting that a deficiency or a tensional
state must be present before a motivating situation can exist.
This view of motivation is not healthy from a management-
employee relations standpoint.

As an example of deficiency motivation, if an employee who
takes on a new level of responsibility is told that failure
will mean dismissal, most of his efforts will be dedicated to
seeking new possibilities against the chance that something
goes wrong. If, on the other hand, the employee is told that a
great deal is expected of him, his efforts will go to fulfilling the expectation, provided that each time he scores a success he is told so and is appropriately rewarded.   In both
cases, a serious failure due to lack of ability could result in
dismissal, but the second case is stated in positive terms instead of in threats.   The employee in both cases is motivated
from a "deficit," that is, he has not yet succeeded while success is expected of him, but in the second case his activities
are channeled in productive lines instead of safeguarding personal interest.

*The hedonistic type of motivation is somewhat transitional
in that it relates to the deficiency or tensional state by suggesting that an organism (a human being) will endeavor to attain or maintain a pleasant state, and strive to change or
leave an unpleasant state.*   This type of motivation also does
not go beyond an infantile consideration of needs or wants.

*Growth motivation, on the other hand, is a more positive
type of motivation that puts emphasis upon the individual and
psychological growth of human resources.* The individual is motivated to grow psychologically and to develop his potentialities.   Furthermore, a deficiency need not be present in order
for the individual to become motivated.

One of the outstanding examples of growth theory or positive motivation is the need hierarchy proposed by psychologist
Abraham Maslow (Maslow, 1954).   Maslow postulates that man's
needs are arranged in a hierarchy including five major steps:
physiological, safety, love, esteem, and self-actualization
needs. He advocates that "man is a wanting animal" and that "a
satisfied need is not a motivator of behavior."   These statements mean that when the physiological and safety needs are
relatively well provided for, the motivational emphasis must be
shifted to the social, psychological, and other higher order
needs.

Two other individuals complement the Maslow hierarchy.
Douglas McGregor has adapted and elaborated upon Maslow's hierarchy as a basis of his discussion of "Theory X" and "Theory Y."

(McGregor, *The Human Side of Enterprise,* 1960).   In essence,
Theory X holds that the average human being is lazy, slothful,
has an inherent dislike of work and responsibility, and must be
coerced or forced to put forth work effort. Theory Y, on the
other hand, assumes that work effort and responsibility are
natural to man, that there are methods other than force for
achieving individual and organizational goals and that human
beings are capable of self-control and self-direction.

Frederick Herzberg has made a distinction between real mo-
tivators and what he calls maintenance (or hygiene) factors in
the job environment (Herzberg, *Work and the Nature of Man,*
1966). Maintenance factors such as working conditions, company
policies, and supervision, and even salary, appear to be related to job conditions and operate as an essential base for
the satisfying or real motivating factors.   The satisfiers are
such factors as achievement, recognition, advancement opportunities, interest in the work itself, and the responsibility for
a job.   According to Herzberg, the maintenance factors must be
relatively well gratified before strong positive motivation by
the satisfier factors can become prepotent. The task is a continuous one, however, since acceptance of challenge and the accomplishment of more difficult tasks lead to the belief that
salary and other maintenance factors are inadequate; thus, the
cycle begins again. One of the senior management's most difficult tasks is to decide how much gratification of the maintenance factors is enough, and to apply the decision with consistency.

*Modern motivation puts emphasis on the satisfaction of
psychological rather than physiological needs.* Human beings, as
such, deserve to be motivated according to the best available
concepts and methods. Fundamental concepts of human motivation
and behavior such as individual differences, caused or motivated behavior, prepotency of needs, and human dignity itself
must be preserved if managers are to motivate successfully.

Managing Teamwork

There are some special considerations in managing professional and scientific personnel who work on public works projects.   Consistent with the foregoing discussion of motivation
and the recognition of psychological needs that must be fulfilled, it is important that project managers recognize these
psychological needs--particularly for the professional person.
There is a need for "real" motivation in the sense of providing
intrinsically interesting work, responsibility, recognition and
advancement opportunities for personnel involved in a project.

There are various styles of leadership which may be employed on a project. At different times, all styles are usually
appropriate in bringing a project to completion.   Styles of

leadership range from an autocratic or dictatorial type, through
the paternalistic or benevolent authoritarian type, to the con-
sultative or real participative type of management. The latter
has been called "supportive management" by Rensis Likert,
(Likert, 1961).

*The integrating principle of Likert's findings is what he
calls the "principle of supportive relationships."* The important point is that a supportive environment be created--suppor-
tive in the sense that:

> ... the leadership and other processes of the organization
> must be such as to ensure a maximum probability that in
> all interactions and all relationships with the organiza-
> tion each member will, in the light of his background,
> values, and expectations, view the experience as suppor-
> tive and one which builds and maintains his sense of per-
> sonal worth and importance.

There are other styles of leadership which advocate complete or
collegial participation but they are more utopian types of
leadership behavior which do not currently enjoy a widespread
working reality.

*Similarly, communications may be categorized into several
styles or types which facilitate managerial effectiveness.*
Considerable attention has been given to the formal types of
communication--both written and verbal--and how to transmit
clear messages. Less attention has been given to the receiving
and understanding functions of communications that are also a
part of every manager's job.

*Of further importance are the informal and interpersonal
communications among all parties involved in a project.* The
necessity for open communications and a climate of mutual respect and trust cannot be over-emphasized. The informal--
perhaps daily contacts--between contractors, the contracting
agency, and all project personnel can prevent costly changes
and disruptions in the project. To be sure, open communications
can be dysfunctional at either extreme--that is, communicating
everything can waste a great amount of time and have a harmful
effect when some matters might be better unsaid. At the opposite extreme, failure to communicate feelings and sentiments
can have harmful effects also. There is a necessity for a rational middle ground in both formal and informal communications
between all parties on a project.

*There are significant differences between a collection of
individual managers comprising an organizational unit and a
real managerial team* (even though the "team" label is frequently
applied indiscriminately to both). An important frame of reference in communicating and managing teamwork is that in addition to the transmitter and receiver of a communique, and in

addition to the individual and the formal organization, there
should be an interactive team effect at work. This is a synergistic effect that two or more people who are properly matched
will be more productive together than they would be if they
worked individually.

McGregor (1967) suggests several characteristics necessary
for an effective managerial team:

1. the unit as a whole has one *primary task;*

2. its *flexible structure* adjusts to the demands of the situation;

3. all members of the unit, including the manager, determine
the structure, responsibilities, and the subgoals of the
unit; the principles and the norms governing its operation;
and the standards of performance;

4. opportunities for *intrinsic rewards* associated with membership in and accomplishments of a cohesive group are deliberately created, while control of *extrinsic rewards* is
maintained by management;

5. *cooperative relationships* exist between the members accomplishing the common primary task;

6. *interdependence* between all the members determines the
survival and successful performance of the team;

7. members must have appropriate skills.

*Once assembled, with these characteristics in mind, the
management development and training of the members of the ad
hoc design team may be accomplished in various ways.* The important point to be made is that all members of the team must
develop acceptance of one another, trust and mutual respect for
others. Stereotyping of others from different disciplines can
and must be avoided by making a concerted effort to communicate
with and understand each team member's contributions. Formal
training programs such as T-groups (sensitivity sessions) are
designed to accomplish these purposes of developing understanding of others. For most public works projects, the team members
are probably capable of self-development, with or without the
help of a behavioral science consultant. To speed up the team
building process, it may be desirable to engage in an organization development seminar where "feedback" on professional and
interpersonal matters begins to emerge spontaneously.

The behavioral science techniques of listening fully to
others' opinions before rendering one's own and perhaps modifying one's own, still apply. Formal management development

training--*tailored for the individual project*--may be conducted
for an organization by management consultants or inhouse personnel. This team effort and the development of trust, mutual
respect, and empathy for others' positions is fundamental for
the successful management and completion of a project.

MANAGING CONFLICT AND CHANGE

The changing nature of organizations, their pursuits and
objectives, and the interactive effects on human needs suggest
that flexibility is needed in organizational structures.   Regardless of size or type of organization--private, government,
or an interdisciplinary team--there is a necessity for creating
a supportive type of environment and integrating organizational
objectives and individual needs dynamically.

The Nature of Change

Change is the end product of all activity.   Situations,
organizations, and humans are in a constant state of change.
Whether the change is detrimental and destructive, progressive
and constructive, or time-filling and meaningless depends upon
the character and motivations of the people involved.   Change
can be a satisfying job factor in the work place--if it is
properly introduced and communicated to the parties concerned.
Understanding the forces for and against change will help the
manager in coping with it.

Land use provides a good example of the change process.
Some of the forces promoting change in land use are a manifestation of the desire for not only more but better planning and
other managerial action.   In 1790, when about five percent of
the total United States population lived in cities, urban problems as we know them today were almost non-existent and planning was a luxury.   However, in the 182 years since the first
census, the urban-suburban population has increased to about
75 percent of a total population over 200 million. The problems
of urban sprawl, pollution, traffic congestion and ecological
considerations have brought much of the population to the conclusion that land use planning is a necessity.

Some of the forces opposed to changes in land use result
from a fear that planning will eliminate the private enterprise
system.   Historically land use in the United States has been
based on the economic worth of the land in a free market.   Land
use planning, it is feared by some, will interfere with this
system and will compromise the property rights of the individual.   In addition, some governmental units sense that much of
their power will be abrogated by comprehensive planning for
land use.   Planning must encompass more than one governmental
jurisdiction and many jurisdictions must relinquish some control if the planning is to be effective.

Thus, the forces for and against change constitute a paradox. The conflict that has arisen from the interaction forces can be resolved; but in order to do so, society as a whole (through representative governments) must reevaluate its objectives and develop new policies to reflect necessary changes. As the American Institute of Architects states in a resolution which was adopted unanimously:

> That because of the growing conflict between our traditional concepts of private property and land use and the already desperate need for a national housing policy, the AIA recognizes that under more and more conditions the public interest must prevail over the interests of private property, and that the development of land is a privilege and not a right. ("AIA Backs More Land-Use Control," Engineering News Record, July 1, 1971, p. 16.)

Any modification of our traditional values must come about through the change process if they are to be accepted. It is vital then that an understanding of the decision making processes and the communications of changes be understood by all parties concerned.

Effectuating Change

By gaining an understanding of the causes of change and the consequent results, managers are in a better position to adapt to and effect changes within their own organizational situations. As noted above, the management of change deals with the control and direction of the forces for change in order to achieve a prescribed objective. To this end, today's manager has at his disposal a wide range of techniques. Generally, the techniques can be classified as either behavioral or structural. Behavioral refers to the individual and the social relations within an organization and structural refers to the physical compositions of activities.

The human side of the system does not deal with static structures but with dynamic interrelationships of individuals, groups, training, evaluation, and physical factors. Humanistic and social values are influencing our views about complex machine systems. Two social psychologists have suggested that:

> Too often changes in the social arrangements of organizations are lagging and fragmentary adjustments to technical changes already accomplished. We automate the equipment first and repair the social dislocations afterwards. The same rational considerations, however, that operate within the framework of technology can be utilized to determine social objectives and thus make the technical system the means rather than the master of social technology. (Daniel

Katz and Ronald L. Kahn, *The Social Psychology of Organizations,* New York: John Wiley and Sons, Inc., 1966, p.
471.)

The behavioral techniques for change are designed to facilitate and increase the motivating power for change in the
organization. There are as many ways of achieving behavioral
change as there are managers. However, the techniques can be
classified by behavioral concept and the methods employed.
Among others, the behavioral concepts and methods for coping
with change include:

1. compulsion: threats and bribery; while these techniques do
    bring immediate results, their long-range effect is a
    greater detriment than the initial advantage gained.

2. persuasion: the use of rewards in order to be effective,
    must match the individual's need and goal structure. Man-
    agement must recognize that it must make some concessions
    in order to achieve mutual objectives.

3. security: guaranteeing variables in the employment situa-
    tion (such as income) through long-term contracts may over-
    come fears of inadequacy but may produce a less-than-
    dynamic work force.

4. understanding: it is fear of the unknown which creates
    anxiety over change. Communications are the means by which
    understanding is brought about.

5. involvement-participation: allowing the parties a chance to
    participate in the formulation of policies and other as-
    pects of the job situation.

There are additional methods including flexibility and
criticism where the amount of change desired and the amount
that can be accepted are important variables. The individual
manager must decide which one or combination of techniques he
will utilize in bringing about change.

*People are not inclined to resist change if they are in-
volved in the changes taking place.* Most human beings want to
be a part of improvement and change. It is through change that
better material and social welfare are achieved. It is generally not technological change which is resisted. Resistance
develops primarily because of psychological and sociological
deprivations and because economic well being is threatened.

*Structural organization of activities is becoming increas-
ingly complex due to the rapid growth, increasing size and diversity of ownership that results in a need for control to be
placed on managers.* In a rapidly changing environment, organization structure becomes of paramount importance as it provides

a basis for an organization's efforts to adapt and provides the vehicle for coping with change. Care must be exercised in establishing and promoting an organizational structure because too rigid a structure restricts change itself and the ability of personnel involved in change.

The organization structure that has flexibility built into it and permits adaptation to change is based on open-systems theory. Three authors writing on the management of systems, have suggested that:

> The successful administrator can foresee that the disruption and trauma resulting from the rapid application of systems techniques are potential sources of difficulty for both the individual and the organization. Because of the many ways of displaying resistance to change and the great difficulty in correcting resistance and moving back into a new, effective equilibrium, it is highly desirable for management to make every effort possible to effectuate the change properly and to facilitate adjustments to it. Giving adequate consideration to the problem of change, both technical and social, prior to its implementation is essential to long run success. (R. A. Johnson, F. E. Kast, and J. F. Rosenzweig, *The Theory and Management of Systems,* 2nd ed., (New York: McGraw-Hill, Inc., 1967, p. 398.)

The next section of this chapter deals further with traditional and adaptive organizational structures.

DESIGN AND OPERATION OF ORGANIZATION STRUCTURES

The tasks of reconciling differences among objectives, other plans, and human activities--of making a multidisciplinary team truly interdisciplinary--must be organized. Organization is the process of assigning authority and responsibility to various components of an organization structure in order that the predetermined objectives and plans may be fulfilled. Authority is the essence of organization. Some general principles of organization are discussed below.

Principles of Organizing

Authority to carry out a job and the responsibility for that job must be equal. No matter where the authority comes from (from the formal organization or from an acceptance of that authority by lower-levels in that organization), it is inconsistent to provide more authority to a unit or its leaders than is necessary for the responsibilities involved. Similarly, it is inappropriate to exact responsibility from individuals if they have not been given the authority over a project.

624

Decisions should be made at the lowest practical level of management where the authority can be exercised most competently—that is, where the individual manager is close enough to a situation to assess properly the needs of the project, client, organization, and employees. Authority should be delegated to individuals at the operational level for efficiency purposes and also as a means of developing managers.

Whether the authority should be centralized or decentralized on a project is a function of the newness of the project, the experience of the managers on the project, and the overall objectives of all parties to a particular project. If the complexities of the project are such that perspective on some issues can only be gained by viewing several areas of the organization simultaneously, then the decision-making authority must be centralized at a higher level of management rather than delegated to the on-site project manager. However, the more vertical space (the more levels of organization) between the top management of an organization and the individuals responsible for carrying out a project, the more probable that there will be miscommunications and misunderstandings about the objectives and the fulfillment of the project.

Unity of command is desirable in that only one individual should have decision-making authority regarding the commitment of resources to the project. The need for unity of command is strengthened where the diversity of objectives and the environmental considerations are present.

The span of control and overall span of management for any one individual is limited. The number of people that one individual can effectively manage is limited by the training and experience of the individuals working on the project, the diversity of functions being managed, and the abilities of the manager himself.

Minimizing the number of interfaces between levels of management serves to cut down on the miscommunications and misunderstandings regarding the project's objectives. Reducing the interfaces between levels and on a particular level reduces the necessity for coordination and unnecessary communications.

Responsibility for measuring outputs in terms of cost, quality, and meeting plans should be placed on one individual—usually a project manager. Thus, he should have the authority with respect to hiring employees, training employees, and other personnel functions.

Departmentation of activities within a firm should be contiguous—that is, similar activities should be under the functional head of one individual. For example, water resources

projects that have similar activities should be under the general supervision of one individual even though there may be
many projects headed by field managers.

*Completed staff work should be expected of headquarters
staff and staff in the field such as data processing technicians, personnel specialists, and other individuals who serve
the project manager as assistants.* The staff personnel should
complete all work from the standpoint of doing a thorough job
of analysis (looking at alternatives to problems and making recommendations to the project manager).

*Compulsory staff service should be required. Compulsory
staff service means that the project manager should use or at
least listen to the staff.* This does not mean that the project
manager must follow staff advice and recommendations but he
should hear them out before making a decision.

Staff independence is a desirable principle which overcomes the tendency of staff personnel to become "yes men." The
staff should be independent with respect to hiring, promotion,
and other personnel actions of the managers who they are advising or serving as staff.

Traditional Forms of Organization Structure

Organization structures usually evolve. Most firms start
as a line organization--that is, where managers in the chain of
command have all decision-making authority. Similarly, the line
has the responsibility and the duties of carrying out all activities.

As an organization becomes larger, it usually develops
staff personnel to carry on some of the specialized activities
within the firm. The line maintains the authority and responsibility for the direct accomplishment of objectives. The staff
serves in an advisory capacity to assist the line as experts
and as service personnel who can specialize in their particular
areas.

A functional type of organization is one in which the management of the firm is divided along functional lines--usually
such lines as product, process, or service being rendered by
the firm. The source of authority here is the perceived expertise of the individual in a unit of an organization. This
authority is present in the organization regardless of line-
staff considerations. The functional organizational form can
violate the principle of unity of command discussed above.
However, the violation of unity of command principle can be
overcome if the incumbents in a particular function adhere
closely to their own areas of expertise and do not exercise
authority outside of their specialty.

Bureaucratic organization structures are a composite of line and staff and functional decision-making forms of organization. There is still a hierarch where each level of management is subject to the control of the next higher level of management. And there is a systematic division of labor among occupants of offices based upon the occupants' technical qualifications. The increasing complexities of projects such as the legal, insurance, financial, and safety requirements indicates the necessity for continued use of specialized staff in each of these areas--notwithstanding the complex, bureaucratic nature of these forms of organization.

Examples of Organization Structures

There are numerous examples of organizational structures in the government, industrial, conglomerate and service organization fields. Nearly all of them follow the basic principle of line and staff organizations, that is, managers in the chain of command are delegated all the decision-making authority, and that line has the responsibility and duties of carrying out all activities. The following section will show some representative examples of the organizational structure employed by agencies and entities engaged in design and control of major public works projects.

Corps of Engineers. Most typical of a federal agency involved in major public works projects is the Department of the Army, Office of the Chief of Engineers. Top management is vested in the Office of the Chief of Engineers, headed by the Chief of Engineers who in turn delegates the responsibilities to a number of divisions such as the Pacific Ocean Division, North Pacific Division, North Atlantic Division, South Atlantic Division, etc. The organizational chart for a typical division is shown in Figure 13-1. Each division in turn supervises the work of districts under its own jurisdiction. The district organization is outlined in Figure 13-2. The division responsibilities are outlined as follows:

a. The Division Engineer is directly responsible to the Chief of Engineers. He commands and supervises District Engineers assigned to his control. This includes reviewing and approving major plans and programs of the districts, transmitting these to the Chief of Engineers, interpreting plans and policies of the Chief of Engineers and transmitting these to the districts, and reviewing and controlling individual acts of the districts.

b. The Division Engineer, within authorities delegated by the Chief of Engineers, assigns missions to his districts, coordinates their execution, combines results, develops cooperative interests with offices on his level and represents the division as a whole. The mission assigned by a Division Engineer to his districts include:

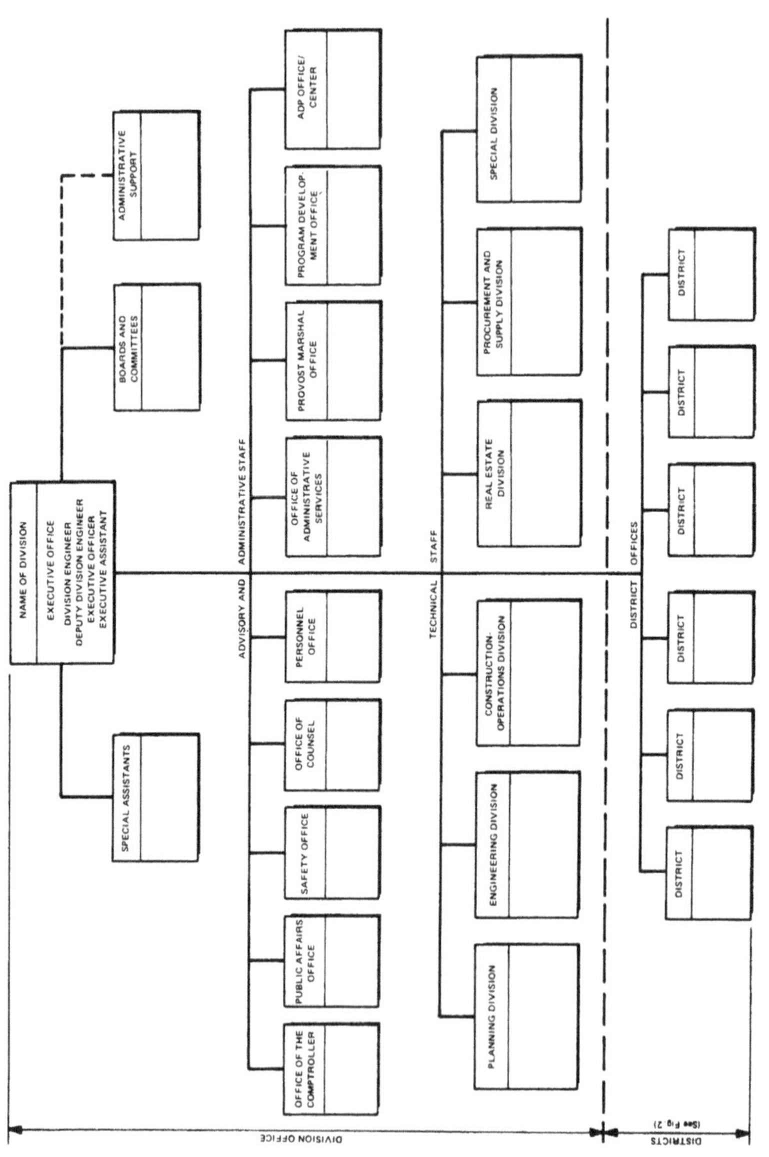

Figure 13-1. Corps of Engineers--division organization.

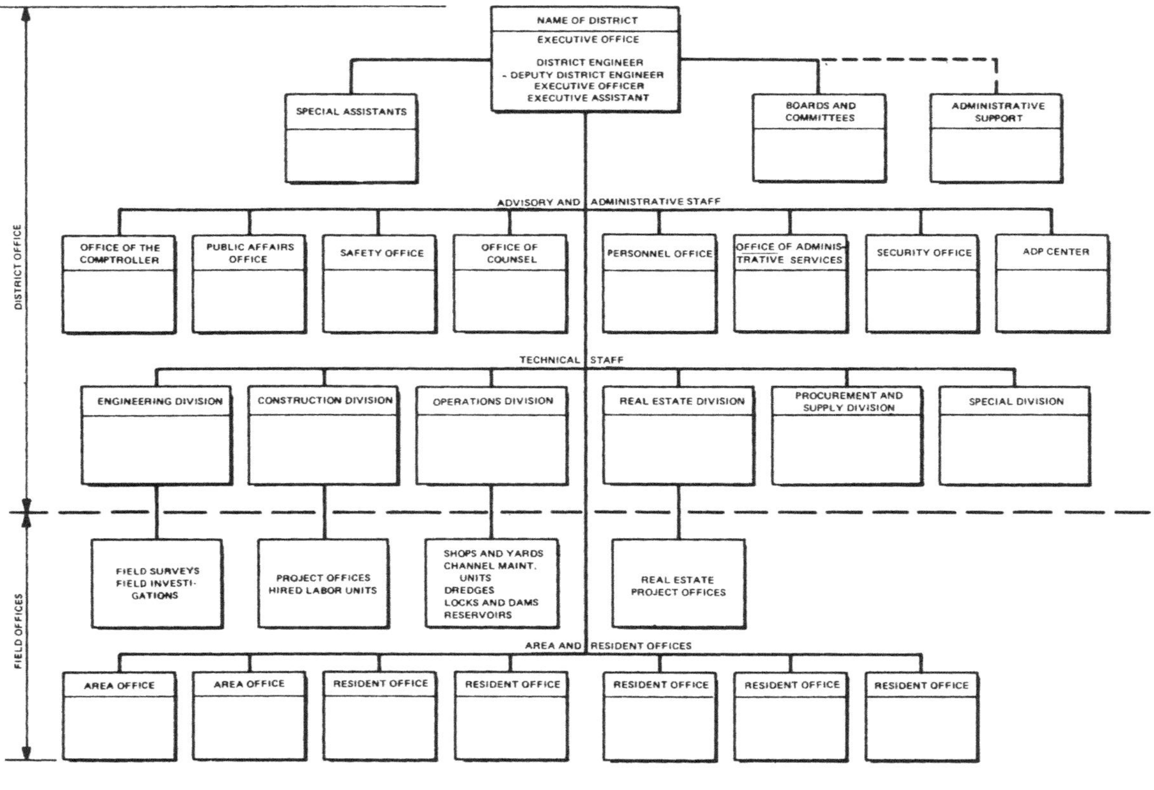

Figure 13-2. Corps of Engineers--district organization.

629

(1) Preparing engineering studies and design.
(2) Constructing military and civil works, and other facil-
    ities.
(3) Operating and maintaining flood control, river and har-
    bor, and other civil works facilities and installations.
(4) Administering the laws on civil works activities.
(5) Acquiring, managing, and disposing of real estate.
(6) Performing other functions assigned by law or directive.

The district is the operating arm of an engineer division
having the broad missions assigned to it for a given geographical area. In addition, the district may set up area engineer
offices as shown on Figure 13-2. For instance, where distance
makes the control and conduct of the work by the district office difficult or costly, or the nature and the complexity of
the work requires another field level, an area engineer is assigned who will often be in charge of one or more construction
projects or operating facilities.

State departments of public works. Many states also maintain highly competent service organizations to provide design
services for their own public works projects. Most state highway departments have that capability. The State of California
also maintains a major capability in the design and operation
of their major water supply facilities. Figure 13-3 shows the
organization structure for the California Department of Water
Resources and Figure 13-4 shows the organization of the Division of Resources Development within the department.

Municipal departments of public works. Most major cities
also maintain competent organizations to deal with municipal
public works projects. These organizations are generally
structured to provide design and contracting services for most
of their projects and will also maintain the capability of administering contracts for services to be provided by private
consultants. Municipal organizational structures vary widely,
with most cities having unique separation of authorities for
the various municipal public works activities such as streets
and highways, sanitary engineering, water supply, solid waste
disposal, construction of municipal buildings, parks and recre-
ations, etc.

Figure 13-5 shows the organization chart of the Bureau of
Engineering, Department of Public Works, City of San Francisco.
It again shows the traditional line organization and indicates
the limits of the activities of the Department of Public Works
of that particular city.

Many municipalities have elected to set up special authorities, for one or several of their public works activities. The
advantages of such authorities are that they are somewhat separated from day-to-day political pressures and other influences.

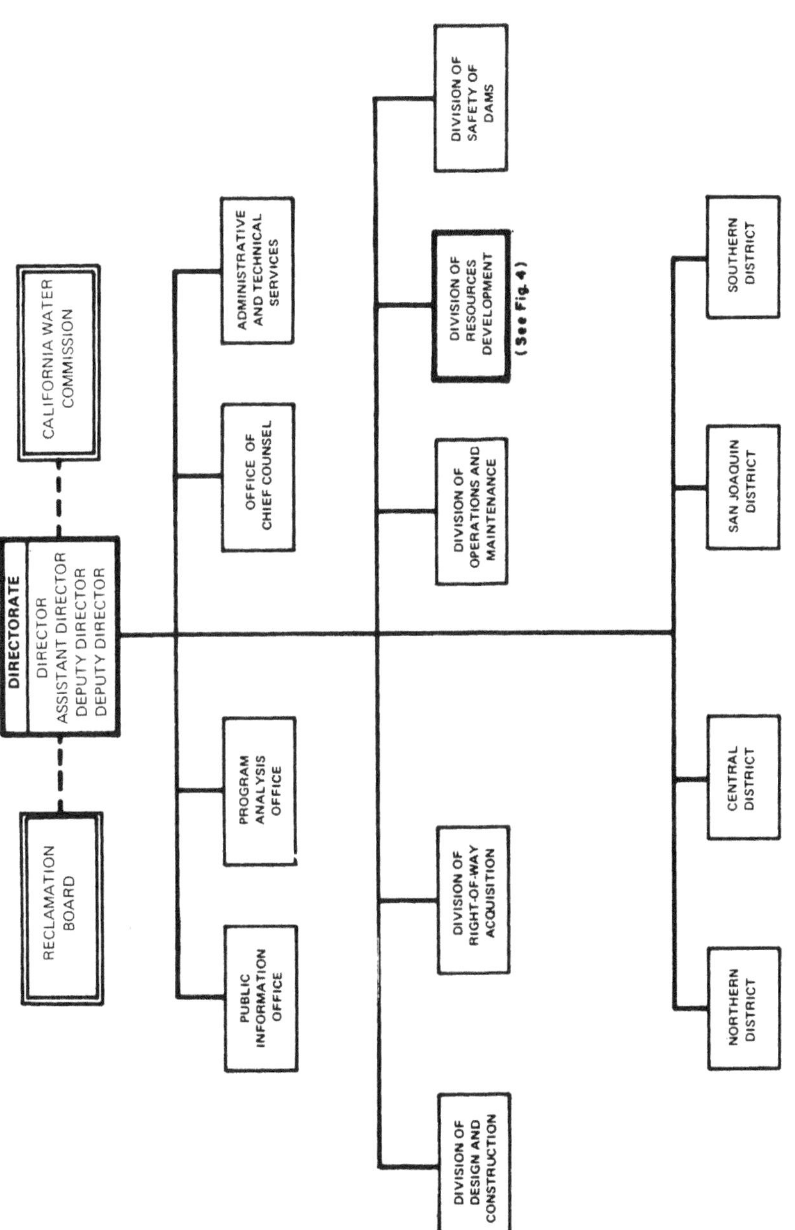

Figure 13-3. State of California, The Resources Agency, Department of Water Resources.

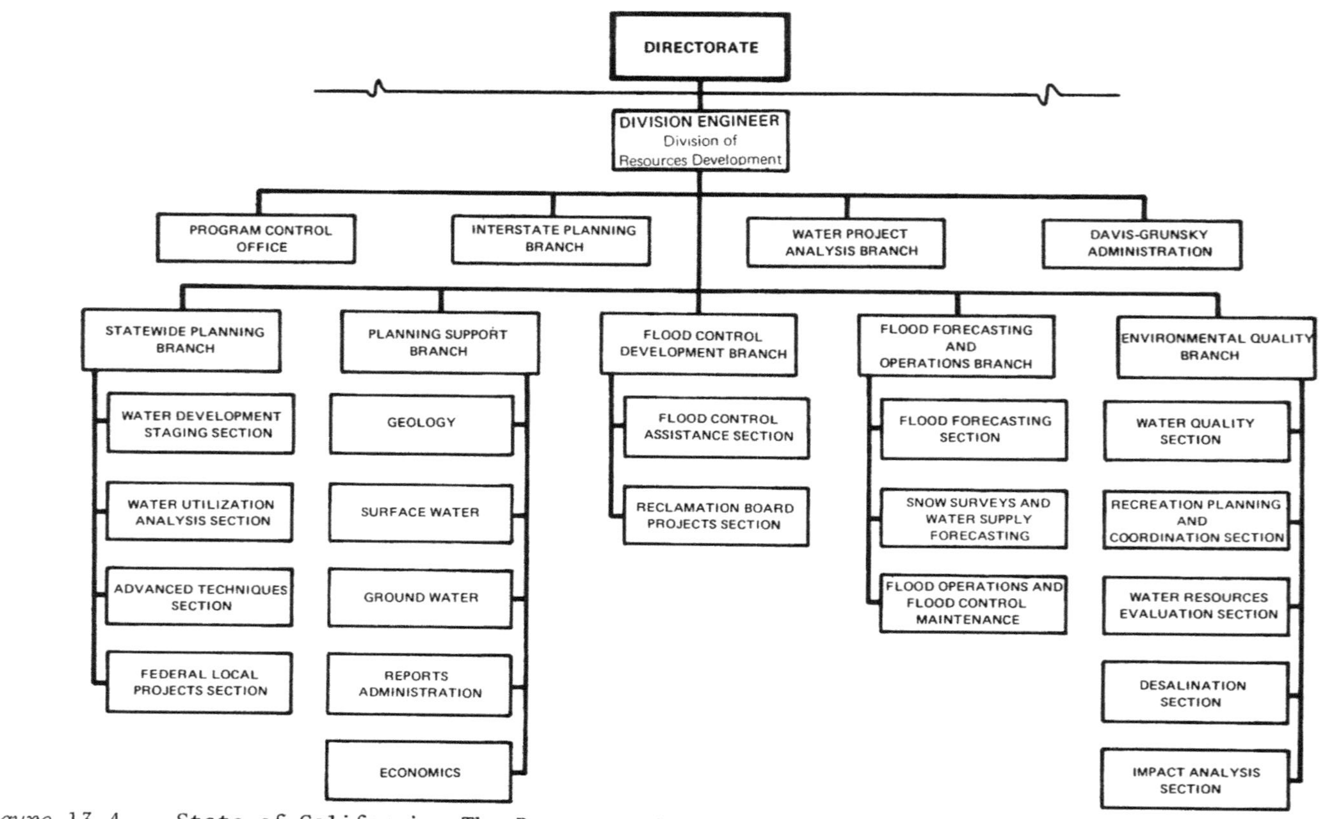

Figure 13-4. State of California, The Resources Agency, Department of Water Resources, Division of Resources Development.

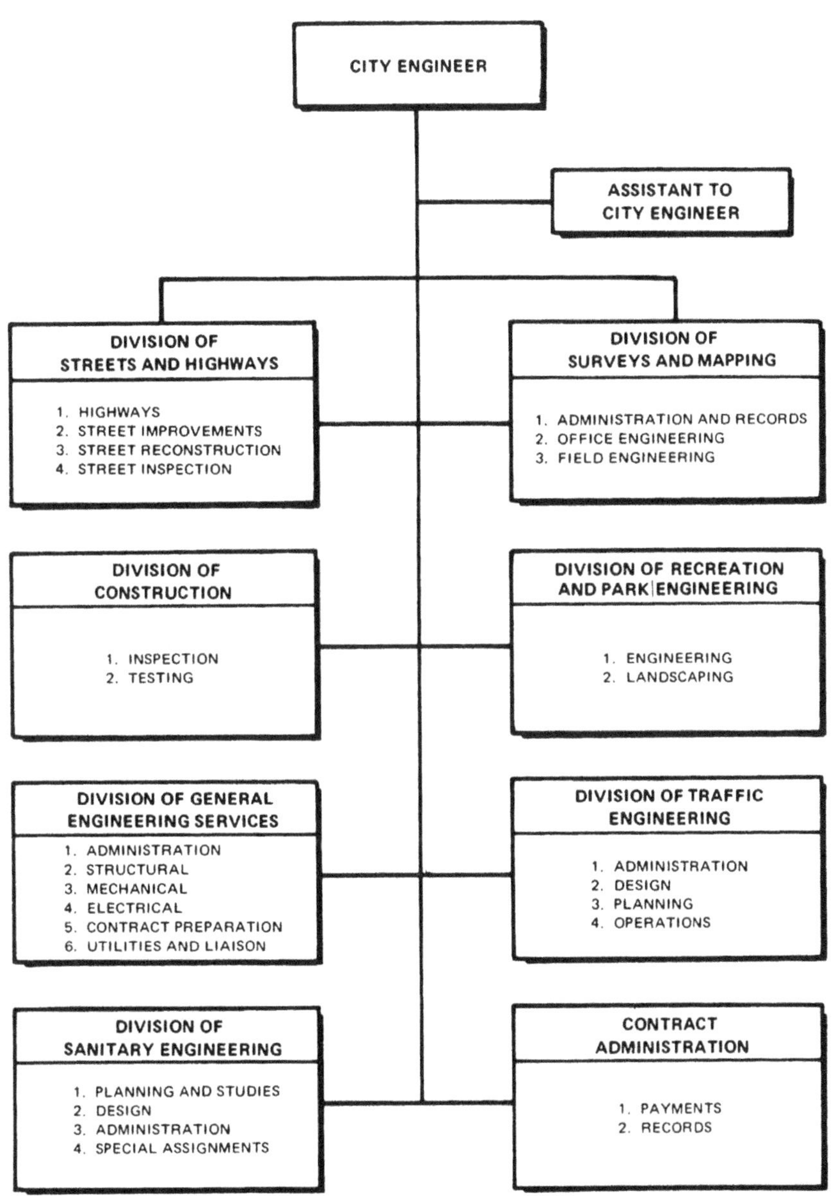

Figure 13-5.   Organization chart,   Bureau of Engineering, Department of Public Works, City of San Francisco.

633

They are generally financially independent and are able to manage their affairs through bond financing without drain on tax dollars.   An example of such an authority is the Denver, Colorado Board of Water Commissioners.   Figure 13-6 shows the management organization of this board and Figure 13-7 shows the organization structure of their engineering division.

Industrial conglomerates.   The growth of conglomerates in our industrial society provided a new dimension to the service industry engaged in the design of public works projects.   The large industrial corporations, principally those originally heavily involved in defense work, have decided to expand and diversify and some have elected to acquire existing design and construction firms. Some examples are the Raytheon Co., TRW, and others. A partial organization chart of a conglomerate engaged in engineering and construction derived from the Raytheon organization chart, is shown on Figure 13-8.   Again the basic principle of line and staff organization is evident. Organizational emphasis is upon strong policy and control responsibility centralized at the corporate level, with responsibility and authority for products and operations decentralized in cognizant divisions and subsidiaries. The decision making occurs at the level where the information is available to support a sound logical decision; however, such a decision must be made against the framework of adequate objectives, policies, and controls which are dictated by the top management of the company and which allow for the proper evaluation of operating performance. Organizational flexibility is intended to enable the conglomerate to respond purposefully to change and to adapt their resources to new opportunities.

Organization of service companies.   A substantial proportion of all public works designs is performed by traditional service companies.   The service companies fall into two basic general categories; the design-construct firm and the architect-engineer-planning firm.   Design-construct firms such as the Ralph M. Parsons Co., The Bechtel Corp., the Austin Co., etc. provide total project development from the conceptual design through the construction documents to the completion ɔf the construction and operation and training of operators and other personnel. A basic organization chart for a design-construction service company is shown in Figure 13-9.

Traditionally, the architect-engineer-planning firms stop their services at the letting of a construction contract and may render construction inspection services through the completion of the work. The design elements available to both design-construct as well as the architect-engineer-planning firms are usually organized in a similar fashion.   Today's major public works projects invariably require a multidisciplinary organization to cover all the necessary planning and design elements inherent in such projects.   Figure 13-10 shows the traditional

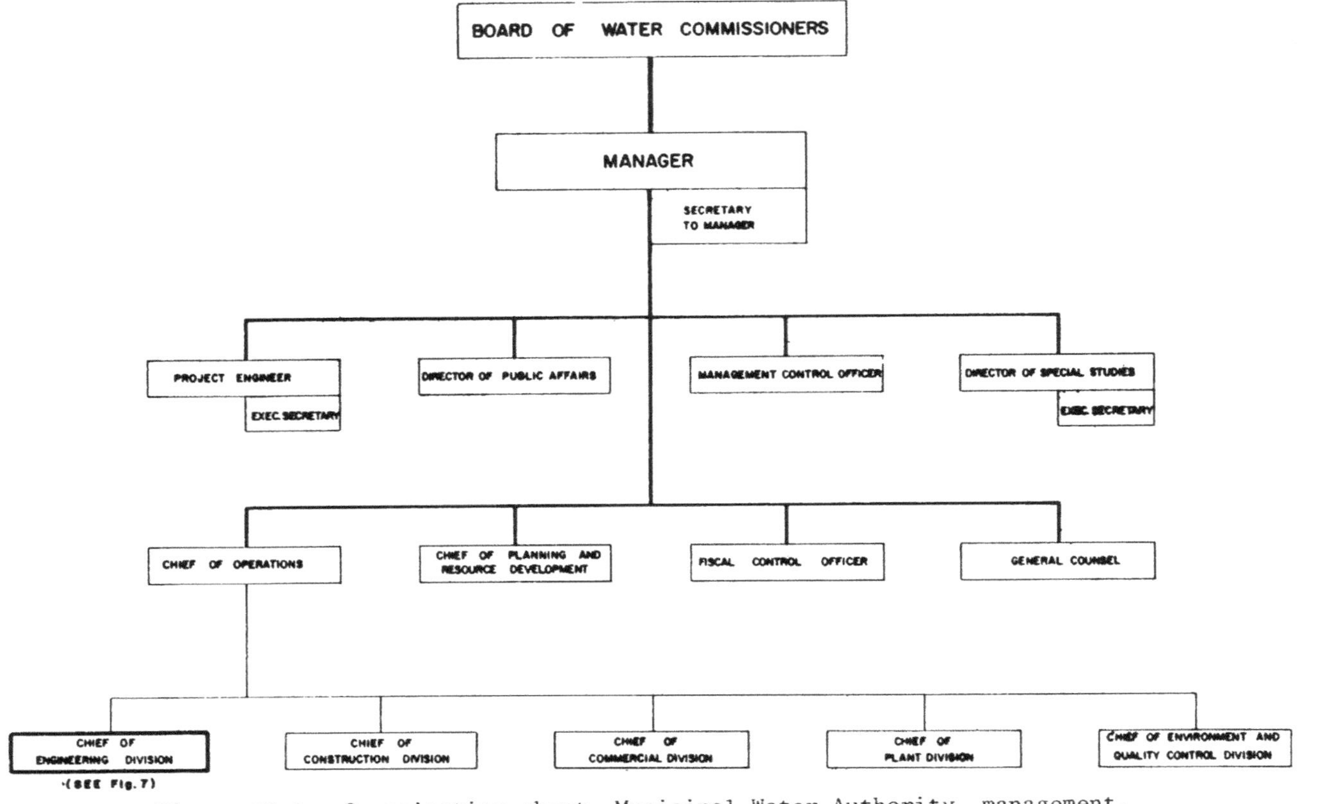

Figure 13-6. Organization chart, Municipal Water Authority, management.

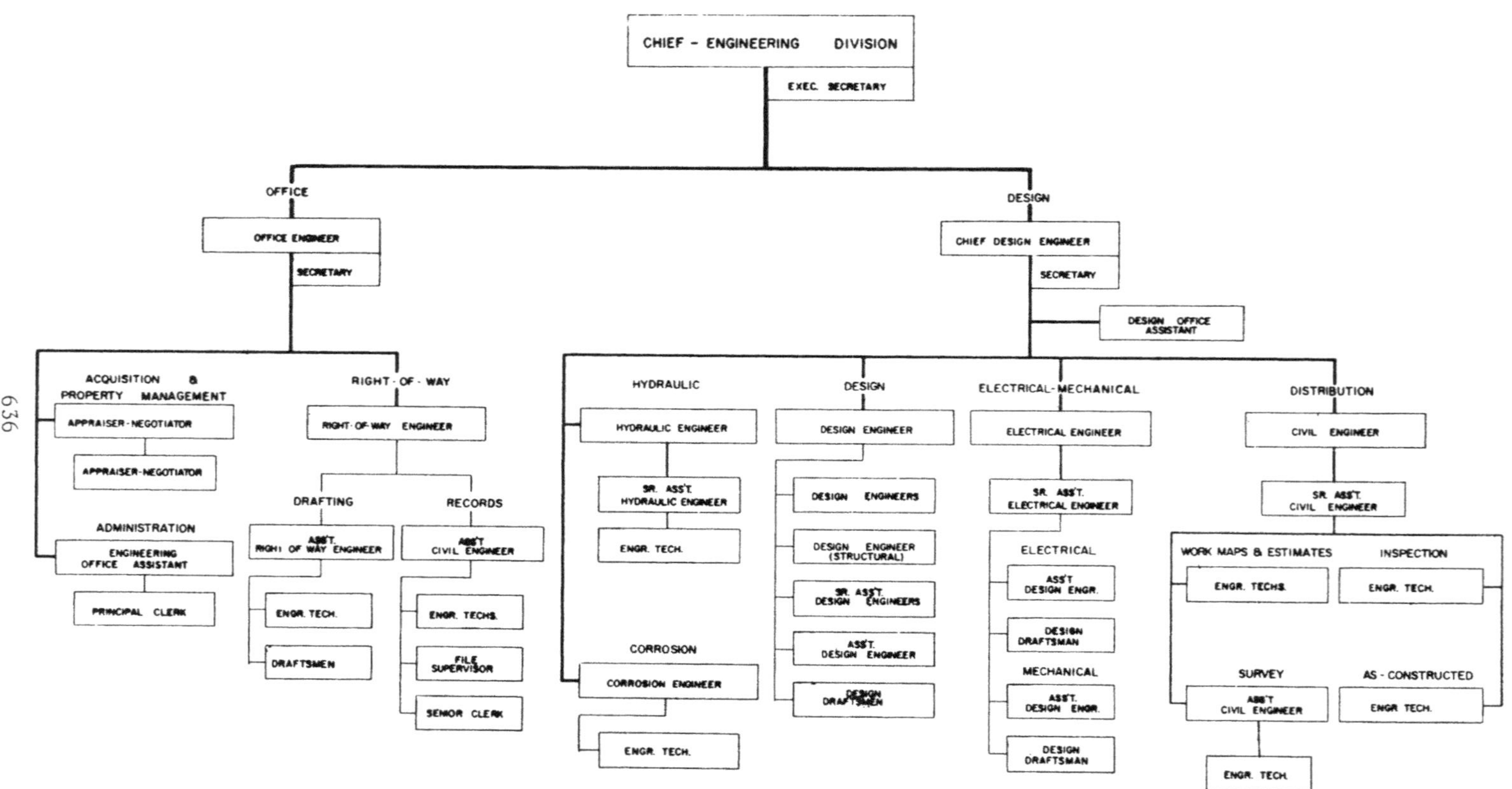

Figure 13-7. Organization chart, Municipal Water Authority, engineering.

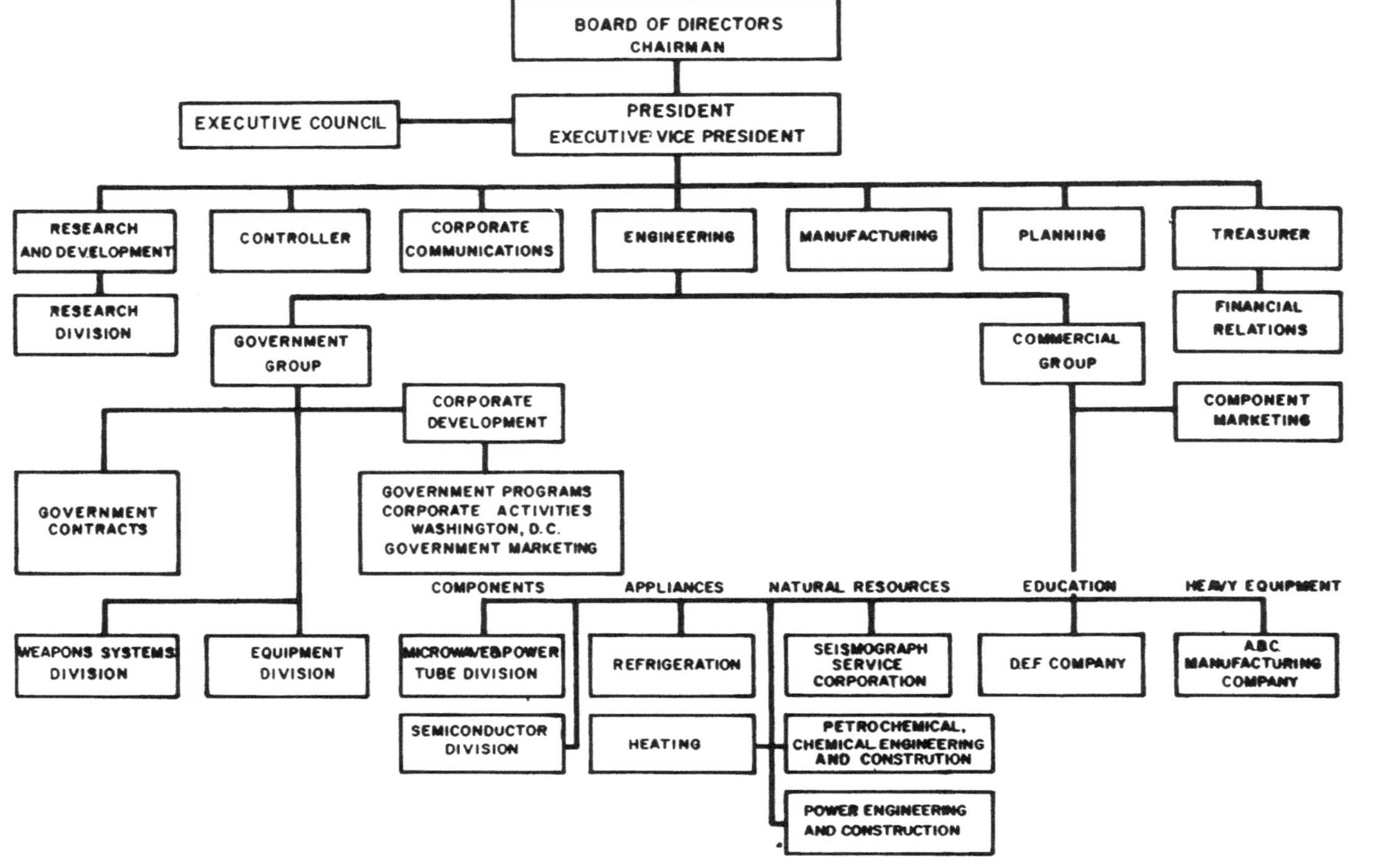

Figure 13-8. Industrial conglomerate, typical organization structure.

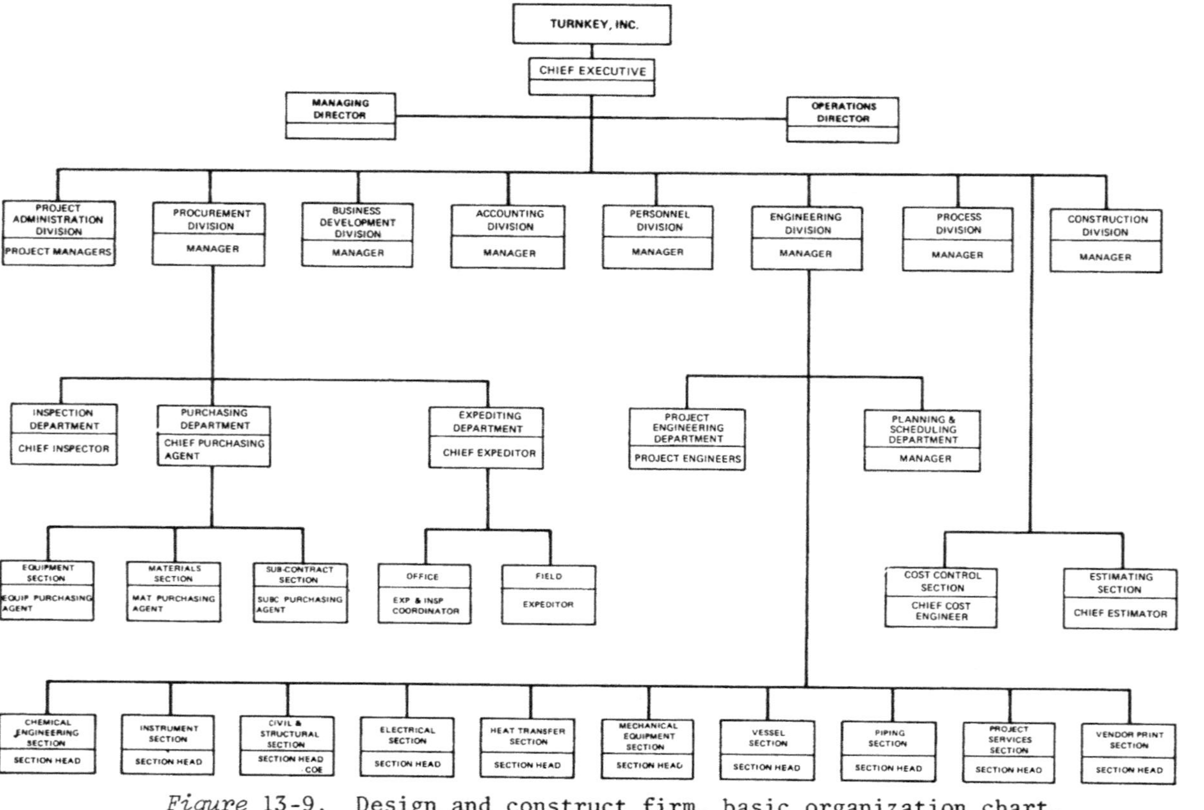

Figure 13-9. Design and construct firm, basic organization chart.

organization of a multi-disciplinary service company. Again, it shows basically a line and staff operation separating the administrative services from the professional services.

Day-to-day management of the activities is divided between technical management and the project management. Technical management is generally by the department or division head who controls the technical activities of his own department. Project management deals with contracts administration, client liaison, coordination, scheduline, programming and other matters related to the implementation of specific tasks.

A modification to the traditional orgamization might be the separation of the operating divisions into broad areas of activity rather than technical disciplines. For example, some firms organize their operating divisions into general areas of activities such as marine terminals, transportation, commercial and industrial buildings, power plants, urban planning and the like, rather than into civil, architectural, mechanical, structural, etc. However, the basic management control structure is unchanged.

Joint ventures. Another form of organization which reflects the realities of today's market place is the typical organization for a multidisciplinary joint venture shown on Figure 13-11. Many public works projects are of such magnitude that very few single service organizations could undertake the total project without internal reallocation of resources, which may be undesirable. A common solution is for two or more service organizations to form a Joint Venture partnership to undertake such assignments. The purpose of such a Joint Venture is to bring together firms with complementary skills, and to divide the risk which is inherent in having to commit a substantial portion of a firm's total resources to one project. Since projects can be stopped, delayed, or cancelled on short notice, such a risk is evident. The organizational structure of the Joint Venture is a typical line type, with certain features which are required to respond to the realities of management and provide the sensitivity necessary to deal with local present-day outside influences.

As shown on Figure 13-11 the Joint Venture may set up a Board of Control consisting of principals of each of the member firms. It assures that each firm is properly represented and it guarantees that the resources of each member firm are properly allocated for the use of the project manager. A board of consultants is usually made up of prominent outside experts who (1) will give a technical overview to all work performed, (2) provide an independent critique of the work done, and (3) offer a safety valve for the senior technical staff members who can find a sensitive and sympathetic ear for their opinions by individuals to whom they are not beholden by employment agreements. The chart also shows a Citizen Advisory Board to perform

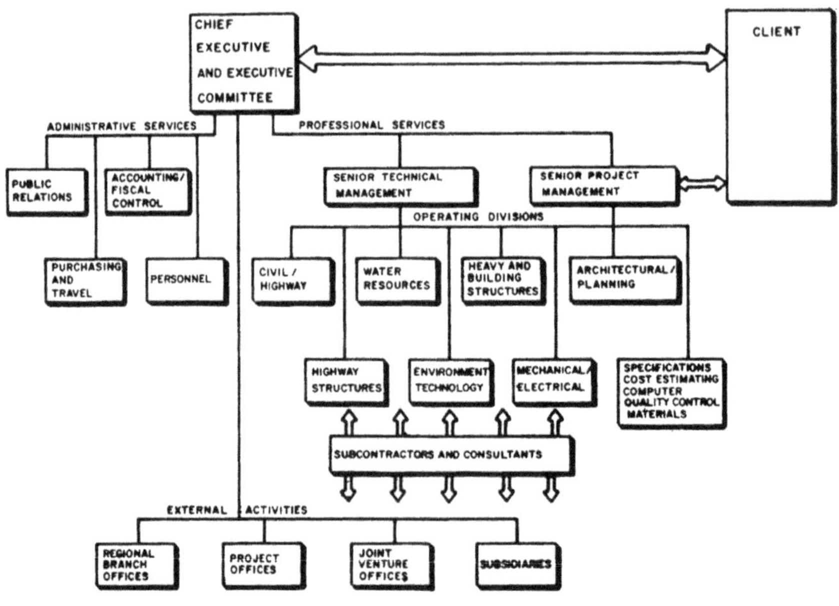

Figure 13-10.   Multidisciplinary service company, traditional organization.

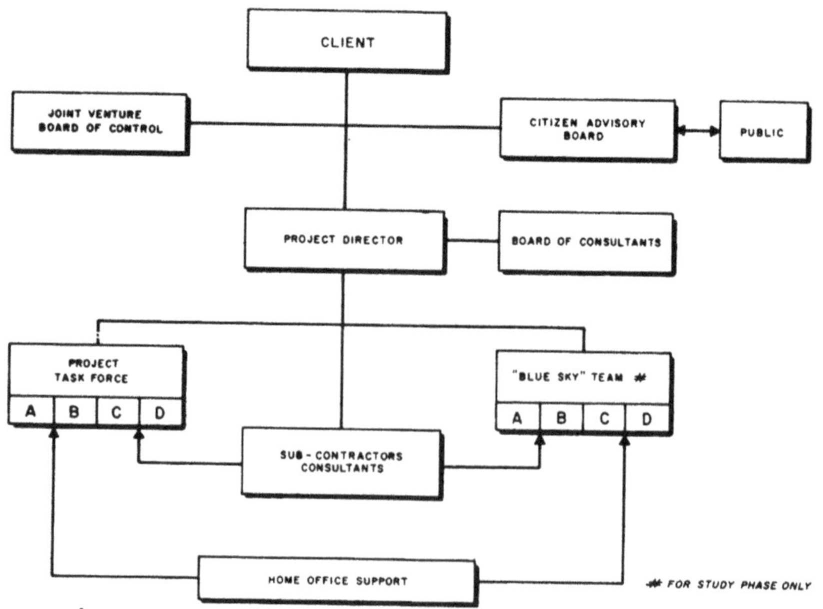

Figure 13-11.   Multidisciplinary joint venture, typical organization.

liaison with the public since participation in major public
works projects by the public is an essential element of today's
undertakings.   A noteworthy feature of the structure displayed
is the concept of two separate teams, or task forces, under the
direction of the manager (project director).   One, the Project
Task Force is in the traditional position of the operating divisions shown on Figure 13-10.   The other is the "Blue Sky"
Team, utilized for the conceptual or study phase only.   The
purpose of this team is to work independent of the traditional
organization and to seek solutions to problems which will
achieve the ultimate objective through non-traditional and
hopefully more imaginative and innovative means.   An example of
an undertaking by a joint venture is the design and construction inspection of a new 75 mile subway system for the Bay Area
Rapid Transit District in the San Francisco-Oakland area by a
Joint Venture of Parsons, Brinckerhoff, Quade & Douglas; Tudor
Engineering Company, and Bechtel Corp.

Matrix (project) form of organization.   A form of organization which is new to the service industry, but is gradually
finding acceptance by multidisciplinary, multi-office service
companies, is the matrix or project form of organization.   This
type of organization has been used with success in the aerospace and defense industries.   The project form of organization
is a distinct type of structure from the others previously discussed.   While the common practice is to refer to a project
manager on a design or a construction job, the job may still be
set up along line and staff concepts.   The project form of organization or matrix form of organization (so named because the
organization may be put into a matrix as shown in Figure 13-12)
suggests that one individual is put in charge of a project including complexities such as the legal, insurance, financial,
and safety responsibilities.

A major potential advantage of the project form of organization is that it centralizes control and other aspects of management at one palce. If the project manager has the knowledge,
the methods, and the data available, it is possible for him to
exercise better control.   There is less coordination time involved and hence, the developmental time involved in a project
is shortened.   One individual is responsible for the external
relationships in relating to the various communities involved
in a project.   An outstanding advantage of the project form of
organization is that it provides a vehicle for accelerating the
development of future managers for larger projects and at higher levels in the organization.

Potential disadvantages of the project form of organization include that too much responsibility is placed on one individual, the internal organization of the company is more complex as shown in the accompanying figures, and there is less
utilization of the total manpower resources of a firm.   The

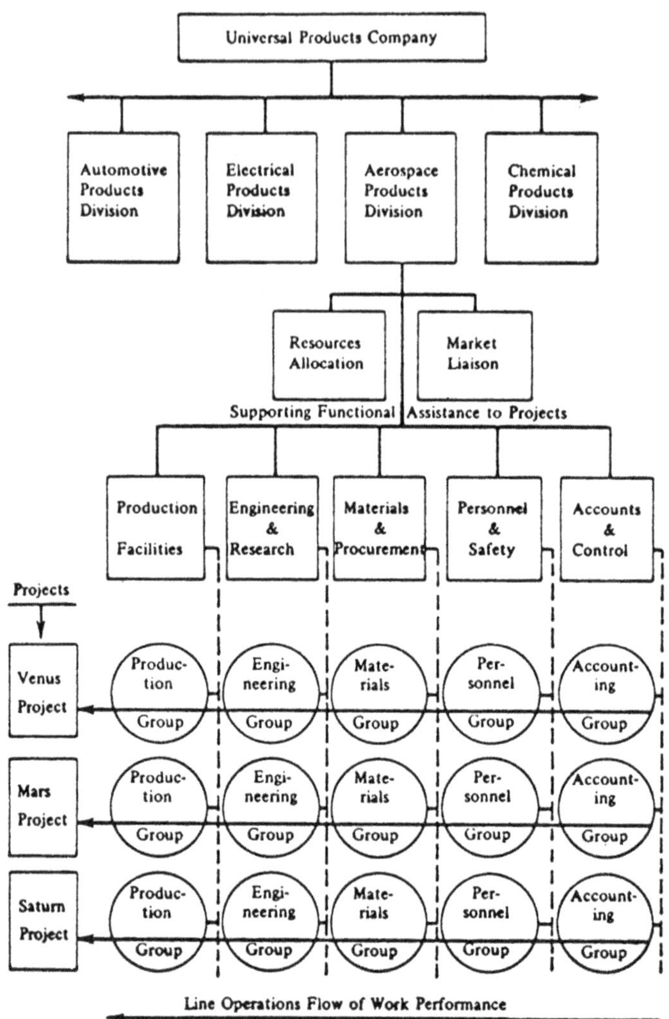

Figure 13-12.   Matrix form of organization (aerospace industry).

unity of command and span of management principles suggest that it is difficult for one manager to manage a large project and the concomitant support group. There is shifting of personnel in order to keep personnel fully utilized. When one project phases out, another project must be found or the project leaders must be reassigned to tasks which may not fully utilize their abilities. These personnel actions create anxieties and frustrations and frequently make-work assignments. The career setbacks and consequent loss of loyalty for human resources within the firm are significant disadvantages which must be overcome for the technique to show continuing success.

An adaptation of the matrix form of organization is shown in Figure 13-13. It shows a corporate form of service company with a number of regional offices where the company is involved in all the basic technical disciplines covering the public works field. Obviously, no service company can maintain all the technical disciplines in all the geographic locations. Under the matrix form of organization, each regional office is headed by a senior officer who manages the affairs of that office. Each technical discipline in turn is headed by an officer-in-charge who has the technical responsibility for his technical discipline irrespective of the geographic location of that activity. With proper control and sensitivity, this form of organization enables sound management control and proper technical control of the activities in any particular area.

Portions of Figure 13-13 are shaded to reflect a hypothetical public works project; a major transportation project in Atlanta under the jurisdiction of the South-East Region. The shaded areas indicate all the elements of the service company that are partly or totally involved in such a project. For example, the senior vice president in charge of the South-East Region has the management responsibility in the vertical column. The assistant vice president in charge of planning has the technical responsibility in the horizontal row. In this particular case, even though there are planning capabilities in five regional offices, as indicated by the black crosses, only two offices will participate in the technical accomplishments of the planning portion of this project; namely the South-East regional office, and the Eastern Seaboard office. The same is true in several of the other disciplines where more than one office will contribute to the success of this project, utilizing skilled manpower, locally available, developed and trained on similar projects within their own region. The advantage of this type of organization is that the project direction rests locally permitting direct access to the client, to the regionally unique information and to the affected "public." The technical supervision, including necessary staffing either by transfer of people or transfer of work elements, is vested in the senior technical management officer of the firm for the discipline involved who is able to bring to bear the total technical resources of the firm, even though the project may be accomplished in many geographically decentralized offices.

Organization for specific projects. Several general forms of organization that are applicable to those engaged in the design of major public works projects have been examined. Any of these organization management structures may be applied, but should be adapted, to specific projects. Figure 13-14 shows a project organization chart for a Metropolitan Transportation Corridor Study. It combines some of the elements shown on Figures 13-10 and 13-11 in a single structure.

Officer in Charge	Office Locations / Technical Divisions	Senior Vice President — Northeast Region (Boston)	Senior Vice President — Eastern Seaboard (New York)	Senior Vice President — Southeast Region (Atlanta)	Vice President — Midwest Region (St. Louis)	Vice President — Rocky Mountain Region (Denver)	Senior Vice President — Western Region (San Francisco)	Senior Vice President — Pacific Basin (Honolulu)	Vice President — International (New York)
Asst. Vice Pres.	Administration	X	X	X	X	X	X	X	X
Asst. Vice Pres.	Civil - Highways	X	X	X	X		X	X	X
Asst. Vice Pres.	Highway Structures	X	X				X	X	X
Vice Pres.	Heavy Structures	X	X	X			X	X	
	Architechture	X	X	X					
Asst. Vice Pres.	Planning	X	X	X		X	X		
Asst. Vice Pres.	Pollution Control	X	X	X		X	X		
Asst. Vice Pres.	Water Resources	X	X			X	X		
Asst. Vice Pres.	Mechanical / Electrical		X	X			X		
	Marine Terminals		X				X		
	Specifications - Estimates	X	X	X			X	X	
Vice Pres.	Airport		X				X		
Asst. Vice Pres.	Advanced Technology	X	X						
Vice Pres.	Transportation	X	X	X	X		X	X	X
	Economics		X	X			X		X

Figure 13-13. Multidisciplinary, multi-office service company, matrix form of organization.

644

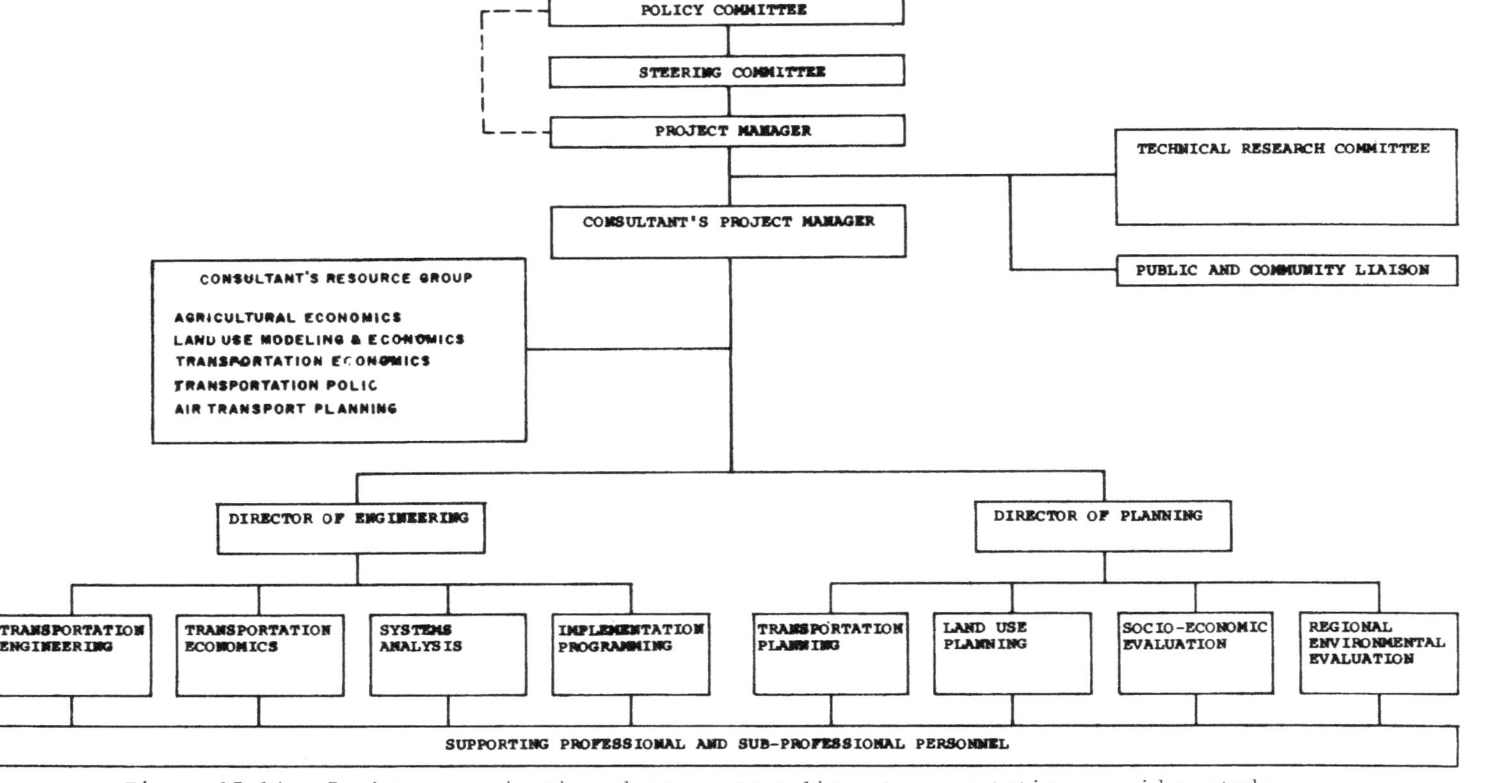

Figure 13-14. Project organization chart, metropolitan transportation corridor study.

The Technical Research Committee is the equivalent to the Board of Consultants shown on Figure 13-11 and the Public and Community Liaison is equivalent to the Citizen Advisory Board shown on the same figure. The director of engineering and the director of planning may be representatives of each of two companies forming a joint venture, where one of the two companies has their main expertise in engineering and the other in the planning field, or they may be individuals who are the department heads in a single office or, in the case of a matrix form of organization (Figure 13-13) they may be the officers in charge of their respective technical environmental and other divisions. The Policy Committee, Steering Committee, and Project Manager shown at the top of Figure 13-14 represent organizational elements of the client's staff.

In summary, what is needed are organization structures that are adaptive to the changing needs of society and individual projects. One author has suggested that:

> ... . organizational analysis must be suitable to dealing with open systems involving a variety of dynamic relationships that are much more complex and subtle than the simple boss-subordinate relationships of more primitive organizations. Similarly, methods of evaluating and controlling organizational behavior must go beyond the simplistic 'one best way of managing' to a view of the individual as a participant in a network of relationships, each strand of which may require different behavior in response to different cues and signals. (Leonard R. Sayles, "Managing Organizations," *Columbia Journal of World Business* (Fall, 1966), p. 86).

The project form of organization, because of its transient and changing nature, probably does not meet all of the needs of any organization. A modified line and staff organization with flexibility and recognition of an individual's contributions to a work project will probably continue to serve the purposes of public project management well. The complexities of environmental design and the ecological considerations necessary suggest the continued use of specialized staff in many areas.

IMPLEMENTATION

Many walls are decorated with elaborate organization and schedule charts but many projects fail because the managers use the charts as ends rather than means to help them achieve their objectives. The purpose of this section is to describe the manager's role in the implementation of the design of a major public works project.

*The project manager must be identified at the very earli-
est stage in the evolution of a program.* Many times a potential
client will insist on meeting and interviewing the designated
project manager before he even invites the consultant to prepare a technical proposal. Once the technical proposal is produced and the client has satisfied himself of the competence of
the consultant to perform his work, he will often require the
designated project manager to make a presentation of the indicated program describing *his* approach in achieving the stated
objectives. This is logical and reasonable from every point of
view since the personality of the project manager and his ability in presenting a program to the client will have a substantial effect on the success of the project.

Consequently, the designated project manager must be completely involved in the earliest phases of a project, including
the preparation of the technical proposal, the definition of
the scope of work, the selection of subcontractors, the possible selection of joint venture partners, the negotiation of the
fee and the accomplishment of the contractual vehicle prior to
commencement of the design of the particular project. He must
also be heavily involved in assembling the staff which will ultimately participate in the work. This is particularly important since the staff has to be compatible with the needs of the
project as well as with the personality of the project manager.

As early as possible and prior to any final contractual
commitments by the client, the project manager must develop a
general outline of the work to be done to achieve the intended
objective. An example of such an outline is the First Stage
Scheduling Precedence Diagram shown on Figure 13-15. This diagram shows the sequence of the major work elements that are to
be produced in the course of the execution of the contract. It
is called a precedence diagram since it is developed to indicate which element is to precede the following element and
therefore what must be achieved before the following element
can be started.

The precedence diagram is somewhat similar to a simplified
PERT/CPM technique but it limits itself to the most basic elements. The traditional elements that have been inherent in
public works projects in the past, including the technical development, project management, client participation and data
control are shaded. The unshaded portions of the diagram show
work elements which have been introduced into the design process of public works projects in the recent past.

Figure 13-15 shows the difference in management of a project after imposing environmental needs. These elements are
basically related to environmental aspects and public affairs
which must be highlighted since they in fact have become controls on the ultimate execution of the project. The National

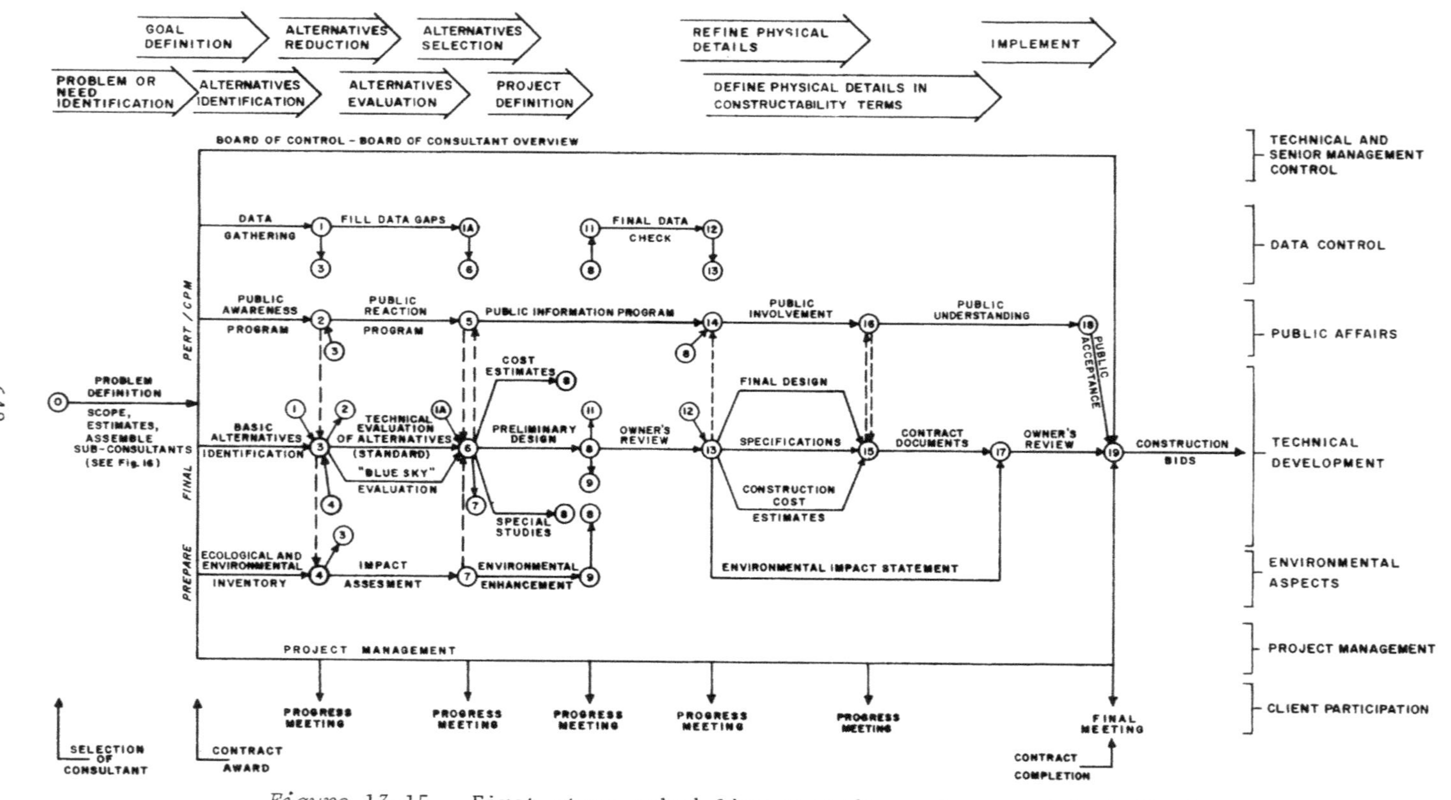

Figure 13-15. First stage scheduling precedence diagram (simplified).

Environmental Policy Act of 1969 makes it very clear that no
major public works project can go forward until all the environmental impacts have been evaluated, and until it can be
satisfactorily demonstrated that no major adverse environmental
impact has been overlooked by the consultant.

The work elements related to public affairs are a response
to the need for a better understanding of the fact that public
works projects must properly serve the public for whose benefit
they are intended. It has become evident that the public has a
clear understanding of its own needs, and consequently it must
be involved in all stages of the program. The various types of
involvements are indicated in Figure 13-15. Consistent with the
precedence approach of this diagram, it is evident that construction bids cannot be taken unless that activity is preceded
by public acceptance of the project. It also shows that inviting construction bids after the owner's review cannot be accomplished unless the owner's review is preceded by the preparation of an acceptable Environmental Impact Statement.

In working through the precedences to the beginning it becomes evident that the programs for environmental aspects and
for public affairs must commence on the first day of the project. This is a substantial departure from traditional engineering practice where the attitude has been, "let's get something down on paper, let's agree on it first internally and
then let's bounce it off somebody." This attitude has become
counter productive and unless this is properly recognized and
dealt with during the programming of a project, the projects
success may be jeopardized.

Another important feature of developing this type of diagram is to insure that all major elements are recognized before
the financial limits of the work are finalized, so that in the
preparation of the fee estimates for the work, allowances can
be made not only for the traditional technical portions of the
work but for the non-traditional elements that have now become
essential parts of the work. Most public contracting agencies,
particularly on the federal level, limit the consultant's fee
to a percentage of the estimated cost of construction or to
published fee curves developed over the past several decades.
These fee calculations do not include the new non-traditional
work elements which contribute heavily to the total cost to the
consultant. Identification of the need for these elements will
allow the consultant to negotiate proper fees and therefore
give the project manager appropriate financial tools to accomplish the project to the complete satisfaction of everyone concerned.

Another feature of the precedence diagram is the problem
definition portion which is further amplified and elaborated in

Figure 13-15. It might be described as the pre-contract precedence diagram. It is not generally recognized that a substantial effort takes place before a contract is consummated between the public works agency and the consultant who will be
required to perform the design of the public works project.

The completeness of the contract, the proper scheduling of
the work, and the inclusion of all appropriate work previously
performed in connection with such a project is essential to the
success of the project, and this investment must be borne by
the client. There are numerous examples where the client has
prepared a reasonably comprehensive scope of work indicating
his definition of the problem and his proposed solution. Frequently, upon thorough review by a competent consultant, the
scope can be substantially modified to reflect the true needs
of the client. This should be done before any contract is
signed and the types of activities that precede the start of
the contract are generally outlined in simplified form on Figure 13-16. An example of the value of this pre-contract analysis is a major flood control project in Yonkers, New York where
the consultant demonstrated to the client (the Corps of Engineers) that the inclusion of a diversion tunnel could be more
economical and environmentally more acceptable than the more
traditional channel lining approach.

After programming the work, and execution of a contract,
proper scheduling of manpower, time, and costs must be accomplished. There are a number of tools used for work schedules.
An example of the Initial Work Flow Diagram for a hypothetical
Metropolitan Transportation Corridor Study is shown on Figure
13-17. This figure is simply an elaboration of the First Stage
Scheduling Precedence Diagram. It also works on the precedence
principles and identifies the major tasks in an orderly fashion
and further separates the tasks into those to be performed by
the consultant, those to be performed by the client and those
to be performed jointly with others. This diagram becomes a
basic control diagram from which the manager can derive a number of other scheduling tools.

The most commonly used tool is the PERT/CPM technique. The
PERT/CPM diagram is usually initially produced by hand and then
converted through the use of computer punch cards into a computer assisted tool.   It can then be regularly updated by computer controlled graphic outputs as well as computer tabulated
printouts. This will enable the manager and the client to perform timely periodic reviews of the status of all the work elements of the project.   Further, it will allow early warning of
delays in the execution of the work, thus enabling the manager
to restructure the work by changing emphasis, or by introducing
additional technical help, after agreement with the client in
accepting changes to the schedule at an early opportunity.

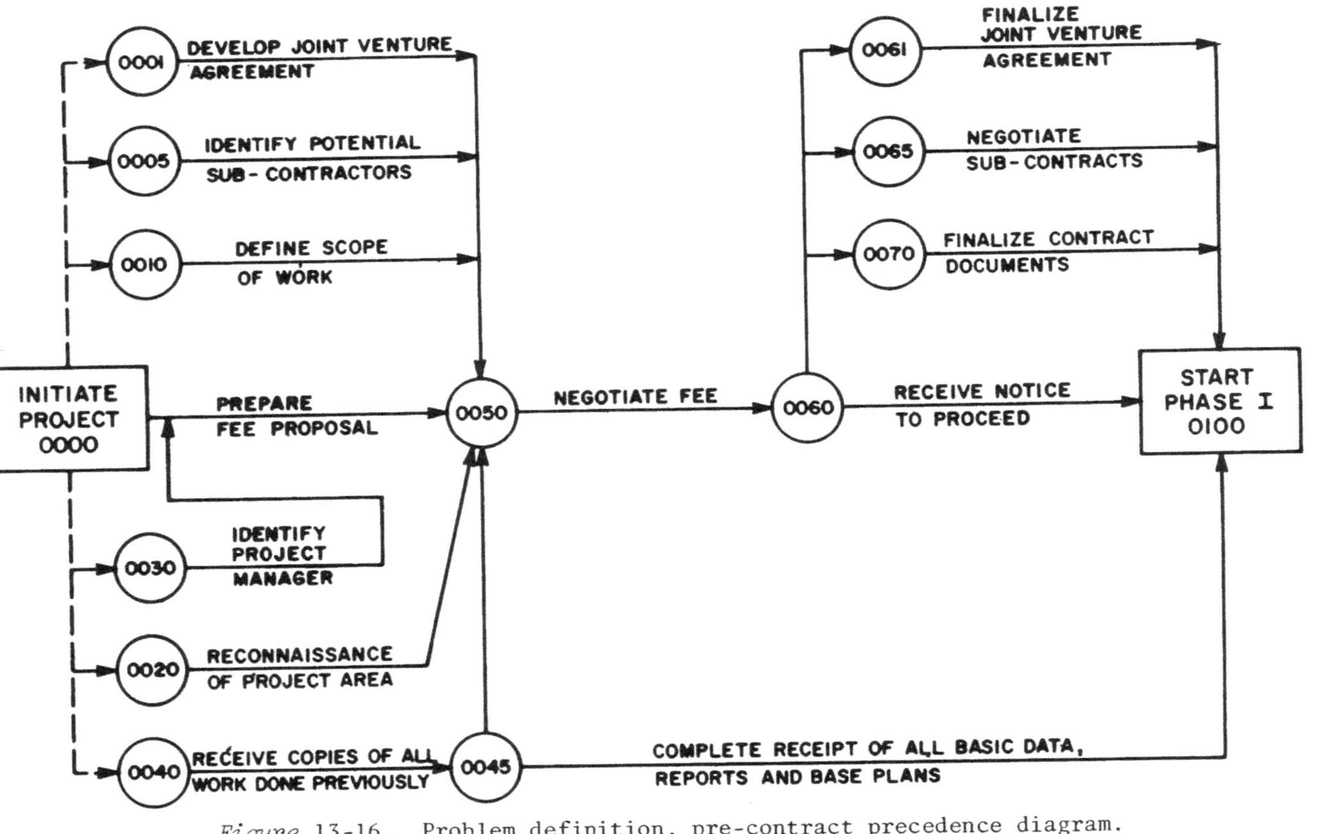

Figure 13-16. Problem definition, pre-contract precedence diagram.

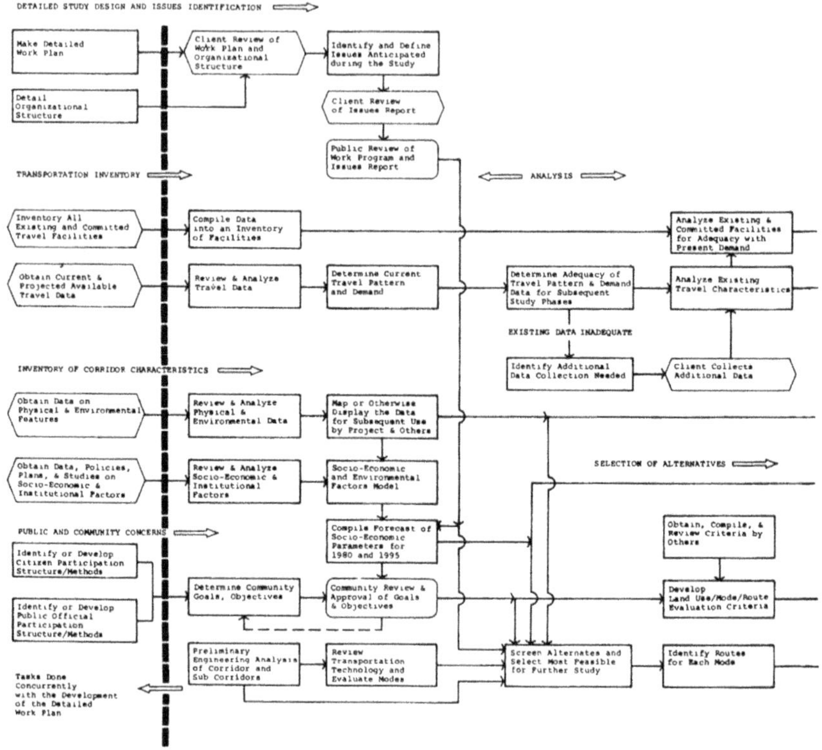

Figure 13-17. Work flow diagram, metropolitan transportation corridor study.

The use of a bar graph in scheduling work is a very common device and is used as a simple representation of the flow of work over a period of time. A Bar Graph showing the work schedule for a Metropolitan Transportation Corridor Study is indicated on Figure 13-18. This bar graph is derived from the Work Flow Diagram on Figure 13-17 and thereby helps to control and check the efficacy of the work flow diagram.

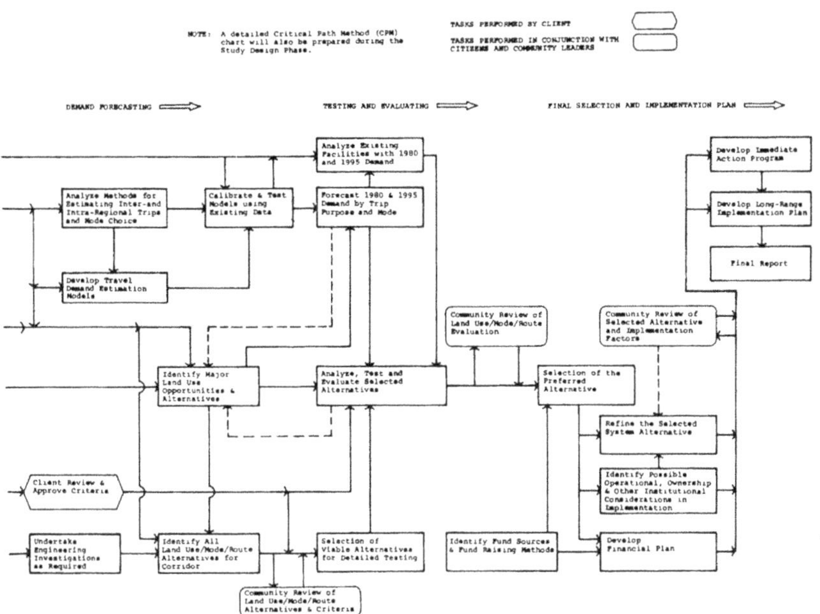

Figure 13-17. Work flow diagram, metropolitan transportation corridor study.

Use of a check list in controlling and managing a project is also widespread.   A well prepared check list is not only a good management tool but it also gives the client a clear indication of all the intermediate products that he can expect to receive at scheduled intervals which he will utilize in reviews or presentations to the public or to his own management.   A check list of items to be produced for a Metropolitan Transportation Corridor Study is shown on Figure 13-19.

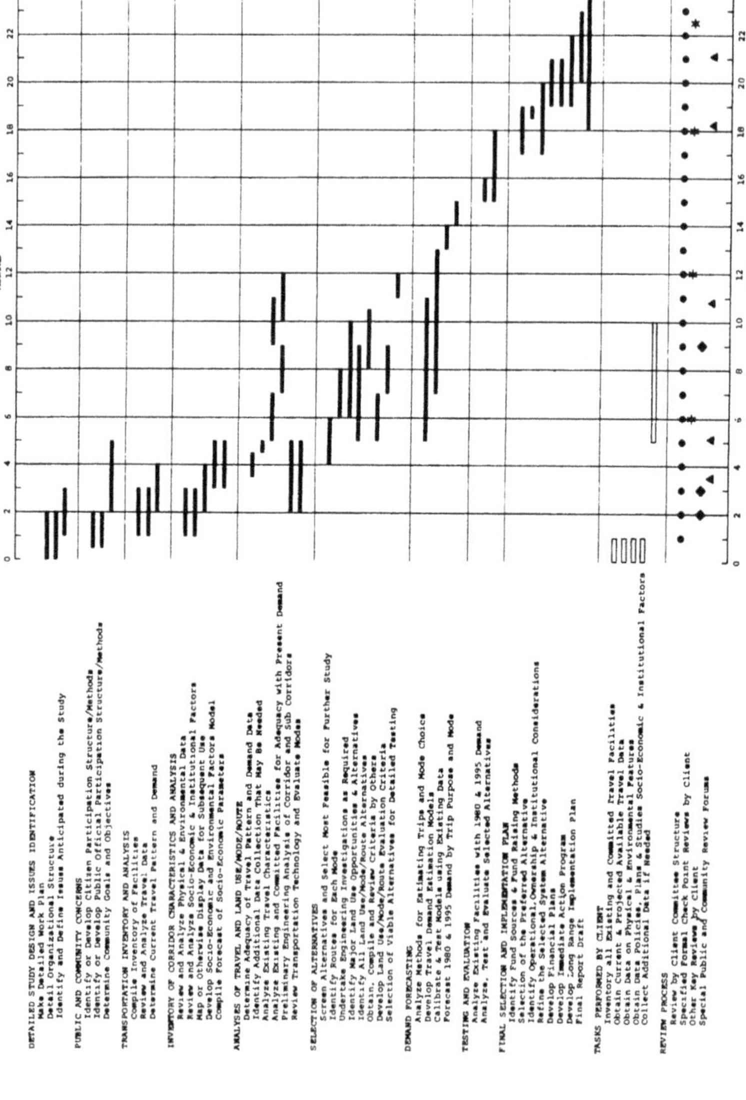

Figure 13-18. Work schedule, metropolitan transportation corridor study.

| ITEM | FORMAT | CLIENT CHECKPOINT PRESENTATION | REVIEW | | COMPLETED (Month Ending) |
			Client's Committees	Public Forum	
Progress Reports	Letter Reports		●		Monthly
Study Design and Organizational Structure	Technical Memorandum, Charts & Diagrams		●		2nd Month
Public and Community Participation Program	Technical Memorandum		●		2nd
Issues Report	Report and Visual Aids		●	■	3rd
Goals & Objectives Report	Report and Visual Aids		●	■	5th
Inventory & Analysis of Existing Systems and Planned Improvements	Technical Memorandum and Tables	✳	●		6th
Preliminary Screening of Alternatives	Memorandum Report, Maps & Charts		●		
Evaluation Criteria	Technical Memorandum & Charts		●		9th
Selection of Viable Alternatives	Memorandum Report, Maps, Tables and Visual Aids	✳	●	■	12th
Travel Demand Estimation Modes	Technical Memorandum, Printout and Tape		●		13th
Travel Forecasts for 1980 & 1995	Maps, Charts, and Printout		●		15th
Analysis Testing and Evaluation of Alternatives	Memorandum Report, Printout, Charts Tables, and Visual Aids	✳	●	■	18th
Selection of Recommended System	Memorandum Report, Maps, Charts and Visual Aids		●		19th
Institutional Considerations	Technical Memorandum		●		20th
Financial Plan	Draft of Final Report (a reproducible copy will be furnished 3 weeks after review)	✳	●	■	21st
Action Program					23rd
Long-Range Plan					
Corridor Studies Procedural Guide	Report, Charts and Tables		●		24th

Figure 13-19. Items to be produced, metropolitan transportation corridor study.

655

An outgrowth of the various scheduling tools is the manpower estimate. Such an estimate must be made for all the work elements indicated as appropriate for the particular assignment. Such an estimate may be shown on the PERT/CPM diagram or in any other convenient form. In addition to producing the estimate for the total effort required for any particular assignment, the time to be devoted must be distributed in a meaningful fashion. Such a distribution can be made by work elements or by disciplines, but it must be related to real time. An example of manpower distribution (by individual) for a Metropolitan Transportation Corridor Study is shown on Figure 13-20 which distributes the manpower by man-months over the 24 months' scheduled total duration of the study. The total man-months per discipline and the total man-months of effort per month give an indication of the total effort to be dedicated at any point of time. It enables the Project Manager to control manpower and also to develop a financial plan for billing of fees and prediction of cash flow.

The manpower estimate, regardless of what tool is used to develop and to describe the control, represents only one part of the total cost of the project. Cost controls must be set up for labor as well as for subcontractors expense, consultants fees, and other direct expenses such as travel, per diem, reproduction costs, computer costs and the like. Numerous manual and computer assisted tools to accomplish cost control are available.

The specific device to be used for a particular public works project should be tailored to the assignment. In many instances, the cost control system that is selected is one that is compatible with the client's cost control system or one that the client may demand. All the work described as elements of the project must then be converted into budgeting terms as a control device to maintain the schedule. Budgets must be set up for the calendar time to be allocated to the work, as well as for all the costs to be incurred. Budgets become the true control device for the manager and no costs should be incurred without his knowledge and prior approval. This concept is consistent with the Project Manager's dual responsibility--that is, the responsibility for a technically sound project to the client and for a financially sound project to his own senior management. One of many budget control devices is a computer print-out Project Status Indicator (PSI) which is shown on Figure 13-21.

The Project Status Indicator (PSI) program is designed to provide the manager with information necessary to aid him in controlling costs and eliminating contract cost over-runs. This system does not provide all the information a manager may require nor is it "management" in itself. It does however require minimum of input burden while maximizing output. The function

POSITION IN TEAM	MONTHS																								TOTAL MAN-MONTHS
	1	2	3	4	5	6	7	8	9	10	11	12	13	14	15	16	17	18	19	20	21	22	23	24	
Project Manager	.8	.8	.8	.6	.8	.8	.8	.6	.8	.8	.8	.8	.6	.8	.8	.6	.8	.8	.6	.8	.8	.8	.8	.6	18
Director of Engineering	.2	.3	.2	.2	.2	.2	.3	.2	.2	.2	.2	.2	.2	.2	.2	.2	.2	.2	.2	.2	.2	.2	.2	.2	5
Principal Transportation Engineer			.4	.4	.4		.4	.4	.4	.4		.2					.4	.4		.4	.4		.4		5
Principal Transportation Economist		.1				.3				.1		.3					.3	.3		.3	.3				2
Principal Systems Analyst					.3	.3	.5	.5	.5	.5	.5	.5	.6	.5	.3	.4	.2	.3				.1			6
Principal Implementations Engineer																				.5	.5	.5	.5		2
Director of Planning	.2	.2	.3	.2	.2	.2	.2	.3	.2	.2	.2	.2	.2	.2	.2	.2	.2	.2	.2	.2	.2	.2	.2	.2	5
Principal Land Planner			.4	.4	.4	.5	.5	.5	.5	.2	.2	.2						.4				.4	.4		5
Principal Transportation Planner		.2	.2	.2	.2	.2	.2	.2	.2	.2	.2	.2	.2	.2	.3	.3	.3	.3				.2			4
Principal Regional Economist		.2	.3	.3	.4	.2		.3	.4			.2					.2	.2				.1	.1	.1	3
Principal Environmentalist		.2	.4	.5	.5	.2		.2	.2			.1			.1		.2	.2				.1	.1		3
Resource Specialists			.2	.2		.2			.2			.2				.2	.2	.2			.2	.2			2
Other Professionals		.3	.4	.4	.4	.4	.4	.4	.4	.4	.3	.3	.3	.3	.3	.3	.4	.4	.4	.4	.3	.3	.3	.2	8
Subtotal Professional Man-Months	1.2	2.3	3.6	3.4	3.8	3.5	3.3	3.6	4.0	3.0	2.4	3.4	2.1	2.2	2.2	2.2	3.4	3.9	1.4	2.8	3.0	3.1	3.0	1.2	68
Subprofessionals and Technicians	0.5	2.0	2.5	2.5	2.5	2.0	2.0	2.0	2.0	2.0	2.0	2.0	2.0	2.0	2.0	2.0	2.0	2.0	2.0	2.0	2.0	3.0	3.0	2.0	50
Total Man-Months	1.7	4.3	6.1	5.9	6.3	5.5	5.3	5.6	6.0	5.0	4.4	5.4	4.1	4.2	4.2	4.2	5.4	5.9	3.4	4.8	5.0	6.1	6.0	3.2	118

Figure 13-20. Man-power distribution, metropolitan transportation corridor study.

```
JOB NUMBER          WATER RESOURCES PROJECT STATUS INDICATOR
2579 F110                                                          PG.    1
TASK WGT.  21.2                         03-03-72

FC SUB-  SUB-TASK DESCRIPTION B U D G E T  E X P E N D    PERCENT      ACT   WGTED
LP TASK                                                  TO DATE      PCT   PCT
A                        MDAY DOLLAR MDAY DOLLAR MDAY DOLLAR COMPL COMPL
G
    1000 REVW HYDGY & HYDICS     10    630   10    559   100     88 100.0    2.9
  C 1005 COLCT ADDTL NEDD DTA     0      0    1     48     0      0   0.0    0.0
    1009 RVW HYDGY & HYCS TNL    10    630    5    303    50     48 100.0    2.9
  * 1010 DEV BWTR PROF FOR 3Q     5    320   14    623   280    194 100.0    1.5
  * 1016 COE PRVD PRPTY US MP     1     60    2     85   200    141 100.0    0.3
  C 1020 FLD CK PRP ANL SRM O     4    250    4    200   100     80 100.0    1.2
    1030 DEV PN 890 1450 2000     3    190    2     66    67     34 100.0    0.9
    1040 DTM PRM OBST FLD FLW     4    250    1     54    25     21 100.0    1.2
    1050 PRP FD PRP LST W/WSE     9    570    7    272    78     47 100.0    2.6
    1060 VR STG DS RLNP CR CD     2    130    2    108   100     83 100.0    0.6
  C 1070 DV HYD 890 1450 2000    20   1260   14    765    70     60 100.0    5.9
    1080 VRY DLHG FREQ RLTNPS     4    250    0      0     0      0 100.0    1.2
                               ---------------------------------------------------
******TOTALS********************  72   4540   62   3083    86     68 100.0   21.2***

JOB NUMBER
2579 F110                       DIRECT  EXPENSES                    02-04-72

                    O U T - O F - P O C K E T   D O L L A R S

        COMPUTER    SUB-CONT.   REPROD.    TRAVEL    OTHER

          0.00        0.00       0.00       12.25      0.00
```

Figure 13-21. Water resources project status indicator.

of PSI is to provide an easy to use display of project status
versus project cost information.

The computer print-out shown in Figure 13-21 breaks the
project into various task numbers and task descriptions derived
from the PERT/CPM diagram. It then lists the budget amounts
both in man-days and dollars, the expended amounts in man-days
and dollars, a computer calculated "Percent to Date" in man-
days and dollars, the actual percent completion calculated by
the manager and the weighted percent completion which relates
this particular task to the total project completion. The inputs in man-days and dollars are taken directly from time cards
and expense reports as well as vendors bills. By simple comparison, the manager can discover major differences between
budgeted and expended amounts. An asterisk on the left hand
column automatically flags lines where the expenditures exceed
the budget by more than ten percent.

Other major elements of implementation of the management
procedures for a public works project are related to the legal
and financial aspects. The legal aspects fall into two categories: first, the legality of the contract to be executed with
the client and second, the compliance with the legal requirements and constraints on the project by numerous regulatory
agencies having control over certain of the aspects of the
project. It is essential that the project manager have competent in-house legal counsel to assist him in these various aspects.

The financial element is also divided into two areas:
first, is the financial problem related to the project itself--
that is, the cash flow which is based on prompt and accurate
billings to the client and equally prompt collections, and
second, the financial problem related to assisting the client
in preparing funding applications through grants and loans that
may be available to him under certain existing local or federal
legislation. Here again, competent legal advice as well as
competent financial counsel are essential to enable the Project
Manager to serve his management and the client adequately.

Finally, in order to properly implement and execute the
contract, sound management requires proper communication. The
establishment of proper communication is a procedural matter.
There are numerous ways of communicating with the client, the
project staff, consultants and sub-contractors. As noted earlier, there is no substitute for inter-personal communications.
Another simple, but indispensable device, is the use of Policy
Memoranda which cover all the procedural matters in writing and
which are circulated to all interested parties in order to
achieve proper understanding of the various policies that relate to contracts management. These include man-power allocation and policies, fringe benefits, distribution of correspondence, notification of meetings and all other matters related

to the procedural elements of managing a multidisciplinary major public works project.

Summary

The manager of public projects will be most successful if he has (1) a philosophy of management--especially human resources management, and (2) the competence to implement that philosophy. There are several approaches to management ranging from the coercive, dictatorial, it-has-to-be-my-way management to the consensus, collegial, let's-do-it-together type of management. In most circumstances, the manager of human resources working on public works projects at the professional level will be more effective if he chooses a path between the two, but leaning toward the consensus approach.

This collaborative-consensus philosophy of management is geared to dynamic rather than stable situations. Organizations are not static but are continually changing to meet external and internal demands. The manager's role is one of developing objectives and workable organizations, meeting change, and helping participants achieve their own personal goals while accomplishing organizational objectives.

REFERENCES

1. Drucker, Peter F. *The Practice of Management*. New York: Harper & Brothers, 1954.

2. Fayol, Henri, *General and Industrial Management*. London: Sir Isaac Pitman & Sons, Ltd., 1949.

3. Herzbert, Frederick W. *Work and the Nature of Man*. Cleveland, Ohio: World Publishing Company, 1966.

4. Likert, Rensis. *The Human Organization*. New York: McGraw-Hill, 1967.

5. McGregor, Douglas (Warren G. Bennis and Caroline McGregor, eds). *The Professional Manager*, New York: McGraw-Hill, 1967.

AUTHOR NOTES

This chapter was authored jointly by Dr. Douglas A. Benton, Associate Professor of Management, College of Business, Colorado State University, and Mr. Henry L. Michel, Partner in the firm of Parsons, Brinckerhoff, Quade and Douglas.

Dr. Benton has taught courses in the general and human resources management areas, including fundamentals of management, human relations, personnel administration, training and development, and labor law. He has served as a management consultant for several organizations and has a wide variety of management oriented research activities.

Mr. Michel has had a long history as a practitioner in accomplishing major engineering projects. Prior to joining Parsons-Brinckerhoff in 1965, Mr. Michel headed up his own firm, ENCONI, SA, headquartered in Rome. He directed projects spanning the geographical area from Northern Europe to Lagos to Tehran.

As a principal in one of the World's major engineering-architecture-planning firms he has been Partner-in-Charge for a wide range of projects. He has had specific charge of the firm's burgeoning water resources program, bringing to bear the modern professional advances in systems engineering and environmental planning technology.

The chapter has also profited through inputs from Mr. Gardner Reynolds, Partner of Dames and Moore, Consulting Engineers in the Applied Earth Sciences; Mr. Joe Kellogg, President of Kellogg Corporation, Multidisciplinary Management Consultants; and Dr. Val F. Ridgway, Chairman, Department of Management, College of Business, Colorado State University.

PART IV

CASES

The last part of the document attempts to complement, through a range of representative case studies, the thrust of the argument concerning comprehensive planning by illustrating the types of problems encountered, typical disciplinary responses, and the complexities of real life situations.

CASES: FORT COLLINS BYPASS

BY WAYNE J. CAPRON

No community or large urban area is unaffected by highway construction. Adding new links to the highway system or changing old ones will profoundly affect a community and is likely to be controversial. The Central Fort Collins Expressway project typifies many of the interactions between highways and community values. The project is a microcosim of issues which may be relevant to a larger area. These include green space, displacement of a minority group, noise and esthetics, and the need to facilitate increasing traffic loads. In addition, the case illustrates aspects of many previous chapters.

THE LOCALE

Fort Collins, Colorado, is a city of approximately 45,000 population lying approximately 50 miles north of Denver. It is within a rich agricultural area just east of the foothills of the Continental Divide. Figure 14-1 is a proximity map. The growth rate has been very rapid. Since World War II, the university in Fort Collins has grown from approximately 3,000 students to 18,000 students. Within the past five years, a new Eastman Kodak Plant located some twenty miles southeast of Fort Collins has further stimulated growth. There is no indication that rapid growth will not continue.

THE HIGHWAY NETWORK

Fort Collins is served by several highway transportation corridors. The main corridor lies along Interstate 25 which runs north from Denver to Cheyenne and passes approximately 5 miles east of Fort Collins. The chief purpose of Interstate Highways is to connect large urban population centers. State Highway 287, which is a primary highway, runs north and south

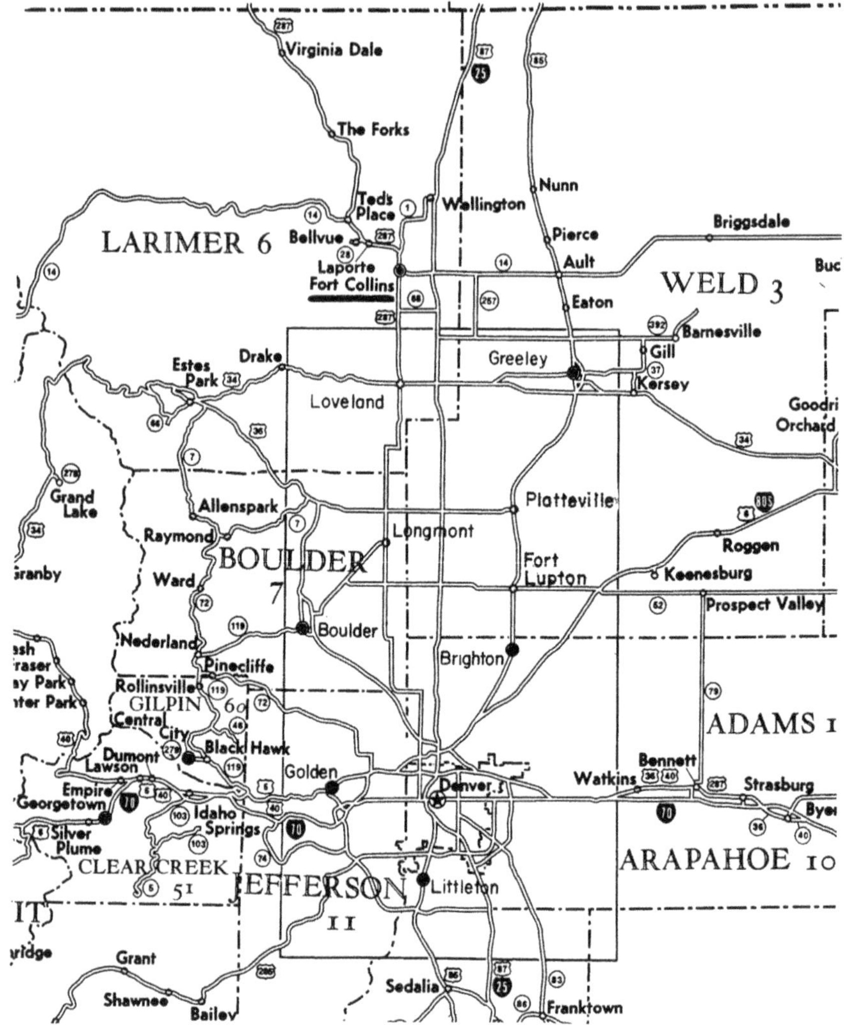

Figure 14-1.   Map showing Fort Collins in relation to Colorado Highway system.

through Fort Collins connecting Fort Collins to the northwest with Laramie, Wyoming, and to the south with Loveland, Longmont, and eventually into Denver. The prime purpose of primary highways is to connect urban areas. Fort Collins is also served by State Highway 14, a secondary highway, which connects Fort Collins with the Interstate to the east by four lanes and then two lanes after leaving Fort Collins running west coincident with U. S. 287 to a point approximately 10 miles northwest of Fort Collins, where it leaves U. S. 287 and goes westerly up the

Poudre Canyon over Cameron Pass into Walden and Colorado's North Park area. This portion serves both as a recreation highway and a local farm-to-market road. U. S. 287 which runs through the heart of the city is also a two-lane highway.

RELEVANT INSTITUTIONAL STRUCTURE

The State Department of Highways is a part of the executive branch of the state government headed by the governor of the State of Colorado. The Department has its own policy making body composed of the Executive Director and the State Highway Commission who act as a Board of Directors with full policy making power. One of the major functions of the executive director and the Highway Commission is the preparation each year of a budget to guide the work of the Division of Highways through the ensuing fiscal period. Funds come from two main sources consisting of: (1) A four-cent federal tax on fuels which goes into the Federal Highway Trust Fund and apportioned to the states by Congress, (2) A seven-cent tax on fuel collected by the State of Colorado and placed in the State Highway Trust Fund. In order for the Highway Commission to determine where these budgeted funds should be appropriated, hearings are held each November with representatives from each county, major cities, highway associations, and other interested persons. Each county is generally represented at these hearings by their county commissioners who have, prior to the hearings, had meetings in their own counties with the representatives of the towns in their county and other interested groups and persons. These requests to the Highway Commission cover all classes of highways in the state including interstate, primary, secondary, urban, and state highways. After the November hearings, the Highway Department prepares a preliminary budget listing all requests made at the hearings. These requests have historically run five to ten times more cost than money available. Therefore, it becomes necessary for the Highway Commission to set priorities for the coming fiscal period.

GENESIS OF THE FORT COLLINS EXPRESSWAY PROJECT

In 1965 at the Highway Commission November hearings, the Larimer County Commissioners requested, with concurrence of the Fort Collins City Administration, that money be appropriated for either the improvement of U.S. 287 between Fort Collins and Laporte or the establishment of a new route to handle the continuously increasing traffic congestion. In August of 1966, the Highway Commission budgeted money for route location studies. As a result of this appropriation, the Highway Department engineering staff, after studying the area, established a proposed new route for U.S. 287 which started approximately one mile west of Laporte where it left existing U.S. 287 traveling

in a southerly direction, crossed the Cache La Poudre River
southwest of Laporte, and paralleled the Colorado and Southern
Railroad towards Fort Collins. This is seen as route C in Figure 14-2.

On September 30, 1969, a Corridor Public Hearing was held
in the Larimer County Courthouse in accordance with the Policy
and Procedure Memorandums established by the Federal Highway
Administration. As a result of that hearing, a recommendation
was made by the Colorado Department of Highways to the Federal
Highway Administration that this route be approved for the design and construction of a four-lane facility through the area.
On November 12, 1969, the Federal Highway Administration approved this route. There was no apparent opposition to the
project; in fact, it was widely endorsed by both local organizations and governmental bodies. There was no further action
taken until July of 1970 when additional funds were appropriated for preliminary engineering. From July of 1970 until March
of 1971, survey data and preliminary design data were collected.

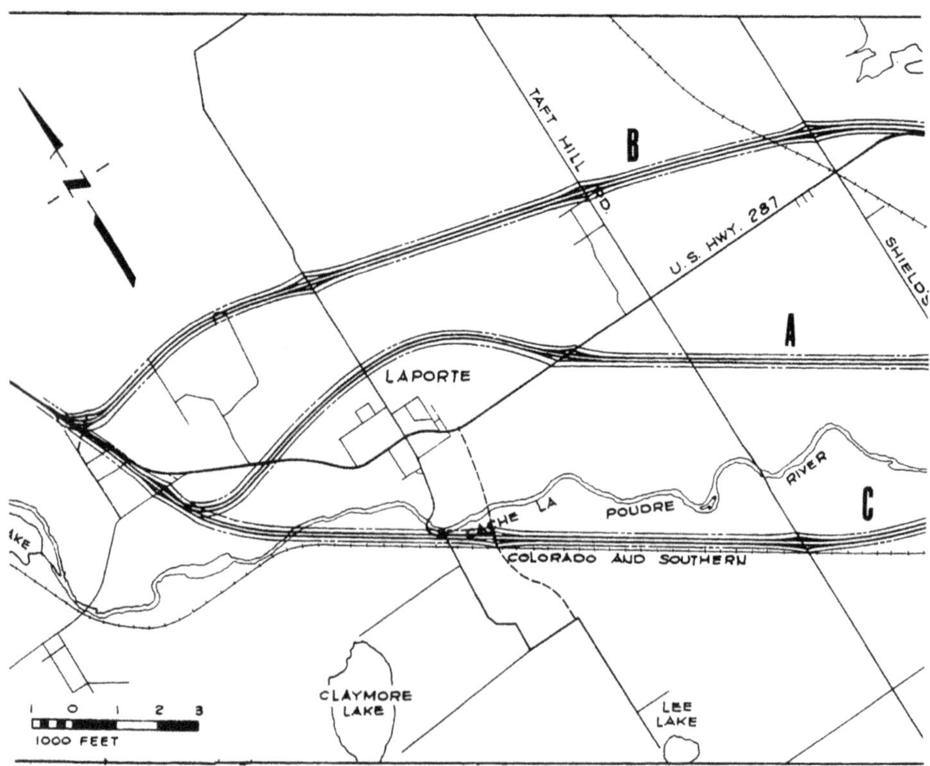

Figure 14-2. Alternate routes, Fort Collins Expressway
(from Environmental Impact Study, Meheen Environmental Consultants).

During this period of planning, the National Environmental
Policy Act, passed by Congress in December 1969, began to shape
decision making in public work projects through regulations,
guidelines and court decisions. This gave some focus to the
increasing sensitivity and awareness toward the environment,
and the effects of large public projects such as the proposed
expressway. Environmental groups began forming. Some of these
groups in the Fort Collins area who became interested in the
project were the Greenbelt Association, Trout Unlimited,
Neighbor to Neighbor, Sierra Club, Colorado Open Space Council,
and others. As a result of the interest of these environmental
groups, the federal government, the state government, the Department, and the Federal Highway Administration held a meeting
concerning the projects future. The main question discussed at
this meeting was whether the project was far enough along at
the time of the enactment of the National Environmental Policy
Act that it would not be affected by the Act, or whether an Environmental Impact Study would be necessary. It was determined
that design approval had not been obtained prior to February 1,
1971, as required by the NEPA Act and, therefore, an Environmental Statement would be required. The CEQ April 23, 1971

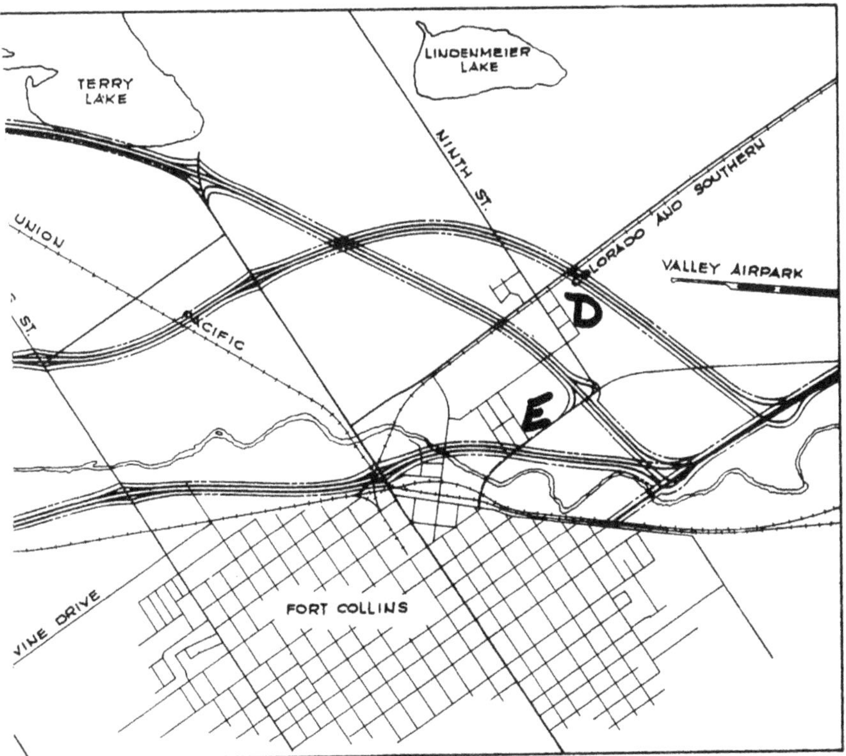

Figure 14-2 (continued). A, B, C = Alternate Routes;
D = Andersonville; E = Buckingham.

Guidelines address this question by stating that any major federal action yet *remaining* required an Environmental Impact Statement. This has been reinforced by numerous court decisions. Requests prior to this time from the various environmental groups has also urged a full environmental study. Therefore it seemed both imperative and advisable that an environmental study be prepared. Meheen Environmental Consultants were retained by the Department of Highways to conduct the study.

Table 14-1 outlines the sequence of major events for the project. It is evident that until February 1971, the events unfolded in a routine manner. However, in February the mandate of NEPA began to take hold and new forces and factors began to shape decision making. This illustrates the contrast between the old and the new in forces and factors that shape highway decision making.

Table 14-1. Chronology of events; Fort Collins Central Expressway.

August 8, 1966	First Supplement to 1966-1967 Budget. P.E.-Route location studies, under CS06-0014-25.
1967	Route location studies were made from aerial maps and by field reconnaissance.
January 11, 1968	Mr. McEldowney, District Engineer, met with Ft. Collins City Planner and Larimer County Planner. Route was basically accepted as well as interchange locations.
January 17, 1968	Mr. Merten and Mr. McEldowney covered the location in the field.
February 21, 1968	Mr. McEldowney and Mr. Shouse, District R. O. W. Engineer discussed location with Larimer County Commissioners, County Engineer, County Planner, and Ft. Collins Planner. Alignment was approved in general and interchange details were discussed.
March 8, 1968	Mr. McEldowney met with Mr. Bower and Mr. Merten in Denver office. Route was agreed on. Interchange location at College Avenue and railroad involvement with this interchange was discussed.
October 16, 1969	The Public Hearing transcript and Public Hearing report was submitted to B.P.R.; location approval was requested.

Table 14-1 (continued). Chronology of events; Fort Collins Central Expressway.

November 12, 1969	B. P. R. approval of the corridor was made. This is the location of what is now known as Route "C" in the environmental studies.
November 1969 December 1970	Field surveys were made and data plotted along the proposed route.
May 18, 1970	1970-1971 Budget included $100,000 P. E. under project F 287-3(3).
January 28, 1971	Survey data for the approved route was submitted to the Denver office, Design Section.
February 1971 March 1971	Numerous groups and agencies were contacted by Resident Engineer Atkins relative to the desires of these groups in connection with the social, economic, and environmental aspects of the project. These included, Division of Game, Fish and Parks, Ft. Collins Landmark Commission, Larimer County Planning Office, Greenbelt Association, and City of Ft. Collins.
March 1971	It was determined that design approval had *not* been obtained prior to February 1, 1971, therefore, an Environmental Statement would be required. Requests were also made by various interested groups for a full environmental study.
April 9, 1971	Meeting held in Larimer County Courthouse with local government agencies and interested groups. The intent to make an environmental study was announced and as a result of this meeting, a Citizens and Officials Task Force was suggested to provide input to the environmental statement.
September 21, 1971	An agreement was made with Meheen Environmental Consultants to make an Environmental Impact Study.
April 21, 1972	The Environmental Impact Study was presented to the Division of Highways by Meheen Environmental Consultants.
May 4, 1972	The Environmental Impact Study was presented to each of the Task Force Committee members.

Table 14-1 (continued).   Chronology of events; Fort Collins Central Expressway.

August 1972	Draft Environmental Statement completed to F.H.W.A. for distribution to all local government bodies and committee members for comment.
January 1973	Corridor Public Hearing held at Poudre High School in Ft. Collins,

CONTENDING FORCES AND PARTIES

It is difficult to comprehend the magnitude and extent that a highway project (or any other public works project) can affect so many groups of people with such divergent values and interests. The Fort Collins Expressway project demonstrates this as various organized and individual groups showed interest in the project. Table 14-2 is a list of some of the groups who became involved and a brief description of their particular interest.

Table 14-2.   Interest groups concerned with Fort Collins Expressway.

Fort Collins Historic Landmarks Commission - Location of project in connection with old Fort Collins fort site.

Neighbor to Neighbor, Inc.- Location of highway in regards to minority.

Trout Unlimited - Highway location in regards to rivers.

Poudre School District R-1 - Location of highway in reference to school boundaries, busing, etc.

Cache La Poudre General Improvement Association - Highway location in reference to town of Laporte.

Landowners - Location in regards to agricultural lands and potential development.

Local City Governments - Relocations to best fit their needs.

Larimer County - Location in reference to best fitting the County's needs.

Poudre Valley Greenbelt Association - Open space and recreation areas.

Fort Collins Chamber of Commerce and North College Avenue
Merchants - Business interests as affected by highway.

League of Women Voters and American Association of University Women - Proper governmental procedures and best plan for
region.

Northern Colorado Rod and Gun Club - Development of sport-
ing areas in regards to new highway location.

Designing Tomorrow Today - Local planning group interested
in total planning for area.

Virtually all of these groups showed strong and active interest in the project.   The Environmental Study attests to the
intensity and extent of interest in fact by some 30 letters of
commentary reproduced as a part of the report.   The element of
public participation was certainly a central concern as the
project unfolded.

In cognizance of the wide public interest in the project
a meeting was held in April of 1971 with local government agencies and interested groups.   From this meeting, a Task Force
Committee was established to help the consultant in the preparation of the Environmental Study. It was the intent of the Division of Highways that this committee would be able to supply
the consultant with adequate input, from the viewpoints of
their respective concerns, to help him complete the Environmental Impact Study.   The committee was made of representatives
from the following groups: Fort Collins Historic Landmarks
Commission, Neighbor to Neighbor, Inc., Trout Unlimited, American Association of University Women, City of Fort Collins,
Poudre School District R-1, Cache La Poudre General Improvement
Association, CSU Environmental Corps, Landowner, Larimer County
Planning Commission, Poudre Valley Greenbelt Association, Fort
Collins Chamber of Commerce, League of Women Voters, Northern
Colorado Rod and Gun Club, North College Avenue Merchants, De-
signing Tomorrow Today, CORAD, and WIN School.

Part of the strategy of the Department of Highways in selecting the Task Force Committee was knowledge that some of the
groups represented had conflicting interests.   Therefore, by

getting representatives from each group together on the committee, it was hoped that they would compromise some of their differences within their committee and, with the help of the consultant, possibly come up with a route recommendation. The Task
Force Committee members provided valuable input into the study,
but did not come up with a recommended route. It was felt by
the district engineer, that this being the first time they had
tried this approach, that this failure was due to not setting
up guidelines and objectives for the committee. This case is a
prime example of the need for some formalized knowledge, as
proposed in Chapter 12, on how to absorb public participation
into the decision making process.

TRAFFIC STUDIES

In order to determine the traffic loads in the study area,
the Division of Highways made an origin and destination study
and a traffic study. The origin and destination study involved
first setting up interview stations along state highways around
the perimeter of the Fort Collins area. A percentage of cars
passing each interview station were stopped. The driver was
asked such questions as: (1) where he is coming from; (2) where
he is going; (3) whether he intends to stop in the area or pass
through and, if he is passing through, where he will be leaving
the area; (4) determine whether it is local or out-of-state vehicle and whether it is a passenger car or truck. Results were
recorded as illustrated in Figure 14-3. Figures 14-4 and 14-5
are flow diagrams for projected 1991 traffic, constructed from
the OD survey information and projected trends. It is however,
difficult to make these projections without knowing the future
land use policies of the area. Highway access will, of course,
influence land use patterns.

ENVIRONMENTAL STUDY

The Environmental Impact Study of the Fort Collins Expressway was initiated formally in the fall of 1971. The charge
given to the consultants was to ascertain the environmental effects of alternate routes for the expressway within a three
mile wide corridor between Fort Collins and Laporte.

The environmental aspects of the problem were structured
into three main categories. These were physiographic features,
wildlife habitat, and land use and community values. These
three basic factors were then broken down into subfactors.
After identifying the geographical location of the various environmental factors, they were shaded in on three different
overlays. The overlay mapping method, as introduced by Ian
McHarg, is outlined in Chapter 11. As these overlays were
placed over a base map of the area, theoretically the darkest

REGIONAL TRANSPORTATION STUDY
EXTERNAL INTERVIEW REPORT
1971

Interviewer's Name

Supervisor's Name

CARD NO.
STATION NO.
TIME PERIOD
INTERVIEW DIRECTION
DATE OF TRAVEL

Vehicle Type	Vehicle Occupancyy	TRIP ORIGIN AND DESTINATION	TRIP — AUTO	PURPOSE — TRUCK	FOR TRUCKS ONLY — Commodity Carried	Load Factor	Tare Wt. of Truck	Ent-Exit Route
1.Car - Local / 2 Car-Foreign / 3.Pickup-Panel-Camper / 4.Single unit Tr. / 5.T-T comb. Tr.	1 / 2 / 3 / 4 / 5 / 6 / 7 / 8 / 9	ORIGIN _______ DESTINATION _______	1 Home / 2Work or work related bus. / 3 Personal Business / 4Shopping / 5 Social-recreational / 6School / 7 Change travel mode / 8Serve Passenger	1.Wholesale,Retail,Ind., Pick-up or Delivery / 2.Const.;Equip.and Tools, Pickup or Delivery / 3.Main-Rep.Service Calls / 4.Personal use (non-delivery) / 5.Work or business use (non-delivery)	1.Processed Food Beverage / 2.Ag.Produce / 3.Bldg.Mtls. / 4.Chem-Fuel Product / 5.Mfg.Products / 6.Tools-Equip. / 7.Mixed Load / 8.Mail / 9.Empty	1. 0 / 2. 1-25 / 3. 26-50 / 4. 51-75 / 5. 76-100		
1.Car - Local / 2.Car - Foreign / 3.Pickup - Panel-Camper / 4.Single unit Tr. / 5.T-T comb. Tr.	1 / 2 / 3 / 4 / 5 / 6 / 7 / 8 / 9	ORIGIN _______ DESTINATION _______	1 Home / 2 Work or work related bus. / 3 Personal Business / 4 Shopping / 5 Social-recreational / 6 School / 7 Change travel mode / 8 Serve Passenger	1.Wholesale,Retail,Ind., Pick-up or Delivery / 2.Const.;Equip. and Tools, Pickup or Delivery / 3.Main-Rep.Service Calls / 4.Personal use (non-delivery) / 5.Work or business use (non-delivery)	1.Processed Food Beverage / 2.Ag.Produce / 3.Bldg.Mtls. / 4.Chem-Fuel Prod. / 5.Mfg.Products / 6.Tools-Equip. / 7.Mixed Load / 8.Mail / 9.Empty	1. 0 / 2. 1-25 / 3. 26-50 / 4. 51-75 / 5. 76-100		
1.Car - Local / 2.Car - Foreign / 3.Pickup - Panel-Camper / 4. Single unit Tr. / 5.T-T comb. Tr.	1 / 2 / 3 / 4 / 5 / 6 / 7 / 8 / 9	ORIGIN _______ DESTINATION _______	1 Home / 2 Work or work related bus. / 3 Personal Business / 4 Shopping / 5 Social-recreational / 6 School / 7 Change travel mode / 8 Serve Passenger	1.Wholesale,Retail,Ind., Pick-up or Delivery / 2.Const.;Equip. and Tools, Pickup or Delivery / 3.Main-Rep.Service Calls / 4.Personal use (non-delivery) / 5.Work or business use (non-delivery)	1.Processed Food Beverage / 2.Ag.Produce / 3.Bldg.Mtls. / 4.Chem-Fuel Prod. / 5.Mfg.Products / 6.Tools-Equip. / 7.Mixed Load / 8.Mail / 9.Empty	1. 0 / 2. 1-25 / 3. 26-50 / 4. 51-75 / 5. 76-100		
1.Car - Local / 2.Car - Foreign / 3.Pickup - Panel-Camper / 4.Single unit Tr. / 5.T-T comb. Tr.	1 / 2 / 3 / 4 / 5 / 6 / 7 / 8 / 9	ORIGIN _______ DESTINATION _______	1 Home / 2 Work or work related bus. / 3 Personal Business / 4 Shopping / 5 Social-recreational / 6 School / 7 Change travel mode / 8 Serve Passenger	1.Wholesale,Retail,Ind., Pick-up or Delivery / 2.Const.;Equip. and Tools, Pickup or Delivery / 3.Main-Rep.Service Calls / 4.Personal use (non-delivery) / 5.Work or business use (non-delivery)	1.Processed Food Beverage / 2.Ag.Produce / 3.Bldg.Mtls. / 4.Chem-Fuel Prod. / 5.Mfg.Products / 6.Tools-Equip. / 7.Mixed Load / 8.Mail / 9.Empty	1. 0 / 2. 1-25 / 3. 51-75 / 4. 51-75 / 5. 76-100		

Figure 14-3. Form used by Colorado Division of Highways for origin-destination study.

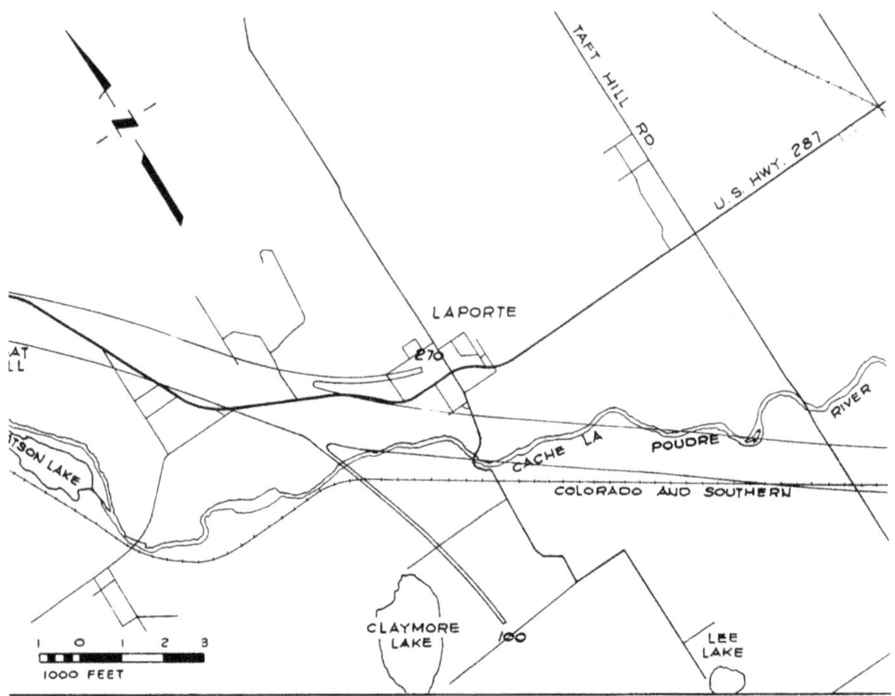

Figure 14-4.   Flow diagram of traffic along corridor from 1991, (from Environmental Impact Study, Meheen Environmental

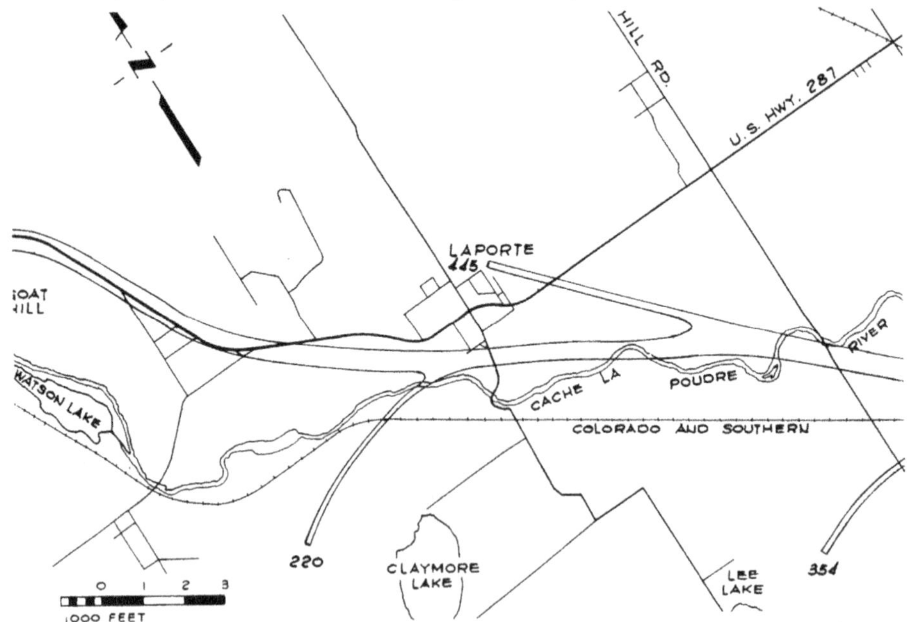

Figure 14-5.   Flow diagram of traffic along corridor from 1991, (from Environmental Impact Study, Meheen Environmental

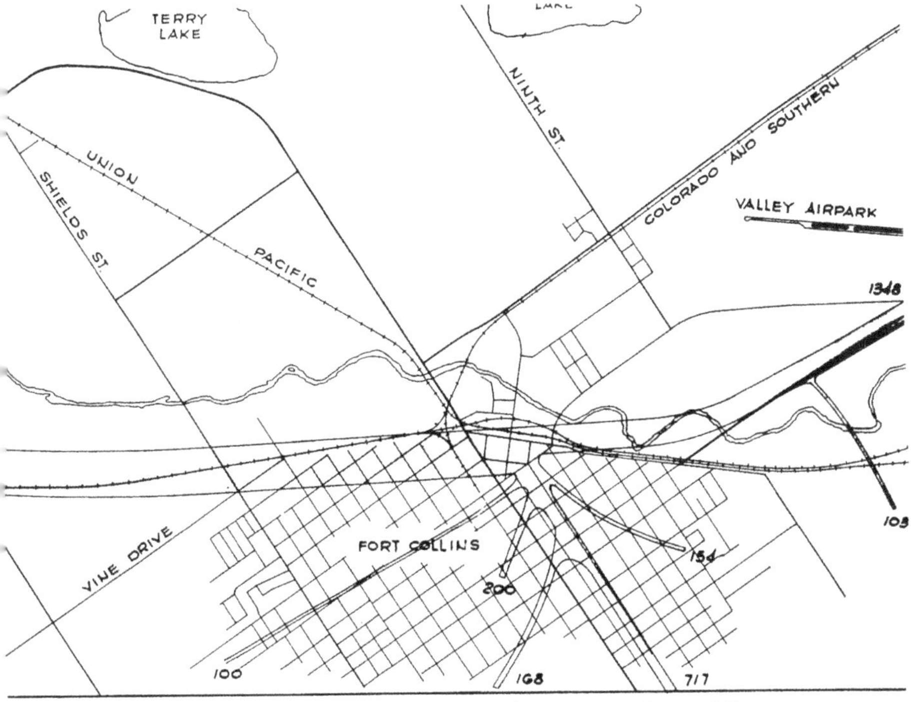

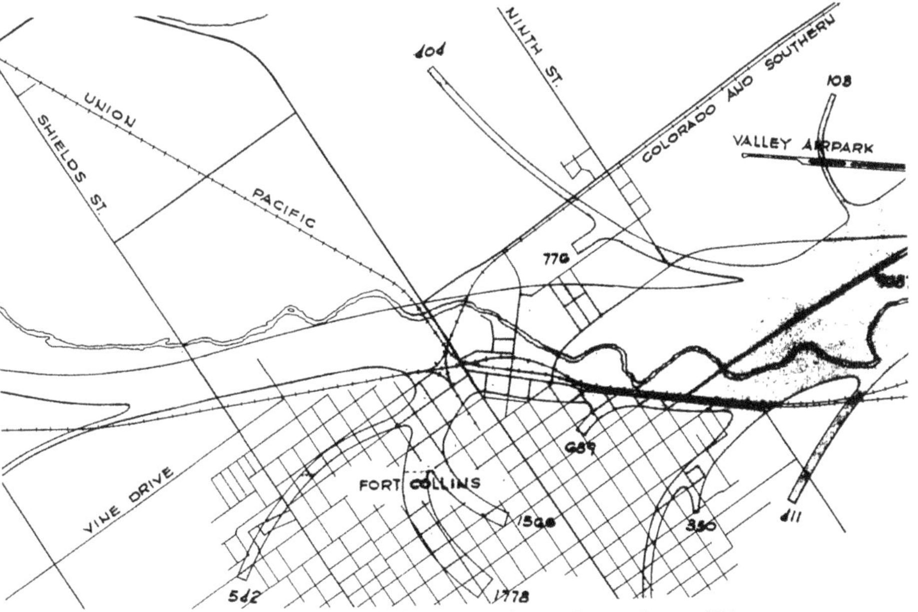

origin and destination studies and projected traffic counts to
Consultants).

origin and destination studies and projected traffic counts to
Consultants).

shades would indicate the least desirable areas in which to place a highway. The lightest areas would indicate where a highway could be constructed, with the least amount of environmental conflict. These studies yielded three main alternate routes which warranted more detailed study. They were considered further in the context of 37 factors. Table 14-3 lists the 37 factors, and attempts to assign a number to each factor relating to each of the three routes.

A grading system of 1 to 5 was used on each of the 37 factors. The lowest total score on a given route would show it to be the best and the highest grading would indicate the least desirable route location. The consultant stated in his evaluation that the lowest grade would show superiority only if it was by a large margin. The scores for the three routes were: Route A - 105, Route B - 97, Route C - 108. The closeness of these scores would indicate that, no one of the alternates was considerably better than the other two. In April of 1972, the consultant turned over to the Colorado Department of Highways its completed Environmental Impact Study of the Fort Collins Expressway.

DRAFT ENVIRONMENTAL IMPACT STATEMENT

The Environmental Impact Study was used by the Department's Environmental Resources Analysis Section in the preparation of the Draft Environmental Statement. By using the information obtained in this report, the draft statement was completed in August of 1972 and was submitted to the Federal Highway Administration for distribution to all federal agencies for their review. The draft statement was issued also to all members of the Task Force Committee and local governmental jurisdictions for their review and comment. The draft statement did not recommend any particular corridor.

Another corridor Location Hearing was held in January of 1973 to afford the public the opportunity to comment on the alternates proposed in the Environmental Impact Study. Since the project had become highly controversial by this time, an outside moderator, an attorney retained by the Department, conducted the hearing. There was considerable response to the Draft Environmental Statement. After a period of ten days in which the Department received additional comments, the Department's Environmental Resources Analysis Section sat down to study and analyze all of the questions and comments made during and after the hearing. The Environmental Resources Analysis Section has the responsibility to produce the Final Environmental Impact Statement for the project; the completed statement is to be transmitted, along with the Division's route recommendation to the Federal Highway Administration and eventually to the Council for Environmental Quality.

Table 14-2

Factor	Title	Routes A	B	C
1	Citizens Aspirations and Desires	3	3	3
2	Neighborhood Identity and Cohesion	3	2	4
3	Education Facilities and Programs	3	3	3
4	Religious Facilities and Programs	4	3	3
5	Health and Welfare Facilities and Programs	3	3	4
6	Displacement and Relocation of Persons and Families	3	2	4
7	Type Condition and Availability of	4	2	3
8	Social and Cultural Facilities and Programs	3	3	3
9	Recreational Facilities and Programs	2	3	4
10	Neighborhood Shopping Facilities	3	3	3
11	Health, Well-Being and Safety of the Public	3	3	3
12	Fire, Police and Ambulance Facilities and Services	2	2	2
13	Employment	2	2	2
14	Property Values	3	3	3
15	Taxes	2	2	2
16	Highway Costs	5	2	1
17	Distribution and Intensity of Economic Activity	2	2	2
18	Effect on Land Use	3	2	3
19	Consistence of Highway with Areawide Plan	3	4	2
20	Traffic Service Levels	3	3	3
21	Adaptability to Future Transportation Needs	3	3	3
22	Coordination and Integration with other Transportation Facilities	3	3	2
23	Conservation of Parks, Historic and Cultural Sites and Landmarks	2	2	4
24	Areas of Natural Beauty, Recreation or Scenic Value	3	2	4
25	Perception of the Road from the Neighborhood	2	2	2
26	Perception of the Road by the Driver	1	1	1
27	Noise	3	2	3
28	Air Pollution	3	2	4
29	Water Quality and Flood Control	3	4	4
30	Neighborhood Growth and Development	2	3	4
31	Regional Growth and Development	3	4	2
32	Joint Development Potential	2	2	2
33	Multiple Use of Highway Rights of Way	3	3	3
34	Conservation of Natural Resources - Ecological Considerations	3	2	4

679

Factor	Title	Routes		
		A	B	C
35	Accessibility as a Factor in the Overall Environment	3	3	2
36	Disruptions Caused During Construction	4	4	4
37	National Defense	3	3	3
	TOTAL GRADE	105	97	108

COMMENTARY

Lack of comprehensive regional transportation and/or land use planning makes highway location and traffic projections difficult. The basic problem in constructing a large facility, whether it be transportation or some other public works, is not with the facility itself but with the lack of any local or regional planning relating to community goals. As a result of questions posed by the Fort Collins Central Expressway the community is in the process of reevaluating its position and developing goals. A Regional Transportation Plan is getting a start. The expressway has opened up the issue of transportation planning, land use planning, and community and regional goals to large public debate and dialogue. Through this planning process, in the future, many of the decisions will be made and problems solved before the Highway Commission is requested to construct new highways. The Department of Highways is also gaining expertise in the relatively new field of environmental evaluation and planning techniques.

The issues and contending forces of the Fort Collins expressway case are rather typical, on a limited scale, of many similar urban situations where large structures, superposed on a community hold the spector of causing an array of new "equilibrium" conditions. The case also illustrates the metamorphosis in planning protocol which public works agencies are now undergoing. The expressway illustrates this transition period. It further illustrates the interlocking of governmental systems and how now the new need is for more active regional planning at the local level. Unless the local communities develop community goals and land use plans the state and federal governments will find it difficult to provide the necessary interregional and interstate transportation systems in a manner compatible with the goals of communities. A new era of planning is indeed portended.

REFERENCES

Meheen Environmental Consultants, Central Fort Collins Expressway Environmental Impact Study, for State of Colorado Division of Highways, April 1972.

Colorado Division of Highways, Draft Environmental Statement for Project F 287-3(3), Central Fort Collins Expressway, August 1972.

AUTHOR NOTES

This case history was reconstructed by Wayne Capron, Assistant Chief Engineer for Operations, Colorado Highway Department. He started his career with the Division after obtaining his B. S. degree in Civil Engineering from Colorado State University. His early years of experience with the Department consisted primarily of field work on highway construction projects in the mountains and plains of western Colorado. As Mr. Capron advanced in the department, he became active in highway administration and has been involved with the development of environmental policies of the Department, public hearings regarding environmental aspects, and preparation of environmental impact statements. He is an active member of the Department's "Action Plan" Committee which has formulated the plan to handle all future projects and their correlation to environmental issues. He has served on the Technical Committee for the Northern Colorado Area Transportation Committee which is a cooperative effort with the Weld-Larimer Regional Planning Commission for regional transportation planning.

CASES: LOGAN CANYON HIGHWAY

BY WILLIAM T. HELM

Logan Canyon is carved into the Bear Mountains by the Logan River; it extends northeasterly from the City of Logan, Utah toward Bear Lake. A Road, U.S. Highway 89, uses this canyon as a route between Logan and Garden City, a distance of about forty miles. The lower twenty miles of road passes through the rather narrow and rugged canyon area. Above, this canyon gives way to broad meadows and more gentle slopes for an additional ten miles, after which the road drops rapidly from the 7800 foot summit to Garden City. Figure 14-6 is a portion of a Utah highway map showing the canyon highway in relation to the state highway system. Figure 14-7 is an aerial view showing about five miles of the lower portion of the canyon.

INTRODUCTION

Logan Canyon, an exceptionally beautiful canyon, contains eighteen campgrounds and nine picnic sites. The river is a popular fishing stream; hunting in the canyon is very popular in fall and early winter and there is a well patronized ski resort near the summit. Quite a number of people bicycle up and down the highway, and picnics and family outings are very numerous during the summer season; vacationing tourists and local outdoor enthusiasts keep the campgrounds nearly filled to capacity. There are many U. S. Forest Service sanctioned summer home areas. Figures 14-8 and 14-9 are indicative of the varied uses and opportunities.

This canyon is specifically identified on state road maps, as shown in Figure 14-6, and is presumably a scenic road or at least has some interesting features. It is one of the possible routes between Salt Lake City and Yellowstone Park. There are a number of alternate routes between these popular tourist

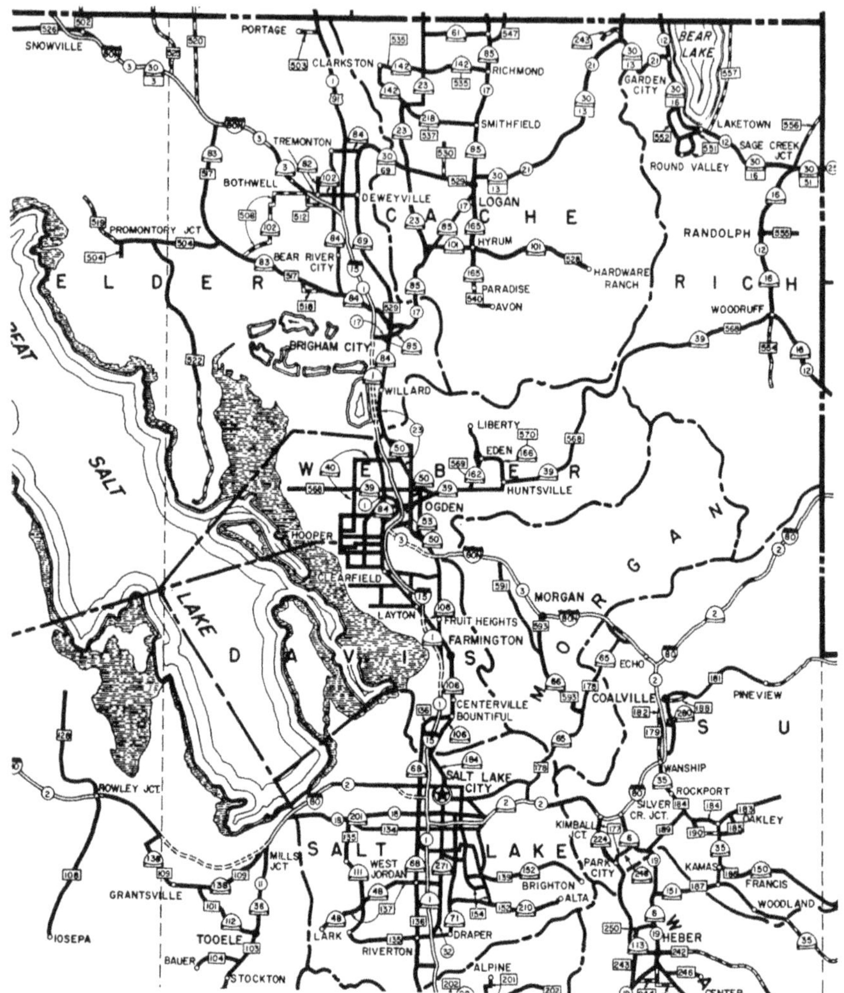

Figure 14-6.   Map of northern Utah and southern Idaho.
Logan Canyon lies between Logan and Garden City, northeast of
Logan.

points, and so Logan Canyon is not the only route to Yellowstone Park from Logan.   Traffic counts obtained from the Utah
Department of Highways indicate an average daily traffic (ADT)
volume of 2300 cars at the mouth of Logan Canyon in 1970 and
1372 some ten miles up the canyon in the area proposed in 1970
for new construction.   An estimated ADT of 600-700 cars cross
the summit of the highway about thirty miles from Logan.   Between Logan and the summit are nearly all of the numerous campgrounds, all of the picnic sites, all of the rivers in which to
fish and the ski resort.   Numerous small mountain roads   are

available for the adventuresome during the warm season and nearly all of these intersect the highway on the Logan side of the summit. There is obvious tourist use of the highway based on the number of out-of-state license plates observed. There is some truck traffic but much of this is of local origin.

Figure 14-8.   Forest Service Campground entrance.   Wood
Camp is but one of the many popular campgrounds in Logan Canyon.
Note recreation vehicle on highway in background.

Figure 14-9.   Forest Service special use area sign.   Each
such area is clearly marked for the convenience of canyon users.

Reconstruction of a 4.3 mile section (section one, milepost 2.6-6.9) of highway at the mouth of the canyon (Figure 14-10) was completed in the early 1960's. Figure 14-7 shows the reconstructed road; the old road can be seen also. There was little public participation in this first project. Public sentiment, however, was expressed during and after the construction of this portion of the highway and there was considerably more public input in the planning of the second section. Extensive environmental damage prompted the faculty of the College of Natural Resources (then the College of Forest, Range and Wildlife Management) at Utah State University to write a treatise, "Road Construction and Resource Use," (J. H. Berryman et al, 1962) on responsible environmental treatment in highway planning. Reconstruction of an additional 4.8 miles (section 2, milepost 6.9-11.7) was completed between 1968 and 1969 after considerable delay and much negotiation among the various agencies and people involved. The construction in these two sections totaling 9.1 miles was, in general, according to AASHO (American Association of State Highway Officials) standards.

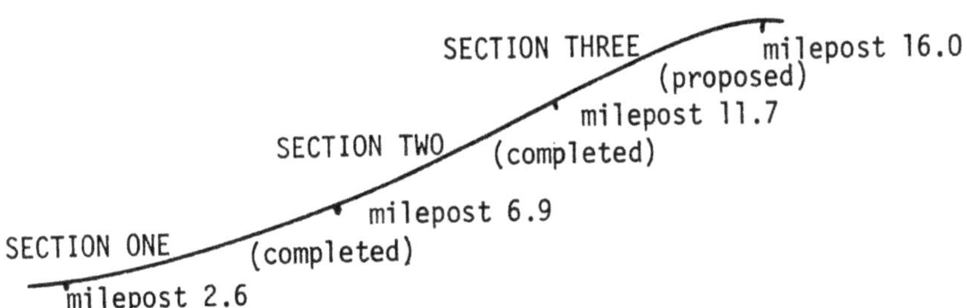

Figure 14-10. Diagrammatic sketch of completed and proposed construction on Logan Canyon highway. Milepost 2.6 is at the mouth of the canyon. For several miles above milepost 16.0 construction of a wide highway would be very damaging to the canyon environment.

The original speed limit in the canyon was thirty-five miles per hour, the present speed limit on the 9.1 miles of reconstructed road is fifty miles per hour, with some curves signed for lower speeds. The current speed limit, on the remaining portions of the old highway in the true canyon section, is still thirty-five miles per hour. Plans for a third section to be reconstructed, a distance of 4.3 miles (milepost 11.7-16.0), call for a speed design of forty miles per hour.

The remaining old road is composed of two lanes each about
nine feet wide, a narrow shoulder in many places and usually no
sculptured ditch.   Trees and   other vegetation frequently grow
very close to the ditch and in a few places form a canopy or
near canopy over the road; Figure 14-11 illustrates. The elevation or the road is not too much higher than the river in many
locations, as shown in Figure 14-12, and there are open vistas
where the river can be observed and others where the cliffs and
upper parts of the canyon are clearly visible.   In the reconstructed area there are generally two lanes of twelve feet each,
a solid painted line at the outside of the driving land, six to
eight feet of paved surface outside of this line and an additional ten feet of ditch.   To achieve this conformation, along
with the gradual slopes called for with today's standards, it
has been necessary to do a great deal of cutting and filling.
Although an attempt was made to revegetate the cut slopes at
the time of construction, most such efforts have been unsuccessful.   The reconstructed highway contains passing lanes in
four locations with limited opportunities for passing in several others.   Section three of the old highway has no passing
lanes and very limited passing opportunities. Accident rates on
this entire highway are somewhat below the state average.

Figure 14-11.   Old 1950's vintage highway above milepost
11.7. Close proximity of vegetation to road is not only scenic,
it may also serve to encourage greater driving caution and thus
reduce the number and severity of accidents on this twisty
mountain road.

Figure 14-12.   Logan Canyon Highway near Logan Cave. The river is readily visible from vehicles traveling in either direction, and most scars from channel alteration by previous highway construction are concealed by vegetation.   This moderately sharp curve is the site of more accidents than any other location, yet it is not marked (William T. Helm).

At the time of the original decision to reconstruct the highway, in the late 1950's and early 1960's, there was of course, no National Environmental Policy Act and no requirements to undertake the types of studies and considerations necessary at the present time.   The first and second portions of the highway which were reconstructed were thus completed during a time when environmental considerations were not required by law, nor where they considered necessary by many citizens.   It is interesting that the delay in reconstruction of section two, from about 1962 until 1968, resulted from pressures to incorporate environmental considerations into highway planning and designing.

FORMATION OF CITIZEN'S ADVISORY COMMITTEE

Concern about various aspects of the proposed continuation of highway reconstruction, this time section three immediately above and contiguous to section two, prompted formation of an informal "citizens advisory committee" during the winter of 1969-70. This committee was composed of thirteen members of the faculty of Utah State University, drawn from the disciplines of Agricultural Economics, Botany, Civil Engineering, Forestry, Humanities, Landscape Architecture and Environmental Planning,

Mechanical Engineering, Philosophy, Political Science, Sociology and Wildlife. The purpose of this citizen's committee was to focus attention on those aspects of planning deemed most necessary for the proper evaluation and development of a highway without undue environmental degradation in this unique setting. In terms of composition and operation, this committee fulfilled the requisites of an interdisciplinary team.

Participation with Public Agencies

Three agencies were involved in the planning and execution of highway construction in Logan Canyon. The Utah Department of Highways planned for and contracted the construction, pending approval by the Federal Highway Administration and the U. S. Forest Service. A special use permit from the U.S. Forest Service is necessary before any use can be made of such federal lands. Communication of concerns and ideas was initiated with the above agencies on June 23, 1970 with the following letter, sent to a Utah State Road Commissioner, the Regional Forester, and the Division Engineer of the U.S. Bureau of Public Roads:

"We are a group of Utah State University professors who are concerned about the proposed additional modification of the Logan Canyon highway. We are expressing ourselves as private citizens and from the vantage point of our respective disciplines.

Because we believe that the value of Logan Canyon ranges far beyond the role it plays in the highway transport system, we are taking this occasion to pose a number of questions, in the hope that a just consideration of them will optimize the benefits of all concerned. We fully acknowledge that the designers and decision-makers in the above three organizations are competent and conscientious professionals--and we certainly do not mean to suggest the contrary in raising the following questions.

1. What are the objectives of the proposed Logan Canyon highway?
   a. As a scenic highway.
   b. As a primary highway.
   c. For access to points in the Canyon.

2. Has a benefit-cost study been done?
   a. If so, what values were considered?
   b. How were impacts on scenic, aesthetic, recreational, and historical values assessed?
   c. Has the economic study been done also in terms of marginal value/marginal cost? or total net benefits?

3. Has the proposed Logan Canyon highway modification been examined in its systems context?
   a. In terms of a transportation system per se? -- how have the alternative primary routes been evaluated?

b. In terms of the ecosystems involved?
   (1) the German brown trout fishery?
   (2) the total recreational fishery?
   (3) other recreational impacts?
   (4) aesthetic impacts?
   (5) wildlife effects?
   (6) river turbidity and channel effects?
c. In terms of social impacts?
   (1) commercial effects?
   (2) recreational enjoyment?
   (3) sightseeing at moderate speeds?

4. Have all of the specifications for the construction of the highway been checked to determine compliance with the new Environmental Policy Act?

5. What were the projections for the future in terms of highway use, number of cars per hour, and requirements for any additional road widening or significant road-bed changes?

6. What basic data were collected to determine whether there was a need for this new road?

7. How were the above gathered and how were they evaluated in terms of relative importance?

While we recognize that few precedents exist for the articulation of these environmental-social considerations in a benefit-cost analysis, and that laws, public attitudes, and values have changed markedly in the past few years, we nevertheless feel that the incorporation of all values in the "design mix" should be attempted. In particular, we feel that the scenic, aesthetic, recreational, and historical values should not be ignored, especially since highway construction is an irreversible act. We believe that such an approach is possible, and we hope that from it some creative and exciting insights in highway construction could result.

If you should wish further discussion of any of these considerations we would be happy to cooperate in any way we can. While we do not see ourselves as providing final answers, we would be able to sharpen the focus on the important issues, and perhaps assist in defining the most feasible alternatives.

As we understand the decision making process, the design is proposed by the Utah State Highway Department upon recommendation by the Highway Commission; the permit to build the highway is issued by the U. S. Forest Service; and approval must be given by the U.S. Bureau of Public Roads. Therefore, we address ourselves to each of these organizations. We look forward to an early response from each of you.

Sincerely,"

Response to the above letter included the scheduling by the Utah Highway Commission of an extra public meeting in late

August, 1970, attended by representatives of the three groups mentioned above, members of the citizen's advisory committee and others. The location and design hearings required by law had already been held. Although there were obvious disagreements about many of the items under discussion at the meeting, areas of contention were more clearly identified and each group had a better grasp of the objectives of the other groups. An inspection tour of the proposed highway construction was agreed upon, subject to selection of a mutually agreeable date. One event foretelling the future changes in highway department procedures was the circulation of an undated highway department memorandum stating their intention to form a highway department environmental advisory committee.

The inspection tour took place about six weeks later, attended by representatives of the same groups. On-site examination of some of the problems further crystallized the issues, and set the stage for definitive studies of some of them.

Identification of Problems

Meetings of the citizen's committee, prior to the public meeting in August, 1970, were devoted to discussions of both general and specific problems related to highway routing and design in northern Utah and southern Idaho. While much discussion centered on problems in Logan Canyon, the role of various roads in moving traffic in the general area was also explored. Little specific information was available about the proposed design for the third section of highway, so it was assumed this section would be designed similar to the previous sections.

After the public meeting in August, the full extent of the situation was evident. Obvious general problems were related to inadequacies of basic assumptions about this type of situation, inadequacies and faults in planning, mistakes in judgment and lack of consideration of problems other than engineering. A partial list of specific problems identified included: lack of consideration of alternative routes for through traffic; designation of Logan Canyon as an important highway link (principal arterial) while ignoring its importance as a recreational road; reference to improved highway safety with a new road without any evidence for such an achievement; refusal to consider alternative construction standards; no involvement of landscape architects at the planning stage; many massive road cuts; locating the extra wide portion of road containing passing lanes in a narrow and particularly beautiful area of the canyon; encroachment of road fill onto the riverbank; potential siltation problems with the river; encroachment into and/or loss of picnic sites and campgrounds; aesthetic degradation by the type and placement of guard rails; and lack of coordination with highway and recreational uses. The lack of interdisciplinary planning was clearly evident.

Solutions to problems were sought by discussion among citizen's committee members, by subcommittee study, and by correspondence with and suggestions to members of the various public agencies involved. A request for sketches of the probable appearance of some areas where huge road cuts were proposed resulted in sketches prepared in time for the inspection tour. A suggestion that development of some alternative route might be preferable resulted in a limited and inadequate report on one such route. The U. S. Forest Service responded to comments on the need for recreational planning in the canyon with consideration of the effect of roads of varying standards upon those plans, by establishing a two year study. Funding for this study, unfortunately, has been held up until fiscal year 1974. A request for information on highway accidents in the canyon was answered with a short memorandum by the highway department (H.A. Guerts, unpuglished data). Questions on the selection and placement of guard rails drew the admission that these were subjective decisions. Here again, lack of input from disciplines other than engineering was clearly evident.

Subcommittee study of some of the problems resulted in the preparation of reports and sketches. Study of other problems culminated in the compilation of data, conversations, phone calls and letters to various members of public agencies about such topics as tourism, road width, encroachment on riverbanks, recreational planning, coordination of highway and recreation uses, guard rails and wildlife habitat and migration patterns.

Sketches were prepared to illustrate such problems as the effects of vegetation removal on recreational use of the river, picnic sites, and campgrounds. Encroachment of the road into recreational areas would eliminate some and reduce the size of other picnic sites and campgrounds. Removing the vegetative screen from between such sites and the highway would expose recreationists to the annoyance of unsightly highway fills, vehicles passing within full sight and increased traffic noise levels. Increasing the speed limit would further increase noise levels. Fishermen, desirous of solitude, would be exposed to the view and noise of passing traffic wherever the screen of vegetation along the riverbank was removed. Even while on the highway the canyon user would be penalized. Removal of roadside vegetation and replacement with eroding road cuts of up to 1000 feet in height would occur in numerous places. While the far-field views would remain unchanged, the highly important near-field views would be extensively altered, resulting in a great reduction in the beauty of the canyon. Thus one of the major uses of this highway, recreation, would suffer reduced values as a result of the proposed design for reconstruction (Figures 14-13, 14-14, 14-15).

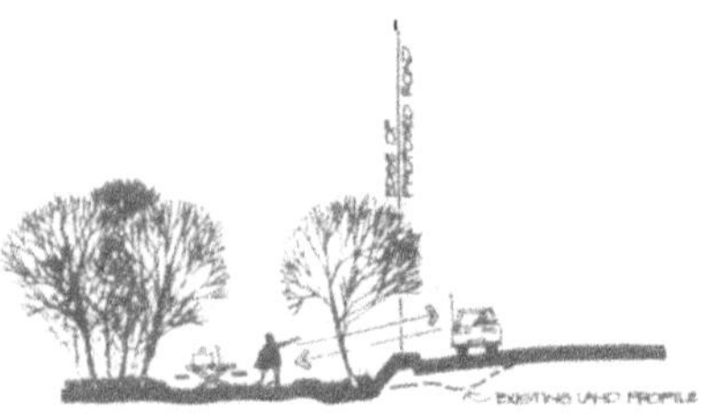

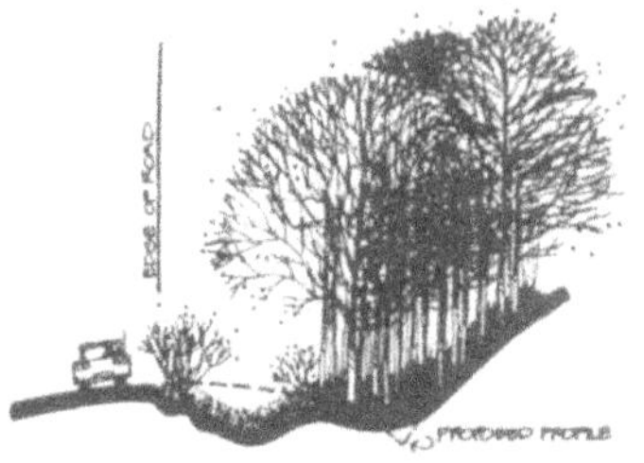

Figure 14-13.   The changing roadside environment   (I).
Under existing conditions (left) recreation areas are screened
from the road; road construction as proposed would leave insufficient room for vegetative screening (right) in many places
(Courtesy of Gerald Smith).

Revegetation of cut slopes has not been very successful or
satisfactory in the lower portion of the canyon, leaving unsightly and eroding slopes.   Cuts were too steep, unstable and
dry in most cases for a dense stand of vegetation to prosper.
Planned replacement of some particularly interesting slopes
near Wood Camp with huge cuts represented a distinct loss (Figure 14-16). Cuts of such magnitude were required because of the
wide driving lanes, wide parking lanes and wide ditches called
for in the design. A reduction in width, as noted in the accident report, could encourage drivers to exercise caution and
also reduce the extent of the cuts, thus decreasing the amount
of scenic damage. Revegetation could also be made more successful and aesthetically acceptable by utilizing native forbs
and shrubs in combination with soil pockets and irregularities
in slope (step slope technique).

Views of the river, part of the enjoyment of driving in
the canyon, would also be diminished by increased road elevation above the river and installation of vision obstructing and
unsightly guard rails.   An examination of the sections of road
previously reconstructed revealed that guard rails, almost entirely absent along most of the old highway, were installed in
numerous locations along the reconstructed segments.   The wide

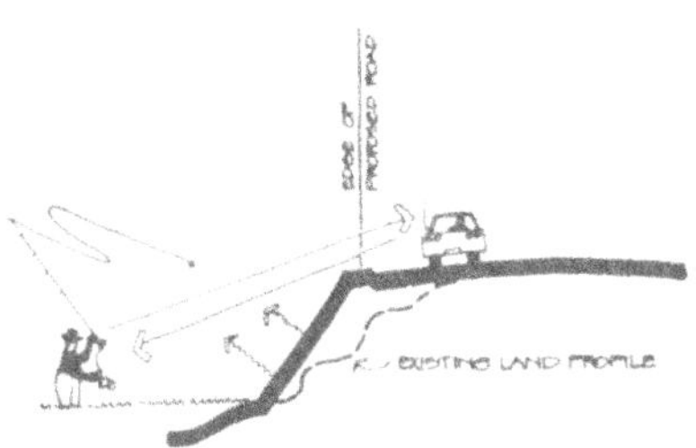

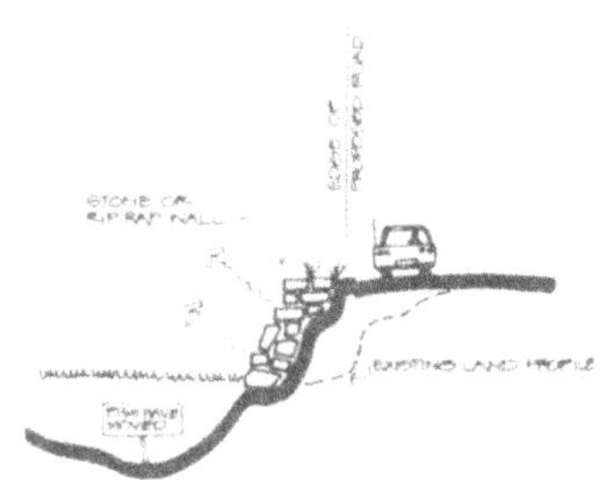

Figure 14-14.   The   changing roadside   environment   (II).
Streambank cover screens fishermen from view and provides cover
for brown trout (left); encroachment of road fill as originally
proposed (right) eliminates such cover, decreasing recreation
enjoyment and reducing the number of trout in the river (Courtesy of Gerald Smith).

pavement, wide ditches and broad galvanized guard rails, all
foreign to the landscape, combined to diminish the perception
and enjoyment of the canyon.   Increased height of the roadbed
above the river and the guard rail along every portion of the
road near the river reduced visibility of the river to only
brief flashes, as seen in Figure 14-17.

According to the plans, road fill would reach or nearly
reach the riverbank in some areas, eliminating vegetation
along the shoreline and between the shoreline and road.   Destruction of shoreline vegetation, alteration of stream banks
and stream beds greatly diminishes the trout carrying capacity
of the river.   Figure 14-18 shows a typical channelized stream
reach which is now poor habitat for trout. Figure 14-19 is an
example of good trout habitat.   Comparison of altered and natural portions of the Logan River adjacent to the highway showed
eighty-seven percent fewer trout and a more limited size distribution of fish in altered than in natural areas.   This is
seen in Figure 14-20 (W. T. Helm, unpublished data).   Although

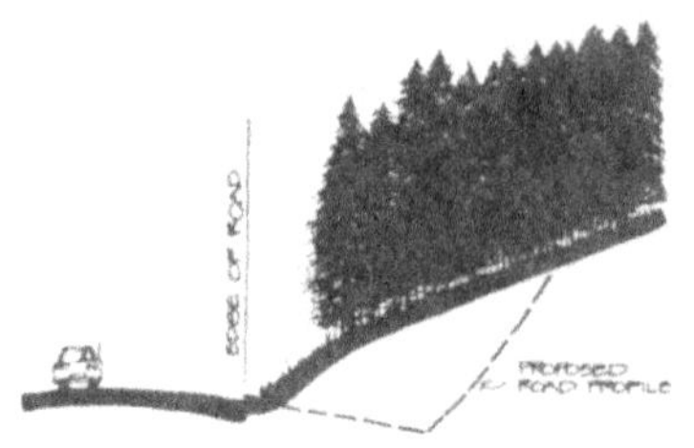

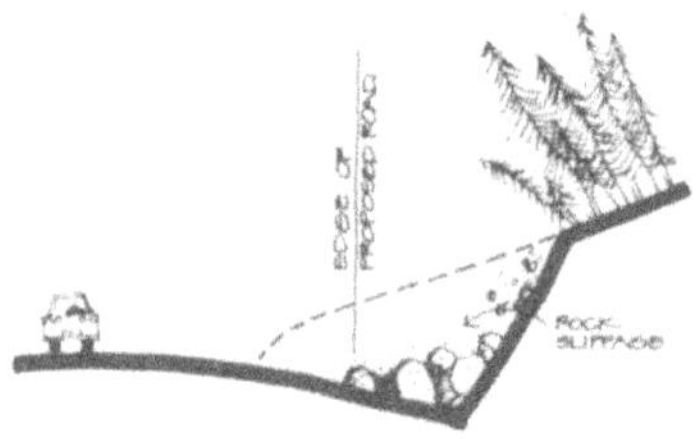

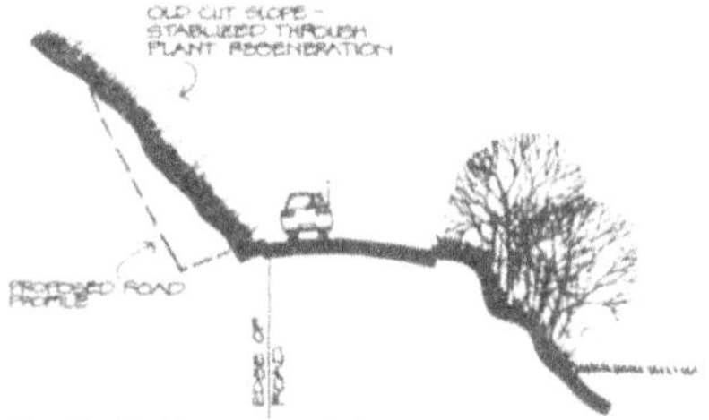

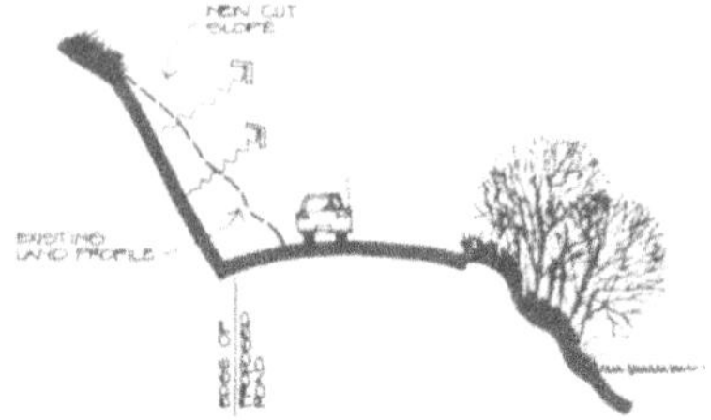

Figure 14-15.   The changing roadside environment (III). Vegetated hillsides and small, vegetated road cuts (left) would be replaced by very extensive road cuts, many too steep and too high (right) to be revegetated (Courtesy of Gerald Smith).

Figure 14-16. Roadside beauty near wood Camp. A mixture of evergreen and deciduous vegetation combined with large steep-like rock outcrops and steep slopes make this one of the very beautiful areas in the canyon. The road here will be straightened and passing lanes installed by cutting deeply into the mountainside.

Figure 14-17. Road and guard rail. The view from the passenger's side on a right curve in section two. All such potentially interesting views were destroyed by the increased height of road surface above the river and the installation of the guard rail.

Figure 14-18. Poor riverbank habitat. Encroachment of highway fill onto riverbank, or replacement of natural channel with a shortened, straight channel destroys desirable brown trout habitat as well as creating a scenic eyesore.

Figure 14-19.   Bank protection   plus habitat.   Brush de-
flects current away from bank, providing protection from erosion.   Quiet water downstream from bush serves as resting area
for trout, and interface between quiet and flowing water is a
feeding area. Diversity in depth and speed of flow is required
for good brown trout populations.

the effects of increased siltation caused by road construction
were not assessed in this area, such effects have been documented elsewhere (Peters, 1965 and 1967).   Siltation of trout
streams is one of the most serious and detrimental of habitat
damages.

Much of the encroachment of road fill into river banks and
recreation areas, denudation of large cut-fill slopes, siltation of the river, elimination of views of the river by greatly
increased elevation of the road and damage to the side canyon
ultimately selected for dumping excess fill material is a direct result of the design for extreme width of the proposed new
road.   Redesigning to reduce this width would decrease much of
this damage and could eliminate excess fill, and thus also
eliminate the impact of disposing of that fill in some side
canyon. The basic problem is that the road is designed too wide
for the canyon terrain.   Figure 14-17 shows how the widened
highway near the mouth of the canyon begin dominates the canyon
landscape in comparison to the old highway which can also be
seen.

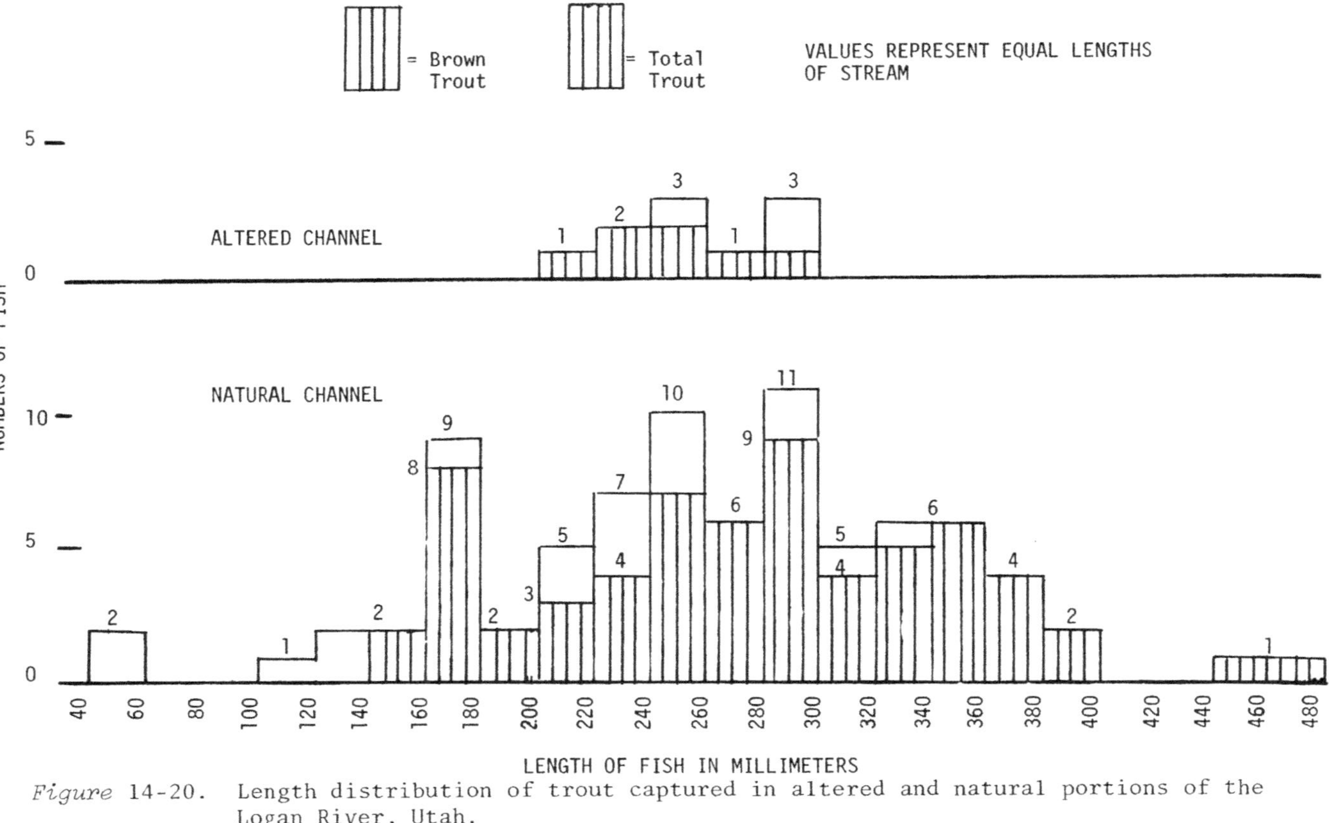

Figure 14-20. Length distribution of trout captured in altered and natural portions of the Logan River, Utah.

In response to the highway department Logan Canyon accidents memorandum, a subcommittee conducted a study of available
records. Computer print-outs of all recorded accidents were
made available by the highway department. Although these records had been available for several years, there had apparently
been no effort to utilize them in determining whether there
were any dangerous locations in the canyon, or to determine
whether the reconstructed highway was any safer (or more dangerous) than the old highway. A twelve-page report (P. B.
Holden, R. M. Lanner and I. G. Palmblad, unpublished data) was
written by the subcommittee, and copies submitted to all interested parties. The report included all accidents.

Accident statistics for a three and one-half year period
(January, 1967 to June, 1970) were available for both the first
section of highway reconstructed (4.3 miles in length) and for
the proposed third section (also 4.3 miles in length). Accidents appeared to be randomly scattered during the twelve-
months of the year, and not related to traffic volume (Figure
14-21). It appears the most dangerous times to travel, in terms
of the probability of an accident, are during the low traffic
volume periods. Heavy traffic volume appears not to be an
accident-breeding factor, and thus any improvement in the ability of the road to carry higher traffic volumes (and improved
flows) would be irrelevant to highway safety. Accidents were
about equally distributed between light and dark hours on both
stretches of road and most accidents in both areas occurred on
dry road surfaces. The various kinds of accidents were quite
similar between the two areas, regardless of the physical differences as shown in Table 14-2. Cars leaving the road in the
first section, where the ditches were wide, were more apt to
hit fixed objects than those in the third section, where the
ditches were very narrow, in spite of the disparity in proximity of fixed objects to the road. The greatest difference,
however, was in severity of accidents. Table 14-3 refers to
reportable injuries, so an additional statistic is required. Of
all accidents, forty-four percent of those in the first section
resulted in bodily injury, while only thirty-two percent of
those in the third section resulted in bodily harm. Among the
conclusions drawn in the report were: there is an apparent lack
of analyses of accidents on canyon highways; the higher speed
encouraged by the broad road surface and lack of obstructions
close to the road caused a higher proportion of people in accidents to be injured, and also caused a higher proportion of
more severe injuries on the reconstructed road than on the old
narrow road. Available evidence suggests that drastic design
changes will make the highway more dangerous rather than less.
Although a small sample size (86 accidents) makes the interpretation of small differences in values difficult, the broad implications appear to be valid.

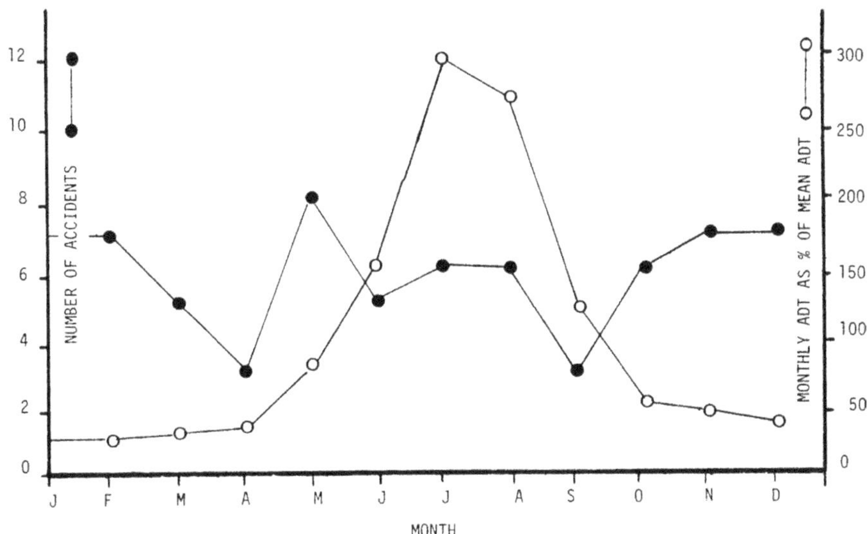

Figure 14-21.   Number of accidents per month and monthly average daily traffic (ADT) in Logan Canyon. Values are for the first section and proposed section combined.

Table 14-2.   Kind of accidents - values in percent.

	First Section of Reconstructed Highway	Third Section (Proposed for Reconstruction)
Run-offs	50	65
Two-car collisions	19	18
Hitting fixed objects	17	6
Over-turns	8	3
Deer	6	3
Undetermined	0	6

Table 14-3.   Severity of accidents   in which injuries were re-
              ported.

	A	B	C
First Project	61	22	17
Proposed Project	45	28	28

A = Visible signs of injury, bleeding wound, or distorted mem-
    ber, or had to be carried from the scene.
B = other visible injury as bruises, abrasions, swelling, limp-
    ing, etc.
C = no visible injury, but complaint of pain or momentary un-
    consciousness.

In addition to statistical comparisons, accident hotspots
on the old road were identified in the accident report. Unfortunately, such information has not been utilized to properly
mark and sign such places to reduce accidents.

Interdisciplinary discussion and planning, by the citizen's advisory committee, had identified numerous problems in
the basic planning of the highway, and had developed proposed
alternatives for most of them. The value of combining the varied talents of widely divergent disciplines was obvious to the
committee and, it appeared, to the U.S. Forest Service and members of the Utah Highways Environmental Advisory Council.
Copies of each of the reports generated had been submitted to
all agencies involved, and communications indicated the reports
were of value.

EFFECTS OF INTERDISCIPLINARY EFFORT

The identification of problems related to the proposed
highway reconstruction, the subsequent elucidation of these
problems and communication with various public agencies resulted in broader consideration of the issues. Formation of the
Utah Highways Environmental Advisory Council provided a vehicle
for conveying broad concerns to the highway department. All
information generated by the citizen's committee was made
available to the Council. Consultation between members of the
two groups resulted in additional clarification of some problems. The Highway Environmental Advisory Council, is composed
of nine members and includes citizens who are an engineer, a
physician, a city planner, a sociologist, an architect, a contracts specialist, a retired mayor. Views of divergent interests were sought and considered in efforts to identify and
solve problems related to road design. The presence of this

committee and its influence on decision making has resulted in
many design changes making the proposed highway more environmentally acceptable.

Subsequent revision of the design for the third section of
highway reveals the value of interdisciplinary planning. Although proper consideration has not been given to alternative
routing and different road standards suitable for this beautiful canyon recreation area, the revised road plans do include
improved consideration for other issues. Since most of the
damage would be caused by greatly increased road width, a reduction in planned road width in some particularly troublesome
locations has been proposed. A combination of retaining walls
on the side of the road toward the river. more modest increase
in pavement width and a minimum of ditch have reduced the magnitude of road cuts. Instead of more than 1000 feet the largest
cut should now be less than 150 feet in height. This has also
reduced the amount of excess fill which must be disposed of
somewhere in the area. Retaining walls would be set back from
the riverbank, with a prohibition on machinery use in the river
or between the wall and the river, with no disturbance of
existing vegetation in this area, thus retaining some of the
natural character of the river and riparian habitat. Planning
guard rail installation has been reduced. Provisions have been
made for parking at numerous and regular intervals along the
highway. Hopefully, revegetation of cut slopes can be more effectively managed with the use of a mixture of plants and soil
pockets, than on the two previous sections.

In spite of the above noted improvements in the design of
the proposed reconstruction, the most valuable contribution of
the citizen's committee has been to identify major weaknesses
in the basic approach to highway location and design. Rather
than considering the nature and role of particular roads in the
regional network of roads and classifying each accordingly,
selection of roads for "improvement" often appears to be made
in a vacuum. In this case easily obtainable information which
should have been utilized in making one initial decision of
whether or not to reconstruct the road, would have revealed
many of the problems identified by the citizen's committee.

Concepts such as scenic parkways and alternate routes for
different types of traffic should have been thoroughly explored.
Identification of the types of highway users, their origins and
destinations would have provided the necessary information to
select a number of possible alternate routes. Blacksmith Fork
Canyon from Myrum to Round Valley, Logan to Preston, Idaho and
via Mink Creek to Montpelier, and Logan Falls to Swan Valley to
Victor to Jackson all represent workable alternate routes. The
best local alternative would probably be via Immigration Canyon
between Preston and Montpelier, Idaho, but development of this
route would require two different state highway departments to

talk to each other and work out a system of priorities. Lack of
lines of communication plus the "pork barrel" aspects of such
projects has inhibited enlightened planning under the highway
management system which has existed until the early 1970's.

So-called "standards" constitute an equally serious impediment to sensible and responsible development of highways.
Federal cut-fill standards have remained unchanged for over 40
years.   Since most road building money comes from federal
sources, states too often use the "excuse" of federal standards
rather than try to change.   Federal standards are apparently
expected to apply to all situations for all time, a patently
unsound concept. New guidelines for cut-fill, highway pavement
width and other problem areas related to mountain roads are
long past due, and would help upgrade these roads without destroying the beauty of the canyons.

In the absence of interdisciplinary planning, the conventional two-hearing system is inadequate as either a means of
informing citizens of the true nature of a proposed highway or
of eliciting helpful citizen comment upon the proposal.   The
location hearing is usually the first presentation of such information to the public, and there is no opportunity for careful study and analysis of the proposal prior to the meeting or
prior to the highway department decision on which route to follow.   Similarly, the design hearing, at which details of the
proposed highway are first made available to the public, does
not satisfy the intent to obtain either citizen contribution to,
or approval of the plans.   It takes time to examine and digest
many pages of maps, to examine the sites and determine the
probable effects of the proposed construction. Good intentions
aside, in actual practice inclusion of the public in any meaningful participation in the planning of such public projects is
very difficult.   Interdisciplinary planning when a tangible reality will greatly facilitate a more sensitive approach to
planning and design and the involvement of public groups.

Examples of the finality of foreclosing options without
adequate consideration of alternatives can be viewed at several
locations including the upper end of Third Dam Reservoir and
near the mouth of the canyon.   Nibbling just a little bit at a
time is seen in Figure 14-22. The original road bed can be seen
by arrow. That a subsequently rebuilt road cut off a portion of
the river, and the last reconstruction of the road cut off an
additional portion of the river can also be seen in Figure
14-22.   While the old road skirted the heavily used reservoir
the new road cuts across a portion of it as seen in Figure 14-7,
and isolated this portion and reduced the scenic and recreational values of the area. A hydro-electric generating plant at
the mouth of the canyon, operated by the Utah Power and Light
Company, has for years drained a portion of the Logan River
during the summer, fall and winter to supply water for its turbines.   When the first section of highway was reconstructed,

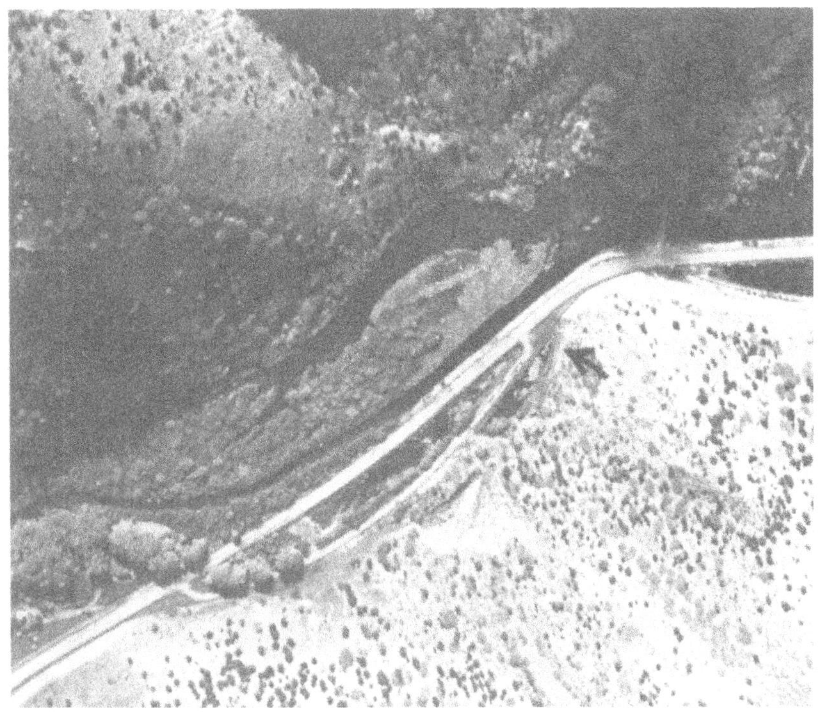

Figure 14-22. Just a little bit. Three stages of highway
reconstruction near upper end of Third Dam Reservoir in Logan
Canyon.   Arrow points to original road. Curving line of trees
in lower left, crossing the road, marks the original river
channel.   Bare cut slopes appear as light areas on toes of two
slopes by curves.   Canyon here is too narrow for a wide, high
speed road (William T. Helm).

much of the river was channeled to make highway construction
cheaper and easier. Since completion of the construction in the
early 1960's the power company has abandoned the generating
plant, returning the normal flow of water to the previously
dried-up channel.   Fishermen now have a low-productivity channelized stream in this area with little chance for any significant improvement.   Banks along this channeled portion are eroding and unsightly in many areas, and the erosion contributes
damaging silt to downstream areas.   Since the normal flow has
been returned to the river, increased velocity due to the
shortened and altered channel has now resulted in severe erosion of a previously stable stream bank at the mouth of the
canyon.

There are enough unresolved issues, related to both the
third section of proposed reconstruction and the remainder of
the highway, for which reconstruction is proposed in increments,
to warrant reconsideration of the entire project.   Many irreversible foreclosures have been made in the 9.1 miles already
reconstructed. In addition, a U.S. Forest Service spokesman has
stated that the present reconstruction efforts are doing all
the damage the environment can stand.   Yet highway department
traffic projections indicate the new road will be obsolete and
in need of additional widening by 1990. By 1973 plans for going
ahead with the Logan Canyon project were suspended.   Anything
not as yet done will have the benefit of an interdisciplinary
team not only through the Environmental Advisory Council, but
through additional added interdisciplinary staff capabilities
within the Utah Highway Department. This is insured further by
the reorganizations of all state highway departments, in accordance with the 1972 Federal Highway Administration Policy and
Procedure Memorandum 90-4,   for the purpose of being more environmentally sensitive.

REFERENCES

J. H. Berryman et al.,"Road Construction and Resource Use',' Extension Circular 297, 1962, Utah State University.

J. C. Peters,"The Effects of Stream Sedimentation on Trout Embryo Survival," in *Biological Problems in Water Pollution,* Third Seminar, 1962, U. S. Dept. Health, Education and Welfare, Public Health Service Pub. 999-WP-25, pp. 275-279.

J. C. Peters, "Effects on a Trout Stream of Sediments from Agricultural Practices," *Journal of Wildlife Mgt.,* Vol. 31, No. 4, Oct. 1967, pp. 805-812.

AUTHOR NOTES

Dr. William T. Helm, author of Chapter 14B, is co-author of Chapter 5, Ecology, which contains a summary of Dr. Helm's professional work.

Mr. Donald M. Turner, Environmental Biologist, Utah State Department of Highways, has reviewed the chapter.   A number of his helpful editorial comments have been incorporated into the text.

15

PERSPECTIVES AND FUTURE CAPABILITIES

BY P. H. MCGAUHEY

This treatise herein presented is based on the rationale that the time has come when large public projects must be subject to a balanced environmental design. In this context an "environmental design" is considered to be one in which practitioners of the full spectrum of disciplines that are, or should be, interested in the environmental effects of the project participate in decisions at every state--conception, planning, system selection, physical layout, functional design, construction and operation. A "balanced design" is defined as one in which the environmental benefits and disbenefits are identified and weighed, and tradeoffs and compromises are consciously made in the decision process. To make a balanced design possible, it is postulated that each of the several disciplines involved must be able to (1) identify the criteria which will be applied in bringing the discipline into the design process, (2) describe the methodology that will be used in applying the criteria, and (3) develop a scale which relates degree of environmental damage to the extent of compromise with the ideal or absolute condition.

It is the objective of this treatise to establish to the degree possible the parameters which make for a balanced integration of disciplinary considerations when decisions are made in the process of the environmental design of large public works. The potential of the volume to be of significant value to practitioners of the several professions involved in a balanced environmental design depends upon many factors. Some of these have immediate application; others are long-term in their response.

Of most immediate concern is the extent to which the authors of the several chapters have successfully delineated by objective analysis of concepts, and by examples, how their respective disciplines are to be applied in the over all design of a project. Here success is measured both by a clarity of language, which makes the considerations understandable to professional in other specialty areas, and by a scientific objectivity which other objective-minded men may accept. In the longer span the book's potential depends not only upon the success of its authors in achieving the immediate goal of understandability and practical methodology, but also upon the degree to which the conclusions of each author are accepted or repudiated by other professionals in his discipline. Each preceding chapter is, therefore, concerned with the factors which professional practitioners in all participating fields should understand relative to the particular discipline discussed in that chapter if a balanced environmental design is to result. It also serves to indicate the considerations, reasoning, and methodology the practitioner of the discipline might describe in supporting his recommendations to a public body holding official hearings on which to base its own decisions on a given project.

It is the purpose of this chapter (Chapter 15) to put the problem of environmental design into further perspective and to assess to some degree the future capabilities of a balanced multidisciplineary decision process for producing public projects which are at the same time harmonious with nature and beneficial to mankind. To this end, three major aspects are considered:

1. The feasibility of a multidisciplinary undertaking.

2. The need for, and viability of, a multidisciplinary approach to the design of public projects.

3. The degree to which individual disciplines can develop viable criteria and compatible methodologies for balanced design.

In exploring these three areas a conscious effort is made herein to avoid moralizing and to minimize editorializing, although the evaluations must necessarily be those of the .uthor. The general approach adopted is to look to the past from time to time to note how project design has traditionally been accomplished. Thereafter, evaluating the present against this past may reveal what progress has been made in bringing a greater number of appropriate disciplines into the development and design processes. The past and present situations, when weighed against the concepts on which this treatise is founded, might then suggest how we ought to go about achieving balanced multidisciplinary decisions. Finally, by comparing present methods and their future potential with some ideal concept, the efficiency of our capabilities might be assessed in at least some qualitative way.

Although it might seem logical to develop the discussion along a temporal scale ranging from the past to the future, a more random approach is herein utilized. This is done because in some particulars the pattern is different for transportation systems than for water resource development and power production even though all suffer from the same lack of design in the environmental sense. Moreover, the various disciplines derive from different bases--some mathematical and some conceptual. In addition, they are at different stages of readiness to contribute to a balanced environmental design by any identifiable common approach.

THE QUESTION OF FEASIBILITY

To establish a perspective from which to assess the future capabilities of professional practitioners from many disciplines to bring together and adjudicate the diverse considerations necessary to insure good environmental design, it is necessary to recall several human and historical factors. In attempting to do this the author has drawn freely upon concepts voiced by various individuals during the deliberations which helped to develop this book, as well as upon personal experience and evaluations.

One of the first difficulties in establishing effective multidisciplinary action is that the degree of specialization necessary to achieve professional competence in any discipline tends to make the individual highly knowledgeable in one field and a near illiterate in others. This problem, which has been recognized and deplored by the academic as well as the professional world, cannot be resolved by abandoning specialty studies in favor of general education. The reason is that the resulting specialist in generalities, who understands everything and can do nothing, is incapable of achieving environmental design although his contributions to a multidisciplinary effort may be significant. It requires little imagination to predict the results of an environmental design in which six professional men, each illiterate in five disciplines, try to make a decision in which each of the six disciplines involved are evaluated reasonably well in scale with each other.

Fortunately, professional men are not generally as narrow as their critics sometimes suggest. Nevertheless, in acquiring the competence needed for professional practice in his field, no man can acquire a similar competence in all fields. It is therefore of utmost importance that the concepts and methodology of one discipline be made understandable to other disciplines--the basic objective of this treatise.

The second roadblock to multidisciplinary project design is the "Tower of Babel" nature of communication between various specialty fields. Each discipline tends to have its own special jargon which may or may not be understandable by its own practitioners, but which is definitely not understood by all men concerned to apply environmental design to public projects. Thus multidisciplinary discussions are immediately handicapped by a language barrier.

Next is a way of thinking about things and situations. Engineers and others who apply the physical sciences, customarily see a job to be done in terms of structures or systems which man needs now, even though these may become obsolete as human knowledge increases. Economists and political scientists often find it less easy to identify any point where a continuously changing situation can be interrupted by action on the basis of today's knowledge, when tomorrow will obviously be different. Biological scientists tend to take a longer view, perhaps because the area of unknowns is so vast that one cannot foresee whether any given course of action is a prelude to ecological disaster or to a benefit to mankind. The social scientist is confronted with an even greater complexity of problems because of the unpredictability of people and of their persistence in behavior that is difficult to define as rational.

A fourth phenomenon that makes it difficult to bring the knowledge of several disciplines into a balanced decision concerning environmental design for public projects is the need that this book was concerned to alleviate. This is, that many disciplines, especially the non-physical sciences, do not yet have a methodology or research approaches sufficiently specific to supply the evidence required for a balanced decision to emerge from a public hearing. The disciplines concerned are not here criticized for such a lack of methodology. The variables are simply far too great for any easy resolution.

The preceding roadblocks to effective multidisciplinary environmental design are substantially based on the assumption that all disciplines approach the project with a similar degree of objectivity. This has by no means always been possible, hence as discussed in more specific terms in sections which follow too few disciplines have been involved in many of our past public projects. Beliefs, prejudices, and orientation to causes or organizations have compounded the illiteracy of many individuals, or colored their approach to even their own specialty areas of knowledge. The growth of affluence and its demand on natural resources triggered a public and official concern for environmental factors. Lacking other approaches, and beset by these beliefs and prejudices, an emotional missionary zeal moved into the vacuum of knowledge as an influencing factor in public decision. In spite of this development one might conclude on the basis of past and present actions that multidisciplinary environmental design of public projects is essentially infeasible. However, as one goes deeper into the need for environmental design, the longer range implications of the public mood, and, especially, the nature of emerging environmental law, it becomes evident that in some manner environmental design of public projects is going to characterize the future. As the author of Chapter 11 has stated: "Conflicting interests must be accommodated regardless of the tenability of their stated positions."

NEED FOR ENVIRONMENTAL DESIGN

Basic Concepts and Assumptions

A need for environmental concern for public projects has often been postulated in terms of resource depletion, threat of ecological disaster, and fragility of the planet's equilibria. However, if it is assumed that public projects are going to be built in the future with appropriate attention to these general factors, the need might better be related to the identifying and overcoming of shortcomings revealed by past projects. Examples have already been cited of highway location where the objective has been engineering for the benefit of users of the roadway without apparent attention to the effects of the facility on other sectors of the community. An example of the failure of a sociologically oriented project is the high rise apartment complex in St. Louis intended to house negro migrants from the rural south. These facilities were later dismantled because the project merely proved what any of its occupants could have told its originators, i.e., that the tenants were neither culturally nor socially prepared to live in such an environment as the project provided. At great economic and social cost it was concluded that dispersed housing was the answer, and a new round of sociological experimentation was begun.

Such examples should not be interpreted as necessarily bespeaking a narrow outlook by the professions of engineering and sociology. It must be remembered that projects are often derived from the zeal and enterprise of one or two enthusiasts; and that enthusiasm is many times a symptom of untested preconceived theories, impractical dreaming, or personal value commitments. The point to be remembered is that regardless of whether failure results from a bias of the principal practitioners involved in a project, or from the limits inherent in any given discipline, it could be prevented by a system which insures a balanced environmental design procedure. It is in this context that the need is best defined.

The validity of the assumption that through public projects the physical world will forever be altered to meet the objectives of man is hard to dispute, even though the citizen in an idealistic mood may applaud those who would obstruct such projects. The increasing number of people to be served, as well

711

as the individual citizen's appetite for transportation, recreation, and power-dependent facilities, is an inescapable fact of life. However, the fact that the citizen grows uneasy at the voice of the "prophet of doom," is constantly dismayed by traffic congestion, is alarmed by smog, and may have suffered a loss of value of his property as a result of some public project evidences his belief that the consequences of public works should be more clearly foreseen.

The assumption that public projects will continue to be built does not imply that any particular benefit man seeks to enjoy must be achieved by constructing the most obvious facility. Alternate locations, systems, or approaches may be capable of achieving the project objectives. And if environmental quality is to be the critical consideraiton, it follows in logic that all elements which define environmental quality should be evaluated in the selection of alternatives, the choosing of tradeoffs, and in the design of the project itself. This means that economists must transcend the limits of the traditional benefit-cost ratio expressed in dollars. It means that political considerations must go beyond the responsibilities, motives, and scope of elected public officials to include those areas in which the political scientist is expert. It means that the sociologist must find a way to harness the energies of people and to evaluate social costs and incorporate them into design decisions. It means that the "ecologist" and "environmentalist" must contribute the most positive inputs of which their disciplines are capable.

Without repeating the details presented in previous chapters there is clearly a need for multidisciplinary environmental design activity in the development of large public projects. The problem is how a viable design procedure can be established and made to function effectively within any time frame scaled to the life expectancy of man. In the paragraphs and sections which follow, the author therefore reflects upon both the need for and the viability of multidisciplinary environmental design within the temporal concepts outlined in the introduction.

Transportation

As emphasized in Chapter 3A, the social or human welfare goals of transportation, as contrasted with the simpler motives of moving people and goods from one place to another, first became widely recognized in the 1960's. This concept of transportation is, however, somewhat poorly reflected in public policy. The result is that transportation projects have often been fully or partially completed before people recognize that they disagree with the design. This has been especially true in smaller cities where businessmen once believed that their businesses would suffer if through traffic was not routed down the main street. Only when resulting traffic density made it impossible for either the traveler or the local resident to patronize the shops did this misconcept of highway location become evident at the level of local politics.

No permanent economic loss occurred during this stage of development of highway transportation except in cases where extensive street widening and setting back of existing structures was involved. Generally, only routing of traffic was concerned and the fallacy of assumption soon became evident to even the most casual observer. The next phase, however, was somewhat more costly but no less disastrous. It involved building bypass highways skirting the city, typically without adequate access controls. The result was that service stations soon clustered on the borders of the highway. They were quickly joined by small businesses, particularly of the "hot dog stand" type. Dwelling houses soon closed the gap between clusters of service stations and it was not long until there arose a local public outcry against the speed of motorists and consequent danger to children. The public demanded, and got, traffic lights and speed zones. Soon what was designed as a 50 mph route for the traveling public became a 25 mph street interrupted by unsynchronized traffic lights and bordered by an unzoned development destined to early decay as highway transportation moved on into its present phase: i.e. to the limited access freeway bypassing the city entirely.

The bypass freeway concept, of course applies to both the small and the large city where space permits. However, the most difficult situation has arisen at all stages of the highway transport program in the large urban metropolitan centers where the project goal was to move commuters in and out of the city while at the same time keeping through traffic off the city streets. This goal gave rise to the sunken or elevated freeway which often traversed neighborhoods already in economic decline, thus giving rise to physical separation, noise, and other effects destructive of the social structure of the community.

Figure 15-1 illustrates the result of the elevated freeway in San Francisco. Here the structure was stopped in mid-air by popular demand only after a great deal of money had been invested on a public project that people had to see before they rejected its further progress. This example does not, of course, mean that freeways in metropolitan areas should not be built in the future, but it does underscore the concept that more than traditional engineering and economic factors and the guidance of planners having only advisory powers should go into the project design.

Because public policy has not in the past required environmental design, examples of environmental damage by transportation facilities, whether by highway, railway, or airline, have created pressures for society to adopt repressive policies.

Figure 15-1.   Embarcadero Freeway in San Francisco.

Such pressures tend to come from organized citizen groups having simplistic concepts of the complexities of the situation.
This simplistic approach has not been confined to citizens concerned over air pollution, rush hour stagnation of traffic,
noise, parking fees, and related hazards and inconveniences.
The spread out of urban communities which the automobile made
possible after World War II hastened the decline of the central
city.   Local government, therefore, has sought ways to lure
people back to the city or, at least, to alleviate the incon-
venience of working there.   San Francisco is a good example of
such a situation. At present commuters can by no means be lured
back once they have survived the evening traffic rush.   The
convenience to the housewife of the local shopping center has
reduced her interest in the city.   In such a situation a cultural decline of the city became inevitable. Women no longer
went to the city to shop, meet their husbands for dinner, and
stay on for the theater.   The central city has therefore moved
in the direction of a financial, banking, and business center
with an overthrust of tourist trade based on convention activity. Meanwhile, the worker comes home to the suburbs and settles
down before the TV set.

Like many areas which dismantled a once viable interurban
streetcar system, the San Francisco Bay Area's governments and
citizens have sought to restore the social and cultural attrac-
tions and to destroy the evils attributed to the automobile by
proposing an extensive rapid transit system.   Unlike others,
however, the area has embarked on a public project to construct
the Bay Area Rapid Transit System (BART). For BART some people
hold the highest of hopes.   As a possible solution to present
transportation difficulties, including the automobile-generated
fraction of smog, and as a possible blueprint of the future it
is receiving international attention.   Significantly, it is
being built by local tax levies and, as is typical of large
public works projects requiring years in the making, will cost
twice the original estimate.   In the context of multidisciplinary project design BART may well provide important evidence.
During the 10 turbulent years between authorization and initial
service runs, numerous changes were made in design in response
to environmental considerations of which the public became conscious. Valuable data on the social and economic costs of pro-
longed disruption of business in construction areas has been
generated by the project. In addition, the effect of its physical structures and its services upon an entire region should
provide information of the type needed by many disciplines in
developing methodologies for environmental design.

The project, however, is not without its critics or its
skeptics. Some see in BART an attempt to go back to an era that
is gone.   They doubt that a system built in 1970 will be more
viable economically and socially than were its 1907 predecessors. They cite the social and cultural benefits of freedom of
movement the automobile provides at other than commute hours
and, especially, in the residential areas.   There is concern
that the problem of origin and destination of the transit system itself may be too inadequately resolved to make rapid transit a convenient or even an acceptable alternative.   Finally,
they cite the added quality of life associated with planting
and tending one's suburban plot--a sense of some control of and
participation in one's own environment that automobile transport has made possible.   Thus it is conceded that should BART
make possible the continued spreadout of the urban community,
it is not likely to restore the central city to its former social and cultural status.

It is not the purpose of the author here to judge the merits of rapid transit or to prejudge its effectiveness. The
point in perspecte is that the BART system was conceived by the
voting public as an escape from the horrors of commuting and
the hazards of air pollution; and by the local government as a
way to arrest at the same time the declining role of the central city. Social and cultural objectives were therefore strong
in the motivation of the project but the kind of balanced multidisciplinary evaluation which this treatise seeks to outline

was not the basis of the original decision. However, significant progress toward that end seems to have accompanied the birth of BART.

The need for the capability of men to apply a balanced decision-making process to the design of transportation systems in the future does not hinge entirely upon the success or failure of sub-systems such as the BART. Unloading the city streets and the commuter routes at peak hours might eliminate the need for elevated freeways within the city and hence make possible better environmental design within the area. It does not, however, overcome the fact that urban centers represent an over-loading of land with people to a degree that leads them to turn outward in search of recreation. Already the highways in the populous states are crowded on weekends, which get longer as affluence cuts down the working hours, and an impressive volume of intercity truck traffic occurs at all times. As America moves from 80 percent of 200 million people on 10 percent of the land to 300 million people on 3 to 5 percent of the land, the outward reach for recreation via highways can be expected to increase. Therefore the need and the opportunity for multi-disciplinary attention to environmental design in the realm of ground transportation will continue in the future and can hard-ly be resolved by repressive policies intended to make life in the central cities indefinitely possible or tolerable.

As noted in Chapter 3A, highways have been viewed as "a despoiler of the physical environment." Designers have there-fore been deplored for overlooking the social costs of tran-secting neighborhoods, or for destroying the visual harmony of the environment. In this context the introduction of an envi-ronmental design which brings sociologists, political scien-tists, and landscape architects, for example, into concert with engineers, economists, and planners should overcome the defi-ciencies of design. This conclusion, of course, presumes that the objectives of design are to provide a transportation system in which the quality of the physical environment is the func-tion to be optimized. However, under a philosophy of government for social ends, such as described in Chapters 1 and 2, envi-ronmental design may offer only a trivial solution to the transportation problem. Such a possibility is suggested in Chapter 3A in relation to the loss of social benefits through a "no project" decision even though the decision be made pursuant to the parameters of several participating disciplines. The possibility is further implicit in the question of weighing the merits of activities made possible by transportation against the virtues of better transportation per se. Here, as the author of Chapter 3A points out, the system's output may be a more important political objective than the system's character-istics, and hence unresponsive to a balanced environmental de-sign procedure.

Factors such as the foregoing vastly complicate the task
of identifying criteria and methodologies which the social sciences, especially, might utilize in participating in a balanced
environmental design.   For example, if we are to relate total
community needs to transportation, total needs must be deter-
minable by some specific method of evaluation rather by anyone's
social theory.   Otherwise action becomes a function of the
transcendent sociological dream of the moment and the result
may be similar to the St. Louis housing fiasco.   Thus the criteria for environmental design become useful only in a secondary sense, the primary problem being that of defining total
community needs.

From a traditional viewpoint mobility may be considered an
inherent quality of life--a heritage--of Americans which should
be reflected in design criteria.   At the same time government
for social ends may impose greater restraints on how and where
vehicles can be used until environmental design becomes a function of individual elements of the system rather than of the
system itself.

The effectiveness of environmental design of transportation systems of the future cannot be considered bright if, for
example, the engineer is to use the criterion of mobility; the
economist, the criterion of cost; the political scientist, the
criterion of social purpose (real or imagined) which transcends
the need for mobility of goods and people; and the environmentalist, the criterion of excluding people from land areas for
old times' sake.

Air transportation is a problem distinct from that of the
highway system. Facilities for an air transport system are more
a problem of location than of balanced environmental design. In
general the airports are experiencing the intermediate stage
cited in relation to highways--the time when their margins are
being occupied by people who then demand that the activities of
the system be modified to suit the whims or convenience of
themselves. There are some sociological and environmental facts
of life to be learned here concerning land use and land development, but they cannot be solved by a multidisciplinary design
of air transport systems and hence are herein considered beyond
the scope of this chapter.

Water Resources Development

Chapter 3B has underscored a need for a multidisciplinary
approach to water resources development, both because of the
nature of the decision and the manner in which it is implemented.   In general, as the author of the chapter notes, water
resources planning has been used to accommodate the future
rather than to shape it.   This approach, which has sometimes
been called "backing into the future," is analogous to peering

at the wake of a boat to see what has been churned up rather
than steering it toward some identified port.   It has characterized water resources development and is responsible for the
most compelling need for environmental design yet recognized in
relation to public projects. This, however, does not mean that
no effort at multidisciplinary design has been made.   What has
happened is best described in the final section of this chapter
in the context of critic versus participant in the project development process.

The development of water resources began moving toward
multi-purpose and hence toward multidisciplinary planning when
single purpose entities began to encroach upon each other. The
result is that the spectrum of disciplines which have been
drawn into the project planning process is already a broad one.
Even more broad, perhaps, is the range of management interests
necessary for optimum operation of a developed facility.

Because the disciplines involved in multi-purpose water
resources development projects has generally been internal to
the agency or group of governmental agencies responsible for
developing the project, their individual inputs have not been
widely identified.   This has contributed to the popular notion
that only engineering and benefit-cost economic considerations
have been involved in public projects related to water resources.   This does not mean that a broader spectrum of interests should not be involved in environmental design, but it
does caution that the environmentally designed project might
not be so much different than the multi-purpose projects of the
recent past.

Chapter 3B calls attention to the conflicts between such
interests as power, flood control, flow augmentation, water
supply, recreation, and fisheries resources.   Such conflicts
appear in two contexts: (1) the people whose major interest is
in one or another of the many purposes of the project, and (2)
the fundamental characteristics which define each purpose. This
latter type of conflict is locked into the very concept of multiple purpose projects and cannot be designed out of the system entirely. Specifically, if flood control is to be a project
objective, the reservoir water level must be at a design elevation when the flood season begins.   If flow augmentation and
minimum stream flows are design objectives, water must be discharged at appropriate rates as the season requires.   If irrigation or other water supply is involved, discharge of stored
water during the dry season is also necessary.   Although these
discharges may take place through a power-generating turbine
power has its own set of additional requirements if it is to
pay the costs it is customarily expected to carry.   All these
factors mean that a shoreline of 150 miles, for example, is not
going to delight 8000 waterfront homeowners each with 100 feet
of fish spawning habitat in his front yard.   Check dams across

branches of the reservoir may maintain the fisheries resource
or keep a constant level for recreational purposes, but to many
people the carrying capacity of the facility for some types of
recreation may be disappointingly small.   Further disappoint-
ment may accrue to the urban citizen when he discovers the inherent conflicts between various water recreation activities
themselves. For example, the rowboat fisherman versus the motor
boat and water skiing enthusiasts.

Despite the history and reality of multidiscipline planning and operational management of water resources projects,
few resource developments were under greater attack than that
of water as the decade of the 1970's began.   To a significant
degree the attack came from people who for various reasons, or
lack of reason, advocate no development at all.   However pure
the sentiment may be, the inherent conflicts in water resources
use cannot be avoided nor the welfare of man enhanced by delegating natural resources to the uncontrolled whims of nature.
But more of this present situation in more appropriate sections
to follow.   For the future, the present system of multidisciplinary inputs can achieve greater capability through a clearer
concept by professional practitioners of the factors within
their individual disciplines which merit consideration.   When
these concepts are accompanied by a methodology for their application to environmental design, water resource development
projects will take on a new and more satisfactory dimension.
Nevertheless, it is too much to expect that all projects will
be more influenced by reason than by political considerations,
or that the eventual effects of the best designed project can
be foreseen. Two examples might be cited in this context--the
TVA and the Bonneville Dam.

The ramifications of either of these two projects are too
vast to catalog here. Nevertheless, it is significant that more
than a generation after the TVA was conceived to bring prosperity to an entire region, southern Appalachia is still considered a depressed area.   Changed, yes, but not prosperous.   In
contrast, the Bonneville Dam was built on quite different sociological dreams; and roundly opposed at the time for the excess
of power it would generate.   As it turned out, the power was
almost immediately needed in a national emergency (World War II)
and established the Pacific Northwest as a center of the aluminum industry. Although the "no development" enthusiast of 1972
might deplore this industry, the project contributed heavily to
the economic prosperity and to the welfare of thousands of
people who wished to live in the northwest.

Although government for social ends may be reflected in
water resource development policies, they are less likely to
confuse the objectives of environmental design than is the case
with transportation. More likely social goals will lead to no-
project decisions than to a confusion of the characteristics of

the systems which are built. Likewise in comparison with
transportation systems the results of narrow concepts of water
resource project development have a much broader impact. In
fact, one of the most startling features of all types of large
public projects is the human multi-ignorance of seemingly unrelated phenomena which has appeared in water resource development projects. As we see yesterday's tomorrow in retrospect the
factors overlooked (such as ecological considerations) as well
as the ones misinterpreted (such as channelization) seem to im-
pugn our intelligence. The implications of these factors are
set forth in detail in previous chapters, particularly in Chapter 5, and hence need not be catalogued here.

Power Production

Of the three types of large public projects herein discussed, power production is the most varied in the nature of
project evolvement. Unlike the highway and water resource development, a power project may be basically a government enterprise, or it may be initiated by a public or private utility.
Therefore the ways in which it is possible for a number of disciplines to participate in environmental design of power projects may differ more widely from case to case than in transportation and water resource projects. For example, if power is to
come from a water resource development, government alone is
likely to be involved in initiating the project. If nuclear
power is proposed, both government and private enterprise are
certain to be involved in the project's beginnings; and if fossil fuels are to be the source of energy, some type of utility
operating under governmental restraints is generally the origi-
nating agent. Although the nature of environmental design considerations may differ little with the type of project, as regards the energy source, the way in which outside groups may
cause their parameters to be considered in project design is
far from constant.

Viewing the situation in the time series framework adopted
for this chapter, water power appears historically as the oldest of the three power systems. Simple diversion of water by
individual industries gave way to single purpose power dams
when electrical energy replaced direct mechanical drive by
water wheels. Such power dams were usually low structures oc-
cupying the most desirable dam sites. Thus the potential of a
stream to serve man was not achieved because the project had to
be profitable at once in terms of salable power alone. Eventually government adopted the policy of maximum benefits from
water resource development and the privately owned power dam
reached the end of an era. It was replaced by the politically
initiated government multi-purpose project, amid a chorus of
opposition to the entry of government into the power business.
By the time the current concept of multidisciplinary participation in public projects for environmental objectives came into

being, power was but one of a number of goals of water resource development projects; and, perhaps, the least important of the lot in terms of environmental design.   For the future, there-fore, it seems evident that if the environmental design objec-tives of other purposes of a water resources development are met, those related to power will have disappeared in the pro-cess.

Fossil fuel power projects have a somewhat different his-tory. Initially, steam driven equipment required that power, or energy conversion, be produced on the site of its use.   Later, electricity became the product of steam generation but the practice of transporting fuel instead of energy continued. Perhaps the limitations of early electric power transmission technology, and political deference to the barge and railway systems, contributed to this custom.   In any event the result was to generate in the areas where people are concentrated such adverse environmental factors as are associated with power plants burning fossil fuels. What the future may bring in this type of power project is obscured by considerations of policy and of the availability of resources, as well as by what may transpire relative to nuclear power.   Of particular importance to environmental design, however, is that the fuel driven power plant typically belongs to a private utility company operating under government controlled rate structures, and subject to the regulations of air and water pollution control agencies.   Pro-posed new facilities are hence subject to approval by agencies established by state and federal environmental law. This means that environmental design will be necessary, although through some arrangement between project owner and environmental de-signers which is not yet fully developed. Probably consultants from the appropriate disciplines will be called in to advise the design firm retained by the power company, which may well have professionals of many disciplines on its own staff as well.

Nuclear power has no past history comparable to that of water or fossil fuel powered projects. Significantly, it did not become feasible until standards for environmental protec-tion against radiation had been developed and technology for power production was at hand. Even then it became available to power companies and utilities only under strict federal licens-ing and control procedures. At present nuclear power projects are multiplying faster than are fossil fuel burning plants. Because heat is the principal waste product, however, location near large bodies of water is essential. For maximum safety a location away from urban centers is also preferred. Siting of facilities, possible consequences of reactor failure, manage-ment of radioactive wastes associated with reactor regeneration, and the ecological effects of heat are the principal environ-mental factors most commonly discussed. Most of these are in reality concerned with policy rather than design.   They are therefore little amendable to resolution by design criteria,

hence multidisciplinary inputs have been largely negative and
counter-productive.

The year 1972 found power production project proposals
under concerted attack by organized citizen groups which often
seem to imply that the use of power is immoral. Water power
projects are opposed because they involve dams; fossil fuel
projects because they consume resources and sometimes pollute
the air; and nuclear power projects because radioactivity and
heated water are deemed environmentally unacceptable. Because
all three of these attitudes are held simultaneously by the organizations involved, 1972 saw the rise of a potential power
shortage of serious implications. Inasmuch as this treatise
emerges from the concept that public projects are going to be
built, there is little point in going deeply into the matter of
opposition to power production in relation to power development
per se. The phenomenon of obstructionism is discussed in another section. Suffice it to say here that it is difficult to
foresee how some disciplines can contribute much that is positive to a balanced environmental design without first rethink-
ing what contribution it is that they wish to make.

DEVELOPMENT OF CONCEPTS AND METHODOLOGY

General Considerations

If one accepts as realistic the two concepts that: (1)
physical facilities will continue to be built for the benefit
of man, and (2) environmental design of large public works must
involve a balance of parameters derived from a number of disciplines, it follows that the needed parameters must represent a
composite of criteria and methodologies for their application
developed by each discipline itself. At the present time the
resolving power of some sciences is considerably greater than
that of others because what the author of Chapter 6 has termed
the "elusiveness of subject matter" prevents the social sciences from producing exact answers and rigid methodologies such
as usually associated with the physical sciences. Morec.er the
time span over which any parameter remains valid is highly variable from discipline to discipline. For example, the design
criteria and the technology for building a dam may be expected
to change relatively slowly as the physical sciences continue
to develop over the years. In contrast, social attitudes and
political expediency may change, or even reverse, with exceeding rapidity because they are matters of the human mind and
hence more dependent upon fancy than upon scientifically demon-
strable fact. The same is true of the biological sciences but
for a quite different reason. Here the extreme complexity of
ecosystems and their environmental equilibria inevitably make
for a slow rate of evolution of environmental design methodology. The result is that men may become impatient and so react

with extreme caution, or from emotion rather than rationality.
In such a situation it only requires the factors of missionary
zeal and simplistic naivete to generate a spirit of obstructionism.   Either of these add to the burden of the biologist
engaged in the development of criteria and methodologies.

If the nature of the subject matter and the life expectancy of its interpretation is to vary from discipline to discipline, it seems logical to consider the subject of development
of concepts and methodology in terms of key factors and of the
status of individual disciplines.   Therefore, in what follows
particular attention is directed to some specific aspects which
confront one or more disciplines or which characterize the discipline itself.

Public Policy and Environmental Law

Because environmental law is the unifying and overriding
factor in all disciplines concerned with environmental design
it is herein selected for first consideration.   Looking to the
past, it might be said that public policy is normally expressed
in legislation which both prescribes the conditions that society is expected to meet and establishes institutions with component administrative agencies to enforce those conditions.   Administrative agencies then, under the authority of the law or
by assumption of power, issue further ukases which have the effect of law.   Should administrative fiat become too oppressive
the matter may be taken to the courts which then interpret the
law in its original context, or reinterpret it in terms of constitutionality or their own social attitudes. Changed public
policy or failure of the courts to uphold administrative interpretations of the law may then lead to additional legislation
to clarify or update the original laws.

At the present time federal spending power makes it possible for federal agencies, often on the basis of administrative
regulation, to control the nature of public works.   An example
is the much debated requirement that sewage treatment achieve
an 85 percent reduction in BOD (Biochemical Oxygen Demand) if
the project is to qualify for a federal grant-in-aid. Because,
for numerous reasons beyond the scope of this chapter, building
of treatment works without a federal grant is impossible from a
practical viewpoint, the requirement achieves the force of superior national law.

In the case of environmental law as such the common pattern cited in paragraph one is subject to considerable variation because this area of legal activity is still in flux.   As
noted in Chapter 9, environmental law as an area of specialization by the legal profession has not yet (1972) clearly emerged.
Nevertheless, there is a rash of laws and legislative bills
aimed at various elements of environmental control which if

gathered together for a single treatise might be expected to discourage a professional practitioner of environmental law.

As noted by a reviewer of an early draft of Chapter 9, "Law, to be effective, must reflect the physical realities of the real world.   Ability to float 10 feet off the ground may very well be in accord with the values, hopes, aspirations, and expectations of many men, but a law establishing this as a requirement for all of us would not be meaningful.   Many knowledgeable people are of the view that some 'environmental laws' now being passed do indeed ignore the physical realities. Such laws will not be implemented because they simply cannot. In the meantime, attempts of implementation can do great harm to society at large." In the context of environmental design, environmental law is relatively recent.   Its principal unifying effect on multiple disciplines is the requirement that all projects submit an environmental impact statement. It is axiomatic that the project design must therefore include the considerations of a number of appropriate disciplines. Several major implications in the matter of environmental design then emerge relative to the impact statement.

1. It precludes design based entirely upon technological and dollar cost considerations.

2. It leads to a diversity of professional backgrounds on the staff of engineering firms, as well as the use of multidisciplinary consultants in project development and design.

3. It makes urgent the matter of development of criteria and methodology by professional practitioners of the nonmathematical disciplines for use in environmental design.

These reform minded provisions of NEPA are self implementing through the requirement of the impact statement.

A somewhat more difficult problem for environmental law to solve in the future is discussed in Chapter 8 in relation to land.   Ownership and general freedom to use land is a strong cultural factor in the U. S.   It is harder to constrain by administrative fiat than are practices which involve the use and abuse of water and air resources.   Nevertheless, environmental design which takes into consideration the relationship of the physical facilities of a community and the social costs of location of public projects cannot ignore realities that go with the ownership of land.   There are always individuals who for reasons of their own do not consider the best developed plan for land use to be in their personal interest.   As the author has often noted, even the Lord had to displace people who did not like the idea in order to create a "Promised Land."

A final comment on the future capability of environmental
design concerns the time factor. One of the principal by-
products of obstructionism is that the planning horizon -for
large public works is lengthened from some 5 to 10 years to
perhaps 20 to 25. In the meantime technology will change and
the initial proposal may become obsolete. Thus either a further
delay is required while the project is updated, or man is
forced to build in the past. If the project is indeed not
needed for 25 years, this should be the first decision of the
multidisciplinary design process and the time and human energy
involved in long term bickering put to more beneficial use.
Although the prospect of interminable delay may delight the opponent of a project, the time span is too long to be acceptable.
Public policy as currently reflected in environmental law leads
to such a time distortion but it is probably safe to predict
that this condition will change as both the refinement of law
and the capability of disciplines increases. However, there is
a popular tendency to overestimate the role of public policy,
the inference often being that citizens must accept drastic
changes in their way of life to accommodate the day to day pronouncements of the administrative agency currently in the as-
cendency. In this matter it is often forgotten that although
government has the power to tyrannize over people, people have
the power to change governments if a policy becomes intolerable,
and they may do so even though the result may be for the worse.
The record of several thousand years of unsuccessful urging of
man to change his inhuman habits may be pertinent to environmental design. One alternative, which history suggests that man
might choose, is to continue to develop public projects which
he deems desirable as long as his ingenuity and the earth's resources permit. This he may prefer to do with the benefit of
environmental design if it becomes a viable tool, but he will
go ahead without it if he must, even though it ultimately ends
in disaster. In the prospect of such an eventuality it seems
appropriate for the several disciplines to get on with the development of their capabilities to participate in a balanced
environmental design, rather than to seek to arrest the momentum of civilization because of their lack of readiness to participate effectively.

The Origin Public Projects

In previous sections it has been assumed that some identified need or desire for a specific benefit is to be met by a
public project and that in the development of that project, in
the choice of alternatives, and in the final decision and design a spectrum of disciplines will cooperate to optimize environmental considerations. To this end environmental law may be
viewed as a catalyst as well as a generator of constraints; and
individual disciplines are expected to develop appropriate criteria of their own as well as practitioners who understand the
criteria of all disciplines concerned. Good environmental design would then result from applying these criteria through

methodologies having sufficient flexibility to permit the compromises necessary for a balance of various viewpoints. Nothing
has been said concerning the sequence of inputs from the various disciplines.   It may, therefore, be useful to remind ourselves how projects originate.   It may be that environmental
design at its optimum involves sequential rather than simultaneous application of disciplinary parameters. If such should be
the case, qualitative inputs might precede quantitative ones
and hence need not be of the order.   Thus methodologies could
differ widely as long as they were defined and flexibility and
interdisciplinary understanding prevailed.

It has often been noted that decisions on public projects
are political decisions and that the role of the professional
practitioner, including the lawyer, is that of determining the
nature, cost, and consequences of alternatives. Quite as truly
it may be said that the origin of many public projects is political and hence from an environmental viewpoint a no-go decision remains one of the viable alternatives and might be the
correct answer. Water resource developments have been particularly susceptible to political origin in the past, often being
constructed without adequate prospect of long-term success. For
example, it has been alleged that the Red River Project in
Oklahoma supplied agriculture with water already too saline for
irrigation purposes; and that the Central Arizona Project depends upon water previously assigned to other public projects
at the time of their origin.   Whether or not such reports are
strictly accurate is not to be debated here.   The significant
factor is that it is an uncommonly inept legislator who does
not generate proposals for public projects within the area of
his constituents.   The compulsion to generate proposals for
public projects is independent of the project's merits. Thus
from an ideal viewpoint good environmental design depends upon
the selecting of a good public project initially, hence a multidisciplinary input at the time of origin of the proposal.
Such a need is more readily identified than implemented because
the multidisciplinary team can hardly be assembled until a public project has been proposed and alternatives, development,
and design decisions become matters of substance.

If multidisciplinary inputs are by circumstance ruled out
until legislative committee or public hearings are held to consider a project proposal, the question arises as to how the
professional practitioner might contribute at the time of origin of the proposal.   Further, what discipline might be most
appropriate to the task.   At first thought it might seem that
the political scientist is the logical one to contribute to a
public project at its very beginning. Such logic is weakened in
practice by the fact that political science is not the disciplinary source of politicians the way engineering, for example,
is the source of practicing engineers. Most elected officials,

who by definition are professional politicians, have a background in law, or some area of business. Thus, they are more concerned with practical politics, which may range from considerations of how to get re-elected to the acquiring of a working knowledge of the customs, routines, and tradeoffs within the elected body which make possible a positive contribution to the welfare of his constituents or to the larger community. The political scientist, in spite of hiw knowledge of how a society governs itself and of the ways in which people influence or are affected by government remains an outsider quite as surely as does the practitioner of any other discipline as far as the origin of public project proposals is concerned. The profession of law, although the background of many politicians, is in but little better status because, as previously cited, environmental law is not yet a solid core in the education of lawyers.

If the foregoing analysis is of reasonable validity it means that the professional practitioner must compete with the impractical dreamer and the self-seeking individual for the attention of the politician concerned to make a project proposal. It is therefore of utmost importance that the professional practitioner who may have entree to the politician in a given situation be well informed as to the valid criteria which other disciplines will introduce in project development and design. The alternative is for the practitioner to find himself identified as a visionary should the project be shown to be unwise after its author has committed his prestige to it and its demerits have been established through public hearing and expert testimony.

Not all public project proposals first surface at the political level. In fact those that do so are generally first conceived by interests which prefer to remain in the background. Many, and possibly most, large public projects originate in some public agency. Best known of these agencies are such entities as departments of highways, and the Corps of Engineers, the Bureau of Reclamation. Although projects prepared by such agencies eventually get into the political arena for final decision, most result from studies made within the agency itself and do relate to some benefit to man which the agency foresees and identifies. Undoubtedly interagency rivalries, motives of self-perpetuation, and other less noble motives have attended the origin of project proposals by public agencies in the past. However, from the standpoint of environmental design, some opportunity for multidisciplinary participation exists during the earliest stages of projects at present, both because of the greater range of disciplines now on the staffs of public agencies and the need to consider the environmental impact of any project. For the future the evaluation and design of agency-initiated projects will be refined in direct proportion to the ability of various disciplines to establish criteria and methodologies which reflect environmental considerations and so insure that solid rationality rather than emotionalism is the parameter of project validity.

The third source of project origin is the private or public special purpose utility, particularly that concerned with power. Such projects are commonly based upon an analysis of the public need for service and the lead time needed to provide facilities at the appropriate date. Opportunity for multidisciplinary inputs during the period of project exploration depends to a significant degree upon the availability of staff who have the working knowledge of criteria within their own discipline and the inter-disciplinary understanding to which this treatise is addressed.

Looking to the future it seems probable that the ability of a multidisciplinary group of professionals to insure good environmental design of public projects can eliminate much public controversy through inputs at the time of origin of project proposals. The critical factor is the development of this multidisciplinary potential.

Ecological and Environmental Factors

In developing an enlarged set of design parameters which includes the ecological, economic, social, and political, as well as the engineering and technological, no professional is confronted with more difficulties than is ecology. Attention has been directed in Chapter 5 to the extreme complexity of ecosystems and of the equilibria on which any system is dependent. This, as noted, has often been the cause for a caution on the part of the professional ecologist which would in effect call for a moratorium on public projects until viable parameters can be developed. The evidence, however, is that public projects will continue to be built albeit with attention to ecological factors to the extent of our understanding. But limited understanding is not the major problem confronting the professional ecologist. There are others which are worth highlighting before looking to the future.

To begin with, ecology as a special branch of biology interpreted in the context of environment, is a quite recent development. Thus the development of parameters and methodologies suited to a balanced environmental design has had little lead time before the need was recognized. This tends, perhaps, to add a sense of urgency or impatience when public projects having ecological implications proceed more rapidly than the implications can be evaluated. Moreover, the importance of ecology has been so recently recognized and so widely acclaimed that the ecologist could hardly be blamed for a sense of new-found importance.

But the true cross which the professional ecologist bears is the great mass of organized and disorganized people who have suddenly discovered "ecology," who speak loudly in its name, and who equate it to something called "environment." To make matters worse for the practitioner who would develop criteria useful in a balanced environmental design, the chorus is too often joined by professional ecologists who have value commitments that make them incapable of working on a multidisciplinary team or accepting the compromises that go into a balanced environmental design.

Looking backward in a somewhat broader context it is clearly that in early years of our nation removing forests to open agricultural land was more important to human survival and to the creating of a nation than was any consideration of the esthetic or environmental attitudes or realities of future generations. And irrigating the 17 western states to make them habitable was more urgent when the family farm was the way of life than were ecological considerations which might become acute when the food problem was solved and industry became a mainstay of affluence.

Looking to the present: It is quite common in the 1972 climate of concern for our poorly defined concept of "environment" to view as a permanent and irretrievable disaster a stress or strain in any element of a system which might signal its failure. This is particularly true when the system is an ecosystem. Here there is a tendency to underevaluate at least three phenomena: (1) the significance of change, (2) the resilience of nature and (3) the ingenuity of man.

Concerning change, not all change is environmentally or ecologically bad. Numerous instances of ecological benefits resulting from public works may be cited. For example, the southern California coast, where the diversity and abundance of aquatic life has been enhanced by discharges of treated domestic sewage. When catastrophy has resulted from toxic or oxygen depleted wastewaters, or from natural floods, the resilience of nature has re-established aquatic life within one or two seasons after the event. Typical examples are the recovery of the fish population of the Mississippi River following the massive fish kill of a few years ago; and the return of oysters to Mobile Bay after a disastrous flood wiped out 90 percent of the crop through lowering the salinity of the overlying water. More revealing, perhaps are such examples as the change in east San Francisco Bay in the vicinity of the city of Emeryville. Here a malodorous anaerobic embayment became a choice area for striped bass within a few years after sewage discharge in the area was discontinued.

Examples of the ingenuity of man predate written history. In his public projects man probably learned how high he could

pile stones only when the bottom stone began to crumble. Far
from triggering the end of high-rise structures, the event led
man to enquire why failure occurred and to set out to discover
what he must do to prevent failure of the foundation stone from
limiting his design.

This does not mean that man cannot create a desert upon
the land or even in some sectors of the hydrosphere, but it
does give cause for optimism. Especially in relation to the
water resource, it suggests that an alertness for signs of
stress will indicate the nature of investigative work to be
done. In the field of ecology, where both the complexity and
durability of ecosystems are but dimly revealed, it suggests
that we may have to fracture the system in order to learn where
to start. This does not mean, however, that beginning today we
should build public works in our accustomed fashion and start
empirically to assess the stresses they impose on ecosystems.
The authors of Chapter 5 have shown clearly that we already
have enough information on both the shortcomings of past projects and the interrelationships of ecosystems and the environment to make possible a wealth of valid design decisions. By
using these intelligently, many past failures can be avoided
and our next design will more correctly evaluate the ecological
criteria which it must meet. In the context of public works
such as those with which this treatise is concerned, intelligence and alertness rather than inaction based upon fear of irretrievable damage is the key factor.

With specific reference to the task of the professional
practitioner of ecology in the future, it seems evident that he
must break free from the obstructionists in order to tell other
professionals: (1) how the ecologist thinks, (2) what factors
must be considered specifically in a given situation, and (3)
how the ecologist evaluates these factors and what methodology
or investigative work is necessary for such evaluation.

The authors of Chapter 5 show that this is not an impossible task. The implications are clear that a public project is
not necessarily to be ruled impossible because of ecological
damage. Instead, the damage potential should be assessed in the
project planning stage and considered as one of the penalties
society must accept in exchange for the benefits to be gained
by the best possible project alternative. If the penalty is
judged to be too high, a no-go decision will result. However,
it cannot follow axiomatically that *no* change in habitat or biological communities can be tolerated. What is important in the
application of ecological parameters to environmental design is
that the change be foreseen and a decision made concerning its
acceptability as an aspect of project formulation.

In the previous sections attention is called to the way in which the practitioner of any discipline contributes to the decisions that go into the development and design of public projects. Of equal importance, perhaps, is the way the individual citizen contributes to, or obstructs, environmental design. On the negative side, obstruction by citizen lawsuits has been cited. Such suits may be the result of initiative by the crusading individual citizen, but more often they are associated with the activities of organized groups. In many situations the attitude of the organization represents that of its leaders, who may be far more biased than the majority of the members. Originally, many individual citizens join such groups because they are in favor of conservation as a basically good philosophy; or because they oppose the removal of a tree from the local park; or because of a personal attitude toward any one of a hundred specific local situations or highly general concept of public good. The result is adequate funds for a few men to advocate their own biases and the ability to make themselves heard because of the presumption that they speak for a large group of citizens.

In the context of environmental design the organized citizen group dedicated to preservation in the guise of conservation has contributed but little. The reason is that they have tended to consider obstructing public projects as a highly positive contribution to the cause of "ecology" and "environment."

To minimize the spirit of obstructionism and to take full advantage of the public opinion on which public policy presumably is based, public agencies and professional practitioners presently advocate bringing the public into the project planning process from the very beginning. So far this has been more of an ideal than a reality. The pressures for the attention of the individual are far too numerous to permit him to respond to more than a few at any one time. Therefore he tends to discard the remote and abstract in favor of the immediate and personal threat. An example, perhaps typical, is the experience of Marin County, California in 1972. Here a land use planning session intended to involve the citizen in the planning process drew two citizens and a dozen real estate developers despite a very concerted campaign involving letters to every possible organization and to some 500 individuals. In such cases, it is only when plans are drawn and the individual's own property, tax rate, or physical access is involved that he responds--and unfailingly in the negative.

Numerous examples can be cited of the folly of asking citizens to decide questions beyond their understanding, and it may well prove to be true that in public project planning, citizen inputs can only be expected when there is something definite to oppose or to support.   But here again people are given

the choice between absolutes: i.e., support or oppose; project
or non-project; acceptable and unacceptable design. As noted by
the author of Chapter 2, the citizen is offered all or nothing.
And he asks the question, "Do *all* billboards have to come down?
Must *all* wires be underground?"

In the more contemporary ethic of "environment" people are
being asked to remake themselves so that the advocate of one or
another movement may become a prophet.   For example, what is
called in Chapter 7 the "do your own thing upon the land" fad-
dist, questions public projects because they are designed by.
professionals rather than by simpletons.

Looking to the future it seems doubtful that the citizen
can effectively function in the planning process except through
organizations which take the time to seek information on all
aspects of the project and to discuss these with its members.
For the immediate future, however, there will be far too few
organizations such as the League of Women Voters, and far too
many militant environmental groups to make the citizens' input
to environmental design a positive rather than a negative contribution.

One especially hopeful development is that agencies which
have customarily generated public projects are turning toward
what the U. S. Army Corps of Engineers' Seattle District has
termed "fishbowl planning." Although its success is by no means
assured, it does permit the citizen to be heard formally at a
sequence of stages during the development of a project rather
than only at its beginning and end.

Cost of an Economic Factor

Economics has long played an important role in public
project design.   Generally the engineer has been required by
the public officials concerned with public works to meet a
least-cost criterion.   The smaller the unit of government involved, and consequently the closer it is to the citizens who
are directly going to pay the bill, the more cost consciousness
becomes a constraint on project design. In large projects such
as might characterize a multi-purpose water resource development involving a dam and power plant, this cost constraint has
been less obvious.   Nevertheless, it has been far from absent,
a cost-benefit study being one of the first factors in selecting the project from among conceivable alternates and establishing its priority.   Such cost-benefit studies generally involve both engineers and economists and both have often later
been accused of insensitiveness to all but the cost of physical
structures and land in their deliberations.   The fact that a
project may be initially publicized as costing, say, 60 million
dollars and end up costing twice that amount makes good political
cal campaign material but it does not mean that the engineer

and the economist lost sight of costs once the project was approved. In multi-purpose water resource projects particularly, the cost-benefit ratio may be favorable or unfavorable, depending upon what portion of the cost is assigned to such non-reimbursable benefits as recreation and wildlife habitat. In turn, the fraction of the cost assigned to the qualitative benefits of a project may depend upon whether or not those who influence political decision want the project to be built. The ultimate cost of the project depends upon several factors such as

1. inflation during the period between authorization and completion;

2. features added during the development process to enhance the overall benefits of the project, particularly the non-reimbursable ones; and

3. misinterpretation of initial requests for authorized funds as representing the total project costs.

In major highway projects and power production facilities, non-reimbursable benefits have not customarily been singled out in cost estimating. Moreover, in all types of major public projects costs have been normally primary costs only, leaving the secondary costs unevaluated. For example, the benefit of flood control represents the losses which might result from a flood such as has occurred in the past or is statistically probable once in say 50 or 100 years. Unless, as is the case in the lower Mississippi Basin, the flood control project itself involves secondary works which exclude residential occupancy of the flood plain, it is essentially impossible to control development of the area protected by the flood control project. The reason is the freedom of people to own and use land; a factor discussed in detail in Chapter 8. The results, however, is that in many instances the flood plain becomes developed to the extent that even the design discharge of the dam is enough to cause a disaster when the anticipated flood occurs. Thus, although the geographic area of risk is reduced, the willingness of men to accept or ignore risks is not, hence the flood control benefit of the project becomes increasingly unrealistic with time. This secondary cost factor is not a fault of the design, nor could it be eliminated by a project design which does not involve control of the flood plain regardless of the spectrum of disciplines involved in the design decisions.

Factors such as the foregoing, albeit in a far more complex way than outlined, define much of our past association of costs with economics. That they have also led to a somewhat arbitrary indictment of practitioners of engineering and economics is pertinent only in that it contributes to a polarization of others who would project the changed social climate of today into the past in search of a scapegoat for real or fancied threats of environmental disaster.

The nature of the present mood of America has been discussed in various contexts in preceding chapters and is further noted in other sections of this chapter. The important facts in the context of costs per se as an economic factor is that: (1) we want to bring both short-term and long-range considerations of economics, sociology, ecology, aesthetics, etc., into our public project designs; and (2) we are currently at the stage of identification of objectives as contrasted with implementation.

In this latter context, it is axiomatic that the goals of design must precede the fact of design, just as design must precede construction of a project. However, the factor of costs in hard dollars is not easily overwhelmed by social costs which we find it hard to evaluate. Currently we are in one of the shear zones or turbulent times of rapid change when goals of environmental design are emerging, when people are demanding new goals on the basis of desirability and idealogical concepts, but when the price tag has not yet been attached. In cases where costs of environmental quality have been estimated, the figures, although astronomical, are as yet unreal. This is because they are not yet on the tax bill, and because the delusion is amazingly persistent that a project will be economically painless as a result of grants from the federal government.

When one considers the frequency with which school and other district bond proposals are being rejected by voters, and the growing demands of the taxpayer for relief from his burden, the possibility that old-fashioned concepts of costs will be smothered by the new environmental ethic does not seem bright. The future capability of the balanced environmental design of public projects may therefore be tempered by the trauma of cost. In the meantime society lingers in that condition of euphoria between the pleasant dream and the rude awakening. How dollar cost per se may affect social attitudes or human responses to project proposals is not yet clarified. Quite likely, it may equal any of the more subjective deterrents to public projects.

The Question of Social Costs

The matter of social costs and the consequences of ignoring them are discussed in several of the preceding chapters, and some very cogent approaches to their evaluation are presented in Chapter 6. Because major emphasis on social considerations in public project design is of recent origin, its past is not too revealing. Nevertheless, there are aspects of the past which may serve as guides to the future and which therefore might be worthy of consideration here. Two factors seem particularly important. They are implicit in the terms "social science" and "social cost."

It might be said with some logic that it is too bad that studies which deal with the way people behave, react, and relate had to be constrained with the title of sciences. Such a thesis would hold that had it not once become necessary for every field to become a "science" in order to achieve respectability in a science-oriented world, disciplines such as sociology and political science might have been freer to make their own unique contributions to human welfare. This would not have precluded the use of the methodologies of science in approaching problems, but it might have saved considerable energy losses in the process of problem solving. The potential of social sciences as a free discipline, as contrasted with scientific ones, is not worthy of lengthy speculation here. However, the consideration of such a possibility by professionals in these fields might be useful in answering the question of how their concepts should be translated into environmental design criteria for public works.

More important in the context of the objectives of this treatise is the matter of social costs. Long before the current question of "what is the value of a sunset?" the planners of multi-purpose water resource developments were struggling with the question of the benefits to be assigned to recreation, wildlife preservation, and similar intangibles in the context of a benefit-cost ratio expressed in dollars. Thus the designers of public projects have for a good many years been aware that some non-dollar scale of values is needed if appropriate consideration is to be given to the qualitative aspects of a project. Currently, the need for such a scale is vastly more urgent than in the past. In previous chapters project design procedures have been properly criticized for insensitiveness to the social cost of bisecting a viable community with a freeway, or even an elevated mass transit system. Such a cost is doubly difficult to assess. First, as previously noted, we have no idea of how the cost can be expressed in dollars in a general situation, and no other scale of values has been adequately developed. Second, and more complex, is the fact that question of social cost can rarely be referenced to the proposed public project alone. Frequently those who oppose routing a freeway through a given neighborhood base their opposition on the rationale that it houses a minority group, or the neighborhood comprises under-privileged or poverty-stricken people. Simultaneously, other sociologists, often through institutionalized programs, are concerned with the social costs of permitting such communities to endure.

In the context of environmental design, with which this treatise is concerned, it is not a question of "who are we to believe?" Instead the question is how can the sociologist develop a cost parameter that he himself can understand and that other professionals can also understand when he explains it to them? How can the sociological approach be systematized and measured along some scale that varies in a way which is assessable in itself as well as responsive to changes in similar variables in other disciplines involved in the project design?

To summarize the present situation, it might be said that we are aware of the need for answers to these questions; we are mindful that social costs, however we define them, must be considered in environmental design of public projects; and that the roadblocks to effective action are now pretty clearly defined (see Chapter 6).

But what of the future? How will men resolve the disparity between what the engineer and economist can provide concerning the cost of systems and hardware, and what the sociologist can provide concerning social costs measured by the same scale? The chances are that the solution will come about through a different route than mathematical calculations. A combination of five possibilities can be foreseen:

1. A non-dollar scale of measurement will be developed.

2. Some questions will cease to be asked.

3. Reasonable men will agree upon the weight of sociological factors on a case by case basis.

4. Environmental law will accelerate agreement.

5. Political decision will continue as a major determinant.

Concerning the first, such approaches as the Delphi method and the evaluative methods described in Chapters 6 and 11 are being developed and utilized in various contexts. These are non-dollar scales but assess the weight of current opinion and existing systems. As noted in Chapter 11 the "elusive parameters" may still be veiled in subjectivity and obscure scoring techniques. Also they are responsive to current opinions and situations hence not predictive in the long term. However, as noted in a previous section where the TVA and Bonneville Dam were cited as examples, long-range social effects are not notably predictable.

The classic example of the second possibility is the ancient theological question, "How many angels can stand on the point of a needle?" Although there were sound reasons for asking this question initially, the whole matter was dropped when it no longer made sense to ask the question. In a less profound vein, many standards of water quality, such as the coliform concentration, came to be accepted and only a few scholars now ever enquire whether they made sense originally or continue to do so today. Conceivably, the dollar index for social decision might do the same.

An example of the third possibility is to be found in Orange County, California. There the seemingly unsolvable legal question of who owns the capacity of the earth to store water if natural stored water belong to the overlying landowner, was solved by men simply getting together and agreeing that it was too costly and foolish to seek an answer by litigation.

Trends in environmental law suggest that the question of social costs may be resolved by requiring the sponsors of public projects to show that its social implications have been considered rather than requiring the sociologist to evaluate his design considerations in dollars. This approach confronts all parties concerned with the same qualitative unknowns and hastens the type of agreement cited under item 3.

Finally, the fact that public project decisions are more political than technical, if technical feasibility exists, does much to reduce the degree to which environmental design precision governs the viability of a project.

A driving force toward a solution is, as cited in Chapter 2, that "conflicting interests must be accommodated regardless of the tenability of their stated positions" applies in the final analysis.

New Factors in Environmental Design

H. L. Mencken once said, "Man has an instinctive love of the ugly..." Others have asked "How ugly is ugly?" Mencken's thesis would not be hard to document. And one need only recall the automobiles of the late 1950's to get some basic index. However, the fact that the automobiles were accepted by the public bears out Mencken and underscores one factor that should be borne in mind in environmental design--i.e., that within certain limits what is ugly and what is beautiful is often topical and subject to rapid change. In architecture, it may be remembered that when scale, color, and form become the dominant considerations, the highly sculptured "gingerbread" house was considered as false and therefore ugly. However, as the new design led to what the author terms "shoe box gothic" an increasing number of people find them less attractive than some of the older sculptured forms.

Dwelling houses and automobiles are generally more transient than large public projects which may endure indefinitely. Therefore, environmental design of large public projects may call for more attention to enduring criteria of esthetics than are applied to lesser projects. Here, it might be worthwhile to differentiate between such concepts as functional design, conceptual design, and enviornmental design. Without becoming too technical it might be said that functional design basically concerns the technological aspects of the structure; those factors by which it is made safe, economical, and technologically

sound. These design objectives may be irrespective of whether
it is ugly or graceful, although economics in terms of dollar
costs may make the ugly cheaper. One example of such a project
is the Richmond-San Rafael Bridge across San Francisco Bay.
Here in full view of two of the most graceful spans in the
world--the Golden Gate and the San Francisco-Oakland Bay
Bridge--a broken-backed roller-coaster structure was built to
save some 3 million dollars on a 90 million dollar project.

Conceptual design may relate to the way the project har-
monizes with other elements of the community. Basically, it is
a concern of good planning and may involve all the factors of
visual harmony cited in Chapter 7. To the extent that environmental design is concerned with esthetics, conceptual design
and environmental design overlap.

Environmental design, as herein considered is something
new in the hierarchy of design.   Not only does it involve the
measures necessary to bring the project's structures into harmony with the natural and man-made physical aspects of its surroundings, but also into harmony, or at least compatability,
with the ecological systems of nature and with the social and
cultural goals of man. Environmental design, therefore, may be
considered as functional design and conceptual design with
added objectives which call for a new set of parameters. These
parameters are not constant for all types of public projects.
For example, a power plant or a dam may be basically an architectural problem in terms of the structure itself and in its
relation to existing physical features.   In contrast the successful design of a transportation system which cannot be
viewed in its entirety from any single point, or the recreational features of a reservoir, may be best achieved by including the professional landscape architect in the multidisciplinary design team.

Project Evaluation

As noted in Chapter 6 large public projects are considered
successful when through the appropriate technology they achieve
valued ends. The concept of environmental design in effect
broadens the spectrum of valued ends which are to be achieved,
and hence the number of factors which make for success.   The
author of Chapter 2 however, poses some critical questions concerning project evaluation. What are we going to evaluate? How
are we going to assess or judge it?   When are we going to pass
judgment; at the pre-planning, planning, formulative, or post
completion stage?

In applying environmental considerations to the conception,
development, and design of public projects these questions
should be borne in mind by the practitioners of all disciplines
involved and answers sought at every stage of progress. Many of

the criticisms leveled at "engineered" projects in the past are
judgments made after the project is completed. This has led to
what is almost an ethic of "environmentalists" that some perversity of men who use technology to build physical structures
or systems inspires them to create something cheap and ugly and
to locate it where the most people will be made unhappy and
all other creatures obliterated.

Why have after-the-fact evaluations been more common than
pre-planning or planning assessments? The complete answer is
complex in the extreme. Various authors in the preceding chapters have presented numerous facets of such an answer. Attitudes toward resources and concepts of what constitutes a resource; the nebulous nature of social sciences in the realm of
measurements; the extreme youth of the discipline of ecology,
the unknown linkages with which it deals, and the caution and
polarization of many of its recent adherents are among the
identified reasons for the neglect of qualitative assessments
in the early stages of public project formulation. Lack of institutionalized or even expressed public policies concerning
environmental objectives of projects is also cited as a cause.

The influence of political decisions is not discounted by
successful implementation of the concept of environmental design. Therefore the future capability, or the efficiency of
environmental design in terms of the ideal, may be lowered regardless of the potential efficiency of an environmental design
developed with good multidisciplinary balance of criteria.

The Critic Versus the Participant

As one views in perspective the past history of public
projects and the limited environmental consideration which went
into their design, the factors which motivated this treatise
become evident; and the need for multidisciplinary inputs at
every level of project decision seems obvious. Nevertheless,
the practitioner of any one of several highly important disci-
plines--e.g., ecology, sociology, political science, and es-
thetics--generally finds himself cast in the role of critic
rather than of participant in the project development process.
So much is this the case that organized groups speaking in the
name of various disciplines, or presuming to speak for them under the catch-all title of "environmentalist," make obstructionism and opposition to public projects a way of life. If one
assumes, as does the author of this chapter, that man is going
to continue to build public projects and that they will be subject to environmental design in the future, it is appropriate
to enquire why some disciplines which should be participants
find themselves in the role of critics instead. An understanding of this phenomenon is the key to combating it in the future.

The most fundamental reason cited herein is the lack of
clearly stated or widely understood criteria, methodologies,
and degree of flexibility which disciplines will apply to the
decision process in an environmental design. When environmental
law requiring consideration of quality factors is absent, projects tend to arise from political and engineering responses to
identified needs or objectives of some segment of society. Thus
regardless of whether the project's advocates are insensitive
to or desirous of an environmental design, some proposal
emerges before it is tested out publicly. The result is that
the advocate of any discipline which did not participate in the
proposal can only ask such questions as: Have you considered
this aspect? How was it evaluated? Who did the evaluation?
What harmful or beneficial effects were found? Whether or not
he has the special tools for making the evaluation, he has no
choice but to appear in a negative posture relative to the
project at this point. His role is that of a critic. However,
if he does not indeed have the tools--criteria and methodolo-
gies--his role as a critic degrades to that of obstructionist.
For reasons of self respect he may then condemn the engineers
and other project participants for not possessing and not using
tools which he himself does not possess, but no useful end is
gained in the process. The engineer is condemned for not having
played the role of a social scientist, and the social scientist
is frustrated because he was either not permitted or not prepared to play his proper role.

Emergence of Environmental Design

Attitudes and necessities such as described above are representative of the past and the immediate present. Environmental law, however, is rapidly making broader multidisciplinary
inputs necessary and inescapable. Necessity brings professional
practitioners into the situation before many are prepared. The
result is that the obstructionist plays a delaying role in normally conceived projects. His influence can, however, be expected to wane as professional practitioners of appropriate
disciplines move from the critic to the participant role in
project design. The readiness of professionals to move into
such a role, and the way in which it can be exercised, is set
out in preceding chapters of this book; Chapters 5 and 6, for
example. In addition, consulting firms specializing in various
aspects of project evaluation are appearing in the professional
field. Many consulting engineering firms now have economists,
biologists, social scientists, lawyers, and other specialists
on their own staffs, or retain consultants from appropriate
disciplines to advise on projects from the very beginning.

There is work yet to be done in identifying the criteria
and methodologies each individual will apply to an environmental design, and in stating them with sufficient clarity that
the proponent of a project may know in advance what parameters

the project must satisfy.   There is much to be done within individual disciplines   before a   sufficient community   of acceptance of criteria   permits   any one   of   many   practitioners to
function effectively   on an environmental team.   This does not
mean that   human judgement   should   become   a fixed or a static
thing.   Engineer may disagree with engineer,   for example, but
there is enough unanimity of acceptance of design criteria that
any one of dozens of engineers can design a structure that will
achieve   the project   objectives   even though   the hardware and
processes may differ widely.   The same must soon be true of all
disciplines participating in an environmental design. And there
is no reason to believe that such will not be the case.

Finally, as various authors of this treatise have noted in
one or another context, no discipline can stand upon absolutes.
There will be   situations in which   the considered decision   of
the participating disciplines is that "no project" is the proper answer.   In most cases, however,   it may be expected that a
balanced environmental design will minimize adverse environmental effects; and risk or certainty of damage to one factor will
be consciously accepted in order to achieve the benefits judged
to be worth such sacrifice.

The important fact   is that the stage   is now set   for the
emergence of environmental   design   of   public projects--a   new
concept of design   in which   a number of   concerned disciplines
function as   participants   at all levels of decision; and which
to a maximum degree produces foreseeable benefits and   disbenefits rather   than unanticipated   problems   and   dislocations of
natural equilibria.

AUTHOR NOTES

P. H. McGauhey   is   Director Emeritus of the   Sanitary Engineering Research Laboratory,   and Professor Emeritus of Sanitary and Public Health Engineering,   of the University of California,   Berkeley.   As a teacher, a researcher,   and a professional engineer his wide interest   in environmental engineering
and its   relationship to the   world about him   has led him into
many fields of study.   Early in his career he developed an interest in aquatic biology;   and more   than 30 years ago   he led
one of the first   comprehensive studies   of stream pollution in
which both engineers   and biologists   participated   as   professionals enjoying mutual respect.   More recently he has been engaged in studies of the eutrophication of Lake Tahoe and a participant in a   multidisciplinary study   of the relation of land
development   to coastal water quality in Hawaii.   Prior to his
retirement in 1969   he led a   multidisciplinary study   of solid
waste management.   Later he served as special assistant for environmental studies to the chancellor of the University of California at Berkeley.   His research interests have included such

areas as water reclamation, recycling, solid waste disposal,
detergents, underground travel of pollutants, and economic
evaluation of water quality. Recognition for his leadership and
contributions in environmental engineering has led to numerous
awards, including the Honorary Doctor of Science from Utah
State University, and election to membership in the National
Academy of Engineering.

Currently he is serving as consultant to several major engineering firms and organizations engaged in studies and evaluations of pollutants in the environment, and continuing his
long list of contributions to the literature of environmental
engineering.